89 Springer Series in Solid-State Sciences

Edited by Peter Fulde and K. Alex Müller

Springer Series in Solid-State Sciences

Editors: M. Cardona P. Fulde K. von Klitzing H.-J. Queisser

Managing Editor: H. K. V. Lotsch

Volumes 1–49 are listed on the back inside cover

50 **Multiple Diffraction of X-Rays in Crystals**
By Shih-Lin Chang

51 **Phonon Scattering in Condensed Matter**
Editors: W. Eisenmenger, K. Laßmann, and S. Döttinger

52 **Superconductivity in Magnetic and Exotic Materials**
Editors: T. Matsubara and A. Kotani

53 **Two-Dimensional Systems, Heterostructures, and Superlattices**
Editors: G. Bauer, F. Kuchar, and H. Heinrich

54 **Magnetic Excitations and Fluctuations**
Editors: S. Lovesey, U. Balucani, F. Borsa, and V. Tognetti

55 **The Theory of Magnetism II**
Thermodynamics and Statistical Mechanics
By D. C. Mattis

56 **Spin Fluctuations in Itinerant Electron Magnetism** By T. Moriya

57 **Polycrystalline Semiconductors,**
Physical Properties and Applications
Editor: G. Harbeke

58 **The Recursion Method and Its Applications**
Editors: D. Pettifor and D. Weaire

59 **Dynamical Processes and Ordering on Solid Surfaces**
Editors: A. Yoshimori and M. Tsukada

60 **Excitonic Processes in Solids**
By M. Ueta, H. Kanzaki, K. Kobayashi, Y. Toyozawa, and E. Hanamura

61 **Localization, Interaction, and Transport Phenomena**
Editors: B. Kramer, G. Bergmann, and Y. Bruynseraede

62 **Theory of Heavy Fermions and Valence Fluctuations**
Editors: T. Kasuya and T. Saso

63 **Electronic Properties of Polymers and Related Compounds**
Editors: H. Kuzmany, M. Mehring, and S. Roth

64 **Symmetries in Physics** Group Theory Applied to Physical Problems
By W. Ludwig and C. Falter

65 **Phonons: Theory and Experiments II**
Experiments and Interpretation of Experimental Results By P. Brüesch

66 **Phonons: Theory and Experiments III**
Phenomena Related to Phonons
By P. Brüesch

67 **Two-Dimensional Systems: Physics and New Devices**
Editors: G. Bauer, F. Kuchar, and H. Heinrich

68 **Phonon Scattering in Condensed Matter V**
Editors: A. C. Anderson and J. P. Wolfe

69 **Nonlinearity in Condensed Matter**
Editors: A. R. Bishop, D. K. Campbell, P. Kumar, and S. E. Trullinger

70 **From Hamiltonians to Phase Diagrams**
The Electronic and Statistical-Mechanical Theory of sp-Bonded Metals and Alloys
By J. Hafner

71 **High Magnetic Fields in Semiconductor Physics**
Editor: G. Landwehr

72 **One-Dimensional Conductors**
By S. Kagoshima, H. Nagasawa, and T. Sambongi

73 **Quantum Solid-State Physics**
Editors: S. V. Vonsovsky and M. I. Katsnelson

74 **Quantum Monte Carlo Methods** in Equilibrium and Nonequilibrium Systems
Editor: M. Suzuki

75 **Electronic Structure and Optical Properties of Semiconductors**
By M. L. Cohen and J. R. Chelikowsky

76 **Electronic Properties of Conjugated Polymers**
Editors: H. Kuzmany, M. Mehring, and S. Roth

77 **Fermi Surface Effects**
Editors: J. Kondo and A. Yoshimori

78 **Group Theory and Its Applications in Physics**
By T. Inui, Y. Tanabe, and Y. Onodera

79 **Elementary Excitations in Quantum Fluids**
Editors: K. Ohbayashi and M. Watabe

80 **Monte Carlo Simulation in Statistical Physics**
An Introduction
By K. Binder and D. W. Heermann

81 **Core-Level Spectroscopy in Condensed Systems**
Editors: J. Kanamori and A. Kotani

82 **Introduction to Photoemission Spectroscopy**
By S. Hüfner

83 **Physics and Technology of Submicron Structures**
Editors: H. Heinrich, G. Bauer, and F. Kuchar

84 **Beyond the Crystalline State**
An Emerging Perspective
By G. Venkataraman, D. Sahoo, and V. Balakrishnan

85 **The Fractional Quantum Hall Effect**
Properties of an Incompressible Quantum Fluid
By T. Chakraborty and P. Pietiläinen

86 **The Quantum Statistics of Dynamic Processes**
By E. Fick and G. Sauermann

87 **High Magnetic Fields in Semiconductor Physics II**
Transport and Optics Editor: G. Landwehr

88 **Organic Superconductors**
By T. Ishiguro and K. Yamaji

89 **Strong Correlation and Superconductivity**
Editors: H. Fukuyama, S. Maekawa, and A. P. Malozemoff

Strong Correlation and Superconductivity

Proceedings of the
IBM Japan International Symposium,
Mt. Fuji, Japan, 21–25 May, 1989

Editors: H. Fukuyama, S. Maekawa,
and A. P. Malozemoff

With 198 Figures

Springer-Verlag Berlin Heidelberg New York
London Paris Tokyo Hong Kong

Professor H. Fukuyama

Institute for Solid State Physics, University of Tokyo, 7-22-1 Roppongi, Minatu-ku
Tokyo 106, Japan

Professor S. Maekawa

Department of Applied Physics, Nagoya University, Nagoya 464-01, Japan

Dr. A. P. Malozemoff

IBM, Thomas J. Watson Research Center, Yorktown Heights, NY 10598-0218, USA

Guest Editor:

Professor Dr. h. c. mult. K. Alex Müller

IBM, Zürich Research Lab., CH-8803 Rüschlikon, Switzerland

ISBN-13:978-3-642-83838-5 e-ISBN-13:978-3-642-83836-1
DOI: 10.1007/978-3-642-83836-1

2154/3150-543210 – Printed on acid-free paper

Preface

This volume contains the proceedings of the IBM Japan International Symposium on Strong Correlation and Superconductivity, which was held in Keidanren Guest House at the foot of Mt. Fuji, May 21–25, 1989.

The purpose of the Symposium was to provide an opportunity for discussions on the problem of strong correlation of electrons in the context of high-T_c superconductivity. Sixty-eight scientists were invited from seven countries and forty-three papers were presented in the Symposium.

Soon after the discovery of high-T_c superconducting oxides, Professor P.W. Anderson proposed that the essence of high-T_c superconductivity lies in the strong correlation among the electrons in these materials. This proposal has stimulated a wide range of theoretical investigations on this profound and difficult problem, which are expected to lead eventually to new concepts describing strong electron correlation. In the Symposium, Anderson himself started lively discussions by his talk entitled "Myth and Reality in High-T_c Superconductivity", which was followed by various reports on theoretical studies and experimental results. Concise and thoughtful summaries of experiment and theory were given by Professors H.R. Ott and P.A. Lee, respectively. It is our hope that this volume reflects the present status of the research activity on this outstanding problem from the viewpoint of the basic physics and that it will further stimulate the effort to understand these fascinating systems, the high-T_c oxides.

We would like to thank all participants, who made the Symposium very stimulating. In the process of identifying its scope and forming the program, we received valuable advice from the members of the advisory committee: Professor P.W. Anderson of Princeton University, Dr. L. Esaki of the IBM Watson Research Center and Professor R. Kubo of Keio University, to whom we are deeply indebted. Last but not least, on behalf of all participants, we would like to express our sincere gratitude to IBM Japan for its financial and organizational support which made this symposium possible, and to its members, Mr. T. Takeda, Mr. S. Ohno and Mr. A. Nemoto, who so attentively and finely steered every detail of the unforgettable gathering. Additional thanks are due to the authors of the manuscripts in this volume, especially for their promptness, which made the rapid publication possible.

June, 1989

H. Fukuyama
S. Maekawa
A.P. Malozemoff

Conference Organization

Advisory Committee

P.W. Anderson	(Princeton University)
L. Esaki	(IBM, T.J. Watson Research Center)
R. Kubo	(Keio University)

Executive Committee

H. Fukuyama	(ISSP, University of Tokyo)
S. Maekawa	(Nagoya University)
A.P. Malozemoff	(IBM, T.J. Watson Research Center)

Secretariat

T. Takeda	(IBM Japan)
S. Ohno	(IBM Japan)
A. Nemoto	(IBM Japan)

Word of Welcome by Leo Esaki

Ladies and Gentlemen,

As a Board Member of IBM Japan and one of the organizers of this meeting, I am honored to welcome all of you coming to this IBM Japan Symposium on Strong Correlation and Superconductivity. I am particularly happy to see many overseas participants.

Looking back on the 1980s one can see a few remarkable achievements which are considered milestones in solid-state physics research. In the semiconductor area, the quantized Hall effect, including the observation of fractional quantization, in the two-dimensional electronic system, and the evolution of synthesized superlattices, quantum wells, wires and dots, perhaps, belong to this category. In surface physics, the development of the scanning tunneling microscope made a profound impact on understanding the surface structure by unprecedented visualization. One of the most recent examples is obviously the discovery of new superconducting rare-earth copper oxides by Bednorz and Müller, which has sparked intensive world-wide investigations of these heavily-doped semiconductor-like materials because they exhibit remarkably high transition temperatures. We notice that the time interval between their 1986 discovery at Zürich and the 1987 Nobel Prize at Stockholm was very short. This indeed reflects the accelerated pace of scientific progress, where international exchange has played a significant role.

Today, one of the most focused efforts in the physics community is the search for new mechanisms to account for such high transition temperatures with anomalously small isotope effects. The scope of this symposium is precisely along this line. In this context, P.W. Anderson first pointed out that superconductivity could occur near a metal-insulator transition, where the insulating phase is the resonating-valence-bond (RVB) state which he hypothesized in 1973. Since then, this inspiring proposal has been at the center of excited theoretical discussions. H. Fukuyama and S. Maekawa, who have been stimulated by this work of Anderson, were instrumental in organizing this symposium.

I must say that Japan is a most suitable country in which to hold this symposium because it is a country that consists of a strongly correlated human system. A few years ago, I wrote a casual article entitled "Why the West Misreads Japan" [The Bridge 14, 2 (1984), a publication of NAE]. In this article, I pointed out that this group-conscious society generates a kind of collective mode in industry and government and even in academia to perform work, which you do not see often in the West. One may call this the Japanese mode. I would like to quote from the article:

"This term may be explained more precisely by using the analogy of phenomena in solid-state physics. Long ago physicists figured out how to explain elementary processes involving a limited number of particles such as interactions among a few charged particles, gravitational motion of the solar system, etc. Contemporary solid-state physicists, however, are more interested in understanding cooperative phenomena caused by a large number of particles – a collective group. Intriguing phenomena such as superconductivity ...cannot occur except in a collective mode. Let us call the former (elementary process) the single-particle effect and the latter the many-body effect. It is the many-body effect that predominates in Japanese society, where the constitutive individuals are modified in their behavior and the whole is more than the simple sum of its parts. The homogeneity and high density of Japan's population certainly help to promote cooperative interactions. In the United States, the single-particle effect plays a dominant role. The United States was built largely by outstanding individual personalities."

In short, I am saying that you can see a room-temperature superconducting mode in the Japanese society.

Now, I should mention that this seminar house is owned by the powerful Keidanren, Federation of Japanese industrial giants, the spiritual center of this superconducting mode. Some Americans say that these firms, with little basic research, are innovative challengers to U.S. companies. Nevertheless, they built this house for their people to discuss difficult problems in their strongly correlated systems. I am sure you will agree with me that this is a most ideal place to have a lively discussion on Strong Correlation and Superconductivity. If they always reach a successful conclusion in discussions here, why shouldn't we too?

Before closing my talk, I would like to express my gratitude to all members of the Advisory and Executive Committees for their efforts, particularly, A.P. Malozemoff on the U.S. side and A. Nemoto on the Japan side for their handling of major and minor technical matters. I hope fruitful discussions will take place during the symposium and also hope all of you enjoy your stay in Japan.

Contents

Part III Experiments

Part IV Summary

Part V Abstracts

Part I

Overview

Myth and Reality in High-T_c Superconductivity

P.W. Anderson

Joseph Henry Laboratories of Physics, Jadwin Hall, Princeton University, Princeton, NJ 08544, USA

Abstract. I will discuss two major themes. First I will discuss various experimental developments which seriously restrict the possible theoretical alternatives, such as the electron-based material $(Nd-Ce)_2CuO_4$, the low spin susceptibility which shows that the magnitude of the specific heat jump is anomalous, the large gap-like structures at $\sim 500\ \mathrm{cm}^{-1}$ observed in photoemission and the infrared, the evidence for small coherence lengths, etc. Second I will discuss some modern trends in the theory of the one-band Hubbard model which is the appropriate theoretical framework for high T_c: separation of charge and spin due to twisted order parameter in the Néel ordered state near half-filling, the topological renormalization due to doping which leads to generalized flux phases, and possible superconducting order parameters in generalized flux phase states.

In deciding what to do here I had the option of trying to summarize the present rather incomplete state of theory or to focus on the many important experimental facts whose consequences limit the possible theoretical ideas very severely. There is not much basic understanding by the generality of physicists of the importance and quality of many of these results, and therefore I think that it would be more useful for me to focus on this aspect and speak only peripherally to theory. You will find, however, that the experimental facts eliminate the possibility of any conventional theory so rigorously that it is essential to say some things about theory.

In fact, that is how I should begin: with a few words about the electronic theory of condensed matter, since the nature of successful theories is so widely misunderstood. For most such systems, there is

This work was supported by a Grant from the National Science Foundation, DMR 8521377 and from the Department of the Air Force, AFOSR 87-0392.

Springer Series in Solid-State Sciences, Vol. 89 Strong Correlation and Superconductivity
Editors: H. Fukuyama · S. Maekawa · A.P. Malozemoff

neither any possibility nor any point of developing a complete, a-priori, all-electrons theory. We have no such even for the best understood insulators (such as Si) or metals (such as Pb or Al). Instead, we are interested in parametrized, model theories which describe the observed behavior correctly in principle, and particularly we always want to know about low-energy behavior on some scale, since what we are after is explanation of the observed static properties: thermodynamics, transport, low-energy spectroscopy, and, of course, superconductivity—even for high T_c, T_c and gaps are small compared to Fermi energies. Thus we are interested not in "bare" electrons, phonons, or other excitations, but in renormalized ones. This is why the spectroscopy of high-energy probes is so seldom useful in understanding complex materials, relative to transport or thermal properties: the former measure the "bare" particles, the latter the renormalized ones. This has led to a great deal of confusion in such systems as NiO, where the question of whether the "Mott-Hubbard" gap is to "d" or "s" or "p"-like states is irrelevant from the renormalized point of view, but is the most direct result of a photoemission measurement.

We have two familiar forms of renormalized theory. One is Fermi liquid theory, which describes most metals very well, and is particularly strikingly effective in mixed valence systems where the lowest-energy excitations have renormalizations by orders of magnitude in mass, etc. Another, less familiar, is the Mott magnetic insulator, where the low-energy excitations are purely magnetic, charge fluctuations have disappeared entirely, *i.e*, the appropriate fixed point is $U = \infty$.

Experimental data tell us that neither case may be used in the high T_c superconductors, since the transport properties are incompatible with Fermi liquid theory. I will discuss electrical and thermal conductivity, Hall effect, thermoelectric power, and magnetic susceptibility.

One of the few unrenormalized quantities in fermi liquid theory is the Joffe-Mott minimum metallic conductivity; by a purely geometric argument this occurs when $k_F\ell \sim 1$. The normal state conductivity of all the layered cuprates perpendicular to the planes is well below the MMC; and localization theory does not contain the possibility of a metal in one direction being an insulator in the other, since localization is a coherence phenomenon. Thus we have to accept that in some way the state is a two-dimensionally ordered electron fluid.

There are some controversies as to the sign of the T-dependence of the c-axis conductivity for YBa_2CuO_7, but I have yet to see data which did not fit the expression

$$\rho_{ab} = A_{ab}T$$
$$\rho_c = B/T \ (+AT)$$

(see figures) with $\rho_{ab}\rho_c > (\rho_{Mott})^2$. This is brought out clearly in plots of ρT *vs.* T^2. This behavior is much more clearly expressed in good single crystals of $(La, Sr)_2CuO_4$ and of $Bi_2\,(Sr-Ca)_3\,Cu_2O_8$. A very striking feature of these expressions is the absence of residual resistivity, which would show up clearly in these graphs. Interestingly, residual resistance reappears when, and only when, non-magnetic substituents are inserted into the Cu sites in the planes.

Zou and I have presented a theory of this dependence which is at best heuristic. I think what is solid about this theory is the observation that the AT and B/T terms have the same source: they both represent a rate of scattering, in the one case in the plane and in the other between planes. Let me say something about the controversies. $1\,2\,3\,O_7$ is notoriously easily twinned in all three directions; and with conductivity anisotropies of 100–1000 to 1, any small percentage of twinning or surface leakage, or electrode misalignment, or even stacking faults can all have enormous effects. In general, good films or crystals of these materials show a ρ_{ab} of $\sim 60\mu - \Omega$ cm at $100^\circ\,K$, and this value is the best criterion for quality. It is also clear that good crystals should have maximal anisotropy, not minimal. By both criteria the selected large crystals which have been measured by the Japanese and Russian groups seem to be less perfect than the small single crystals measured at IBM and Princeton.

Hagen et al have recently measured anisotropic thermal conductivity, as Ong will discuss. There are many striking features; I say here only that the c-axis conductivity is $< 20\%$ of that in the ab planes and does not have any T_c anomaly, which argues strongly for electronic nature of the latter. That conductivity obeys Wiedemann-Franz with an anomalous factor of 2. It is very striking that the thermal conductivity jumps, below T_c, by an amount of the order of the c-axis phonon conductivity. There appears to be no phonon term in the normal state, in this direction, judging by the T dependence. But we are free to suppose that the phonon term is very flat and jumps below T_c, if that is our preference.

The Hall effect data are puzzling and complex but two common features are clear: the sign of the carriers is compatible only with a Mott-Hubbard view in which Cu^{++} is insulating, and a fairly universal observation is a roughly linear T-dependence of n_H.

$$n_H \sim a + bT$$

with "a" vanishing approximately for $1\,2\,3\,O_7$, the composition for which the thermoelectric power seems to be particularly small also. This and other indications suggest to me that this is the case for which $n_h = 1/2$ in the planes, and that $n_h = 1/2$ represents an approximately compensated material as would be expected if holes were spinless fermions.

One model which does not represent a viable low-energy limit is the so-called "generalized" or "two-band" Hubbard model. In this model the Cu^{++} spins are assumed to coexist with independently moving holes in the oxygen p-band. Aside from the fact that all experience with Kondo and mixed valence systems shows us that hybridization renormalizes up, not down, and that the strength of the semicovalent $Cu^{++} - O^{--}$ bond, which measures this hybridization, is clearly one of the biggest parameters in the problem, several experimental facts exclude this model. The most important is the magnetic susceptibility as measured directly, and the magnetic scattering as measured by neutrons and by NMR. Clearly the carriers deeply affect the magnetic state—in the case of the 214 superconductor, actually shifting the wavelength of magnetic scattering. Of course, the recent observation of electron superconductors also excludes the 2-band model. But actually, to me, the fact that the cuprate phenomenology is so unique was always strong evidence that the one-band model was operative. From a theoretical point of view, it seems almost certain that the 2-band model is not a stable fixed point.

Next let us think about the energetics of superconductivity. Here there are a number of misconceptions current, although the numbers, as far as I can see, are not in real dispute, and everyone gets the same value—about .1 cal/gm $^\circ K$—for the specific heat "jump" at T_c (though it is never clear whether it is a true jump). It is often stated that this is "compatible" with BCS based on values for "γ", the specific heat coefficient, of 45 mj/mol $(^\circ K)^2$ or so (I give figures for $123\,O_7$; other substances are similar). What is not often discussed is that this value is gigantic; if true it would be a heavy-electron γ. The specific heat jump, in entropy units (dimensionless) is .1–.2/carrier! Every indication suggests that this is an underestimate, not an overestimate, of the total entropy in the transition. This seems to be the only major resemblance to heavy electron superconductivity: the large total entropy.

There is, of course, no way of measuring a true "γ", but certainly the bare γ, or that which is estimated from magnetic susceptibility, are nowhere near as large as 45. Two numbers related to ΔC are $\frac{dH_c^2}{dT}$ and the coherence length. The entropy estimate suggests that the coherence lengths must be very short — 3–6 unit cells transversely, one $\perp$ c, *i.e*, $15 \times 15 \times 4\text{Å}^3$ at most for the coherence volume, and $\frac{dH_{c_2}}{dT}$ must be essentially vertical: it seems that all measurements of "H_{c_2}" so far are essentially of irreversibility effects.

The anisotropy of the penetration depth—which in Ginsburg-Landau is about equal to that of coherence length—is a matter of controversy; Malozemoff's group gets ~ 3 while others have proposed 5 or 6. Either is more or less compatible with a reasonable estimate of anisotropy of band mass, and much less than the anisotropy of conductivity: an important confirmation of the non-Fermi liquid behavior.

Next we need to discuss spectroscopy of the low excited states. The most straightforward measurement is single-particle tunneling, which has been carried out on the normal state of $1\,2\,3\,O_7$ by Dynes, Beasley and Ong and on 2 2 1 2 by Greene, at least. The characteristic and universal observation is of a V-shaped curve, $\frac{dI}{dV} = a + b|V|$. (In the department of horrid examples, I note a recent Japanese paper in which many small phonon wiggles were superposed on a gigantic and very clean example of the basic curve: the authors totally failed to mention this major result of their work.) $\frac{dI}{dV}$ measures $ImG_1(V)$ (Schrieffer) and indicates that the density of states averaged over momentum contains a constant term and a larger term linear in energy. Again, a heuristic holon-spinon interpretation has been given by Zou and myself, the essential physics of which seems unique but the details adjustable. The linear term is the superposition of two constant densities of states, assuming the electron is a composite of two excitations. We then assume that some of the pre-existing holons are in a peak of energy states $\sim kT$ wide (as they would be if they were classical or Bose particles, or Fermions in a very flat band). This gives the constant overall density.

$\rho_1(E)$
(Holon)

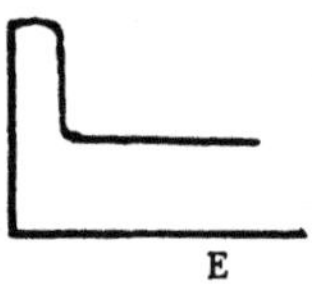

$\rho_2(E)$
(Spinon)

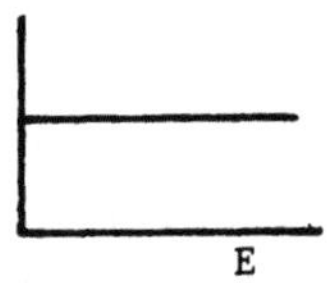

$$\rho(V) = \int \rho_1\,(V - E)\,\rho_2\,(E)\,dE$$
$$= a + b\,|V|$$

In a photoelectric spectrum averaged over momentum one should see the same as in tunneling:

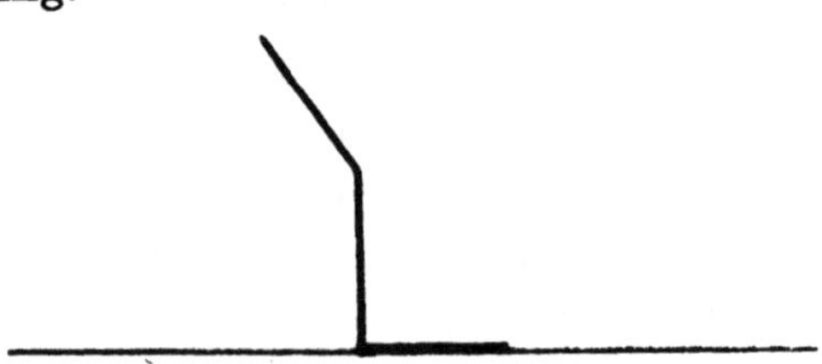

a Fermi edge with a slope below it. In a photoelectric spectrum angle-resolved, one should see at each momentum either nothing or a fermi edge:

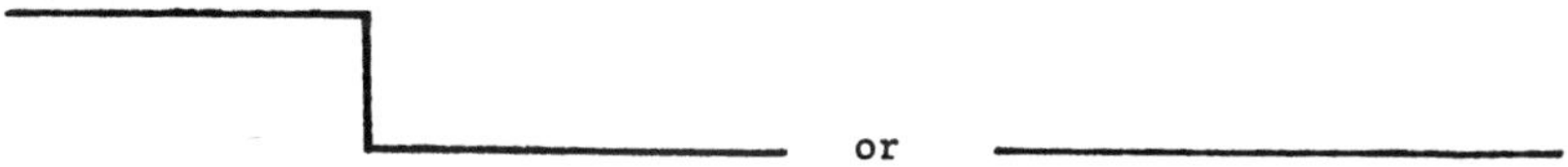

This may resemble what is seen.

Single-particle spectroscopy in the superconducting state is much less well-developed. Most characteristically, in tunneling one sees what Dynes has characterized as a multiple gap structure (or gap-like structure). Ginsburg emphasized that with the very short c-axis coherence lengths, there is no reason to suppose that the order parameter is constant in space, and, in fact, one might expect that it would fall off sharply as it approaches the surface. Thus the top two or three layers should show small order parameters and it is reasonable to suppose that the weak structure which usually appears at $2\Delta \sim 10kT_c$ may be the bulk "gap". This speculation is confirmed by the fact that by now two independent groups have seen large "gaps" in the photoemission spectrum.

Infrared and Raman spectroscopy have so far given very interesting information. The infrared spectrum has been analyzed in many different ways, and much has been made of features which turned out to be artifacts. As far as the normal state spectrum is concerned, it is characterized by a very large frequency dependence of the effective relaxation time in the infrared region up to a few tenths of an ev in the $a\,b$ plane. It is very significant that ir confirms that the c-axis direction is transparent at low frequencies. Various parametrizations can be given but to me the best is the simplest:

$$\sigma_{ir} = \frac{n\,e^2}{m^*\,(i\omega + \frac{1}{\tau})}$$

with $\frac{1}{\tau} = \max\{{kT \atop \hbar\omega}\}$ and m^* being of reasonable order, ~ 2 or $3\ m_e$. The striking fact is the frequency dependence when $\hbar\omega > kT$, which gives a reflectivity qualitatively different from conventional metals. Whatever is causing the striking temperature dependence of the $d.c.$ resistivity must be causing the same effect as a function of frequency. One is obviously called upon to identify the (V) dependence of tunneling, the linear T dependence of ρ, and the linear ω dependence of $\tau^{-1}(\omega)$, as all caused by the composite character of the electronic excitations. Incidentally, the magnitude of τ^{-1} is so large as to preclude Fermi liquid theory in yet another way (Lee): the quasi particles are diffusive, not propagating.

When the material goes superconducting, there is a large shoulder which appears in the *ir* spectrum at a large gap value, but no true gap appears until a much lower one. There is some disagreement as to the T dependence of the large gap structure, but its magnitude checks the two other best determinations of the location of the major structure. It is impossible to accept a simple Fermion BCS gap picture as describing the observed structure, but whether we are merely seeing anisotropic gaps, or more exotic possibilities are forced upon us, is not at all clear.

Raman scattering has provided a number of fascinating surprises. The cleanest and most obvious is the large "two-magnon" peak which appears in the magnetic, insulating phases at least, and appears to persist into superconducting phases at least to some degree. This peak's shape appears to be greatly affected by zero-point motion in the antiferromagnet, and it is reasonably well described by simple phase-space considerations in a spinon description of the system. Recently T. Hsu has been working on a unification of the RVB Fermionic description of the antiferromagnetic state with the Néel ordered state. He shows that Néel order introduces a mass term into the Fermion Hamiltonian and produces a smallish energy gap in the spinon spectrum. Higher energy states such as those comprising the two-magnon peak are little affected and so the spinon and Néel descriptions should coincide if properly worked out.

The Illinois group has seen a continuous Raman scattering even in the superconducting state, and one very significant result is that this shows a strong, large energy gap on going into the superconducting state. This kind of electronic low-energy continuum is perhaps one of the best experimental demonstrations that conventional Fermi liquid theory is not viable. While such scattering is not unknown in other cases, the fact that it exhibits Fano resonances with Raman-active phonons is pretty good evidence that, in this case, it is electronic in origin.

Finally, I would like to talk about the phenomenology of T_c and chemical occurrence of superconductivity. The early work on $(La, Sr)_2CuO_4$ and to some extent that on $1\,2\,3\,O_{7-x}$ supported the idea that transition temperatures were a continuum, the higher the more holes in the CuO_2 layers. It almost seemed that there was a universal curve of T_c *vs* n. This no longer seems viable, although a very influential group has recently revived this suggestion in terms of a universal connection between n_s and T_c, where n_s is the coefficient in the supercurrent, or inverse penetration depth $1/\lambda^2$, which in old-fashioned clean Ginsburg-Landau theory is just n/m^*, the number of carriers divided by their mass. On the other hand, as I will discuss shortly this is clearly not the case here and the arrow of causation may go the other way. (By a sum-rule argument.)

One fact and one speculation stand out when one looks at the systematics of T_c *vs* composition. The fact is that for hole-type superconductors, T_c is most straightforwardly related to the number of CuO_2 layers which are in relatively close contact, *i.e.*, separated only by a layer of cations. We may classify substances as 1-layer, 2-layer, and 3-layer, and T_c ranges from 0–40° for 1-layer, 60–95° for 2-layer, 110–125° for 3-layer. The $(La, Sr)_2CuO_4$ which gives the highest 1-layer value of 40° is actually anomalous in that the interlayer contact is particularly strong in this case; otherwise there is no sound evidence for $T_c >\sim 10°$K.

The speculation is that T_c is not very continuous as a function of doping, but that, in fact, it exhibits a "plateau" structure. The bulk of 1 2 3 materials seem to be stable either 55–60° or 90–95°, with little detailed dependence on doping percentage. As studies continue, it seems likely that the onset T_c in almost every 2 1 4 sample is near 37–40°, which is achieved at a rather sharp doping percentage near .16. Too many T_c measurements report a 50% or zero-resistance point, which is theoretically meaningless; if onset temperatures are reported, they seem to be very sharply grouped indeed.

It is possible, by doping with substitutional elements for Cu in the plane like Ni, Zn, Ga to reduce T_c's very rapidly, but this seems to have to do with disrupting the normal state in the planes, and is almost always signalled by appearance of residual resistance, large changes in Hall effect, etc.

This phenomenology suggests to us the following picture.

(1) Single layers are not superconducting: not at all, in a mean field sense.

(2) Tightly-coupled groups of layers have a mean field T_c controlled primarily by the size of the interlayer pair tunneling matrix elements. The pairs are not bound but relatively neutral or repulsive within the layers. This layer pair structure is responsible for the 2-d fluctuation behavior observed in 1 2 3 and Bi 2 2 1 2, which leads to a large $\frac{1}{T-T_c}$ fluctuation conductivity. This is not observed in the c-direction, nor is it obvious in 2 1 4 crystals.

(3) At a temperature slightly above T_c (mean field) the intergroup tunneling which is rather weak, corresponding to a $T_c \sim$ 5–10° K—takes over and causes a crossover to 3d.

How can T_c be dominated by interlayer pair tunneling? This can be the case only if the holon spectrum is extremely massive with either a roton-like or a diffusive behavior in k-space. We believe, thus, that it is charge fluctuations which are very massive and are responsible for the large entropy and low coherence length in these materials.

What is puzzling, and in this very much resembles the corresponding situation in heavy electron materials, is the small extent to which the heavy fluctuations couple to lattice degrees of freedom. This is evident in the lack of any evident phonon structure in the resistivity and thermal conductivity, and in the absence of self-trapping effects which one would expect from such heavy carriers at such low resistivities. Somehow the magnetic and Mott-Hubbard couplings seem to dominate the situation completely. In fact, the weakness of phonon coupling may be a requirement for high T_c; in the electron superconductors as in $Ba - KBiO_3$, conventional phonon effects seem to be present and, correspondingly, T_c is reasonably low, and coherence lengths normally large. This weakness of coupling is reasonable if one supposes that the carriers are in a Haldane-Anderson type of band dominated by $O\ p\sigma$ character, as seems to be the case and as was suggested by Sawatsky.

The "plateau" structure strongly recalls the commensurability behavior of the FQHE. It seems likely that $1\,2\,3\,O_7$ is $n_H = 1/2$, the 60° K material $n_h = 1/4$ or $1/3$. $2\,1\,4$ is $n_h = 1/6$, and the Bi and Tl materials also have larger doping (possibly 1/3, according to Sleight). On the other hand, we cannot quite rule out the possibility that commensurate Néel ordering could be playing a role (or both, if Hsu's speculations are correct.)

In conclusion, the experimental picture, while complex, is full of hints which, if we ever are able to read them, will undoubtedly tell us the full story. I, personally, believe that what is going on is so complex that purely theoretical analysis is very unlikely to give a satisfactory resolution.

Aside from financial help from IBM, NSF and the AFOSR, I must acknowledge very fruitful discussions with a wide variety of people. Only a sample can be given here.

Students: Zou, Hsu, Wheatley, Kan, Ren, Wen, Marston

Colleagues: Baskaran, Shastry, Affleck, Krishnamurthy, John, Fisher, Coppersmith

Experimentalists: Ong, Thomas, McGuire, Mehran, Holzberg, von Molnar, Dynes, Birgeneau, Shirane, Sawatsky, and many others.

Theoretical Postscript

Let me, as a final brief note, describe recent developments in pure theory. One of the most important is the discovery of spiral or "twisted" Néel phases for lightly doped Mott-Hubbard layers. Early unpublished speculations by John et al (Princeton group) were that holes would be associated with stable "meron" configurations, with a complete half-hedghog topological defect associated with each hole, analogously to the fact that in 1d each hole carries a domain wall. It turns out that the relevant amount of twist per hole is much smaller, of the order of 1/L where L is the lattice size. Thus a large number (but not $O(N)$) of holes may be accomodated in a slightly twisted Néel state. The effect is qualitatively, but not quantitatively, equivalent to the fact that the caliper of the nested Fermi surface changes with occupancy.

Various mean field theories of such states have been produced by Shraiman and Siggia, Lee et al, Krishnamurthy et al, and others. But a problem ensues when there are large numbers of holes, in that one derives a "double twist" as in the infamous "blue phase" of cholesteric liquid crystals, and this leads to topological difficulties. One suspects that topology may be the force which breaks up Néel order and mobilizes the holes, but so far no neat way of doing this has surfaced.

There are several hopeful, semi-variational attempts to identify the critical concentrations of holes with my idea of "generalized flux phases", but so far, again, nonc of this work has reached a sufficiently firm conclusion to discuss at length here.

Supplemental Comments at the IBM Conference
Keidanren Guest House
Philip W. Anderson
June 12, 1989

Everyone has the experience of wishing he had thought of the appropriate response long after a conversation. In science, unlike life, one has the opportunity to come back and complete the dialogue, and I would like to submit three belated discussion remarks to the organizers of the IBM symposium which, if they agree, might be included in the Proceedings.

(1) The first of these has to do with the measurements of Dr. Suzuki of NTT, in relation especially to Dr. Newns' claims about the sudden appearance of a full-fledged hole band as soon as doping is carried out.

It seemed to me, at the time, that the .85 μm band, proportional to doping, undoubtedly was an interband transition into the holes in the lower Hubbard band. I think it is not very relevant from which of the many bands in the top of the hybridized Cud-Op antibonding bands the transition is occurring; major components will be the antibonding $d_{Z^2} - O_{p\sigma}$ and $d_\pi p_\pi$ combinations. These are forbidden at high symmetry points but quite allowed at general k. The frequency of about 1.5 ev is very reasonable. The observed Fermi level in the gap is somewhat less than this, but we have to remove the intraband, $d_{x^2-y^2}$ portion of the top of the lower edge of the Mott gap, and 1 ev is a reasonable number for $d_{Z^2} - d_{x^2-y^2}$ splitting.

The relevance to Newns' talk is that this is, in fact, a transition into the empty part of the Op-$d_{x^2-y^2}$ band to which Newns ascribes a full-fledged Fermi surface leaving a very capacious upper band (which he likened to the gigantic DOS seen in Ce mixed valence systems just above the Fermi level). He suggests a magical process whereby this empty band develops full-blown and almost discontinuously as soon as doping occurs. But we find a DOS linear in doping δ, not in $1 + \delta$ as it would be in a Fermi liquid picture. The slow appearance of this band can, of course, be described by a picture in which Z and hence $1/n$ grow linearly with δ, but this is not Newns' picture, nor is it compatible with other data.

(2) I want to discuss the $1/T_1$ results in the normal state emphasized by Slichter and, especially, Yasuoka. I had not really gotten my thinking on NMR relaxation rates clear but I now think the implications of these data are unequivocal.

The point I had missed is that a $1/T_1$ process is purely inelastic and must represent a real, low-frequency scattering of some excitation, hence a real density of states for excitations. There is almost no possibility that there is any coupling other than simple hyperfine, and the neutron and susceptibility data leave us preferring not to postulate fluctuations of a local Néel structure. It is simplest to suppose that there is some additional low-frequency (heavy mass) excitation which does not carry a spin moment (since it does not contribute to Pauli susceptibility) but is capable of coupling to nuclear moments.

Fortunately I have already postulated such a degree of freedom, and I believe it also exists in the Laughlin-type theories in the normal state. In my version it is the heavy (quasiclassical) holons of Wheatley and Wheatley, Hsu and Anderson. If we give holons Bose character it is easy to see that either (or both) of their strong interactions with each other and with the spinon gas will make them effectively classical. Various authors have shown that the choice of Bose or Fermion statistics for holons is arbitrary.

In the Laughlin type of theory (and, in fact, in ours) the $U(1)$ gauge field coupling spins and holes is assumed to take on a finite mean field flux value. But this field is actually dynamic and will fluctuate at any T about mean field. I postulate that in the normal state this field represents a soft (heavy mass) degree of freedom to which the nuclear spins may couple. I do not believe this picture is really distinct from mine.

How does it couple to nuclear spins? At the very least, spin-orbit coupling at the $\sim .1$ level can couple spin current with orbital current, which the holon certainly can carry.

The T-dependence: if T is bigger than natural gaps or band widths, we get T-independence. It may be that the band is roton-like, in which case one expects $T^{1/2}$, or it can develop a mass gap as Laughlin prefers.

Yasuoka emphasizes the large magnitude of $1/T_1$, and (since short of Néel structure there is no extra coupling possible) this low-frequency charged mode seems the only solution to that dilemma as well.

Another comment on $1/T_1$ measurements: the chain layers in the normal state of 1 2 3 O_7 have been postulated by my group to be insulating, magnetic Cu^{++}. This is not incompatible with the Korringa T_1 observed in this case, but in fact strongly supported by that. The effective Hamiltonian for these layers will be $\sum_i (J_{||} S_i \cdot S_{i+x} + J_{\perp} S_i \cdot S_{i+y})$, with $J_{||} >> J_{\perp}$. Magnetic coupling to the plane layers is probably weak. This quasi-one dimensional coupling surely leads to an RVB state with Fermionic spinons with velocity $\propto J_{||}$. These will disappear as O vacancies appear in the chains, accounting for the rapid change in T_1.

(3) A major topic for discussion at the conference were the angle resolved photoemission measurements. Neither those of Takahashi et al, reported at the conference, nor those of Arko et al, of which an informal preprint is available, seem to resolve the crucial question discussed in my paper: is the energy spectrum at fixed angle (as opposed to angle-averaged) a peak or an edge? Any theory which agrees with tunneling measurements of angle-averaged $Im\ G_1(E)$ gives an edge at the Fermi level, a comment which seems to have escaped the Arko group. But edges in the angle-resolved spectrum are a new phenomenon. My personal reading of the data is that they depict edges.

Part II

Theory

Recent Advances in the Stability and Properties of Commensurate Flux Phases

P. Lederer, D. Poilblanc, and T.M. Rice*

ETH-Hönggerberg, Theoretische Physik, CH-8093 Zürich, Switzerland
*On leave of absence until August 31, 1989 from
Physique des Solides, Univ. Paris Sud, F-91405 Orsay, France

Recent progress in flux phases is reviewed. First numerical evidence favors a conjecture, due to Wiegmann, that a state with a flux per plaquette equal to the electron density has the lowest energy for a n.n. tightbinding model on a square lattice. The $t-J$ model is examined using a renormalized mean field theory and shown to have an instability towards flux phases at large enough J/t. These flux states have a collective gauge variable and exhibit superconducting properties.

1. Introduction

This lecture will cover some very recent theoretical developments [1,2] on the stability and properties of flux phases which make them attractive candidates to explain high-T_c superconductivity in the CuO_2-planes. The name flux phases has been coined by Affleck and Marston [3] who found that one of the many equivalent forms of the quantum spin liquid state [4] for the undoped plane, is as a projected determinantal fermion wave function in which the one electron wave functions are solutions of electrons moving in a magnetic field with flux per plaquette of $\Phi = 1/2$ (in units of the universal flux quantum hc/e). Here we will be concerned with the case of finite doping and discuss the stability and intriguing properties of generalizations of this flux state to values of Φ commensurate with the electron density. Using a renormalized mean field theory [5] we will examine the stability and properties of such states, starting from a model with a finite density of mobile holes in a Heisenberg spin system.

2. Noninteracting Electrons in a Magnetic Flux.

Our interest in this approach was sparked by P. Wiegmann during a recent visit to Zürich. He proposed the following conjecture. Consider the case of spinless non-interacting electrons moving on a square lattice in a nearest neighbor tightbinding model. If we introduce a phase φ_{ij} with each bond $< ij >$ then the Hamiltonian reads (in dimensionless units)

Springer Series in Solid-State Sciences, Vol. 89 Strong Correlation and Superconductivity
Editors: H. Fukuyama · S. Maekawa · A.P. Malozemoff

$$H_0 = \sum_{<ij>} e^{i\,\varphi_{ij}}\, c_i^+ c_j + \text{h.c.}\,. \tag{1}$$

The conjecture is that for a filling of ν electrons per site then the ground state is a uniform flux state i.e. a state in which the sum $\sum_{ij} \varphi_{ij}$ over each plaquette (or elementary lattice square) is nonzero and equal to $2\pi\ \Phi$. The value of Φ which minimizes the energy $E_0(\nu,\Phi)$ in a square lattice is $\Phi = \nu$.

The Hamiltonian, H_0 has been studied by Hofstadter [6] and others. He found a complex spectrum of eigenvalues. The case of rational values of $\Phi(= p/q)$ is easiest to study since then the problem may be made periodic. The spectrum of eigenvalues for flux Φ shows a large gap with exactly ν states below the gap corresponding to filling exactly the lowest Landau-Hofstadter level. This gap then causes a cusp minimum at $E_0(\nu,\Phi)$ at $\Phi = \nu$. There is at present no analytic proof but numerically as can be seen from Table I of Hasegawa et al. [1] the evidence is convincing.

This flux instability is in a way a generalization of the Peierls instability in 1-dim. to the 2-dim. square lattice. The energy in 1-dim. is lowered by a potential whose period is related to the density while in 2-dim. it is a flux which is determined by the density. In 1-dim., it is easy to realize such a potential, simply make a periodic displacement of the ionic lattice. It is less easy to realize a microscopic flux and so one of the questions which will be addressed in the next section will be the physical mechanism by which such a microscopic flux can be realized. In addition we will examine the properties of a self consistently generated flux phase.

3. Flux Phases in the $t - J$ Model.

We use the $t - J$ model to describe the strongly correlated CuO_2 planes [7,8]

$$\begin{aligned} H &= P_0 \sum_{<ij>,\sigma} [t\ c_{i\sigma}^+ c_{j\sigma} + \text{h.c.}]\, P_0 + J \sum_{<ij>} \vec{S}_i \cdot \vec{S}_j \\ &= H_t + H_J\,. \end{aligned} \tag{2}$$

Note in writing the kinetic energy term with a real hopping matrix element, t, we have implicitly chosen a gauge for the atomic states created by $c_{i\sigma}^+$. We can change to another gauge by the transformation $c_{i\sigma}^+ \to e^{i\,\theta_i}\, c_{i\sigma}^+$ on the operators and transform $t_{ij} \to t\ e^{-i(\theta_i - \theta_j)}$. The $t - J$ Hamiltonian is of course invariant under such a transformation. The projector $P_0(= \Pi_i(1 - n_{i\uparrow}\, n_{i\downarrow}))$ is on to the subspace with no double occupation.

According to the previous section, energy is gained by introducing a flux which is commensurate with the density. Actually earlier this year generalizations of the Affleck-Marston 1/2-flux state were proposed as interesting states by Anderson, Shastry and Hristopoulos [9] and related proposals were made previously by Wiegmann [10], Laughlin [11] and others [12]. We consider then the wave function at density ν (els./site spin) of the form

$$\mid \Psi >= P_0 \mid \Psi_0 > \; = \; P_0 \prod_l^{occ} \tilde{c}_{l\uparrow}^+ \prod_l^{occ} \tilde{c}_{l\downarrow}^+ \mid vac > \tag{3}$$

where $\tilde{c}_{l,\sigma}^+ (= \sum_k \psi_{l,k}^* (\{\varphi_{ij}\}) \; c_{k,\sigma}^+)$ are creation operators for electrons. in eigenstates of H_0. The eigenvalues of H_0 are dependent only on the flux Φ but the eigenstates are gauge dependent (i.e. they depend on the phases $\{\varphi_{ij}\}$). The choice of a flux $\Phi = \nu$ in (3) was suggested on general grounds by Anderson, et al. [8].

A simple and reasonably accurate way to treat the projection operator P_0 is following Zhang et al. [5] to introduce Gutzwiller renormalization factors i.e. set $< H >=< \psi \mid H_t + H_J \mid \psi > \; = g_t < H_t >_0 + g_J < H_J >_0$ where $g_t = 4\delta/(1-2\delta)$ and $g_J = 4/(1+2\delta)^2$ in terms of the hole concentration $2\delta (= 1 - 2\nu)$. In the calculation of the exchange term we can view it as arising from correlated hopping processes i.e. terms of the form $c_{i\uparrow}^+ \; c_{j\uparrow} \; c_{j\downarrow}^+ \; c_{i\downarrow}$. The trick in evaluating such terms is to introduce two phase factors and factorize

$$< c_{i\uparrow}^+ \; c_{j\uparrow} \; c_{j\downarrow}^+ \; c_{i\downarrow} >_0 \; = \; < e^{i\varphi_{ij}} \; c_{i\uparrow}^+ \; c_{j\uparrow} >_0 < e^{-i\varphi_{ij}} \; c_{j\downarrow}^+ \; c_{i\downarrow} >_0 \; . \tag{4}$$

Provided we choose the phases φ_{ij} to be the same as in (3) then this term is simply the square of a sum over the Hofstadter [6] eigenvalues $E^n(\vec{k})$. This is possible only if we choose the <u>same</u> phase factors for $\uparrow$ and $\downarrow$ spins i.e. **the flux and the gauges for $\uparrow$ and $\downarrow$ spins in (3) must be the same.** This choice gives the lowest energy. In other words the correlated hopping processes described by H_J allows us to introduce phase factors so that $\uparrow$ and $\downarrow$ spins see effective one electron Hamiltonian of the form (1) but the phases for the $\uparrow$ and $\downarrow$ are linked by the correlated motion.

The introduction of phase factors corresponding to a finite flux is however not favored by the kinetic term, H_t. This term is diagonalized by Bloch states and it favors therefore a state without flux. However the number of hops involving H_t varies as the number of holes or empty sites i.e. $\propto \delta$. Therefore we can see that the correlated hops can dominate over single particle hops roughly if $J \gtrsim \; \delta t$. In this case we can write the energy as

$$< H_t + H_J > \; = \; 2tg_t \Big(\frac{q}{L} \sum_{n,\vec{k}}^{occ} E_{\vec{k}}^n\Big) K(\Phi) - \frac{3J}{16} g_J \Big(\frac{q}{L} \sum_{n,\vec{k}}^{occ} E_{\vec{k}}^n\Big)^2 \tag{5}$$

with the function, $K(\Phi) = Max\{\overline{\cos\varphi_{ij}}\}$ which represents the maximum in the single particle term given that we have a flux, Φ in each plaquette. If $\Phi = 0$ then $K = 1$ since we have simply the Bloch state but if $\Phi \neq 0$ we can use the gauge degree of freedom to achieve the best possible value. This gauge is chosen relative to the gauge in the definition of H_t. For $\Phi = 0$ then $\varphi_{ij} \equiv 0$ and $K = 1$ is the obvious answer but when $\Phi \neq 0, K < 1$ and there is an inherent competition with the H_J term which favors a choice $\Phi = \nu$. The function $K(\Phi)$ was studied by various authors in the context of arrays of Josephson junctions and investigated numerically in Ref. [2]. The choice of a Landau gauge gives $K(\Phi) = 1/2$ but is possible to do better. In addition to the gauge degree of freedom one can also add a flux ± 1 to individual plaquettes with changing the Hofstadter spectrum and therefore the H_J-term. We found the best value of $K(\Phi)$ corresponding always to a state in which the total flux enclosed in a supercell is always zero although the flux in individual plaquettes is ν or $\nu \pm 1$. This means that there will be no net angular momentum in the bulk of the plane but there can be surface currents.

The values of $K(\Phi)$ are $\gtrsim 0.6$ but this value represents a considerable reduction of the H_t-term in the renormalized meanfield theory and therefore states with $\Phi = \nu$ are stable only for critical ratios $J/t > J/t \mid_{crit} \gtrsim 1$ (the exact value depends on ν). Nonetheless it is clear an instability to a flux phase can take place for certain values of J/t and we turn now to an examination of their properties.

4. Superconductivity from Commensurate Flux Phases.

These states which are generalization of the Affleck-Marston 1/2-flux to values $\Phi = \nu$ have intriguing properties when following Ref. [2] we look at their response to external fields. Consider first applying a small external magnetic field $\vec{H}_{ext}$. Will the wavefunction respond? Since the local value of the magnetic flux in the bulk of the sample has been chosen to optimize the H_J-term this will not change. So the wavefunction is rigid in a magnetic field and since, as mentioned above, it is a wavefunction with no net magnetic flux in a supercell it will correspond to a Meissner state. In other words the ground state of the system even if we apply a small $\vec{H}_{ext}$ remains field free in the bulk and this can be achieved when $\delta \neq 0$ by modifying the surface currents.

Current carrying states are easy to make by a gauge transformation of $\mid \Psi >$ to a wavefunction $\mid \Psi_S >= exp(i \sum_{j\sigma_j} S(r_j, \sigma_j)) \mid \Psi >$. This state will correspond to a uniform current carrying state in the x-direction when $\vec{\nabla}_x S = q_x$. $< H_J >$ is unchanged so long as we take S independent of the spin direction. These current carrying states can be viewed as displacements of the lowest filled Hofstadter band in $\vec{k}$-space and this is achieved as in usual BCS superconductors by a gradient of the collective gauge variable of the wavefunction. In the BCS superconductors it is by forming Cooper pairs of $\uparrow$ and $\downarrow$ spin electrons that the gauge coupling between them is obtained. In the commensurate flux phases it is achieved through the coherent correlated hopping motion of the $\uparrow$ and $\downarrow$ spin electrons. The effect is similar in that the gauge collective coordinate S is the same for both spins. This in turn means that in a closed ring geometry the variable $S(\vec{r})$ need not be periodic $(mod(2m\pi))$ but only the sum over spin i.e. $2S(\vec{r})$ need be periodic $mod(2m\pi)$. This gauge coupling between $\uparrow$ and $\downarrow$ spins then leads to flux quantization for the ring geometry with an Onsager value, $hc/2e$, as in the case of usual BCS superconductors.

In a similar way we can examine the response of the system to an external vector potential $\vec{A}_{ext}$. This will affect the energy through the $< H_t >$ term only and for small external fields one can expand the $< H_t >$-term and obtain a form similar to a Ginzburg-Landau expansion for a BCS superconductor. Details can be found in Ref. [2]. The dependence of the energy on the gauge and the vector potential $\vec{A}_{ext}$ comes from the dependence of the function $K(\Phi)$ on the individual phases $\{\varphi_{ij}\}$.

5. Conclusions.

The physical picture that comes from the mean field theory of the flux phases with a flux commensurate with the electron density is quite different from that of a usual BCS-superconductor. The key to the flux phase is the correlated motion of the $\uparrow$ and $\downarrow$ electrons which we usually interpret as a Heisenberg spin-spin coupling. If we look at the origin of this coupling in Andersons's superexchange theory [13] then it is clear that it is a form of correlated motion. If we wish to set up extended states for the $\uparrow$-spin electrons say then this motion will be a combination of one electron and correlated electron hops. These latter allow us to make coherent one-electron states which are not Bloch states but "Hofstadter states" which are solutions of (1). The electrons undergo quite a different motion to that in usual metals, where Bloch coherence dominates.

The stability of these commensurate flux phases is a consequence of their gap structure and the gaps are linear in the effective hopping matrix element in this case, J. Therefore although at the moment we do not have a full finite T mean field theory, the energy scale would seem to be linear in (J, t) and a high value of T_c can be expected. The extension to finite T is only one of a series of open questions.

First we need to examine the relative stability more carefully. In the dilute electron limit ($\nu \rightarrow 0$ or $2\delta \rightarrow 1$) clearly one electron hops will dominate and Bloch coherence will predominate.

In the limit $\nu \to 1/2$ or $\delta \to 0$ there can be a close competition with projected BCS states which were called RVB states by Anderson [7]. Without doping [$\delta = 0$] the d-wave RVB and Affleck-Marston states are identical. Further for small δ, the correction linear in δ to $< H_t >$ is the same for both states if the phases $\varphi_{ij} = \pi/4$ in the Affleck-Marston state. Choosing a commensurate flux then gains an additional energy but only of order δ^3.

A further open question is the effect of terms beyond the $t-J$ model on the relative stability. Also we need to understand better the correct order parameter. A third set of questions concerns the relation to other approaches such as Laughlin's [10]. If we regard low energy excitations as involving continuous variations in the lowest filling Hofstadter spectrum then it is clear that it will involve changing the flux when we change the density ν. Thus a low energy singlet excitation in which 2 electrons and a flux quantum are added is possible. On the other hand if a single electron and a half flux quantum is added then there is a long range disturbance in the vector potential (or phases φ_{ij}). In this sense there is a close relationship to the ideas of Laughlin and others who concentrate on the fractional statistics that would accompany half-flux quasiparticles [11,12].

Finally we need to understand better how our solution, which is closely related to the Quantum Hall Effect (our uncorrelated trial wavefunction would have, in a real external field, Quantum Hall number 1) is related to BCS-theory. Poilblanc [14] has shown how (3) can be written as a form of paired wavefunction with constant particle number, so there may be close parallels although the underlying microscopic mechanism is very different.

Acknowledgments. We are very grateful to P. Wiegmann, G Blatter, Y. Hasegawa and B.S. Shastry for stimulating discussions and to the Swiss Nationalfonds for financial support.

References

1. Y. Hasegawa, P. Lederer, T.M. Rice and P. Wiegmann, ETH-preprint (1989).
2. P. Lederer, D. Poilblanc and T.M. Rice, ETH-preprint (1989).
3. I. Affleck and J.B. Marston, Phys. Rev. B37, 3774 (1988).
4. I. Affleck, Z. Zou, T. Hsu and P.W. Anderson, Phys. Rev. B38, 745 (1988).
5. F.C. Zhang, C. Gros, T.M. Rice and H. Shiba, Super. Sci. Tech. 1, 36 (1988).
6. D.R. Hofstadter, Phys. Rev. B14, 2239 (1976).

7. P.W. Anderson, Science 235, 1196 (1987) and in "Frontiers and Borderlines in Many Body Physics", Varenna Lectures, ed. R.A. Broglia and J.R. Schrieffer, North-Holland (1988).
8. F.C. Zhang and T.M. Rice, Phys. Rev. B37, 3759 (1988).
9. P.W. Anderson, B.S. Shastry and D. Hristopoulos, Princeton-preprint (1989)
P.W. Anderson, Cargèse Lecture Notes (1988).
10. P.B. Wiegmann, Phys. Rev. Lett. 60, 821 (1988); Physica Scripta T27 160C.
11. R.B. Laughlin, Science 242, 525 (1988) and these lectures.
12. X.G. Wen, F. Wilczek and A. Zee, ITP-preprint (1988).
13. P.W. Anderson, Phys. Rev. 115, 2 (1959).
14. D. Poilblanc, ETH-preprint (1989).

U(1) Gauge Field, t′-J Model and Superconductivity

P.A. Lee

Department of Physics, Massachusetts Institute of Technology,
Cambridge, MA 02139, USA

We argue that the physics of the oxide superconductors may be represented by the t-t'-J model, where J is the exchange, t_{ij} is the hopping between opposite sublattices and t'_{ij} is the hopping between the same sublattice. Here the sublattices A and B are introduced as labels only, without assuming long range antiferromagnetic order. The hopping terms t and t' have very different physical consequences. The t term disorders the local antiferromagnetic ordering as the hole hops. This leads to an enhancement of the hole mass for the coherent motion of the hole. At the same time, the t term may lead to a spiral structure, as described by Shraiman and Siggia. Thus the t term may be responsible for the destruction of long range order upon doping. The t' term, on the other hand, allows the hole to hop without disordering the spin. Nevertheless, as the hole makes a closed loop on one sublattice, it is subject to a slowly varying spin quantization axis and the hole wavefunction picks up a phase equal to half the solid angle subtended by the spin orientations around the loop. The phase can be represented by an Aharonov-Bohm flux, resulting in a U(1) gauge theory, as first pointed out by Wiegmann. There is a natural attraction between holes on opposite sublattices because they experience opposite Aharonov-Bohm fluxes. We treat the gauge theory in the presence of a finite concentration of holes and describe the resulting superconductivity. The gauge field also enhances coupling to particle-hole excitations, leading to a $T^{4/3}$ law for the normal state resistivity.

It is commonly accepted that the essential physics of the copper oxide superconductors can be modelled by two dimensional sheets of copper and oxygen orbitals. The insulating compounds consist of spin 1/2 local moments on the copper sites and doping introduces holes on the oxygen orbitals. Anderson[1] and Zhang and Rice [2] argue that a symmetrized orbital of the oxygen hole forms a singlet with the copper moment. The resulting complex can be considered a hole in a square lattice of copper orbitals. The resulting one band model can be mapped onto the so-called t - J model

$$H = J\mathbf{S}_i \cdot \mathbf{S}_j + t \sum_{\langle ij\rangle\sigma} (1 - n_{i,-\sigma})\, c^{\dagger}_{i\sigma} c_{j\sigma} (1 - n_{j,-\sigma})$$

where t describes the hopping of a particle between nearest neighbors in a square lattice, subject to the constraint of no double occupation on any site, and $n_{i\sigma} = c^{\dagger}_{i\sigma}c_{i\sigma}$ and $\mathbf{S}_i = c^{\dagger}_{i\alpha}\sigma_{\alpha\beta}c_{i\beta}$. There has been much discussion concerning the applicability of the one band model. In particular, recent estimates seem to indicate a substantial overlap between neighboring oxygen p orbitals,[3] which was not considered in the Zhang-Rice derivation. We take the point of view that when the hole concentration is small it is difficult to imagine a heavy Fermi liquid ground state consisting of a uniform compensation of all the copper moments. Instead, we assume that the formation of a local singlet between the copper and oxygen spins is a good starting point. The inclusion of direct oxygen-oxygen overlap leads to a substantial further neighbor hopping between the singlet complexes due to direct overlap and to an expansion of the size of the singlet. Thus we propose that the one band model should be extended to a t - t' - J model where t_{ij}, t'_{jk} denote hopping on opposite and same lattices respectively. (We divide the square lattice into A, B sublattices as labels only, without assuming Neel ordering.) t and t' are strongly renormalized and are very difficult to compute from bare values given by band calculations. In this sense the one band t - t' - J model is not derivable from a one band Hubbard model and, depending on parameter range, may have different physics.

In a locally antiferromagnetic (AF) environment, the t and t' terms have very different physical consequences. A hole hopping from A to B sublattice leaves behind a mis-oriented spin and therefore destroys the local AF order. This effect is absent for the t' hopping term. We believe the t term is responsible for the destruction of Neel order upon the introduction of a small concentration of holes.

On the other hand, when the hole makes an even number of hops using t, the overturned spins it leaves behind can be repaired by the spin-flip term J. We have shown elsewhere[4] that in both the ordered and disordered AF phases, this process generates a coherent hopping matrix element of size J on the same sublattice, provided that $t \gg J$. Thus the major effect of t is actually to generate a $t' \approx J$. In this paper we assume that the t term has done its job in disordering the spins and we focus on the effect of t' which we expect to be $\gtrsim J$ with contributions from both direct hopping and via t as mentioned earlier.

A number of workers [5,6,7,8] have pointed out that the spin fluctuation and its coupling to the holes should be described by a compact U(1) gauge theory. In particular, Weigmann [7] has studied the t' - J model with disordered spin (which he denotes as a quantum paramagnet). He pointed out that holes on opposite sublattices couple to a U(1) gauge field which represents spin fluctuations with opposite charges, resulting in superconducting pairing. Wen [8] recently developed a phenomenological description with similar results. The treatment so far is restricted to a few holes moving in a spin background. In this paper we first re-consider this limit and provide a description of the gauge field in terms of the instantaneous sublattice magnetization. This physical picture will prepare us for the second part of this paper, which treats the t' - J model with a finite concentration of holes.

We employed the Schwinger boson-slave Fermion formalism to handle the constraint of no double occupation, i.e. in the t and t' term the Fermion operator $c^{\dagger}_{i\sigma} = \hat{f}_i \hat{b}^{\dagger}_{i\sigma}$ where $b_{i\sigma}$ is a boson operator which carries the spin index and $\hat{f}^{\dagger}_i$ is a spinless Fermion operator which creates a hole on site i relative to half filling. We restrict the Hilbert space to the constrained subspace satisfying[9]

$$\sum_{\sigma} \hat{b}^{\dagger}_{i\sigma}\hat{b}_{i\sigma} + \hat{f}^{\dagger}_i \hat{f}_i = 1 . \qquad (1)$$

The quantum partition function can be written as a functional integral $Z = \int db_i df_i d\lambda_i \exp(-\int_0^\beta L d\tau)$ over complex b_i, grassmanian f_i and time independent Lagrange multiplier field λ_i and $L = L_J + L_{t'} + L_\lambda$

$$L_J = \frac{1}{2} \sum_j (b_j^\dagger \frac{\partial}{\partial\tau} b_j - \frac{\partial b^\dagger_j}{\partial\tau} b_j) + J \sum_{<ij>} b_i^\dagger \sigma b_i \cdot b_j^\dagger \sigma b_j \tag{2a}$$

$$L_{t'} = \frac{1}{2} \sum_j (f^*_j \frac{\partial}{\partial\tau} f_j - \frac{\partial f^*_j}{\partial\tau} f_j) + \sum_{ij} t'_{ij} f_i f_j^* b_i^\dagger b_j + \text{c.c.} \tag{2b}$$

and $L_\lambda = \sum_j i\lambda_j(b_j^\dagger b_j + f_j^* f_j - 1)$. We have used the complex spinor notation $b_j = (b_j\uparrow, b_j\downarrow)$ so that $\hat{\Omega}_j = b_j^\dagger \sigma b_j$ represents the instantaneous spin orientation (or quantization axis) on site j. Note that the gauge transformation $b_j \rightarrow b_j e^{i\theta(j,\tau)}$ leaves $\hat{\Omega}_j$ invariant and indeed leaves L_J invariant if holes are absent and $b_j^\dagger b_j = 1$. In this case we can fix the gauge and parametize $b\uparrow = \cos\theta/2$, $b_\downarrow = \sin\theta/2\, e^{-i\phi}$ where θ, ϕ are the Euler angles of $\hat{\Omega}$. For any pair of spinors we have the identity[10]

$$b_1^\dagger b_2 = e^{i\hat{\omega}_{12}/2} \left|\frac{1}{2} (1 + \hat{\Omega}_1 \cdot \hat{\Omega}_2)\right|^{1/2} \tag{3}$$

where $\hat{\omega}_{12}$ is the solid angle subtended by the unit vectors $\hat{z}$, $\hat{\Omega}_1$, $\hat{\Omega}_2$. Using Eq.(3) we can readily see that the first term in Eq.(2a) when integrated over τ corresponds to the Berry phase for adiabatic motion of a quantum spin, i.e. it is the solid angle swept out by the unit vector $\hat{\Omega}_i(\tau)$ when it executes a closed loop from $\tau = 0$ to β. The second term in Eq.(2a) is simply the classical spin energy $J\hat{\Omega}_i \cdot \hat{\Omega}_j$. Thus we have recovered Haldane's prescription for a path integral representation of spin 1/2. Haldane [11] has shown that L_J reduces to the nonlinear σ - model in the long wavelength limit, i.e.

$$L_J = \int dr \frac{c}{g} [\frac{1}{c^2} (\frac{\partial}{\partial\tau} \hat{\Omega})^2 + |\nabla \cdot \hat{\Omega}|^2]$$

where $\hat{\Omega}(r)$ is a continuum unit vector corresponding to the local sublattice magnetization, c is the spin wave velocity and g is an effective coupling. For $g > g_c$, the σ-model has a disordered ground state[12,13] with a gap $2\Delta_s$ for triplet excitations and a correlation length $\xi_s = c/2\Delta_s$. Haldane has pointed out that there are subtle corrections to the σ-model when hedgehogs are present. We shall not consider these complications here.

Now we are ready to interpret the Lagrangian in the presence of a few holes. We note that $L_{t'}$ corresponds to a tight binding Hamiltonian on A and B sublattices separately with a time dependent hopping matrix element $t'_{ij} b_i^\dagger(\tau) b_j(\tau)$. Since i and j are on the same sublattice, b_i is slowly varying and corresponds to $\hat{\Omega}(r)$ and $-\hat{\Omega}(r)$ on A and B sublattices respectively. Using Eq.(3) we see that $b_i^\dagger(\tau) b_j(\tau) \approx e^{i\hat{\omega}_{ij}/2}$. It is natural to associate a lattice gauge field A_{ij} on the bond ij equal to $\hat{\omega}_{ij}/2$. The associated time dependent magnetic flux through an elementary plaquette in sublattice A is then equal to the solid angle on the unit sphere subtended by the four $\hat{\Omega}(r,\tau)$ on the corners of the plaquette. Thus we see that the fluctuation of the $\hat{\Omega}$ field gives rise to an effective electromagnetic field which acts on the holes. It is also clear that the holes on sublattice B couples to the gauge field with a "charge" of opposite sign since the solid angle swept out by $-\hat{\Omega}$ is of opposite sign.

We remark that while L obeys local gauge invariance $b \to b\, e^{i\theta}$, $f \to f e^{i\theta}$, $L_{t'}$ does not. Using the constraint the first term of Eq.(1) becomes $-iA_0(f_j^\dagger f_j + b_j^\dagger b_j)$ where $A_0 = i\, b_j^\dagger \frac{\partial}{\partial\tau} b_j$. The $A_0\, f_j^\dagger f_j$ term can be added to Eq.(2) to produce $\partial/\partial\tau \to \partial/\partial\tau - iA_0$ and a gauge invariant Lagrangian in the limit of zero hole density. The time component A_0 can be related to the spin structure around a space-time plaquette in the same way as we did for a plaquette in space.

Our relation of the flux of the gauge field with the space and time dependence of the spin quantization axis on a lattice allows us to give a physical picture of the pairing mechanism discussed by Wiegmann. Consider the Feynman paths of two particles on opposite sublattices which begin and end close to each other. An Aharonov-Bohm phase factor exp $(\pm i \int \mathbf{A} \cdot d\mathbf{l})$ is associated with each path and the propagation amplitude will be suppressed by the factor $\exp(i\oint \mathbf{A} \cdot d\mathbf{l}) = \exp(i\phi)$ where ϕ is the total flux enclosed by the loop formed by the two paths. If the spins are disordered we expect $<\exp(i\phi)> \approx \exp(-\text{area})$,[14] leading to strong suppression of the propagation amplitude, unless the two paths are always within ξ_s of each other. Thus the pair on opposite sublattices are bound and propagate essentially as a free boson. A finite density of these pairs will Bose condense. Note that the pairing mechanism relies on coherent propagation on a length scale $>> \xi_s$ and is inherently associated with the disordered state. However, the superconducting state corresponds to a dilute gas of tightly bound pairs, analogous to the bi-polaron limit. It does not describe the experimental situation in oxide superconductors in that, among other things, it predicts a gap in the single particle excitation above T_c. Clearly this picture is valid only if the average distance between holes is larger than ξ_s, because the exchange between holes has been ignored. Experimentally it appears that in order to suppress the Neel ordering, sufficient density of holes must be introduced so that this criterion is violated.

In the remainder of this paper we treat the opposite limit $k_F \xi_s >> 1$. Using the relation between spin solid angle and flux, the typical "magnetic" flux within an area of ξ_s^2 is unity, so that $k_F \xi_s >> 1$ implies the "magnetic" flux per particle is small, so that a perturbative treatment is possible.

Unlike the t-J model, the t'-J model given by Eq. (1) and (2) can be treated systematically in a 1/N expansion. We invert the quantization axis on B sublattice by introducing $\bar{b}_\sigma = \varepsilon_{\sigma\sigma'} b^*_{\sigma'}$ so that b_i and $\bar{b}_{i+1}$ form a slowly varying field. Following Arovas and Auerbach, we extend the σ sum from 1 to N and introduce a mean field $D_{ij} = \sum_\sigma < \bar{b}^*_{\sigma i} \, b_{\sigma j} >$ where i,j are nearest neighbors. For N = 2 this corresponds to the formation of a singlet bond. For $2S/N < 0.19$ the mean field theory corresponds to a short range RVB state [16] with an energy gap Δ_s. We shall assume that the effect of t and doping has stabilized this mean field state. Fluctuations around this mean field were treated by Read and Sachdev[15], who produced the effective Lagrangian.

$$\tilde{L}_J = \int d^2r \, \{|(\partial_\mu - iA_\mu) z_\sigma|^2 + \frac{\Delta_s^2}{c^2} |z_\sigma|^2\} \tag{4a}$$

where $z_\sigma(r_i) = \frac{1}{2} (b_{i\sigma} + \bar{b}_{i+1,\sigma})$ corresponds to the slowly varying sublattice magnetization and A_μ is related to the phase fluctuation of D_{ij}. By integrating out z from Eq.(4a) we obtain the Lagrangian which controls the fluctuation in the gauge field,

$$\tilde{L}'_J = \int d^2r \, c^2 F_{\mu\nu}^2 / 16\tilde{e}^2 \tag{4b}$$

where $F_{\mu\nu} = \partial_\mu A_\nu - \partial_\nu A_\mu$ and $\tilde{e}^2 = \Delta_s/8\pi$ (for N=2).

This treatment can readily be extended to include Eq. (2). Suppose we start with second and third neighbor hopping t'_2 and t'_3. The mean field D_{ij} implies $< b_i^\dagger b_j > \neq 0$ so that hopping matrix elements $\tilde{t}'_2$ and $\tilde{t}'_3$ are generated resulting in a band structure $4\tilde{t}'_2 \cos k_x \cos k_y + 2\tilde{t}'_3 (\cos 2k_x + \cos 2k_y)$. Depending on the sign and relative size of $\tilde{t}'_2$ and $\tilde{t}'_3$, the band minimum may be at (0,0), [(π,0), (o,π)] or (±π/2, ±π/2). For simplicity we treat the case with a single minimum (otherwise addition band labels are needed) and expand about it. Including fluctuations about the mean field following Read and Sachdev produces short range interaction terms plus the important coupling to the gauge field,

$$\tilde{L}_{t'} = \int d^2\mathbf{r}\{f^\dagger_A(\frac{\partial}{\partial x_o} - iA_0)\, f_A + f^\dagger_B(\frac{\partial}{\partial x_0} + iA_0)\, f_B \tag{4c}$$

$$+ \frac{1}{2m}\, [f^\dagger_A(\frac{\partial}{\partial x_i} - iA_i)^2 f_A + f^\dagger_B(\frac{\partial}{\partial x_i} + iA_i)^2 f_B]\} .$$

Eq. (4a-c) were first written down phenomenologically by Wen[8].

We emphasize that Eq.(4) is valid only for ω, $q < \Delta_S$. In the absence of holes Eq.(4b) predicts a mode with linear dispersion . It is a singlet excitation created by $F_{\mu\nu}$ which in spin language is a rather complex operator corresponding to the topological charge density $\hat{\Omega} . \partial_\mu \hat{\Omega} \times \partial_\nu \hat{\Omega}$,[14] i.e., the solid angle subtended by three spins as discussed earlier. This mode does not enter $< \mathbf{S}(\mathbf{r},\tau)\, S(o) >$ but should contribute to specific heat.[16] We shall now see that this mode is strongly modified in the presence of a finite areal density $n = n_A = n_B$ of holes.

As mentioned earlier if $k_F\xi_s >> 1$, we can start with the Fermi sea and treat gauge fields perturbatively. We redefine A_μ with a factor of $\tilde{e}/c$ to conform with electromagnetic convention and adopt the transverse gauge $\nabla \cdot \mathbf{A} = 0$. The longitudinal part A_0 decouples and gives rise to a Coulomb potential $4\pi\tilde{e}^2/q^2$. It is convenient to include the A^2 term in Eq.(4c) in defining the bare transverse field propagator $D^0_{\mu\nu}(\mathbf{x}) = <T(A_\mu(\mathbf{r}, \tau)\, A_\nu(0))>$,

$$D^0_{\mu\nu}(\mathbf{q}, \omega_n) = -\frac{4\pi c^2(\delta_{\mu\nu} - q_\mu q_\nu/|q|^2)}{\omega_n^2 + c^2q^2 + \omega_{sp}^2}, \tag{5}$$

where $\omega_{sp}^2 = 4\pi\tilde{e}^2 n/m \approx \Delta_s\varepsilon_F/2\pi$. The coupling to the transverse field **A** is given by the usual **A.J** and $(f^\dagger_A f_A - n_A)\ A^2$ terms. Unlike the usual electron gas where the coupling to transverse fields is small by v^2_F/c^2, here the spin wave velocity $c \approx J$ is comparable or less than $v_F \approx n^{1/2}t'$ and new physics emerges. Fortunately if $c < v_F$ Migdal theorem is obeyed and a systematic diagrammatic analysis analogous to the phonon problem is possible.

The first step is to write down the screened propagator $D_{\mu\nu}$ which is obtained from $D^0_{\mu\nu}$ by replacing ω_{sp}^2 in Eq.(5) by $\omega_{sp}^2 - 2(\tilde{e}/mc)^2\, 2k_F^2\, \Pi_o(q, \omega)$ where Π_o is the usual 2D free Fermion polarization buble. For $\omega << v_F q$ and $q < 2k_F$, the correction term is a constant which exactly cancels ω_{sp}^2 and we have

$$D_{\mu\nu}(q, \omega + i\eta) = 4\pi c^2\, (\delta_{\mu\nu} - q_\mu q_\nu/|q|^2)\, [\omega^2 - c^2q^2 + i\omega_{sp}^2 \frac{\omega}{v_F q}]^{-1} \tag{6}$$

For $\omega >> v_F q$, Π_o is negligeable and $D_{\mu\nu} \approx D_{\mu\nu}^0$. For a given q, it is clear that for both $\omega > v_F q$ and $\omega < cq$, $\mathrm{Re}D(q, \omega + i\eta)$ is negative for $\omega < \omega_{sp}$ and exchange of $D_{\mu\nu}$ will lead to pairing of the gauge invariant pair propagation $<f_B^\dagger(r, \tau)\, f^\dagger{}_A(r, \tau)\, f_A(0)\, f_B(0)>$. In the spirit of BCS , we construct a retarded interaction by replacing $D_{\mu\nu}$ by $D_{\mu\nu}^0\,(\omega = q = 0)$ for $\omega < \omega_c$ and zero otherwise. Clearly ω_c is at least Δ_s but its determination requires a Lagrangian valid for $\omega > \Delta_s$. Physical considerations lead us to set $\omega_c \approx \Delta_s$ because excitations with ω, $cq > \Delta_s$ are indistinguishable from those of a Neel ordered state which we assume to be not superconducting. In analogy with phonon exchange, we obtain a dimensionless couping constant $\lambda \approx (\tilde{e}/mc)^2 k_F^2 m/\omega_{sp}^2 \approx 1$. (The restriction $q\xi_s < 1$ may reduce λ by $(k_F\,\xi_S)^{-1}$ upon averaging over Fermi surface.)

In addition to the exchange of transverse gauge field, we also have to include the longitudinal field, which in the static limit is simply a screened Coulomb interaction $-4\pi\tilde{e}^2/(q^2 + \tilde{\kappa}^2)$ where $\tilde{\kappa}^2 = 4\pi\tilde{e}^2 dn/d\mu$. Note that the dimensionless coupling constant $\tilde{\mu} = -1$ in the $q \rightarrow 0$ limit, and exactly cancels $\mu = 1$ of the ordinary screened Coulomb repulsion. Because of the different q and ω dependence, there will probably be some residual attraction left. An intrinsic difficulty of all previous attempts to raise T_c by exchanging high frequency excitation ω_0 such as plasmons or excitons, is that the renormalized Coulomb repulsion $\mu^* = \mu/(1 + (\mu \ln(\varepsilon_F/\omega_0))$ grows with increasing ω_0. Our mechanism avoids this problem by cancelling out the coulomb repulsion, leaving $\lambda \approx 1$ so that our theory predicts a uniform energy gap Δ equal to a fraction of $\omega_c \approx \Delta_s$. For sufficiently large Δ_s, a substantial part of the Fermi sea may participate in the pairing, with a relatively short coherence length ξ given by $\xi k_F \approx \varepsilon_F/\Delta_s$ or $\xi/\xi_s \approx v_F/c$ which can be in order of unity.

The coupling to low lying transverse gauge fields has a profound effect on the normal state properties. Very recently, Reizer [17] pointed out that in ordinary metal, coupling to transverse electromagnetic field leads to a scattering rate for electrons of order $(v_F/c)^2 kT$. His argument can be directly applied to the present problem in the normal state. For simplicity let us consider a single Fermion with energy ω above the Fermi energy. It decays by emitting Bose excitation with spectral density given by $\mathrm{Im}D(q, \nu)$ where $\nu < \omega$. In Eq.(6), $\mathrm{Im}D^{-1} \approx \nu/v_F q$ is the usual density of particle-hole excitations. The important point made by Reizer is that for transverse excitation, $\mathrm{Re}D^{-1}$ vanishes as q^2. Thus coupling to the transverse gauge field effectively enhances the matrix element for the excitation of long wavelength particle-hole pair. A simple dimensional analysis yields the results that in 2D, the decay rate $\tau^{-1} \approx \lambda\varepsilon_F(\omega/\varepsilon_1)^{2/3}$ where the energy scale $\varepsilon_1 = c^2 k_F^2/\Delta_s$. We recall that the f Green's function is not gauge invariant and has no physical meaning. The physical quantity is the conductivity $\sigma(\omega)$ which is computed with an f bubble with self energy and vertex correction, and the latter converts τ^{-1} to the transport time τ_{tr}^{-1} which requires an additional factor of 1 - cosθ. This leads to $\tau_{tr}^{-1} \approx \lambda\varepsilon_F(\omega/\varepsilon_1)^{4/3}$. At finite temperature $T > T_c$, we expect ω to be replaced by T so that the conductivity should be given by $ne^2\tau_{tr}/m$, with $\tau_{tr}^{-1} \approx \lambda\varepsilon_F(T/\varepsilon_1)^{4/3}$. A long standing mystery about the normal state resistivity of oxide superconductors is that τ_{tr}^{-1} is approximately kT. We have identified a source of strong inelastic scattering and, while our result is limited to $\Delta_s > T > T_c$, it may well be consistent with the experimental data.

A second effect of the strong inelastic scattering is that by Kramers-Kronig relation, the real part $\Sigma'(\omega)$ of the f self energy also goes as $\omega^{2/3}$, so that the spectral weight $a = (1 + \partial\Sigma'/\partial\omega)^{-1} \approx \omega^{1/3}$. The physical Green's function $<c^\dagger_\sigma c_\sigma>$ is a convolution of the f and z Green's function in ω space, and the $\omega^{1/3}$ dependence should show up in the single particle density of states observed by tunnelling for ω above the energy gap. The convolution will smear the usual square root singularity at the gap edge, and increase the gap for quasi-particle excitation to $\Delta + \Delta_s$. We should also include the attraction between $f^\dagger$ and z_σ mediated by the gauge field using Eq.(4). Our preliminary conclusion is that the joint density of states is constant so that a bound state is always formed below $\Delta + \Delta_s$ and the low lying quasi-particles carry both spin 1/2 and charge. Nevertheless, the tunnelling density of states should show substantial deviation from BCS theory as well as usual Fermi liquid behavior far above the gap.

The superconducting state we find has s symmetry [18] and the energy gap is isotropic. Nevertheless, it differs from BCS theory in an important way, in that Anderson's theorem regarding non-magnetic impurities is not obeyed. The pairing is between holes on A and B sublattices; the two partners experiences different random potentials and the compensation leading to Anderson's theorem do not occur. Experimentally it is known that the copper oxide superconductivity is destroyed by non-magnetic doping in the plane. This distinguishes the copper oxide system from $Ba_{1-x}Pb_xBiO_3$ which is in the dirty limit.

The spin correlation $<s(\mathbf{r}, \tau)\, s(0)>$ is a convolution of z Green's function and exhibit at the gap of $2\Delta_s$ corresponding to triplet excitation. [13] The coupling to gauge field and to holes should modify this quantitatively but not qualitatively so that neutron scattering should be mainly sensitive to the spin gap $2\Delta_s$. Since the onset of superconductivity affects only Δ_s/ε_F of the holes, it should effect the neutron scattering only as a small correction.

Our model also raises an interesting possibility concerning the difference between a single layer and multi-layer copper oxide structures. Even if the electronic hopping between the layers is weak, the exchange between the spins on different layers may be sufficiently strong to correlate them if they are not frustrated as in La_2CuO_4. In this case our mechanism will lead to an additional pairing channel between holes on neighboring layers. The interlayer exchange may also reduce Δ_s, so that a quantitative treatment may be difficult. A second comment is that our pairing mechanism per se is not restricted to 2D. In 2D it may operate over a wider range in parameter space within the t - t' - J model because the 2D Heisenberg AF is easy to disorder. We speculate that the 3D copper oxide systems are not superconducting because the hole concentration is so large that the ground state are Fermi liquids. It will be interesting to reduce the hole concentration in these systems by doping to see if an AF phase boundary can be reached and whether superconductivity can be found in a narrow region near this phase boundary which exhibits short range AF order.

In summary, we believe the U(1) gauge theory [5-8] is the language to describe the original RVB picture of Anderson,[1] which envisions a liquid of holes in a background of singlets on opposite sublattices. Indeed, singlet pairing of holes on opposite sublattices is a natural consequence of this picture. Of course our theory is still incomplete in that the role of t is assumed and not fully elucidated. Our main worry is that the t term may produce a spiral state which induces coherent admixture of the two subtlattices and singlets on the same sublattice.[19] This latter effect is pair-breaking for the superconductivity discussed here. Fortunately, the hopping matrix element is expected to be small, $\approx$ Jn and there should be significant parameter space in a t - t' - J model where the superconductivity survives. It will be interesting to study the t - t' - J model by Monte Carlo, because it may be far richer than the Hubbard model itself.

I am grateful to Charles Kane, Gabriel Kotliar, Naoto Nagaosa, Tai-Kai Ng, and Nick Read for many helpful discussions and collaborations which lead to this work. I am particularly indebted to Paul Wiegmann for patiently explaining his work to me. I also thank R. Shankar for a preprint in which the relation between the gauge field and spin orientation is also discussed. This work is supported by NSF through the Materials Research Lab under DMR 87-19217.

References

1. P.W. Anderson, Science 235, 1196 (1987).
2. F.C. Zhang and T.M. Rice, Phys. Rev. B 37, 3759 (1988).
3. A.K. McMahan, R.M. Martin, S. Satpathy, Phys. Rev. B 38, 6650 (1988); M.S. Hybertsen, M. Schluter, and N.E. Christensen, preprint.
4. C. Kane, P.A. Lee, and N. Read, Phys. Rev. B 39 6880 (1989); S. Schmitt-Rink, C.M. Varma and A. F. Ruckenstein, Phys. Rev. Lett. 60 2793 (1988).

5. G. Baskaran and P.W. Anderson, Phys. Rev. B 37, 580 (1988); G. Baskaran, Physica Scripta, to be published.
6. A. Nakamura and T. Matsui, Phys. Rev. B 37 7940 (1988).
7. P.B. Wiegmann, Phys. Rev. Lett. 60, 821 (198); Physica C 153-155, 103 (1988).
8. X.G. Wen, Phys. Rev. B 39 7223 (1989).
9. Note that our constraint differs from Wiegmann's in that he fixed $b^{\dagger}b = 1$ in which case the J term requires an additional factor $f_i f_i^{\dagger} f_j f_j^{\dagger}$. This mult-point vertex is unimportant when the hole concentration approaches zero, and has been ignored by Wen.
10. See E. Fradkin and M. Stone, Phys. Rev. B 38, 7215 (1988).
11. F. D. M. Haldane, Phys. Rev. Lett. 61, 1029 (1988).
12. S. Chakravarty, D. Nelson, and B.I. Halperin, Phys. Rev. Lett. 60, 1057 (1988).
13. D. Arovas and A. Auerbach, Phys. Rev. B 38, 316 (1988).
14. A.M. Polyakov, Gauge Fields and Strings, Harwood 1987.
15. N. Read and S. Sachdev, Phys. Rev. Lett. 62, 1694 (1989).
16. A spinless collective mode has been suggested by D. Rokhsar and S. Kivelson, Phys. Rev. Lett 61, 2376 (1988).
17. M. Reizer, Phys. Rev. B 39, 1602 (1989).
18. If there are multiple band minima, the symmetry of the order parameter between different bands has to be considered, but the full gap survives.
19. B. Shraiman and E. Siggia, Phys. Rev. Lett. 62, 1564 (1989).

Spectroscopic and Theoretical Estimates of Parameters in Model Hamiltonians Describing the High and Low Energy Scale Physics of Doped CuO_2 Planes

L.H. Tjeng, H. Eskes, and G.A. Sawatzky

Department of Applied and Solid State Physics, Materials Science Center, University of Groningen, Nijenborgh 18, NL-9747 AG Groningen, The Netherlands

Abstract. In this paper we describe the determination of relevant model Hamiltonian parameters for high T_c materials using high energy spectroscopies, model compounds and model theories. With this information we discuss the validity of simplified model Hamiltonians.

1. Introduction

In order to understand and describe the physical properties of exotic materials like the high T_c's [1] we usually resort to the study of intelligently chosen model Hamiltonians. These model Hamiltonians are usually too complicated to solve exactly so the next step is to find approximations whose validity is checked by comparison to experimental results. Since the physical properties are determined by low energy scale excitations we are in the end interested in the low energy scale parts of the problem. However, the model Hamiltonians like for example the Hubbard model or the Anderson impurity or lattice model are described with parameters which usually belong to a high energy scale and can be determined, often quite directly, using high energy experimental probes. These high energy scale parameters are not expected to be strongly dependent on the details of the crystal structure and therefore we are justified in choosing stable model compounds with well defined chemical composition, and structure similar in well chosen aspects to the actual materials we are interested in. This is of special importance for the high T_c's where a combination of surface sensitive electron spectroscopies with unknown or uncontrolled surface composition and structure have led to quite varying results.

Having obtained the high energy scale parameters in this way, we then try to distil out of the general Hamiltonian a hopefully simpler new Hamiltonian for describing the low energy scale properties. Examples of this procedure are the Schrieffer-Wolff transformation [2] of an Anderson [3] impurity Hamiltonian containing Coulomb interactions (U) of 2 - 10 eV and Fermi energies and band widths of order several eV's, to the Kondo [4] Hamiltonian containing the impurity-conduction electron exchange interactions of order 0.01 - 0.1 eV. Another even more relevant example for the high T_c's is the Heisenberg exchange Hamiltonian of a half-filled Mott-Hubbard [5] insulating compound in which the superexchange [6] ($J \approx 0.1$ eV) is determined by the high energy scale parameters of a generalized Hubbard or an Anderson lattice Hamiltonian involving $U \approx 10$ eV, the charge transfer energy (Δ) ≈ 3 eV and the one electron transfer integrals $T \approx 1$ eV. The following step is to study the properties of the low energy scale Hamiltonian and compare with known physical properties and predict measurable properties. This shows how high energy scale parameters play a fundamental role in the understanding of high T_c superconductors.

It is generally accepted that the action is on the CuO_2 planes as far as the low energy scale electronic properties are concerned. This does not mean that the other ions and CuO chains are not important. In fact the ions in the other layers of these layered compounds are of utmost importance in setting up the Madelung potential and in setting the stage for the hole or electron doping of the CuO layers. Although it is possible that the out of plane apex oxygens could play a more direct active role we will neglect them to start with. It is also generally accepted that a generalized Hubbard like Hamiltonian is a good starting point for describing the charge and spin degrees of freedom and the low energy scale physics. This is evidenced by the insulating antiferromagnetic character of the undoped "stochiometric" materials with local moments on the divalent Cu ions. This type of behaviour is typical of a Mott-Hubbard insulator. It is also generally accepted that the Hubbard Hamiltonian to be used must include both the 10 fold Cu 3d states and the 6 fold O 2p states leaving us with a complicated multiband problem. It would be considerably more pleasant if the experimental parameters lead to a simplification of the ultimate reduced Hamiltonian valid for the low energy scale physics.

2. Towards a general Hamiltonian

We start with neglecting the small deviations from square planar symmetry, a good approximation for the high energy scale parameters. The parameters we aim to determine are divided into one electron and two electron terms. The one electron terms are the on-site Cu 3d energy ε_d, the on-site O 2p energy ε_p, or the charge transfer energy $\Delta = \varepsilon_d - \varepsilon_p$, and the Slater-Koster two center integrals $(pp\sigma)$, $(pp\pi)$, $(pd\sigma)$ and $(pd\pi)$ from which we obtain the O-O transfer integrals $T_{pp} = (pp\sigma) - (pp\pi)$ and in terms of the D_{4h} irreducible representations the Cu-O transfer integrals $T_{pd}(b_1) = \sqrt{3}(pd\sigma)$, $T_{pd}(a_1) = T_{pd}(b_1)/\sqrt{3}$, $T_{pd}(b_2) = 2(pd\pi)$, $T_{pd}(e) = T(b_2)/\sqrt{2}$ in which the appropriate linear combination of O orbitals are taken in each case. Here b_1 refers to x^2-y^2, a_1 to $3z^2-r^2$, b_2 to xy and e to xz,yz 3d orbitals. The two electron terms we need can be represented by the Racah parameters or the Slater F^n integrals. For Cu 3d these are the Racah A, B and C parameters where B = 0.15 eV and C = 0.56 eV are taken from the free ion optical data [7] since they are known not to be screened [8] and for O these are the Slater F^0, F^2 integrals with again F^2 = 6 eV is taken from free ion data. There remains then the Racah A and Slater F^0 monopole integrals to be determined which are known to be strongly screened in the solid. The energies of the 19 different $3d^8$ states [9] in D_{4h} symmetry can all be written in terms of A, B, and C Racah parameters. For example, the Hund's rule ground state of $3d^8$ consists of for instance $^3B_{1g}$ ($x^2-y^2\uparrow$; $3z^2-r^2\uparrow$) which has an effective Coulomb interaction energy of $U_{dd}(^3B_1)$ = A-8B. The $^1A_{1g}$ forms a 4-fold IR of which the lowest energy state is approximately $\approx$ A - 3B + 2C and the average $U_{dd}(^1A_{1g}) \approx$ A + 4B + 3C. In addition we will be able to determine the interatomic O 2p - Cu 3d Coulomb interaction U_{pd} for nearest neighbours.

In terms of electronic configurations these parameters are represented by

$$\Delta = E(d^{10}p^5) - E(d^9p^6)$$

$$U_{dd}(\Gamma_i) = E(d^{10}) - E(d^9) + E(d^8, \Gamma_i) - E(d^9)$$

$$U_{pp}(\Gamma_i) = E(p^4) - E(p^5) + E(p^6) - E(p^5)$$

where Γ_i denotes the appropriate irreducible representation of the point group. We note that we have neglected the point charge contributions to the crystal field splitting so that the various d^9 as well as the p^5 states are degenerate to start with and the net splitting is in the end solely a result of the different hybridizations yielding the so-called ligand field splittings. McMahan et al. [10] have shown that the point charge crystal field splittings of the d states are at most 0.05 eV and therefore negligible.

As pointed out by Zaanen et al. [11] the most crucial parameter which determines if the materials are Mott-Hubbard or charge transfer insulators is $U_{dd} - \Delta$ since in the absence of hybridization this is the difference in the energy required to remove a Cu 3d or a O 2p electron from the $Cu^{2+}O^{2-}$ ionic ansatz. As discussed in considerable detail elsewhere [9,11], for $U_{dd} - \Delta \geq 0$ the material has a charge transfer gap and the lowest energy hole states are O 2p states whereas for $U_{dd} - \Delta \leq 0$ the material is a Mott-Hubbard insulator and the lowest energy hole state is of Cu 3d character leading to a d^8 like state.

3. The Parameters

There have been a large number of parameter estimates made from experimental studies like PES, core level XPS, Auger, X-ray absorption [for instance see 12-15]. These estimates have however very large ranges especially in the charge transfer energy $0 \leq \Delta \leq 3$ and $5 \leq U_{dd} \leq 13$. The reason for this is that in most cases a detailed analysis was usually not done so that large shifts due to hybridization were incorrectly put in U and/or Δ. Also the multiplet structure of the d^8 states which ranges over 8 eV was not included resulting in a rather arbitrary value of U. The most important problem however is that the valence band PES spectrum is sensitive to $U_{dd} - \Delta$ and insensitive to Δ itself. Also core level spectroscopies introduce (at least) one new parameter, the core hole potential (Q). In the Cu 2p core XPS then for example one basically has two experimental numbers; the satellite position and relative intensity to determine Q, Δ, and T_{pd}. U_{dd} cannot be obtained because the d^8 states are not involved. A third reason for the large spread is that experiments done on the high T_c materials themselves are often of questionable usefulness because of the badly defined materials and surfaces. For example, in most XPS spectra of $YBa_2Cu_3O_{7-\delta}$ system the O $_{1s}$ core line consists of (at least) two dominant components [16] separated by several eV in energy. It is now known that the higher binding energy one is almost certainly due to contamination making previous analysis of O Auger data questionable.

Theoretically the situation is somewhat better. In the two detailed studies done by MacMahan et al. [10] and Hybertson et al. [17] using density functional calculations and frozen configurations to calculate the U's, Δ, and T's, the results are quite consistent except again for $1 \leq \Delta \leq 4$ eV which seems to be difficult to pin down. Ab initio quantum chemical cluster calculations are in our opinion mostly unreliable in estimations of Δ and U because of the well known problems of the long range screening effects due to polarizable ions and other embedding effects as discussed by Jansen et al. for NiO [18].

As suggested some time ago by Zaanen et al. [11] CuO is a charge transfer insulator ($U_{dd} - \Delta > 0$) and it is by now well established that Emery's suggestion [19] that this is also the case for the high T_c materials is correct. The most direct evidence for this is the O K-edge electron energy loss [20] and X-ray absorption [21,22] spectra which probe, because of dipole selection rules, the empty O 2p states. Not only did these data show that the holes are of primarily O 2p character but, at least for the Bi compound, also of p_x or p_y symmetry [23, 24].

The simplest approach to obtain Δ and U is to use the free ion ionization potentials I, electron affinities A and Madelung potentials (V_M) since

$$\Delta = I(O^{2-}) - A(Cu^{2+}) + |V_M(Cu)| + |V_M(O)| + E^{\Delta}_{corr} \quad \text{and}$$

$$U_{dd} = I(Cu^{2+}) - A(Cu^{2+}) + E^{U}_{corr}$$

where E^{u}_{corr} is the correction due to screening, polarization, covalency, etc. which turns out to be very large since these are quite covalent materials containing highly polarizable O ions. For example, E^{U}_{corr} is in the simplest approximation given by $-2E_{pol}$ where E_{pol} is the polarization energy lowering one gets upon placing a charge in a polarizable medium. $E_{pol} = \Sigma\frac{1}{2}\alpha_i F_i^2$, where α is the polarizability of an ion and F_i is the electric field due to the extra point charge. With E_{pol} between 3 to 5 eV for a charge on Cu in a CuO lattice, we obtain reasonable values for $U_{dd} = U_{dd}(\text{ion}) - 2E_{pol}$, using $U_{dd}(\text{ion}) = 16.3$ eV. The factor of 2 appearing in the effective screening of U_{dd} arises because the total polarization stabilization goes as $(F_i)^2$ and therefore is $4E_{pol}$ for two charges on the same site as compared to $2E_{pol}$ for two infinitely separated charges [25]. The same type of argument is used for the formation of bipolarons [26]. It is impossible to estimate Δ in this way since we don't know the charges to use in the Madelung sum, O^{2-} is unstable as a free ion, and errors do not cancel as in the case of U_{dd}. (For a discussion, see Zaanen's thesis). We therefore must resort to experiment, model and ab initio types of calculation.

In principle all these parameters can be determined from a combination of photoelectron (PES) and inverse photoelectron (IPES) spectroscopy. This is quite obvious since PES gives us the spectrum of all ionization states in the actual solid and IPES gives us all the electron affinity states of the solid. Unfortunately we have to solve the many body system described by our model Hamiltonian in order to extract the parameters. However, as mentioned above the high energy scale parameters are expected to be relatively weakly dependent on the details of the crystal structure and also on the translational symmetry. This suggests two approaches. First we can choose systems for which many body effects of the type described by a Hubbard Hamiltonian are expected to be unimportant for the ground state, as well as for the first ionization states. Secondly we could replace the solid by a cluster of suitable choice so that the many body problem has a small number of particles for which the model Hamiltonian can be solved and compare with experiment.

With regard to the first approach we could choose the compound Cu_2O which has Cu linearly coordinated by oxygen ($d_{Cu-O} = 1.84$ Å) and oxygen tetrahedral coordinated by Cu. It is a semiconductor with a band gap of 2.4 eV [27] which is also obtained from a combination of PES and IPES [28]. This is a closed shell system $Cu(3d^{10})$ $O(2p^6)$ and therefore in the model Hamiltonian U_{pp} and U_{dd} are inoperative for the ground state as well as for the ionization (PES) and affinity (IPES) states provided we are justified in not explicitly including the Cu 4s 4p and higher lying states in our model Hamiltonian.

The PES states will be the ligand field split Cu 3d bands (d^9 like) and the O 2p bands (p^5 or $d^{10}\underline{L}$) where $\underline{L}$ denotes a ligand (O 2p) hole. Since the PES spectrum at various photon energies is sensitive to different portions of the partial density of states (DOS) we can with X-ray energies (XPS) extract the d^9 like states and compare this with tight binding band structure and/or model Hamiltonian cluster calculations to extract quantities like the T_{pd}'s and even Δ as it is defined for CuO. This has been done in detail for Cu_2O. The parameters obtained from this are given in table I. Note that LDA band theory gives a very satisfactory description of the PES and IPES data [28].

Since Cu_2O is a full band system for which one electron theory describes the "one hole" states obtained with PES very well we can use the powerful Auger spectroscopy and related theory to obtain quite directly U_{dd}, U_{pp} and U_{pd}. This can be done because in Auger spectroscopy the final state now contains 2 holes for which exact solutions for the Hubbard like Hamiltonian can be found [29]. The Auger spectrum can then be described in terms of the self-convolution of the one hole DOS and the Coulomb interactions as described by Cini and Sawatzky [30]. Alternatively one can also solve the

Table I.
Nearest neighbour Cu-O distance, experimental bandgap energy, one-particle and two-particle energies for Cu_2O and CuO. The free ion optical values are used for the Racah B (0.15 eV), C (0.58 eV) and Slater F^2 (6.0 eV) parameters. The quantities $T_{pd}(b_1)/2$ and $T_{pp}/2$ are the commonly used hybridization parameters in, for instance, an extended Hubbard like Hamiltonian.

	Cu_2O	CuO	
d_{Cu-O}	1.84	1.95	Å
E_{gap} (experimental)	2.4	1.4	eV
Δ	2.50	2.75 - 3.50	eV
$T_{pd}(b_1)/2$	1.47	1.25	eV
$T_{pp}/2$	-	0.50 - 0.65	eV
$U_{dd}(^1A_{1g})$ (averaged)	9.7	8.8	eV
$U_{pp}(^1D)$	5.7	-	eV
U_{pd}	< 1	-	eV

two hole problem for the same finite clusters as those used to describe the photoelectron spectrum and compare to experiment.

The $Cu(L_{23}M_{45}M_{45})$ Auger spectrum probes the local (because of the local core hole) projected Cu d^8 DOS involving U_{dd} directly and the O $KL_{23}L_{23}$ Auger spectrum probes the local O p^4 or $d^{10}\underline{L}^2$ DOS involving U_{pp}. These states are of course mixed because of T_{pd} provided they occur on nearest neighbors. T_{pd} mixes in states of $d^9\underline{L}$ character with a hole on a nearest neighbor ligand which involves U_{pd}. The exact procedure used is described elsewhere [31] and we suffice here by showing the final result in Fig. 1 which shows the experimental O derived local two hole DOS and the comparison to theory. The shaded region arises from mixing in of the the d^9p^5 or $d^9\underline{L}$ states and the higher binding energy region from the atomic $d^{10}\underline{L}^2$ or p^4 states. From this we conclude that $U_{pd} < 1$ eV - screened down from the bare point charge value of 7.8 eV. The on site U values obtained from this are also listed in table I.

A better comparison with the high T_c's is CuO where the CuO distance is comparable (d_{Cu-O} = 1.95 Å as compared to 1.89 Å in La_2CuO_4 or 1.95 in $YBa_2Cu_3O_7$) and the Cu is in a nearly square planar coordination of oxygen. Also here Cu is divalent (d^9) and the superexchange interaction (J = 0.1 eV) is comparable to the high T_c's. CuO is a semiconductor with a gap of 1.4 eV as found by PES/IPES [28].

Of special interest is the XPS valence band spectrum shown in Fig. 2 together with the LDA band structure calculation. Of special importance is the large energy spread of the d spectral weight as measured in XPS as compared to band theory. The states around 10 - 15 eV are due to d^8 final states which we get upon removal of <u>one</u> electron. These states occur at similar energies as the d^8 states in Cu_2O where they could only be reached by <u>two</u> electron removal. That they are indeed d^8 states is clear from their resonance behaviour for photon energies at the Cu 3p edge [32, 33]. Also shown in Fig. 2 is a spectrum at resonance (74 eV) and out of resonance (70 eV) where we also see the large spread due to the Racah B and C parameters in the d^8 multiplets including a $^1A_{1g}$ state at 16 eV binding energy. Also shown in Fig. 2 is a simulation of the Cu 3d (XPS) spectral weight from a CuO_4 cluster calculation [34] using the model Hamiltonian with the parameters listed in table I. We make a special point of the calculated gap since the PES spectral shape is rather insensitive to the choice of Δ

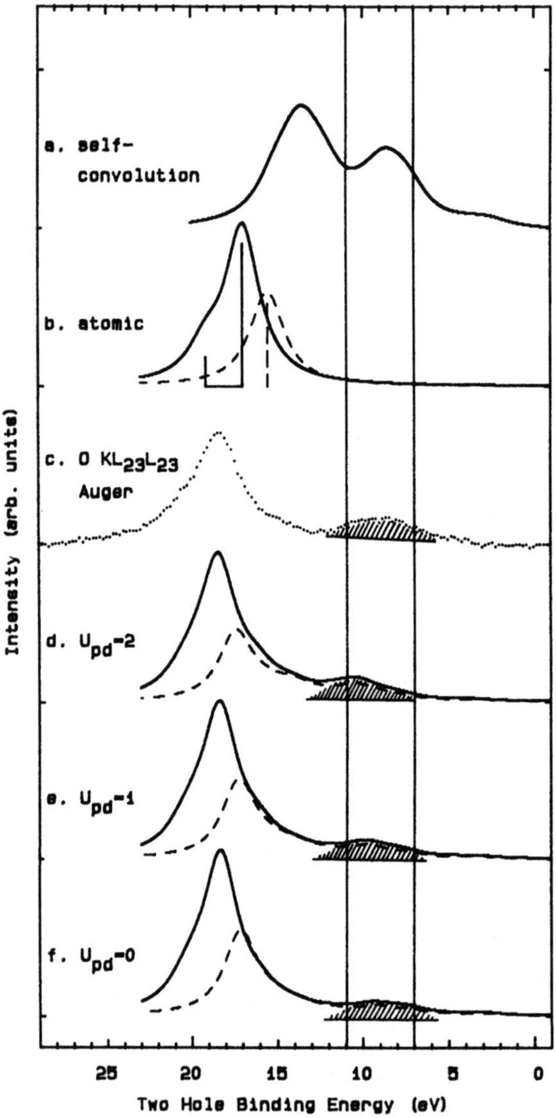

Figure 1.
O $KL_{23}L_{23}$ Auger spectrum and O $2p^4$ local DOS from Cu_2O. (a) self-convolution of the O 2p partial DOS; (b) atomic O $2p^4$ DOS, calculated for a screened F^0 and atomic F^2 as listed in table I; (c) experimental O $KL_{23}L_{23}$ Auger spectrum, for which the kinetic energy scale has been transformed into a two hole binding energy by subtracting the O 1s binding energy; (d),(e) and (f) O $2p^4$ local DOS, calculated using a $(Cu_4O_5)^{6-}$ cluster with an O atom in the centre, using U_{pd} = 2, 1 and 0 eV and parameter values listed in table I. The solid line in (b),(d),(e) and (f) are the singlet DOS, and the dashed lines the triplet DOS, reduced by a

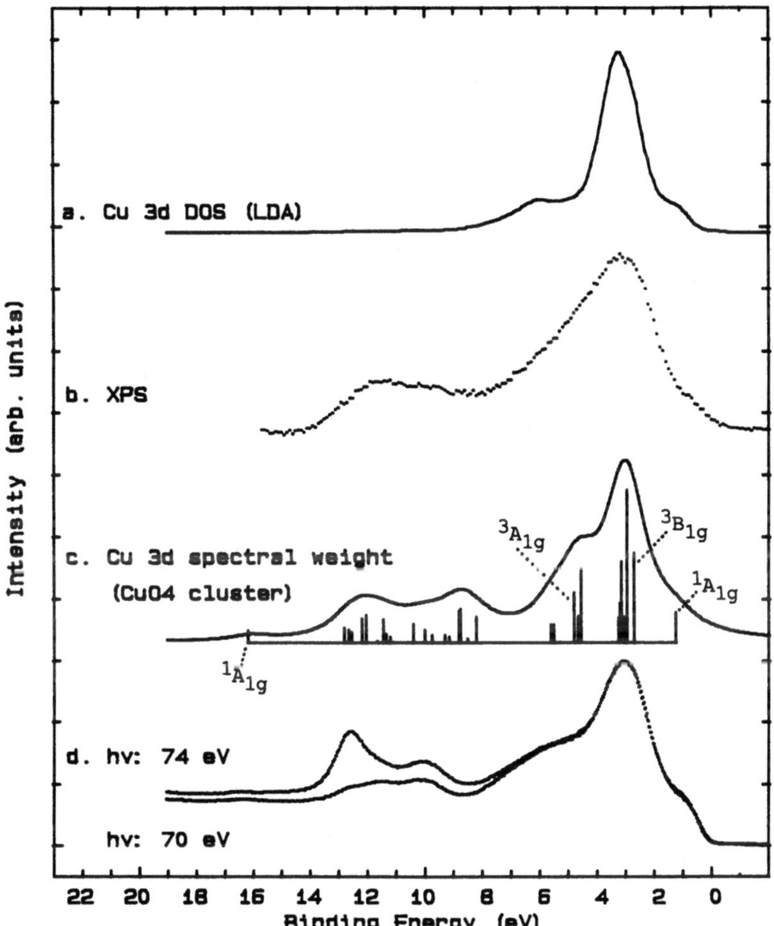

Figure 2.
Valence band of CuO. (a) Cu 3d partial DOS, from LDA band structure calculations [28]; (b) experimental valence band photoemission spectrum using the Al-Kα (XPS) source; (c) Cu 3d spectral weight, calculated using a CuO_4 cluster with model parameters listed in table I; (d) experimental valence band photoemission spectrum at the Cu 3p resonance (hν = 74 eV) and out of resonance (hν = 70 eV).

provided we keep $U_{dd} - \Delta$ constant. However, for $U_{dd} - \Delta > 0$ the gap is determined by the Δ and the oxygen band width or T_{pp}.

It is at this point where we notice that the translation symmetry is important because in the measurement of the gap with PES/IPES the top of the valence band and the bottom of the conduction band are used, but these include a dispersional width $W_v/2 + W_c/2$. Our cluster gap therefore should be taken larger than the measured gap by

$$E_{gap}(\text{cluster}) = E_{gap}(\text{exp}) + \frac{W_v}{2} + \frac{W_c}{2}.$$

To obtain this we must know the extra broadening due to dispersion of the first ionization states and the first affinity states. This dispersional broadening on the other hand will depend on the choice of Δ and T_{pp} so we must determine these self consistently.

Before we look into this a few comments about the parameters of Cu_2O as compared to CuO. The values of T_{pd} in Cu_2O are somewhat larger consistent with the smaller Cu-O distance. U_{dd} is somewhat larger in Cu_2O consistent with the reduced screening due to the lower coordination number.

4. Influence of dispersion

Up to very recently we had not expected the banding due to dispersion of the lowest ionization state shown in Fig. 2c as a $^1A_{1g}$ state to be large enough to cause a significant renormalization of Δ obtained from a cluster [34] or an impurity [35] model. Recently however calculation on larger clusters by Kuramoto [36] and Fukuyama [37] and ourselves [38] have demonstrated that this dispersion is really quite large so we really must reevaluate the Δ values. An estimate of the dispersional width can be made from a Cu_2O_7 cluster calculation. This dispersional width is also of great importance in judging the validity of our small cluster or impurity ansatz. We have recently shown from an exact model Hamiltonian calculation of the Cu_2O_7 cluster shown in Fig. 3 that the lowest energy states are very well represented by neglecting everything except the $3d_{x^2-y^2}$ and $O\text{-}2p_\sigma$ orbitals. The ground state or neutral state has 2 holes in a singlet or triplet with an energy difference due to superexchange as given in table II for several parameter sets. The experimental value for La_2CuO_4 is $J = 0.12$ eV and for CuO somewhat smaller. The difference is primarily caused by the 146° band angle in CuO as compared to 180° in La_2CuO_4 together with the expected $\cos^2\theta$ dependence. The electron addition states are one hole states with a bonding-antibonding splitting of about 0.8 eV resulting in a transfer integral of about $t \simeq 0.4$ eV.

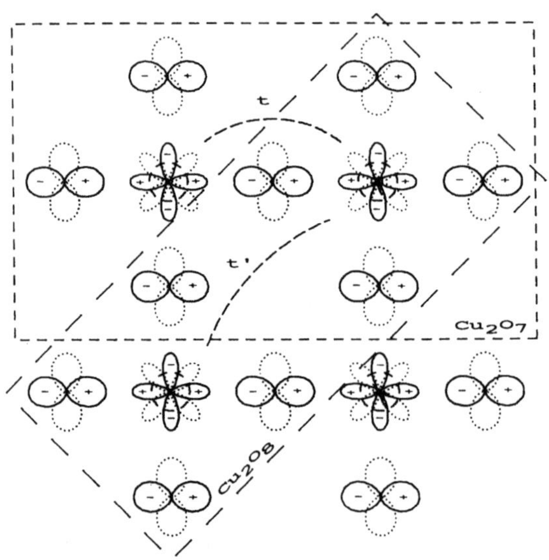

Figure 3.
The Cu_2O_7 and Cu_2O_8 clusters used in the calculations. t and t' are the effective singlet transfer integrals for nearest neighbours and next nearest neighbours respectively.

Table II.
Effective singlet transfer integrals t (nearest neighbours) and t' (next nearest neighbours), and superexchange energies J (nearest neighbours) and J' (next nearest neighbours), calculated using parameter values listed in table I.

	t	t'	J	J'
U_{pd} = 0 eV	-0.44 eV	0.18 eV	0.24 eV	0.024 eV
U_{pd} = 1 eV	-0.40 eV	0.17 eV	0.17 eV	0.018 eV

The first electron removal state is a $^2A_{1g}$ state with an excited $^2B_{1u}$ state as its antibonding partner. The splitting is ≈ 0.9 eV and the states are very well represented by a bonding and antibonding combination of a state with two holes on the left in a $^1A_{1g}$ state and one hole on the right in a x^2-y^2 state and the reverse with an effective transfer integral $t(^1A_{1g})$ = 0.45 eV as given in table II. This would cause a band width of the local singlet described by Zhang and Rice [45] and Eskes and Sawatzky [35] of $8t(^1A_{1g})$.

It has been recently pointed out by Lee [39] that the next nearest neighbor hopping $t'(^1A_{1g})$ of the local singlet state could lead to new physics. Even for the localized (i.e. central Cu^{2+} nearest neighbour), it is easy to see that $t'(^1A_{1g})$ can be quite large by looking at the Cu_2O_8 cluster shown also in Fig. 3. The large transfer integral is due to the T_{pp} integral connecting the four oxygens in the centre. We calculated an effective transfer integral ($t'(^1A_{1g})$ ≈ 0.2 eV) showing that it is indeed not negligible. The fact that t and t' have opposite signs leads to a density of states which is strongly peaked at the top of the band. Taking both t and t' into account in determining W_V and W_C we required Δ ≈ 3.5 eV to yield the experimental band gap.

5. Exchange interactions

Since numerous mechanisms for high T_c superconductors involve magnetism in one way or another it is of interest to determine the exchange interactions with the parameters of table I. The superexchange J has already been discussed having obtained it from the Cu_2O_7 cluster. This value of J = 0.15 - 0.24 eV is in good agreement with that determined by Jennison and Stechel [40] in a similar calculation but taking into account a direct Cu-O exchange and the more sophisticated imbedded cluster calculations of Annett et al. [41]. Annett et al. [41] have also calculated the next nearest neighbor Cu-Cu superexchange (J'= 0.02 eV), in close agreement with the value we obtain from the Cu_2O_8 calculation J' = 0.02 eV. Since this is so small we do not expect a significant frustration effect from this alone.

The Cu-O exchange interactions have been discussed by us in a previous paper [42] although we remark that these should be multiplied by 2 for an exchange defined as the singlet-triplet splitting and the parameters used were slightly different. The most important interaction is the exchange interaction between the $d_{x^2-y^2}$ hole and a hole of the same symmetry on the neighbouring O atoms. This exchange interaction is >3 eV as can be seen from the $^1A_{1g}$ - $^3A_{1g}$ splitting in the CuO_4 cluster calculation (Fig. 2c). We must be careful however since there is a triplet $^3B_{1g}$ between these corresponding to a hole in an x^2-y^2 orbital and the other in a $3z^2-r^2$ orbital. This triplet state is 1.5 eV above the singlet and as pointed out by Fujimori [43] will be degenerate with the singlet if the apex oxygen on-site energy in the high T_c is 1.7 eV or more lower than the on site energy of the in plane oxygen 2p orbitals. Also if the $^1A_{1g}$ band width becomes large it will cross the $^3B_{1g}$ state and the latter could become populated upon increased doping in the high T_c's.

6. A model Hamiltonian

Perhaps the most important result reported here is that the lowest energy states for 3 holes in a Cu_2O_7 cluster can be well described by a bonding and antibonding combination of a state with a singlet on the left and a doublet on the right and vice versa. This demonstrates the stability of the local singlet and gives credence to identifying it as the quasi particle state occupied upon doping. In the absence of doping we apparently have a charge transfer insulator with an extraordinarily large super exchange interaction. The Hamiltonian describing this situation is a so-called t,J-model [44] which basically is a single band Hubbard model but with extra freedom since in the Hubbard model $J = t^2/(U)$ whereas here it is an independent parameter. Within this model t is identified with our $t(^1A_{1g})$ and J the superexchange as in the Zhang-Rice [45] description. We have found however that the next nearest neighbor transfer integral is also large so the t,t',J-model proposed by Lee [39] is even more realistic.

This single band like Hubbard model is based on the assumption that the dispersional broadening of the singlet state due to t and t' is not so large that the identity of the singlet quasi particle may not survive because of strong delocalization and mixing with other states. For example if the singlet state band width is much larger than the energy by which the singlet in the impurity or cluster calculation is split of the other band like states [Eskes 38] one may question the validity of the impurity or cluster approach in describing the quasi particle. Upon examining the parameters it seems that we are still in the regime in which a singlet quasi particle and a t,t',J-model is applicable. This is evidenced by the survival of the singlet description in the Cu_2O_7 cluster as well as by the fact that the singlet in the CuO_4 cluster is about 1.5 eV below the other states whereas the dispersional half width is also 1.5 - 2 eV. The top half of this quasi particle band at least is well described by a t,t',J-model. Also as mentioned above t and t' have opposite signs and compete for the states at the top of the band leading to narrower bands than expected at first glance. At too high dopings $x > 0.5$ this model will probably break down. The long range magnetic order or for that matter the short range order caused by J will further stabilize this singlet quasi particle since any antiferromagnetic correlations will tend to reduce the singlet band width further. Numerical calculations by Imada [46] indicate that as long as $t/J < 10$ the identity of the singlet state survives and we can think of it propagating in an antiferromagnetically coupled lattice of spins.

In conclusion then the model Hamiltonian parameters obtained support a t,J or t,t',J single band model with $t = -0.4$ eV, $t' = 0.2$ eV and $J = 0.12$ eV. We all wonder what the low energy scale physics of this model Hamiltonian is and whether or not it will lead to high T_c be it via exotic flux phases Fermi liquid like behaviour or something resulting in a Fermi energy but a sort of smeared out Fermi surface in k space as the coupled holon spinon picture might give us.

At this point we still prefer a picture as presented earlier [42] in which the local singlet creates around it a large region (a RVB bag) which could be described by a singlet liquid because of the near degeneracy of the singet liquid and Neél state for a 2D spin 1/2 Heisenberg system. This would lead to a new even lower energy scale Hamiltonian in which attractive coupling between the by doping induced holes is caused by the overlap of their singlet liquid like polarization clouds. However at this point this is pure speculation and we will have to wait for the outcome of theoretical studies of the t,t',J-model.

7. Acknowledgements

This investigation was supported by the Netherlands Foundation for Chemical Research (SON) and the Foundation for Fundamental Research on Matter (FOM) with financial support from the Netherlands Organization for the Advancement of Pure Research (NWO).

8. References

1. J.G. Bednorz and K.A. Müller, Z. Phys. B64, 189 (1986).
2. J.R. Schrieffer and P.A. Wolff, Phys. Rev. 149, 491 (1966).
3. P.W. Anderson, Phys. Rev. 124, 41 (1961).
4. J. Kondo, Prog. Theoret. Phys. (Kyoto) 28, 846 (1962); 32, 37 (1964); 34, 204 (1965).
5. J. Hubbard, Proc. Phys. Soc. Ser. A. 277, 237 (1964).
6. P.W. Anderson, Phys. Rev. 115, 2 (1959); Solid State Physics 14, 99, (1963).
7. The atomic values can be obtained from atomic spectroscopic data. C.E. Moore, in Atomic Energy Levels, Natl. Bur. Stand. (USA), Circ. No. 467.
8. D. van der Marel and G.A. Sawatzky, Phys. Rev. 37, 10674 (1988).
9. J. Zaanen, Thesis, University of Groningen, 1986. G.A. Sawatzky, Int. J. of Modern Physics B, Vol. 1, 243 (1988).
10. A.K. McMahan, R.M. Martin, S. Satpathy, Phys. Rev. B38, 6650 (1988).
11. J. Zaanen, G.A. Sawatzky and J.W. Allen, Phys. Rev. Lett. 55, 418 (1985).
12. W.E. Pickett, Reviews of Modern Physics (in press).
13. K. Hass, Solid State Physics (in press)
14. G. Wendin, J. Physique C9, 1157 (1987).
15. J.C. Fuggle, P.J.W. Weijs, R. Schoorl, G.A. Sawatzky, J. Fink, N. Nücker, P.J. Durham and W.M. Temmerman, Phys. Rev. B 37, 123 (1988).
16. J.C. Fuggle, in Proceedings of the Adraitico Research Conference and Workshop on "Towards the Theoretical Understanding of High T_c Superconductors", ICTP Trieste , World Scientific (1988).
17. M.S. Hybertsen, M. Schlüter and N.E. Christensen, Phys. Rev. B 39, 9028 (1989).
18. G.J.M. Janssen and W.C. Nieuwpoort, Phys. Rev. 38, 3449 (1988).
19. V.J. Emery, Phys. Rev. Letters 58, 2794 (1988).20.N. Nücker, J. Fink, J.C. Fuggle, P.J. Durham, W.M. Temmerman, Phys. Rev. B37, 5158 (1988).
21. J.A. Yarmoff, D.R. Clarke, W. Drube, U.O. Karlsson, A. Talab-Ibrahimi and F.J. Himpsel, Phys. Rev. B36, 3967 (1987).
22. P. Kuiper, G. Kruizinga, J. Ghijsen, M. Grioni, P.J.W. Weijs, F.M.F. de Groot, G.A. Sawatzky, H. Verweij, L.F. Feiner and H. Petersen, Phys. Rev.B38, 6483 (1988).
23. N. Nücker, H. Romberg, X.X. Xi, J. Fink, P.S. Durham and W.M. Temmerman, Phys. Rev. B (in press).
24. P. Kuiper, M. Grioni, G.A. Sawatzky, D.B. Mitzi, A. Kapitulnik, A. Santaniello, P. de Padova and P. Thiry, Physica C157, 260 (1989).
25. For details see the calculations done for Ni dihalides, J. Zaanen, C. Westra and G.A. Sawatzky, Phys. Rev. B33, 8060 (1986).
26. B.K. Chakraverty and C. Schlenker, J. Phys. (Paris) Colloq. 37, C4-353 (1976).
27. P.W. Baumeister, Phys. Rev. 121, 359 (1961).
28. J. Ghijsen, L.H. Tjeng, J. van Elp, J. Westerink, G.A. Sawatzky and M.T. Czyzyk, Phys. Rev. B38, 11322 (1988)
29. Kanamori, Prog. Theor. Phys. 30, 275 (1963).
30. M. Cini, Solid State Commun. 24, 681 (1971); G.A. Sawatzky, Phys. Rev. Lett. 39, 504 (1977).
31. L.H. Tjeng, J. van Elp, P. Kuiper and G.A. Sawatzky (to be published).

32. M.R. Thuler, R.L. Benbow and Z. Harych, Phys. Rev. B26, 669 (1982).
33. J. Ghijsen et al. (unpublished).
34. H. Eskes, L.H. Tjeng, and G.A. Sawatzky (preprint).
35. H. Eskes and G.A. Sawatzky, Phys. Rev. Lett. 61, 1415 (1988).
36. Y. Kuramoto et al. (these proceedings).
37. H. Fukuyama et al. (these proceedings).
38. H. Eskes, G.A. Sawatzky and L.H. Feiner (preprint)
39. P.A. Lee (these proceedings)
40. E.B. Stechel and O.R. Jennison, Phys. Rev. B38, 4632 (1988).
41. J.F. Annett et al (preprint).
42. H. Eskes, H. Tjeng and G.A. Sawatzky, in: Mechanisms of High Temperature Superconductivity", p.20. 1989. Springer-Verlag. Eds. H. Kamimura and A. Oshiyama.
43. A. Fujimori, Phys. Rev. B 39, 793 (1989).
44. P.W. Anderson, Science 235, 1196 (1987).
45. F.C. Zhang and T.M. Rice, Phys. Rev. B37, 3759 (1988).
46. Imada (these proceedings).

Effective Hamiltonian for High-T_c Cu Oxides

H. Fukuyama[1] *and H. Matsukawa*[2]

[1]Institute for Solid State Physics, University of Tokyo, 7-22-1 Roppongi, Minato-ku, Tokyo 106, Japan
[2]Department of Physics, Nara Women's University, Nara 630, Japan

Abstract. Effective Hamiltonian has been derived for CuO_2 layers in the presence of extra holes doped mainly into O-sites by taking both on-site and intersite Coulomb interactions into account. A special case with a single hole has been examined in detail. It is seen that there exist bound states below the hole band continuum with various types of the spatial symmetry and spin multiplicity. The spatial extent of Zhang-Rice singlet and the effective transfer integrals between these two singlet states have been found to be sensitive to the relative magnitudes of the direct and indirect transfer integrals between O-sites. Effective Hamiltonian in the case of electron doping is also derived.

1. Introduction

It is now well accepted that the CuO_2-layers are essential parts of crystal structure for high T_c superconductivity in Cu-oxides. Various physical parameters characterizing the electronic properties of the structure consisting of Cu:$3d_{x^2-y^2}$ and O:$2p_\sigma$ are shown in *Fig*.1. Here we have taken the convention that the sign of the transfer integral between Cu and O atoms are $-t_0$ ($t_0 > 0$) everywhere and hence the direct transfer integral between the nearest neighbor oxygens and that between the next nearest neighbors across Cu sites are defined as $+t_1$ and $-t_2$ ($t_1 > 0, t_2 > 0$), respectively. Matrix elements of the Coulomb interaction are considered both at the site of Cu (U_d) and O(U_p) and intersite between Cu and O (V_0).

Estimation of the values of these parameters has been made by band theoretical calculations and with the help of photoemission spectroscopy [1-4]; $U_d \sim 8$, $U_p \sim 4$, $V_0 \sim 1$, $\Delta \equiv \epsilon_p - \epsilon_d \sim 2$, $t_0 \sim 1$, $t_1 \sim 0.6$, $t_2 \sim 0.3$, all in unit of eV and for holes. The large value of U_d is the obvious cause of the existence of local moment at Cu-sites in the undoped systems. Such knowledge on the values of typical parameters is very important and will be the basis for the full understanding of the electronic states of

Editors: H. Fukuyama · S. Maekawa · A.P. Malozemoff

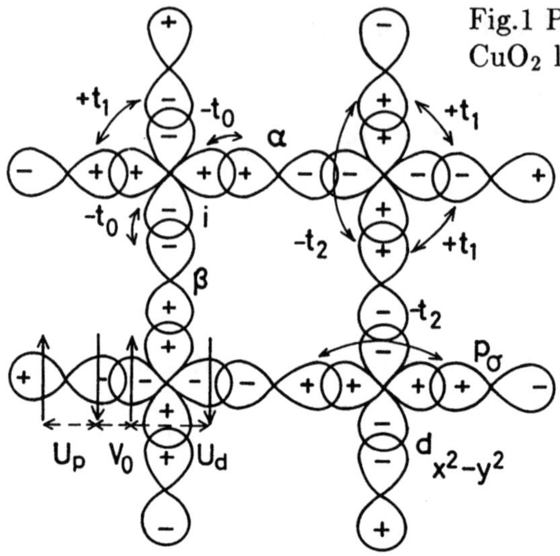

Fig.1 Parameters characterizing the CuO_2 layers.

this remarkable systems. These parameters by themselves, however, determine the electronic states in large energy scale, but for the description of the low-energy excitations, which are crucial to transport properties, one needs the effective Hamiltonian consistent with these large energy scale properties.

2. Effective Hamiltonian

The effective Hamiltonian to describe the low-lying excitation has been derived by assuming that Cu^{++} state is stable against doping and by treating t_0 as a perturbation [5]. By examining all possible (twenty-nine) processes in the order of t_0^2 associated with the introduction of extra holes and by retaining contributions involivng up to two extra holes, we obtained the following;

$$\begin{aligned} H_{eff} = J_d \sum_i{}' [\sum_\alpha{}' \{ S_i \cdot p^+_{\alpha s} \vec{\sigma}_{ss}, p_{\alpha s'} (A_1 + A_2 \sum_{\beta \neq \alpha} n_\beta) \\ + A_3 n_\alpha + A_4 n_{\alpha\downarrow} n_{\alpha\uparrow} + A_5 n_\alpha \sum_{\beta\neq\alpha} n_\beta \} \\ + \sum_{\alpha\neq\beta}{}' \{ S_i \cdot p^+_{\alpha s} \vec{\sigma}_{ss'} p_{\beta s'} (A_6 + A_7 \sum_{\gamma\neq\alpha,\beta} n_\gamma + A_8 (n_\alpha + n_\beta)) \\ + (p^+_{\alpha\uparrow} p_{\beta\uparrow} + p^+_{\alpha\downarrow} p_{\beta\downarrow})(A_9 + A_{10} \sum_{\gamma\neq\alpha,\beta} n_\gamma + A_{11}(n_\alpha + n_\beta) \}] \\ + J_s \sum_{(i\neq k)} S_i \cdot S_k + H_0, \end{aligned} \quad (1)$$

Here $J_d = t_0^2/U_d$ and A_i ($i = 1 \sim 11$) are constants dependent on U_d, U_p, V_0 and Δ; S_i and σ being the Cu-spin at the i-th site and the Pauli matrix, respectively. The summations over α, β and γ are over all the oxygen sites around the i-th Cu-site. The term with J_s, where $\sum_{(i\neq k)}$ implies the summation over pairs, represents the superexchange interaction between Cu sites which is present even in the absence of extra holes. The last term in eq.(1), H_0, represents the direct transfer integrals of holes between O-sites.

The explicit evaluations of A_i have indicated that these parameters are very sensitive to the choices of U_p, V_0 and Δ, especially to the relative magnitude of V_0 and Δ. Our basic assumption of the stability of Cu^{++} put the restriction on the actual values of these parameters. The implication of eq.(1) has not yet been explored and we have examined only some of the limiting cases [5,6].

3. A Case of a Single Hole

Equation (1) in the case of a single extra hole is given as follows [6],

$$H = J_K \sum_i{}' \sum_{\alpha,\beta} \vec{S}_i \cdot p^\dagger_{\alpha s} \vec{\sigma}_{ss'} p_{\beta s'} + J'_K \sum_i{}' \sum_{\alpha} \vec{S}_i \cdot p^\dagger_{\alpha s} \vec{\sigma}_{ss'} p_{\alpha s'}$$

$$- \sum_i{}' \sum_{\alpha,\beta} T_{\alpha\beta} p^\dagger_{\alpha s} p_{\beta s} + J_s \sum_{i\neq k}{}' \vec{S}_i \cdot \vec{S}_k, \tag{2}$$

where

$$J_K = t_0^2 \left(\frac{1}{\Delta} + \frac{1}{U_d - \Delta - 2V_0}\right), \tag{3}$$

$$J'_K = -t_0^2 \left(\frac{1}{\Delta} - \frac{1}{U_p + \Delta}\right), \tag{4}$$

$$T_{\alpha\beta} = -T_0 + \frac{1}{2}\{(t_1 - t_2) - m_{\alpha\beta}(t_1 + t_2)\}, \tag{5}$$

$$T_0 = \frac{t_0^2}{2}\left(\frac{1}{\Delta} - \frac{1}{U_d - \Delta - 2V_0}\right), \tag{6}$$

$$J_s = \frac{2t_0^4}{(\Delta + V_0)^2}\left(\frac{1}{U_d} + \frac{2}{U_p + 2\Delta}\right), \tag{7}$$

where T_0 represents the indirect transfer integral via Cu-sites and $m_{\alpha\beta} = -1$ for nearest neighbor oxygen sites and $m_{\alpha\beta} = 1$ for two oxygen sites across the Cu-site.

The model eq.(2) has recently been examined by Ramsak and Prelovsek [7] mainly without the direct transfer integrals, t_1 and t_2, who have found based on a similar argument by Zhang-Rice [8] that the t-J model represents very well the low energy properties of this model. They have not, however, fully explored effects of direct transfer integrals, which are of our interest in the following.

In order to study eq.(2), we first introduce the Fourier transform of $p_{\alpha s}$ by $p_{\alpha s} = \frac{1}{\sqrt{N}} \sum_k e^{ikR_\alpha} a_{ks}$, where N is the number of the oxygen sites located at (m, n), m, n being integers. Here the lattice spacing between oxygens is taken as unity, and k is in the first Brillouin zone defined in *Fig*.2 by solid lines. In this definition, Cu atoms are located at $(m + \frac{1}{2}, n + \frac{1}{2})$ with $m + n$=even. By use of this Fourier transform for the parts associated with holes, *eq*.(2) is written as follows,

$$
\begin{aligned}
H = {} & 8J_K \sum_{k,k'} \cos\frac{k_x}{2} \cos\frac{k_y}{2} \cos\frac{k'_x}{2} cos\frac{k'_y}{2} \vec{S}_{k-k'} \cdot \vec{\sigma}_{ss'} a^\dagger_{ks} a_{k's'} \\
& + 2J'_K \sum_{k,k'} \cos\frac{k_x - k'_x}{2} \cos\frac{k_y - k'_y}{2} \vec{S}_{k-k'} \cdot \vec{\sigma}_{ss'} a^\dagger_{ks} a_{k's'} \\
& + \sum_{I_\alpha} [\epsilon(k) a^\dagger_{ks} a_{ks} + \epsilon(k + G_\alpha) a^\dagger_{k+G_\alpha s} a_{k+G_\alpha s} \\
& \qquad + (m(k, G_\alpha) a^\dagger_{ks} a_{k+G_\alpha s} + h.c.)] \\
& + \sum_q J(q) S_q \cdot S_{-q},
\end{aligned}
\tag{8}
$$

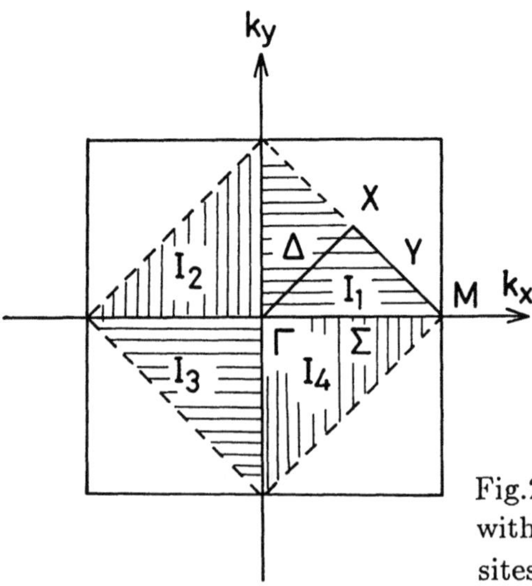

Fig.2 Brillouin zone of O-sites (region within solid lines) and that of Cu-sites (shaded region).

$$\equiv H_K + H_0 + H_S, \tag{9}$$

where $S_q = (2/N)\sum e^{-iqR_i} S_i$, and H_K, H_0 and H_S denote the exchange interaction between Cu-spins and hole spins, the kinetic energy of holes and the superexchange interaction between Cu-spins, respectively. In this equation $\epsilon(k)$, $m(k, G_\alpha)$ and $J(q)$ are given by

$$\epsilon(k) = 2(T_0 - t_1)(\cos k_x + \cos k_y) + 2(T_0 + t_2)\cos k_x \cos k_y, \tag{10}$$

$$m(k, G_\alpha) = 2(T_0 + t_2)\cos(k_x + \frac{G_{\alpha x}}{2})\cos(k_y + \frac{G_{\alpha y}}{2})e^{\frac{i}{2}(G_{\alpha x}+G_{\alpha y})}, \tag{11}$$

$$J(q) = NJ_s \cos q_x \cos q_y, \tag{12}$$

Here $G_1 = \pi(-1,-1)$, $G_2 = \pi(1,-1)$, $G_3 = \pi(1,1)$, $G_4 = \pi(-1,1)$, I_α is defined in *Fig.*2 and the summation over q in the last term in *eq.*(8) is for the shaded region in *Fig.*2.

Equation (8) is difficult to solve due to the periodic array of Cu-spins. In order to get some insight into the electronic states described by this equation, we have examined a case with a single Cu-spin and then investigated that of two Cu-spins in the framework of the tight-binding approximation. To study electronic states in the periodic lattice by treating a single Cu-spin shares a common feature with the study by Eskes and Sawatzky [9] based on the Anderson impurity approach to a Cu-atom. Here in this paper, however, we study the case where there exist only one Cu-spin in a whole crystal but Cu-atoms are located periodically in the sense that we retain the indirect transfer integrals between oxygen sites via Cu-sites.

As will be seen, this study has afforded us interesting features of a hole, which are briefly explained in the following.

(a) Hole bands

If $J_K = 0$, $J_{K'} = 0$, the hole bands are determined by the third term of r.h.s. of eq.(8), and are shown in *Figs.3(a)-(c)* for some choices of T_0, t_1 and t_2. The lower band in *Fig.*3(*a*) without the k-dependence is due to the non- bonding state. For finite t_1 and t_2 the band structures have qualitatively different features depending on the relative magnitudes of T_0, t_1 and t_2, which are classified as follows i) for $t_1 < T_0$, the bottom of the band lies along line M-X with energy $-2(T_0 + t_2)$ and the top of the band at Γ with energy $6T_0 - 4t_1 + 2t_2$. It is to be noted that the location of this bottom coincides with the Fermi surface of the half-filled band of the

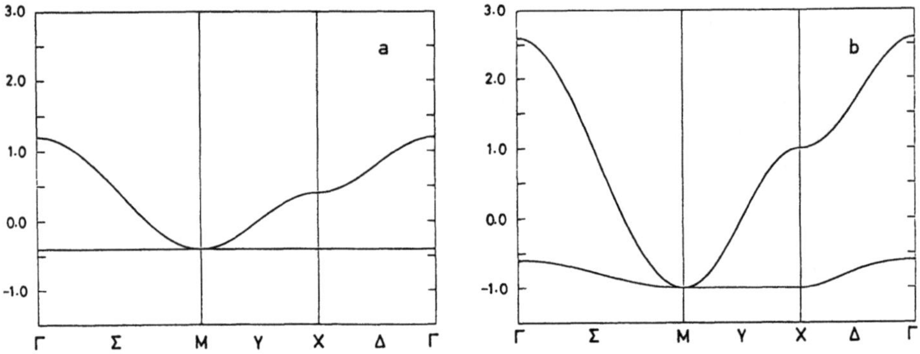

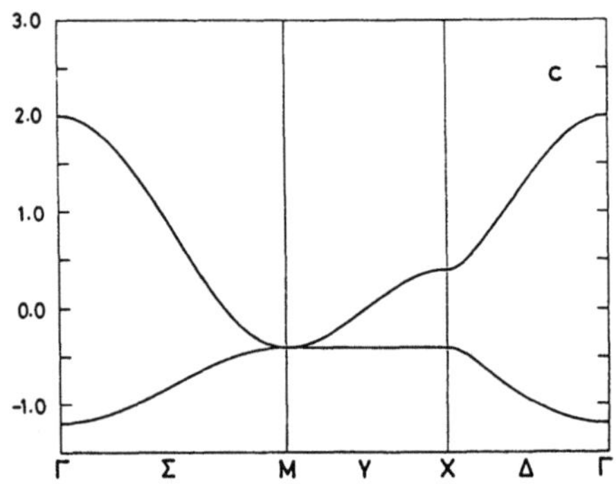

Fig.3 Energy dispersion of hole bands. (a) $T_0 = 0.2$, $t_1 = 0$, $t_2 = 0$, (b) $T_0 = 0.2$, $t_1 = 0.6$, $t_2 = 0.3$, (c) $T_0 = 0.2$, $t_1 = 0.6$, $t_2 = 0$.

oxygen square lattice with the nearest neighbor transfer integral. ii) for $T_0 < t_1 < 2T_0 + t_2$, the bottom of the band lies also along line M-X with the energy $-2(T_0+t_2)$ and the top of the band at Γ with energy $2(-T_0+2t_1+t_2)$. An example of this case is shown in *Fig.3(b)*, where $T_0 = 0.2$, $t_1 = 0.6$, $t_2 = 0.3$. iii) for $t_1 > 2T_0 + t_2$, the bottom and the top of the band are both located at Γ with energy $6T_0 - 4t_1 + 2t_2$ and $-2T_0 + 4t_1 + 2t_2$. An example of this case is also shown in *Fig.3(c)*, where $T_0 = 0.2$, $t_1 = 0.6$ and $t_2 = 0$. In a special case of $t_1 = 2T_0 + t_2$, the k-dependences of the bands are the same as those in the case of $t_1 = t_2 = 0$, but with the different band width. It is seen that the direct transfer integrals, t_1 and t_2, affect the band structure in the opposite way to T_0; t_1 and T_0 counteract whereas t_2 and T_0 cooperate. Hence for small values of t_1, (and with even smaller value of t_2) the total band width gets narrow as t_1 is increased.

(b) Bound states

Once J_K and $J_{K'}$ are turned on bound states with different values of spin multiplicity (singlet and triplet) and different spatial symmetries are possible. The latters are schematically shown in *Figs.4(a)-(d)* in the

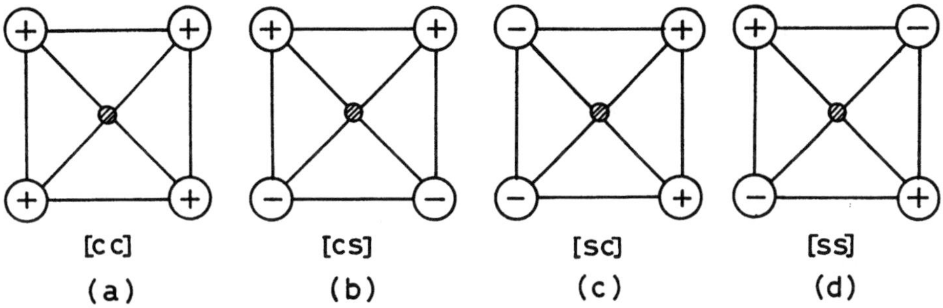

Fig.4 Wave functions in the absence of kinetic energy for (a) F_{cc}, (b) F_{cs}, (c) F_{sc}, (d) F_{ss}.

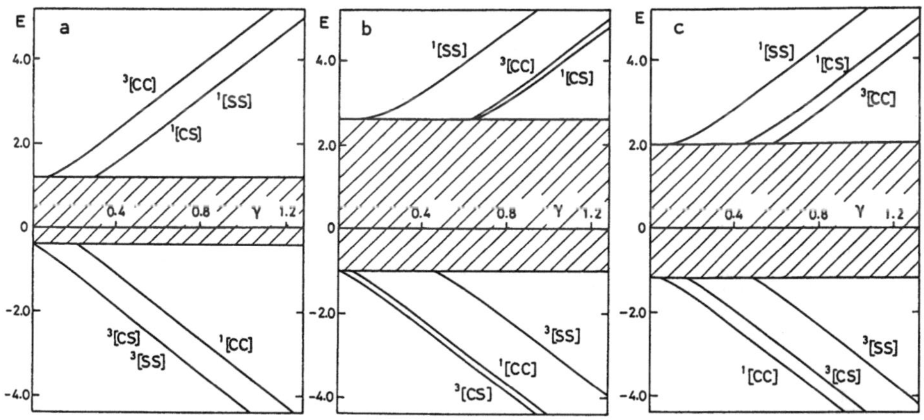

Fig.5 Energies of singlet (*e.g.* ^{1}[cc]) and triplet (*e.g.* ^{3}[cc]) bound states, as a function of the coupling energy. Shaded regions are hole band continuum. (a) $T_0 = 0.2$, $t_1 = 0$, $t_2 = 0$, (b) $T_0 = 0.2$, $t_1 = 0.6$, $t_2 = 0.3$, (c) $T_0 = 0.2$, $t_1 = 0.6$, $t_2 = 0$.

absence of kinetic energy, where Cu spin is located at the center and circles are O-sites with + or − being the phases of wave functions. The binding energies of these states for the same choices of T_0, t_1 and t_2 as in *Figs.3(a)-(c)* are shown in *Figs.5(a)-(c)* as a function of the coupling energy, γ, which takes different values depending on the spin multiplicity and the spatial symmetry.

The state, ^{1}[cc], is the symmetric singlet state, which has been examined by Zhang and Rice [8] in the absence of direct transfer integrals. The coupling energy for this state, γ_s, is considered to be $\gamma_s \sim 1.0 - 1.2$ and then this state is seen to be strongly stabilized.

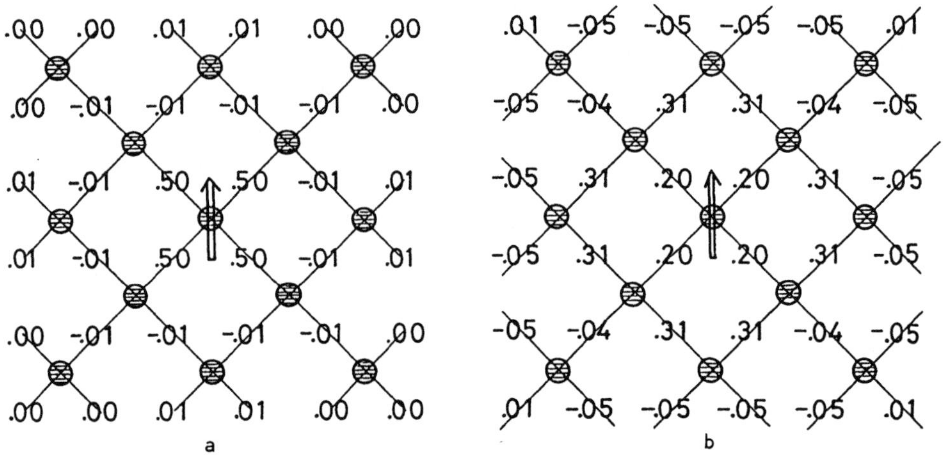

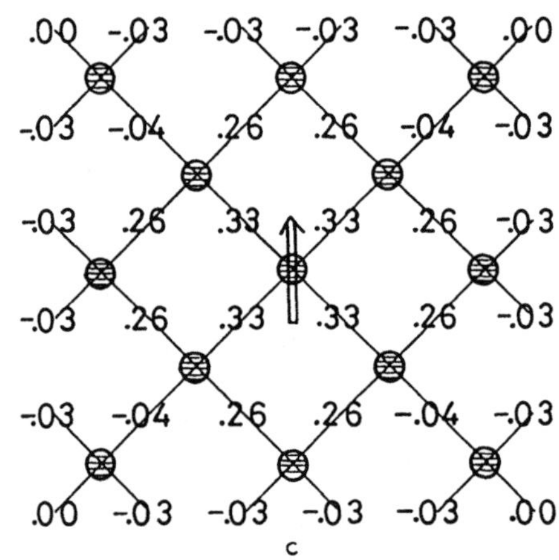

Fig.6 Wave functions of the singlet bound state ^{1}[cc] with $\epsilon_B = -4.0$.
(a) $T_0 = 0.2$, $t_1 = 0$, $t_2 = 0$,
(b) $T_0 = 0.2$, $t_1 = 0.6$, $t_2 = 0.3$,
(c) $T_0 = 0.2$, $t_1 = 0.6$, $t_2 = 0$.

It is interesting to note that triplet states are also possible below the hole band once $J'_K < 0$, *i.e.* $U_p \neq 0$, though their stabilization energies are far less than the singlet state ^{1}[cc].

The spatial extent of the Zhang-Rice singlet state is shown in *Figs.6(a)-(c)* for three choices of parameters. It is to be noted that the way how the singlet state extends is very sensitive to the relative magnitudes of the direct and the indirect transfer integrals between oxygen sites.

(c) Effective transfer integral of the singlet state

The singlet state discussed in the preceding section for a case of a single Cu-spin will have a finite dispersion in the presence of the periodic array of Cu-spins. In order to estimate the effective width of this singlet band, we investigated the case of two Cu-spins located at the i-th site by treating them in the framework of the tight-binding approximation; Let H_i and Ψ_i be the Hamiltonian of a single Cu-spin at the i-th site in the presence of the hole band and its singlet eigen-function; $H_i\Psi_i = \epsilon_B\Psi_i$, and we denote the total Hamiltonian of two Cu-spins at the i- and j-th sites as H_t, *i.e.* $H_t = H_i + H_j - H_0$ where H_0 is the kinetic energy of a hole. Here we ignored the superexchange interaction between Cu-spins. By use of $< \Psi_i|H_t|\Psi_j > \equiv h_{ij} = h_{ji}$ and $< \Psi_i|\Psi_j > \equiv u_{ij} = u_{ji}$ the effective transfer integrals, t_{ij}, between two singlet states located at the i-th and j-th is defined by $t_{ij} = (\epsilon_1 - \epsilon_2)/2$. Here ϵ_1 and ϵ_2 are solutions of the following eigen-value equation,

$$det \begin{bmatrix} h_{ii} - \epsilon u_{ii} & h_{ij} - \epsilon u_{ij} \\ h_{ji} - \epsilon u_{ji} & h_{jj} - \epsilon u_{jj} \end{bmatrix} = 0, \tag{13}$$

and given by $\epsilon_1 = (h_{ii} - h_{ij})/(1 - u_{ij})$ and $\epsilon_2 = (h_{ii} + h_{ij})/(1 + u_{ij})$. In $Figs.7(a) - (c)$ are shown the effective transfer integrals, t_{ij}, thus obtained. It is seen that t_{ij} are sizable between the nearest neighbor pairs. A similar conclusion, but smaller values of t_{ij}, has also been obtained recently by Sawatzky [4] based on a different approach. It is to be noted that in the case shown is $Fig.7(b)$ t_{ij} between the third neighbor is very large. This is due to the particular feature of the spatial extent of the wave function in that direction as seen from *Fig.6(b)*.

As demonstrated these values of t_{ij} are very sensitive to the choice of original parameters, especially to those of the direct and indirect transfer integrals.

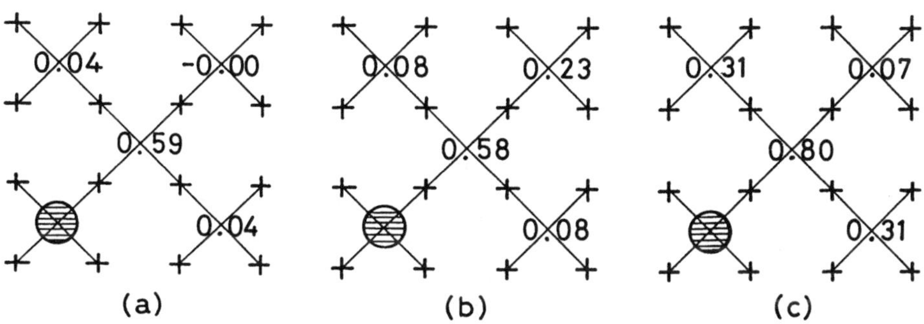

Fig.7 The transfer integrals, t_{ij}, between two singlet states with energy $\epsilon_B = -4.0$ located around the i-th and j-th sites. (a) $T_0 = 0.2$, $t_1 = 0$, $t_2 = 0$, (b) $T_0 = 0.2$, $t_1 = 0.6$, $t_2 = 0.3$, (c) $T_0 = 0.2$, $t_1 = 0.6$, $t_2 = 0$.

4. Case of Electron Doping

In contrast to the case of hole doping the derivation of the effective Hamiltonian for electron doping is very simple in the present context, since extra electrons are considered to go into only Cu sites (resulting in locally Cu^+ (d^{10}) states) because of the completely filled O:p-shelles. The result is [5]

$$H_{eff} = -\frac{t_0^2}{\Delta}\sum_{i\neq j}(1-n_{j,-s})d^{\dagger}_{js}d_{is}(1-n_{i,-s})+(\frac{t_0^2}{\Delta}-\frac{t_0^2}{\Delta+V_0})\sum_{i\neq J}n_i n_j$$
$$+ J_s \sum_{(i\neq j)} S_i \cdot S_j, \qquad (14)$$

where i and j refer to Cu sites and the summations are over nearest neighbor pairs and $n_{is} = d^{\dagger}_{is}d_{is}$, $n_i = n_{i\uparrow} + n_{i\downarrow}$. If $V_0 = 0$, *eq.*(14) is nothing but the *t-J* model, but for finite V_0 there exists a repulsive interaction between two adjacent d-holes.

5. Conclusion

We have derived the effective Hamiltonian to describe the low-lying excitations in the presence of extra holes and examined the energy spectrum of a particular case with a single hole. We have found that various types of bound states are possible below the hole band continuum. Among them the singlet state with symmetrical orbital configuration, which was first examined by Zhang-Rice, is most stabilized. The spatial extent of this Zhang-Rice singlet is found to be very sensitive to the relative magnitudes of the direct and indirect transfer integrals, and is spread out over several lattice spacings for a reasonable choice of parameters. Beside this Zhang-Rice singlet state, triplet states with different spatial symmetries are found to be possible once $U_p \neq 0$.

In order to pursue the possible consequence of the periodic array of Cu spins we have also estimated the effective transfer integrals, t_{ij}, between two Zhang-Rice singlet states around different Cu-sites by use of the tight binding approximation. It is found that t_{ij} is apprecilable for the nearest neighbor Cu-sites and that it takes appreciable values over several lattice sites reflecting the spatial extent of the state. Especially those between the third neighbor can be very sizable for some range of parameters, which we think not so unrealistic. However absolute magnitudes of these effective transfer integrals based on the tight-binding approximation should not be taken literally, since within the spatial extent of the singlet state there are several Cu spins in actual periodic array of Cu atoms, whose participation will reduce the overlap integral of the two singlet states.

From these results we conclude that the low-lying excitations of CuO_2 layers are well described by the singlet band composed of local Cu spins and O hole spins. The essence of this state will be described by the *t-J* model as Anderson [10] has proposed even from the early stage of theoretical studies. The existence of the appreciable spatial extent of the singlet state, however, will result in some modifications of the *t-J* model, *e.g.* finite spatial extent of the transfer integrals and the exchange interactions. Especially that the transfer integrals not only between the nearest neighbors but also those between the third neighbors can be sizeable is note-worthy. This is particularly interesting in view of the fact that the transfer integrals between the same sublattices have very different features from those between the different sublattices as has been indicated by Lee [11].

This observation of the possible importance of spatial extent of the singlet state may have some bearing with the finding by Imada *et al.* [12] of the particular features associated with the extended singlet state in stabilizing the two-hole bound state.

Acknowledgement

This present research was financially supported by Grant-in-Aid for Scientific Research on Priority Areas, Mechanisms of Superconductivity (63631007) and New Functionality Materials-Design, Preparation and Control (63604014) from the Ministry of Education, Science and Culture.

References

[1] K.T. Park, K. Terakura, T. Oguchi, A. Yanase and M. Ikeda, J. Phys. Soc. Jpn. **57**, 3445 (1988).
[2] A.K. McMahan, R.M. Martin and S. Satpathy, Phys. Rev. B **38**, 6650 (1988).
[3] F. Mila, Phys. Rev. B **38**, 11358 (1988).
[4] G. Sawatzky, this proceeding.
[5] H. Fukuyama, H. Matsukawa and Y. Hasegawa, J. Phys. Soc. Jpn. **58** 364 (1989); ibid **58** No.8 (1989).

[6] H. Matsukawa and H. Fukuyama, submitted to J. Phys. Soc. Jpn. (ISSP Tech. Rept. No.2141.) (1989).
[7] A. Ramsak and P. Prelovsek, preprint.
[8] F.C. Zhang and T.M. Rice, Phys. Rev. B **37**, 3759 (1988).
[9] H. Eskes and G.A. Sawatzky, Phys. Rev. Lett. **61**, 1415 (1988).
[10] P.W. Anderson, Science **235**, 1196 (1987); this proceeding.
[11] P.A. Lee, preptint; this proceeding.
[12] M. Imada, N. Nagaosa and Y. Hatsugai, J. Phys. Soc. Jpn. **57**, 2901, (1988); M. Imada, this proceeding.

The Madelung Energy of Cu-O Based Compounds

J. Kondo

Electrotechnical Laboratory, Tsukuba, Ibaraki, Japan

Abstract. Based on the ionic crystal model, we approximate the energy of Cu-O based layered compounds by the sum of the Madelung energy and the energies of the isolated ions. Carrier distribution is determined so as to make this energy a minimum. We have considered compounds like $Bi_2Sr_2(Ca,Y)_{n-1}Cu_nO_{2n+4}$ and found several different types of carrier distribution:

A: holes in the oxygens in the Cu-O planes with Cu surrounded by six or five oxygens

B: holes in the apical oxygens

C: electrons in the Cu-O planes with Cu in square-planar coordination

D: electrons in the Cu-O planes as well as in the Bi-O or Tl-O planes

E: holes in the Cu-O planes and electrons in the Bi-O or Tl-O planes

F: eléctrons in the Bi-O or Tl-O planes.

1. Introduction

Recently discovered Cu-O based layered superconductors consist of many components and can be studied by alloying with other metal elements. Here we consider three series of such compounds: $Bi_2Sr_2(Ca,Y)_{n-1}Cu_nO_{2n+4}$, $Tl_2Ba_2(Ca,Y)_{n-1}Cu_nO_{2n+4}$ and $TlBa_2(Ca,Y)_{n-1}Cu_nO_{2n+3}$. Our concern is what carriers are and which sites they occupy. We take a standpoint that these oxides are essentially ionic crystals and assume that the total energy of a compound is the sum of the energies of the isolated ions E_{ION} and the Madelung energy between point charges E_{MAD}. We will find a minimum of the total energy by varying the valencies of some ions and answer the above-mentioned questions. We have found several different types of carrier distribution including electrons on planar Cu-O sheets.

2. Madelung Potential

First I will mention some algebra of the Madelung potential. Suppose there are N different sites in the unit cell. Let Q_i be the average valence of the i site. Here we assume that ions of different valences are randomly distributed on each site (if there are such ions). Let α_i be the number of the i site in the unit cell. Then, from neutrality condition we have

$$\sum_{i=1}^{N} \alpha_i Q_i = 0. \tag{1}$$

When Q_i's are given which satisfy eq. (1), the Madelung potential at the i site is expressed as

$$U_i = \sum_{j=1}^{N} c_{ij} \alpha_j Q_j. \tag{2}$$

Note that c_{ij} is not unique because of eq. (1). Let us define a unique $\tilde{c}_{ij}$ as follows. We place a unit charge on the i site and a minus unit charge on the j site:

$$\alpha_i Q_i = 1 \text{ and } \alpha_j Q_j = -1. \tag{3}$$

For this choice of Q's the Madelung potentials at the i and j sites are uniquely determined and are set as

$$\begin{aligned} U_i &= -\tilde{c}_{ij} \\ U_j &= \tilde{c}_{ji}. \end{aligned} \tag{4}$$

With this definition of $\tilde{c}_{ij}$ and $\tilde{c}_{ii}=0$, U_i is expressed by the same expression as eq. (2):

$$U_i = \sum_{j=1}^{N} \tilde{c}_{ij} \alpha_j Q_j. \tag{5}$$

One has only to calculate at most N(N-1) $\tilde{c}_{ij}$'s to treat the Madelung energy of a substance. Then, the total Madelung energy per unit cell is obtained as

$$E_{MAD} = \frac{1}{2} \sum_{ij} \tilde{c}_{ij}\, \alpha_i \alpha_j Q_i Q_j. \tag{6}$$

Next let us consider the energy of isolated ions, which is the sum over all the sites of the ionic energy averaged over the valence states involved in that site. Since the fraction of different valence states in the i site can be deduced from the average valence Q_i, the total energy of isolated ions is a linear function of Q_i's. Thus the sum of the Madelung energy and the ionic energy is a quadratic function of Q_i's. We will find a minimum of this energy by varying some of Q_i's. In general in doing such variation we fix the valences of Ca, Ba and Y to 2, 2, and 3, respectively but vary those of Cu, O, Tl and Bi. Since recent experiments indicate that holes reside on oxygen and not on copper, we restrict the ranges of the valences of these elements as follows:

$$
\begin{aligned}
&-2<Q_O<-1 \\
&1<Q_{Cu}<2 \\
&2<Q_{Bi}<3 \\
&2<Q_{Tl}<3.
\end{aligned}
\qquad (7)
$$

Here we are assuming that oxygen accepts holes, whereas Cu, Bi and Tl accept electrons. We use the following values of the ionization potentials of metal elements and the second electron affinity of oxygen.

	Cu	Y	Tl	Bi	La
I_2	20.3(eV)		20.4	16.7	
I_3	36.8	20.5	29.8	25.5	19.2

	O
A_2	-7

3. Results

We now go onto the results. Some of the results have already been reported [1].

3.1 $Bi_2Sr_2CuO_6$

Fig. 1 shows a part of the unit cell of this compound. The table shows the nominal valences for the ions involved in this compound and the corresponding Madelung potentials.

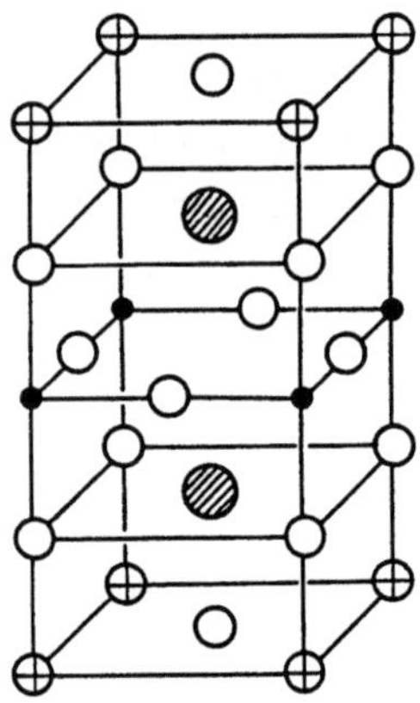

Fig. 1 A part of the unit cell of $Bi_2Sr_2CuO_6$.

	Bi	O4	Sr	O3	Cu	O2
Q	3	-2	2	-2	2	-2
U	17.2	-28.3	27.7	-14.2	42.2	-5.8(eV)

The potentials are for the electron in units of eV. O4 denotes oxygen in Bi-O planes, O3 that in Sr-O planes and O2 that in Cu-O planes. One sees that the potentials of the Cu-O planes are very high compared with others especially those of the Bi-O plane. If one takes an electron from O2 to Bi, one needs an energy equal to

$$U(Bi) - U(O2) - I_3(Bi) + A_2 = -9.5eV ,$$

which is negative and quite large. So we conclude that the charge assignment of the above table is unstable and such a charge transfer should actually occur. To find the amount of the charge transfer we set

$$Q_{Bi} = 3-n$$
$$Q_{O2} = -2+p_2$$

with $n=p_2$. By calculating the $\tilde{c}_{ij}$ of this system we find

$$E_{MAD} = 20.1p_2^2 + 46.0p_2 + \text{const. (eV)}$$

$$E_{ION} = 2p_2[A_2 - I_3(Bi)] + \text{const.} = -65.0p_2 + \text{const. (eV)}$$

We find that the minimum of the sum is obtained when $n=p_2=0.47$. This substance involves a large amount of electron-hole pairs without any doping.

3.2 $Bi_2Sr_2Ca_{1-y}Y_yCu_2O_8$

Two Cu-O planes are stuck together with an intervening Ca sheet. For y=0 the nominal valence assignment and the corresponding Madelung potentials are shown in the table.

	Bi	O4	Sr	O3	Ca	Cu	O2
Q	3	-2	2	-2	2	2	-2
U	16.1	-29.2	23.8	-19.4	34.5	35.2	-11.1

Again the CuO plane is very unstable for the electrons, some of which may spill off to Bi-O planes. We have determined the amount of such charge transfer by the minimum priciple. The result is shown in Fig. 2. Here p_i denotes the content of holes on the i-th oxygen site ($Q_{Oi}=-2+p_i$) and n that of electrons on Bi ($Q_{Bi}=2-n$). As y is increased, some electrons are introduced, which annihilate holes which already existed. However, one sees that the number of electrons on Bi also decreases. This is because Y^{3+} ions now on Ca sheets pull electrons from Bi back to Cu-O planes.

After all the holes have been annihilated, the electrons enter Bi sites up until y=1. Here all the Ca has been replaced by Y, and half of Bi is Bi^{2+}. The valence assignment and the Madelung potentials of this case are shown in the table.

	Bi	O4	Sr	O3	Y	Cu	O2
Q	2.5	-2	2	-2	3	2	-2
U	23.6	-17.3	19.5	-17.2	26.6	23.8	-24.5

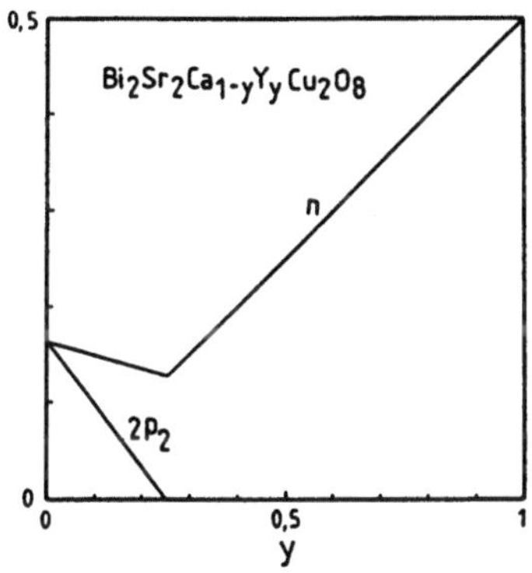

Fig. 2 The number of electrons (n) and holes (p_2) vs y for $Bi_2Sr_2Ca_{1-y}Y_yCu_2O_8$

Looking at this table one notes that these values are quite even and represent a stable valence configuration.

3.3 $Bi_2Sr_2Ca_{2-2y}Y_{2y}Cu_3O_{10}$

There are two kinds of Cu-O planes. The outer ones will be denoted by Cu2 and O2 and the inner one by Cu1 and O1. The table shows valence assignment and the Madelung potentials for y=0. Again the Cu-O planes are very unstable for the electron.

	Bi	O4	Sr	O3	Ca	Cu1	O1	Cu2	O2
Q	3	-2	2	-2	2	2	-2	2	-2
U	13.6	-31.7	21.2	-22.0	32.6	37.2	-9.1	32.6	-13.7

It is interesting to note that the (Cu-O)1 plane is more unstable than the (Cu-O)2 plane, even though the Cu in the former plane does not have an apical oxygen. We note that the difference of the potentials between O and Cu is almost the same for the two planes.

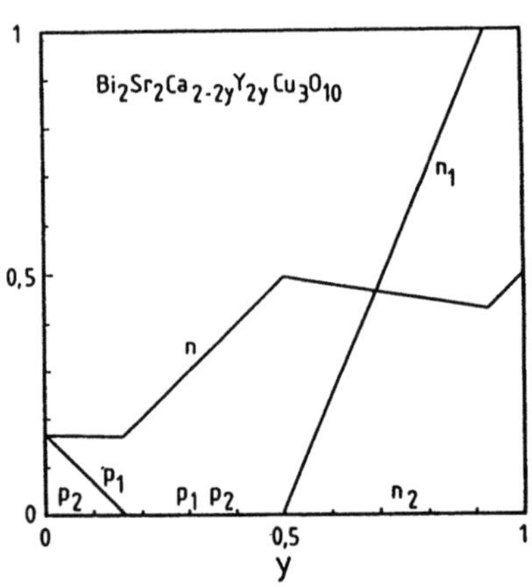

Fig. 3 The number of electrons (n, n_1, n_2) and holes (p_1, p_2, p_3) vs y for $Bi_2Sr_2Ca_{2-2y}Y_{2y}Cu_3O_{10}$

We found that some electrons spill off O1 into Bi. The result is shown in Fig. 3. As y is increased the doped electrons annihilate some holes and for y>0.17 all the holes have been annihilated. For further increase of y all the doped electrons enter Bi, and at y=0.5 half of Bi is Bi^{2+} and half of Ca has been replaced by Y. The table shows the valence assignment and the Madelung potantials for this case.

	Bi	O4	Sr	O3	CaY	Cu1	O1	Cu2	O2
Q	2.5	-2	2	-2	2.5	2	-2	2	-2
U	24.9	-16.1	20.8	-16.0	21.2	19.5	-27.5	24.6	-23.6

Now the potentials of the Cu-O planes have been lowered and those of the Bi-O plane raised. It is interesting to note that the (Cu-O)1 plane is now more stable than the (Cu-O)2 plane. This is because (Cu-O)1 is sandwitched by two Ca sheets which now involve a large amount of excess positive charge due to Y^{3+}.

Since the order of the levels has been reversed between Cu1 and Bi, one might wonder if a reverse electron transfer from Bi^{2+} to $Cu1^{2+}$ would take place. The energy necessary for this transfer

$$U(Cu1) - U(Bi) + I_3(Bi) - I_2(Cu) = -0.23eV$$

is negative but small. So the amount of the transfer will be small. However, as y is increased beyond 0.5, more electrons are introduced which will occupy Cu1. The minimum principle calculation shows that most of the added electrons enter Cu1 and not Cu2, while the number of electrons on Bi remains almost constant (~0.5, see Fig. 3). Finally at y=1 all the Cu1 are Cu^+ and half of the Bi is Bi^{2+}. The Madelung potentials of this case are shown in the table. One finds that the charge configuration of this table is stable.

	Bi	O4	Sr	O3	Y	Cu1	O1	Cu2	O2
Q	2.5	-2	2	-2	3	1	-2	2	-2
U	24.1	-16.9	19.9	-16.8	26.7	14.5	-22.9	24.2	-24.2

We have seen that for the compounds of n=3, electrons are doped into the inner Cu-O plane of triple-layered Cu-O planes. A crucial factor responsible for this fact is an excess positive charge on the CaY sheets in both sides of the Cu-O plane.

The carrier distribution of the series $Tl_2Ba_2(Ca,Y)_{n-1}Cu_nO_{2n+4}$ is quite similar to the series just studied and will not be mentioned here (see [1]).

3.4 $TlBa_2Ca_{1-y}Y_yCu_2O_7$

This compound involves one extra hole for y=0, which mainly occupy the O2 sites, namely the oxygens in Cu-O planes. As Y is introduced, the total number of holes decreases as 1-y. The Y enter Ca sheets which now involve a large excess positive charge due to Y^{3+}. Since the Cu-O sheets are next to the Ca sheets, the O2 sites become unfavorable for the hole. As the content of Y is increased further, the hole occupation switches from O2 to O3 sites (oxygens in Ba-O planes) and finally to O4 (oxygen in Tl-O planes). At the same time the hole level of O2 has been pushed up very much for y ~ 1, due to Y^{3+} in nearby Ca sheets. So, the conduction in this case, if any, will be by holes in the Tl-O or Ba-O planes. It will be interesting to investigate properties of this hole conduction.

3.5 $TlBa_2Ca_{2-2y}Y_{2y}Cu_3O_9$

This substance involves two kinds of Cu-O planes. The outer ones will be denoted by 2 and the inner one by 1. For y=0 there exists one hole, which mainly occupy O2 sites with less than 10% being on O4. As in the previous case hole occupation switches from O2 to O3 and finally to O4 as y is increased from 0 to 0.5. For y ~ 0.5 the hole level of O2 is very high compared with those of O3 and O4, so the conduction in this case will be by holes in the Tl-O and/or Ba-O planes. At y=0.5 no carriers exist.

The hole level of O1 always remains high compared with those of other O sites and O1 will not be populated by holes in any case. However, this means that the electron level of O1 is low. This is because the (Cu-O)1 planes are sandwitched by two Ca sheets, where there exist some Y^{3+} ions. When half of the Ca is replaced by Y, the stabilization of the (Cu-O)1 sheets is enormous. The table shows the Madelung potentials of this case.

	Tl	O4	Ba	O3	CaY	Cu1	O1	Cu2	O2
Q	3	-2	2	-2	2.5	2	-2	2	-2
U	31.0	-16.0	21.5	-16.5	20.4	18.4	-27.9	25.1	-21.6

One notes that the potential of Cu1 is very low. As y is increased beyond 0.5, electrons are added to this system. We have found that the electrons occupy Cu1 sites but not Cu2 sites.

4. Conclusion

Our main conclusion in this talk is that for compounds with triple-layered (or more than triple-layered) Cu-O planes, electrons may enter the inner Cu-O planes when an enough amount of Y is substituted in the interveneing Ca sheets. There is also a case where holes enter Ba-O (or Sr-O) sheets. Furthermore, there is a case where electrons enter Bi-O (or Tl-O) sheets. It will be interesting to study properties of carriers in these cases.

References

[1] J. Kondo: Submitted to J. Phys. Soc. Jpn.

Electronic Properties in the Cu-Oxide Superconductors

S. Maekawa, J. Inoue, and T. Tohyama

Department of Applied Physics, Nagoya University, Nagoya 464-01, Japan

Abstract. We calculate the electronic states in the electron- and hole-doped Cu-oxide superconductors in the cluster model approach, and examine the validity of the strongly correlated one-band Hubbard model, i.e., the so-called $t - J$ model for the oxides. Then, applying the many-body tight-binding method to the $t - J$ model, we discuss the electronic properties of the oxides.

1. Introduction

Since Anderson's proposal [1] that the strongly-correlated one-band Hubbard model near half-filling in the square lattice includes essential physics of the Cu-oxide high T_c superconductors, much study has been done along this line. The simplest Hamiltonian based along this line, i.e., the so-called $t - J$ model, is written as

$$\begin{cases} H_{tJ} = -t \sum_{(ij)\sigma} \tilde{c}^{\dagger}_{i\sigma}\tilde{c}_{j\sigma} + 2J \sum_{(ij)} \vec{S}_i \cdot \vec{S}_j, \\ \tilde{c}_{i\sigma} = (1 - n_{i-\sigma})c_{i\sigma} \, , \end{cases} \tag{1}$$

where $c_{i\sigma}$ is an annihilation operator of an electron with spin σ at site i, $n_{i\sigma} = c^{\dagger}_{i\sigma}c_{i\sigma}$, and $\vec{S}_i$ is a spin operator with S=1/2 at site i. In the Hamiltonian, the first term denotes the hopping of holes (electrons) to the nearest-neighbor sites on condition that no double occupancy is allowed. The second denotes the nearest-neighbor antiferromagnetic (AF) exchange interaction ($J > 0$).

For applying the $t - J$ model to the oxides, there is an assumption that electron- or hole-doping causes the spin defects with electric charge on Cu sites so that the defects can be looked upon as *holes* on Cu sites. Such an assumption may be realized if the doping induces the singlet states localized near the Cu sites [2-8]. However, there are still controversies concerning how good the assumption is and what the physical parameters are in the real oxides when the model is applied to them. Here, we will examine these problems.

Springer Series in Solid-State Sciences, Vol. 89 Strong Correlation and Superconductivity
Editors: H. Fukuyama · S. Maekawa · A.P. Malozemoff

In Sec. 2, we calculate the electronic states induced by electron- and hole-doping in the cluster model approach, and discuss the validity of the $t-J$ model for the oxides. In Sec. 3, we apply the so-called many-body tight-binding method to the $t-J$ model, and derive the dispersion and density of states of holes in the AF state. We will find that the holes make coherent states together with quantum spin fluctuations. This fact reminds us of quantum spin liquid or resonating valence bond (RVB) state with holes. Finally, the stability of the RVB state is discussed in Sec. 4.

2. Electronic States

In the hole-doped superconductors such as $La_{2-x}Sr_xCuO_4$ with T-structure, an extra hole sits on O-2p orbitals and exchange-interacts with a Cu-3d spin. As a result, a singlet state made of Cu-3d and O-2p orbitals may carry the electric current [2-8]. In the electron-doped superconductors such as $Nd_{2-x}Ce_xCuO_4$ with T'-structure [9,10], although an electron is mainly introduced into the Cu-3d orbital, it is also possible for the electron to be introduced into the Cu-4s orbital. Since the Cu-4s orbital strongly mixes with the neighboring O-2p orbitals [8], the doping causes a composite state with spin-singlet, which is made of Cu-3d, Cu-4s and O-2p orbitals. Therefore, the detailed difference between the electronic states will demonstrate the characteristic features in the superconductors.

In the cluster model approach [8,11-14], the strong electron correlation at Cu sites and the charge-transfer between Cu and O sites are taken into account on the equal basis. To elucidate the electronic properties and the motion of the singlet states in the electron- and hole-doped superconductors, we examine the properties in the coupled clusters given in Fig.1. The Hamiltonian is written as

$$
\begin{aligned}
H = & \sum_{mm'}\sum_{ll'\sigma} T_{pd}\binom{m\ m'}{l\ \ l'}\left(p^{\dagger}_{ml\sigma}d_{m'l'\sigma} + d^{\dagger}_{m'l'\sigma}p_{ml\sigma}\right) + \sum_{mm'}\sum_{ll'\sigma} T_{pp}\, p^{\dagger}_{ml\sigma}p_{m'l'\sigma} \\
& + \sum_{mm'}\sum_{l'\sigma} T_{sp}\left(s^{\dagger}_{m\sigma}p_{m'l'\sigma} + p^{\dagger}_{m'l'\sigma}s_{m\sigma}\right) \\
& + U_d \sum_{m}\sum_{\substack{ll'\sigma\sigma' \\ (l,\sigma)\neq(l',\sigma')}} d^{\dagger}_{ml\sigma}d_{ml\sigma}d^{\dagger}_{ml'\sigma'}d_{ml'\sigma'} + U_p \sum_{m}\sum_{\substack{ll'\sigma\sigma' \\ (l,\sigma)\neq(l',\sigma')}} p^{\dagger}_{ml\sigma}p_{ml\sigma}p^{\dagger}_{ml'\sigma'}p_{ml'\sigma'} \\
& + \epsilon_d \sum_{ml\sigma} d^{\dagger}_{ml\sigma}d_{ml\sigma} + \epsilon_p \sum_{ml\sigma} p^{\dagger}_{ml\sigma}p_{ml\sigma} + \epsilon_s \sum_{m\sigma} s^{\dagger}_{m\sigma}s_{m\sigma} \,, \qquad (2)
\end{aligned}
$$

where the operators $d^{\dagger}_{ml\sigma}$ and $s^{\dagger}_{m\sigma}$ create holes in the l-th Cu-3d orbital and Cu-4s one with spin σ at site m, respectively, and $p^{\dagger}_{ml\sigma}$ creates a hole in the l-th O-2p orbital with spin σ at site m. In the numerical calcula-

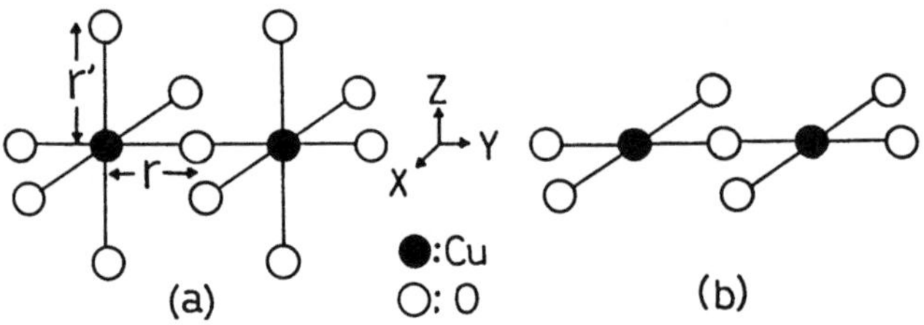

Fig. 1. Coupled clusters used in the cluster model approach. (a) La_2CuO_4, (b) Nd_2CuO_4. In (a), the ratio r'/r is taken to be 1.28.

tion, we take the Cu-3d orbitals with $(x^2 - y^2)$ and $(3z^2 - r^2)$ symmetries and the O-2p orbitals which cause σ-bonds with Cu-3d orbitals. $\epsilon_{3d}, \epsilon_{2p}$ and ϵ_{4s} are the energies of Cu-3d, O-2p and Cu-4s orbitals, respectively. U_d and U_p are the correlation energies in the Cu-3d and O-2p orbitals, respectively.

Let us consider the La_2CuO_4 system, the model cluster of which is given in Fig.1(a). The $(Cu_2O_{11})^{18-}$ cluster describes the undoped system and the $(Cu_2O_{11})^{17-}$ cluster the hole-doped one. In these cases, the Cu-4s orbitals are neglected, since they cause only the shift of the O-2p orbitals. The spins with $S = 1/2$ around Cu sites couple each other antiferromagnetically in the undoped system, so that the ground state is singlet and the first excited state is triplet. From the energy difference between the singlet and triplet states, we obtain the exchange parameter $2J$ in eq.(1) as $2J = 0.13eV$. The parameter values used in the calculation are listed in Table 1(a).

In the $(Cu_2O_{11})^{17-}$ cluster for the hole-doped system, we find that the ground state may be written as the anti-bonding state of the singlet state given in the $(CuO_6)^{11-}$ cluster when the parameter values in Table 1(a) are used. The first excited state mostly consists of its bonding state. Thus, we may take the hopping parameter of the singlet, t, in eq.(1) to be the half of the energy difference between the bonding and anti-bonding states. We obtain $t = 0.29eV$. We note that although two parameters in the $t - J$ model depend on the parameters listed in Table 1(a), they are consistently determined and thus are not independent each other.

In Nd_2CuO_4, it has been shown that the Madelung energy acting on an electron at Cu site is by 3.0eV [17] lower than that in La_2CuO_4, whereas the Madelung energy at the O sites is not much different between the oxides. The $(Cu_2O_7)^{10-}$ cluster is a model of the undoped Nd_2CuO_4 system as shown in Fig.1(b). Taking the Madelung energy, we have $\Delta = 0eV$ and the energy difference between Cu-4s and O-2p orbitals $\Delta_S = \epsilon_{2p} - \epsilon_{4s} = 6.0eV$ [18]. When the parameter values listed in Table

Table 1. Typical parameter values (in eV) in the cluster model approach. T_{pd} stands for the transfer integral between Cu-3d orbital with $(x^2 - y^2)$ symmetry and O-2p orbital and T_{pp} the integral between the nearest-neighbor O-2p orbitals [8]. For the integral between Cu-4s and O-2p orbitals T_{sp} in (b), the value given in **ref.**18 is used. Other transfer integrals in eq.(2), which are not listed in the Table, are estimated following the Harrison's relations [15]. For U_d and U_p, the values given in **ref.**16 are used.

	cluster	T_{pd}	T_{pp}	T_{sp}	ϵ_{3d}	ϵ_{2p}	ϵ_{4s}	U_d	U_p	t	2J
(a)	Cu_2O_{11}	1.17	0.63	—	0.0	3.0	—	8.5	4.1	0.29	0.13
(b)	Cu_2O_7	1.17	0.63	2.5	0.0	0.0	-6.0	8.5	4.1	—	0.08

1(b) are used, we obtain the energies of the ground (singlet) and first excited (triplet) states and then $2J = 0.08eV$. In the Raman scattering experiments, the exchange interaction $2J$ in Nd_2CuO_4 has been observed to be similar to that in La_2CuO_4 [19]. Noting the accuracy of the parameter values used in the calculation for Nd_2CuO_4, we may conclude that this value of $2J$ is in accord with experiment. From the calculation for La_2CuO_4 and Nd_2CuO_4, we also find that the apical oxygens in La_2CuO_4 do not contribute significantly to the exchange interaction.

The $(Cu_2O_7)^{11-}$ cluster describes the electron-doped systems. We have examined the electronic states in the cluster for various values of Δ_S. When Δ_S is large enough, the ground state is the anti-bonding state of the singlet given by the $(CuO_4)^{7-}$ cluster [8] and the first excited state is its bonding state. On the other hand, when Δ_S is taken to be 6.0eV as assumed above, the ground and first excited states are not expressed as the anti-bonding and bonding states each other. If this is the case, the singlet in the electron-doped system is not local and the motion of the singlet do not always describe the electric current in the electron-doped system. This fact suggests that the contribution of the Cu-4s on the electronic states will provide the unique features in the electron-doped superconductors. Having the validity of the $t - J$ model in mind, we study the dynamics of the singlets in the model in the next section.

3. Motion of Holes in the $t - J$ Model

As discussed in the previous section, holes in the $t - J$ model may be identified as the singlets with electric charge introduced by the doping. The holes propagate in the lattice by exchanging positions with spins. Let us consider a hole introduced into the Néel state in the square lattice.

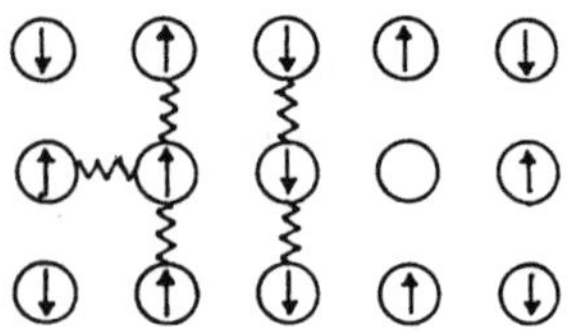

Fig. 2. When the hole propagates in the classical Néel state, it leaves behind a string of the wrong spins which increase the magnetic energy.

When the hole propagates straight as seen in Fig.2, it leaves behind the string of the wrong spins in the Néel state, preventing it from propagating. However, there are two possible ways for the propagation : (i) A hole traverses a closed loop and then propagates to the next nearest-neighbor site in the loop [20,21]. (ii) The string is erased by applying the quantum term of spins in the exchange Hamiltonian, $J\sum_{(ij)}(S_i^+S_j^- + S_i^-S_j^+)$. In the previous paper [22], we have studied the second mechanism for the motion of holes by examining the Green's function of holes in the continued fraction method (or the density of states of holes in the moment expansion method).

Here, we examine the motion of holes in the Néel state in the so-called many-body tight-binding method. The similar method has been used by Trugman [21] and Edward and Miyazaki [23]. A hole is introduced into site i in the Néel state (See, Fig.3(a)). This state is expressed by φ_{i1}, $\varphi_{i1} = \tilde{c}_{i\uparrow}|S>$, where $|S>$ denotes the Néel state at half- filling. φ_{i1} is orthogonal to the state φ_{i2} in which a hole is at site i and the spin direction at one of the nearest-neighbor sites (j) is reversed from the Néel configuration ; $\varphi_{i2} = \tilde{c}_{i\uparrow}S_j^+|S>$. The state φ_{i3} given by $\varphi_{i3} = \tilde{c}_{i\uparrow}S_j^+S_l^-|S>$ is also orthogonal to φ_{i2} where l denotes one of the next nearest-neighbor sites of i. In the same way, we can introduce orthogonal states as many as we can.

Once a number of these orthogonal states are taken to be the basis functions at site i, the trial Bloch function with wave number $\vec{k}$ is given by

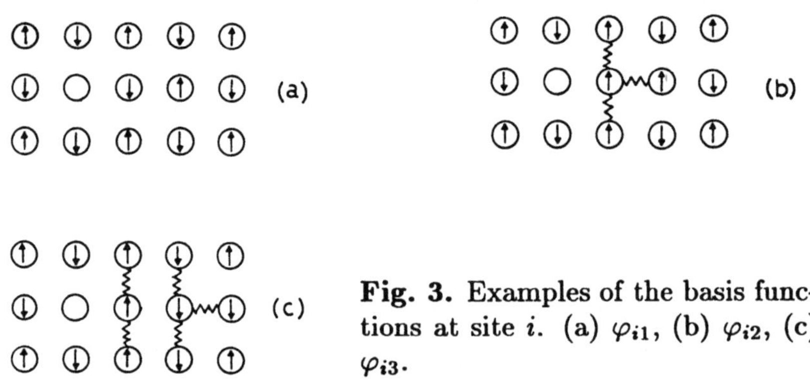

Fig. 3. Examples of the basis functions at site i. (a) φ_{i1}, (b) φ_{i2}, (c) φ_{i3}.

$$\psi_k^{(\mu)} = \frac{1}{\sqrt{N}} \sum_i \exp(i\vec{k} \cdot \vec{R}_i) \sum_\nu a_{\mu\nu}(\vec{k}) \varphi_{i\nu} \; , \tag{3}$$

where N is the number of the lattice sites, $\vec{R}_i$ is the position vector at site i, and $a_{\mu\nu}(\vec{k})$'s are the variational parameters. The energy eigenvalue $E_k^{(\mu)}$ and the wave function $\psi_k^{(\mu)}$ are determined by the Schrödinger equation,

$$H_{tJ}\psi_k^{(\mu)} = E_k^{(\mu)}\psi_k^{(\mu)} \; . \tag{4}$$

The density of states of holes is given by the imaginary part of the Green's function,

$$\begin{aligned} \rho(\omega) &= -\frac{1}{\pi} Im \sum_{\sigma\sigma'} < S|\tilde{c}_{i\sigma}^\dagger \frac{1}{(\omega - H_{tJ} + i0)} \tilde{c}_{i\sigma'}|S > \\ &= \sum_{k\mu} | < \varphi_{i1}|\psi_k^{(\mu)} > |^2 \, \delta(\omega - E_k^{(\mu)}) \\ &= \sum_{k\mu} |a_{1\mu}(\vec{k})|^2 \delta(\omega - E_k^{(\mu)}) \; . \end{aligned} \tag{5}$$

In the numerical calculation, we take 389 basis functions, which include all possible spin configurations up to the 5-th neighbors of the hole in the Néel state. Then, we obtain 389 dispersion curves. In Fig.4, the low energy region of the spectrum is shown for $J/t = 0.5$. The spectrum consists of the almost continuous part and the lowest dispersion curve split from it. We note that the lowest dispersion curve is not sensitive to the number of the basis functions for $J/t = 0.5$. Examining the wave function, we obtain that this curve expresses the motion of a hole

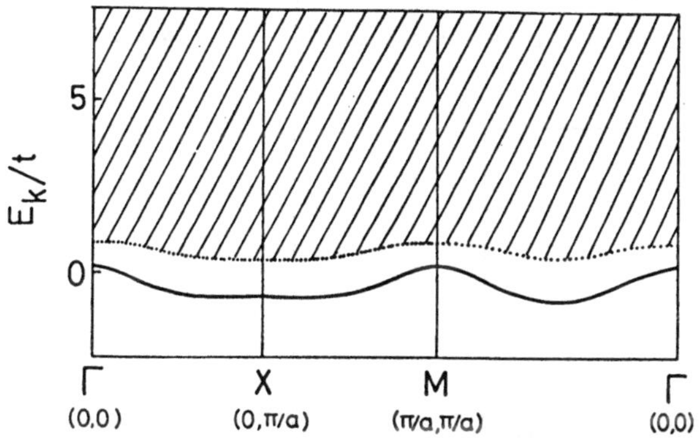

Fig. 4. The energy dispersions E_k/t along Γ-X-M and Γ-M lines calculated for $J/t = 0.5$ by using 389 basis functions. The lowest energy state is split off from the almost continuous part with higher energy.

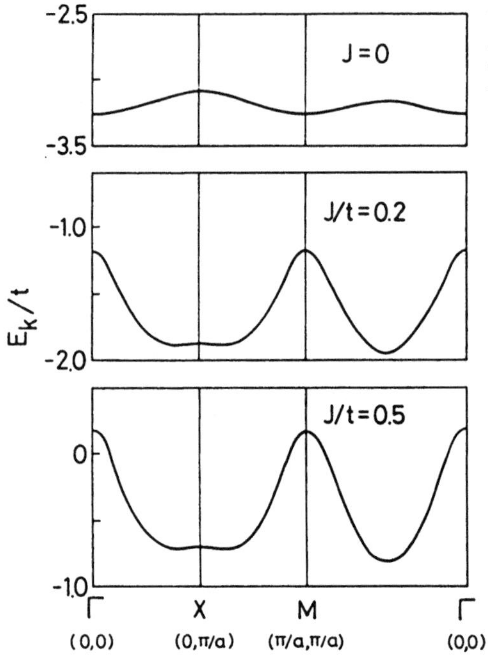

Fig. 5. Dependence of the lowest energy state on the value of J/t.

accompanied by some overturned spins and thus describes the coherent magnetic polaron states. In Fig.5, the lowest dispersion curves are shown for several choices of the J/t value. We find the following characteristic features : (i) The dispersion is a function of $|t|$. (ii) The energy minimum appears at $\vec{k} = (0,0)$ for $J/t \lesssim 0.02$ and at $\vec{k} = (\pi/2a, \pi/2a)$ for $J/t \gtrsim 0.02$ [21,24,25], where a is the lattice constant.

The density of states of holes are shown in Fig.6 for $J/t = 0.5$, where the lowest coherent states cause a sharp peak in the density of states. Such a peak has already been pointed out in our previous study

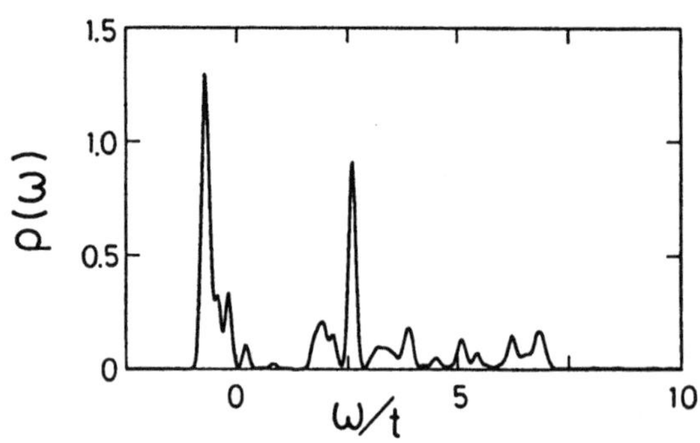

Fig. 6. Density of states of holes calculated for $J/t = 0.5$ by using 389 basis functions.

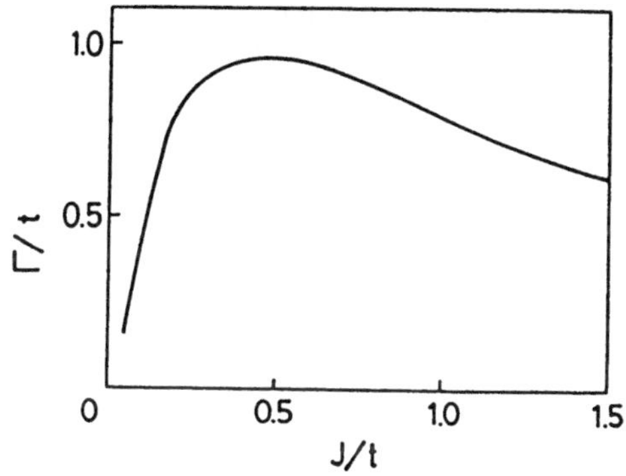

Fig. 7. Band width of the lowest energy state as a function of J/t.

[22]. In Fig.7, the width of the lowest dispersion Γ is plotted as a function of J/t. When $0.1 \leq J/t \leq 0.25$, the width is about $4J$ which is consistent with the results by Kane et al. [25]. On the other band, when $J/t > 1$, the width is proportional to $1/J$ [26].

We have found that the motion of holes in the Néel state is mainly due to quantum spin fluctuations when $J/t > 0.02$. The low energy states are coherent magnetic polaron states. In other words, holes resonate with quantum spin fluctuations and propagate. This fact suggests that holes prefer the quantum spin liquid or RVB state to gain the kinetic energy. In the next section, we compare the energy of holes in the RVB state with that in AF states.

4. Energy of Holes

In order to calculate the energies of a hole introduced in the Néel, quantum AF (QAF), and RVB states at half-filling, we define the on-site Green's function of holes as [27,28],

$$G_{ii}(\omega) = \sum_{\sigma} < S|\tilde{c}^{\dagger}_{i\sigma} \frac{1}{(\omega - H_{tJ})} \tilde{c}_{i\sigma}|S> , \tag{6}$$

where $|S>$ is one of the magnetic states at half-filling. It is convenient to express the Green's function as a continued fraction expansion as follows :

$$G_{ii}(\omega) = \cfrac{1}{\omega - a_0 - \cfrac{b_1^2}{\omega - a_1 - \cfrac{b_2^2}{\omega - a_2 \cdots}}} . \tag{7}$$

The calculation of the coefficients a_n and b_n^2 for the Néel state is straightforward. For the QAF state, the spin wave approximation in the infinite lattice is used to evaluate the coefficients by neglecting the effects of a hole on the spin waves which are found to be small [29]. As for the RVB state, we consider a 5×5 square lattice with a free boundary (Fig.8). A spin at the center of the lattice is excluded by taking the center to be the origin where the hole is introduced. Let Φ_α be a state generated by a product of dimers (nearest-neighbor singlet pairs) which cover the whole lattice sites. Then, the RVB state is given by a linear combination of all Φ_α's generated from a product of dimer coverings of 24 sites [30] ;

$$\Phi_{RVB} = \sum_\alpha \Phi_\alpha \,. \tag{8}$$

Using Φ_{RVB}, we calculate the coefficients up to $n = 2$ in the RVB state. This corresponds to the calculation of the first 5 moments of the density of states of holes. We note that although Φ_{RVB} is made of the nearest-neighbor singlet pairs (dimers), the hole induces the longer pairs when it moves [27,31].

In Fig.9, the lowest pole of the Green's function, ω_0, is plotted as a function of J/t, which is a measure of the stability of a hole in each magnetic state. As seen in the Figure, holes are most stable in the RVB state for $J/t > 0.1$ [28].

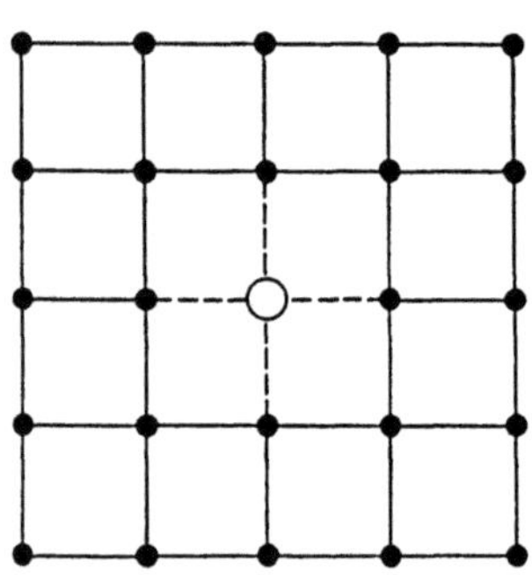

Fig. 8. A 5×5 lattice with a hole at the center of a lattice.

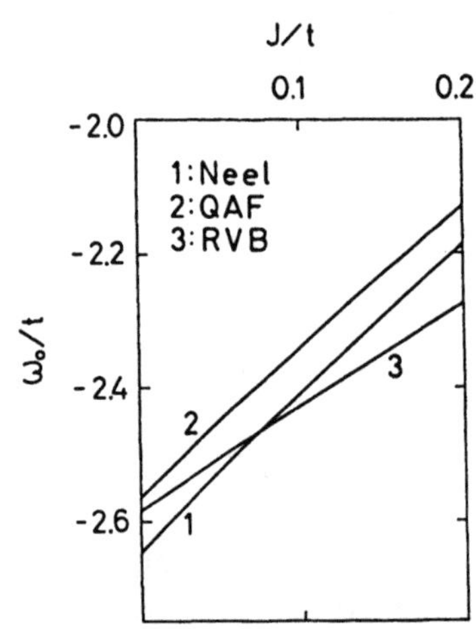

Fig. 9. The energy of a hole in the Néel, QAF and RVB states calculated as a function of J/t.

5. Conclusion

In conclusion, we have discussed the electronic properties in the electron- and hole-doped superconductors. In the hole-doped ones, the ground states consist of the local singlet states and the low energy properties may be described by the $t-J$ model. We have derived the values of the parameters t and J, and studied the motion of carriers in the $t-J$ model. In the electron-doped ones, however, the local singlet states and the validity of the $t-J$ model depend crucially on the Cu-4s states.

Therefore, the detailed comparison between both superconductors will provide valuable information for examining the the mechanism of the superconductivity. In particular, the contribution of Cu-4s orbitals to the electronic states in the electron-doped ones should be experimentally evaluated in more detail.

Acknowledgments

This work has been supported by Grant-in Aid for Scientific Research on Priority Areas "Mechanism of Superconductivity".

References

1. P.W. Andeson, Science **235**, 1196 (1987); Frontiers and Borderlines in Many-Particle Physics, Proceeding of the Varenna Summer School, Varnna, Italy, 1987 (North-Holland, Amsterdam, in press); P.W. Anderson, G. Baskaran, Z. Zou and T. Hsu, Phys. Rev. Lett. **58**, 2790 (1987).
2. F.C. Zhang and T.M. Rice, Phys. Rev. **B37**, 3759 (1988).
3. S. Maekawa, T. Matsuura, Y. Isawa and H. Ebisawa, Physica C **152**, 133 (1988).
4. H. Fukuyama, H. Matsukawa and Y. Hasegawa, J. Phys. Soc. Japan **58**, 364 (1989).
5. H. Eskes and G. A. Sawatzky, Phys. Rev. Lett. **61**, 1415 (1988).
6. N. Andrei and P. Coleman, Phys. Rev. Lett. **62**,595 (1989).
7. M. Imada, "Mechanism of High Temperature Superconductivity", eds. H. Kamimura and A. Oshiyama, (Springer-Verlag in Material Science **11**, 1989) p. 53.
8. T. Tohyama, Y. Ohta and S. Maekawa, Physica C **158**, 525 (1989).
9. Y. Tokura, H. Takagi and S. Uchida, Nature **337**,345 (1989).
10. H. Takagi, S. Uchida and Y. Tokura, Phys. Rev. Lett. **62**, 1197 (1989).
11. A. Fujimori, Phys. Rev. **B 39**, 793 (1989).
12. Z.-X Shen, J. W. Allen, J. J. Yeh, J.-S. Kang, W. Ellis, W. Spicer, I. Lindau, M. B. Maple, Y. D. Dalichaouch, M. S. Torikachvili, J. Z. Sun and T. H. Geballe, Phys. Rev. **B 36**, 8414 (1987).
13. H. Eskes, L. H. Tjeng and G. A. Sawatzky, to be published in Phys. Rev. B.

14. F. Mila, Phys. Rev. **B 38**, 11358 (1988).
15. W. A. Harrison "Electronic Structure and the Properties of Solids", Freeman, San Francisco, 1980.
16. A.K. McMahan, Richard M. Martin and S. Satpathy, Phys. Rev. **B38**, 6650 (1988).
17. J. Kondo and S. Nagai, to be published in Jpn. J. Appl. Phys; J.B. Torrance and R.M. Metzger (preprint) and Y. Ohta and S. Maekawa (unpublished) have also calculated the Madelung energy taking the experimental values of the lattice constants and obtained this value.
18. Although the value Δ_S has not yet been obtained in Nd_2CuO_4, the BIS experiments in $YBa_2Cu_3O_7$ (D. van der Marel, J. van Elp, G.A. Sawatzky and D. Heitmann, Phys. Rev. **B37**, 5136 (1988)) and the band calculation in La_2CuO_4 (M.J. DeWeert, D.A. Papaconstantopoulos and W.E. Pickett, Phys. Rev. **B39**,4235 (1989)) have, respectively, observed the value of Δ_S as about $8eV$ and $8.9eV$. Using these values, $\simeq 8eV$, and the Madelung energy, $3.0eV$, we have $\Delta_S = 6.0eV$.
19. S. Sugai, T. Kobayashi and J. Akimitsu, to be published.
20. Y. Nagaoka, Phys. Rev. **147**, 392 (1966).
21. S.A. Trugman, Phys. Rev. **B37**, 1598 (1988).
22. S. Maekawa, J. Inoue and M. Miyazaki, "Mechanism of High Temperature Superconductivity", eds. H. Kamimura and A. Oshiyama, (Springer-Verlag in Material Science **11**, 1989) p. 68.
23. D.M. Edward and M. Miyazaki, J. Phys. F: Met. Phys. **17**, L311 (1987).
24. B.I. Shraiman and E.D. Siggia, Phys. Rev. Lett. **61**, 467 (1988).
25. C.L. Kane, P.A. Lee and N. Read, Phys. Rev. **B39**, 6880 (1989). See also E.L. Nagaev, Zh. Eksp. Theor. Fiz. **58**, 1269 (1970). In the present work, the value of J is defined as a half of that defined in Kane et al.
26. S. Schmitt-Rink, C.M. Varma and A.E. Ruckenstein, Phys. Rev. Lett. **60**, 2793 (1988).
27. J. Inoue, M. Miyazaki and S. Maekawa, Physica C **157**, 209 (1988).
28. M. Miyazaki, J. Inoue and S. Maekawa, to be published.
29. N. Nagaosa, Y. Hatsugai and M. Imada, J. Phys. Soc. Japan **58**, 978 (1989).
30. P. Lederer and Y. Takahashi, Z. Phys. B **71**, 415 (1988).
31. K. Takano and K. Sano, Phys. Rev. **B39**, 7367 (1989).

Core-Level Spectroscopy in High-T_c Superconducting Materials

A. Kotani, K. Okada, S. Tanaka, and Y. Seino

Department of Physics, Faculty of Science, Tohoku University, Sendai 980, Japan

By theoretically analyzing the core-level spectra, such as X-ray photoemission (XPS) and X-ray absorption (XAS), with the use of the impurity Anderson model, we discuss the electronic structure of high T_c superconductors and some related materials. From Cu 2p-XPS we obtain the information on the charge transfer energy from O 2p to Cu 3d states and on the hybridization strength between these states. The polarization dependence of Cu 2p-XAS and O 1s-XAS provides us with the most direct information on the site and the symmetry of doped holes. An analysis of X-ray emission spectra is also made using the same model system.

1. Introduction

Core-level spectroscopy, such as X-ray photoemission (XPS), X-ray absorption (XAS) and X-ray emission (XES), provides us with important information on the electronic structure of d or f electron systems/1,2/. In the final state of XPS, a core hole is left behind and it couples strongly with d or f electrons through the Coulomb interaction (also through the exchange interaction in some cases). If the ground state of the relevant system is in the mixed-valence state, as in the cases of high T_c superconductors, the different d electron configurations in the ground state manifest themselves as a spectral splitting in XPS through the coupling with the core hole/3-6/. Therefore, by analyzing the XPS spectra we can obtain important information on the mixed-valence character of the ground state, such as the charge transfer energy, the hybridization strength and the average d electron number.

In the XAS process, a core electron is photo-excited to the states above the Fermi level. In hole-doped high T_c superconductors, XAS gives the most direct information on where the holes are, i.e., which site of Cu or O the hole enters, and which state of five 3d orbitals of Cu and three 2p orbitals of O the hole enters. According to the selection rule of the electric dipole transition, a Cu 2p core electron can be excited to the hole states of Cu 3d orbitals and an O 1s core electron can be excited to the hole states of O 2p orbitals. Since the Cu 2p and O 1s core levels have the different binding energy, the Cu 2p-XAS and the O 1s-XAS can be observed selectively in the different X-ray energy region/7,8/. Furthermore, the polarization dependence of Cu 2p-XAS and O 1s-XAS gives important information on the symmetry of the doped hole state/9-12/. For instance, when the electric field of the incident X-ray is in the z direction, the Cu $3d(3z^2-r^2)$ state can be excited in Cu 2p-XAS but the Cu $3d(x^2-y^2)$ state cannot be, and O 2p(z) state can be excited in O 1s-XAS but O 2p(x) and 2p(y) states cannot be, because of the dipole selection rule.

In the present paper, we analyze theoretically Cu 2p-XPS and Cu 2p-XAS of high T_c superconductors and some related materials using the impurity Anderson model/13,14/. From comparison between theoretical and experimental results, we study the electronic structure of Cu 3d and O 2p states in these materials, and discuss where the holes are doped. We also analyze the Cu 2p-XES with the use of the same model/15/, and show that the XES can be explained in a manner consistent with the interpretation of XPS and XAS.

2. Impurity Anderson Model

We concentrate on the Cu site in the so-called CuO_2 plane which is commonly included in La-Sr-Cu-O, Y-Ba-Cu-O and Bi-Sr-Ca-Cu-O high T_c systems. We describe the system with the use of the single-site impurity Anderson model/13/, whose Hamiltonian is given by

$$H = \sum_{\Gamma,\sigma} \sum_{k=1}^{N} \varepsilon_{k\Gamma} n_{k\Gamma\sigma} + \sum_{\Gamma,\sigma} \varepsilon_{d\Gamma} n_{d\Gamma} + (U_{dd}/2) \sum_{(\Gamma,\sigma)\neq(\Gamma',\sigma')} n_{\Gamma\sigma} n_{\Gamma'\sigma'}$$

$$+ \sum_{\Gamma,\sigma} \sum_{k} \frac{V_{k\Gamma}}{\sqrt{N}} (a^{\dagger}_{k\Gamma\sigma} d_{\Gamma\sigma} + d^{\dagger}_{\Gamma\sigma} a_{k\Gamma\sigma}) - U_{dc} \sum_{\Gamma,\sigma} n_{\Gamma\sigma} (1 - a^{\dagger}_{c} a_{c}) , \qquad (1)$$

where Γ represents the symmetry of the Cu 3d orbital states which are taken as irreducible representations a_{1g}, b_{1g}, b_{2g} and e_g in the D_{4h} point group. The first term on the right-hand side of Eq.(1) represents the O 2p valence band states, each of which is assumed to be coupled to the Cu 3d states with the symmetry Γ, and the index k describes the energy distribution of the valence band. The second, third and fourth terms describe the 3d states, the d-d Coulomb interaction and the hybridization, respectively. The last term is the potential of the 2p core hole acting on 3d electrons. In this Hamiltonian we disregard the multipole Coulomb and exchange interactions and the spin-orbit interaction, for simplicity. However, when we analyze the multiplet structure of Cu 2p-XPS in Sec. 3, we take account of these interactions. Furthermore, when we analyze the polarized Cu 2p-XAS of Y-Ba-Cu-O in Sec. 4, we also take account of the contribution from the Cu site on the CuO chain.

The spectra of Cu 2p-XPS and Cu 2p-XAS are expressed, respectively, as

$$F_{XPS}(E_B) = \sum_{f} |\langle f|a_c|g\rangle|^2 L(E_B - E_f + E_g) , \qquad (2)$$

and

$$F_{XAS}(\omega) = \sum_{f} |\langle f|d^{\dagger}_{\Gamma\sigma} a_c|g\rangle|^2 L(\omega - E_f + E_g) , \qquad (3)$$

where $|g\rangle$ and $|f\rangle$ represents the initial and final states with energy E_g and E_f, respectively, E_B is the binding energy, ω is the incident photon energy, and L(x) is the Lorentzian function by which the spectral broadening due to the core hole lifetime τ_c and the experimental resolution is introduced. The Cu 2p-XES is expressed, by applying the formula of the second order optical process/15/, in the form

$$F_{XES}(\omega) = \sum_{f} |\langle f|a^{\dagger}_{c} d_{\Gamma\sigma} \frac{1}{H - E_f - \omega - i(\hbar/\tau_c)} a_c|g\rangle|^2 . \qquad (4)$$

Here we consider the process where a 2p core electron is first excited to a high energy continuum by a monochromatic incident X-ray, and then a 3d electron makes a transition to the 2p level by emitting an X-ray photon with energy ω.

In the case of undoped systems such as La_2CuO_4, the Cu ion is nominally divalent, so that the ground state of our system includes one hole (denoted by <u>one hole-system</u>). There is a general consensus that the symmetry of the hole is b_{1g}. By taking account of the hybridization between d^9 and $d^{10}\underline{L}$

configurations ($\underline{L}$ being a hole in the valence band), the ground state $|g\rangle$ is described as a linear combination of the basis states $d(b_{1g})|0\rangle$ and $a_k(b_{1g})|0\rangle$, where $|0\rangle$ is the state in which all of the valence band states, d states and core states are completely filled. In Fig. 1 we show schematically the processes of Cu 2p-XPS, XAS and XES in the case of the undoped system. When a hole is doped, the nominal valence number of Cu ion becomes trivalent. Then there are two holes in our system (denoted by two-hole system) and the ground state is a mixed state among d^8, $d^9\underline{L}$ and $d^{10}\underline{L}^2$ configurations. Therefore, $|g\rangle$ is represented by a linear combination of the following basis states: $d(\Gamma)d(b_{1g})|0\rangle$, $d(\Gamma)a_k(b_{1g})|0\rangle$, $a_k(\Gamma)d(b_{1g})|0\rangle$ and $a_k(\Gamma)a_{k'}(b_{1g})|0\rangle$. Here, Γ represents the symmetry of the doped hole, and two possibilities of $\Gamma = b_{1g}$ and a_{1g} will be discussed later.

The spectra of XPS, XAS and XES are calculated by numerically diagonalizing the Hamiltonian (1) for a finite system where $\varepsilon_{k\Gamma}$ are taken as

$$\varepsilon_{k\Gamma} = \varepsilon^0_{v\Gamma} - W/2 + W(k - 1/2)/N \,, \tag{5}$$

Here, $\varepsilon^0_{v\Gamma}$ is the center of the valence band with symmetry Γ, W is the width of each valence band, and k = 1~ N. The k dependence of the hybridization is assumed to be elliptical and is given by

$$V^2_{k\Gamma} = \frac{8V(\Gamma)^2}{\pi W}\left((W/2)^2 - (\varepsilon_{k\Gamma} - \varepsilon^0_{v\Gamma})^2\right)^{1/2} \quad , \tag{6}$$

where $V(\Gamma)$ is the strength of the hybridization. The value of N is taken to be so large that the calculated spectra converge well. The key parameters of our model are $V(\Gamma)$, U_{dd}, U_{dc} and the charge transfer energy $\Delta(\Gamma)$ which represents the energy difference between $d^{10}\underline{L}(\Gamma)$ and $d^9(\Gamma)$ configurations in the limit of vanishing V. We fix W = 3.0 eV and disregard the Γ dependence of $\varepsilon_{d\Gamma}$.

3. Cu 2p Photoemission Spectra

We first consider the Cu 2p-XPS of undoped La_2CuO_4 and CuO. As shown in Fig. 1, the final states of XPS are $\underline{2p}3d^9(b_{1g})$ and $\underline{2p}3d^{10}\underline{L}(b_{1g})$ configurations, whose energy difference is $U_{dc} - \Delta(b_{1g})$ in the limit of vanishing $V(b_{1g})$. The latter configuration gives the main peak of the Cu 2p-XPS, while the former gives the satellite. We show in Fig. 2 the calculated Cu 2p-XPS for La_2CuO_4 (the solid curve) and CuO (the dotted curve)/13/, and show in Fig. 3 the corresponding experimental data for comparison/4/. So as to reproduce the intensity ratio and the energy separation of the two peaks, as well as to reproduce the Cu 3d-XPS/13/, the parameter values are estimated as follows: $\Delta(b_{1g})$ = 2.0 eV, $V(b_{1g})$ = 2.5 eV and U_{dc} = 8.0 eV for La_2CuO_4, and $\Delta(b_{1g})$ = 2.8 eV, $V(b_{1g})$ = 2.5 eV and U_{dc} = 9.0 eV for CuO. With these values the average 3d electron number is estimated to be 9.34 for La_2CuO_4 and 9.27 for CuO. From the analysis of the Cu 3d-XPS and Cu 2p-XES (see Sec.5), the value of U_{dd} is estimated as 7.0 eV for La_2CuO_4 and 8.0 eV for CuO/13/. Therefore, the value of Δ is much smaller than U_{dd}, and these systems are found to be the charge transfer type insulator/16/. This means that when holes are doped in these systems they enter mainly the O 2p valence band instead of the Cu 3d states. In order to see the effect of the doped hole, we calculate the Cu 2p-XPS for the two-hole case of our impurity Anderson model. The result is plotted with the dashed curve in Fig. 2, where we use the parameter values of La_2CuO_4 estimated before, and the doped hole is assumed to be in the a_{1g} state with $\Delta(a_{1g})$ = 1.0 eV and $V(a_{1g})$ = 2.0 eV. The result is almost the same as that of the undoped case, except for the high energy shift of the whole spectrum and the broadening of the sattelite, in consistence with the

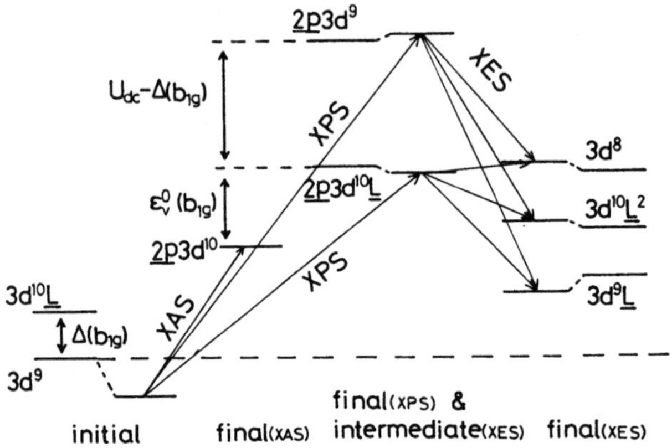

Fig. 1 Total energy level scheme of 2p-core XPS, XAS and XES of nominally divalent Cu in high T_C systems. A limiting case of vanishing valence band width is shown here.

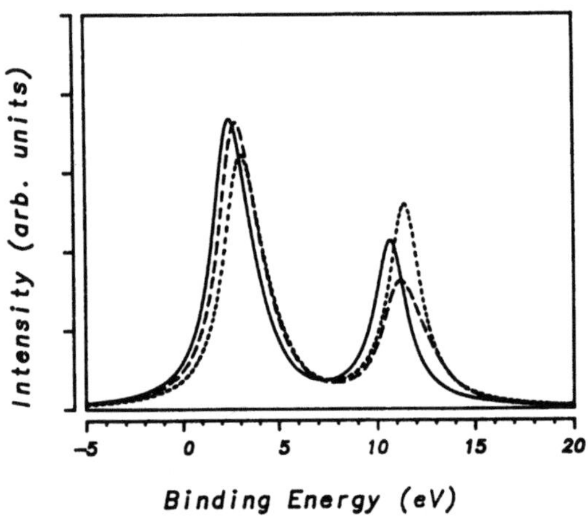

Fig. 2 Cu 2p-XPS of nominally divalent (solid curve) and trivalent (dashed curve) Cu in La-Sr-Cu-O. Cu 2p-XPS in CuO is shown with the dotted curve.

experimental data/3,17/. The result is not very differnt when we assume that the doped hole is in the b_{1g} state. The reason that the effect of the doped hole is very small is that the hole is doped mainly on the O site but the Cu 2p-XPS detects mainly the electronic states on the Cu site. With the parameter values mentioned above, for instance, the ground state of our two-hole case contains only 4.4 % of the $3d^8$ component, so that the doped hole is in the O 2p band with the weight of 95.6 %. According to experimental data, the structure of the Cu 2p-XPS in the Y-Ba-Cu-O system is very close to that in the La-Sr-Cu-O system, and thus we consider that the parameter values concerning the b_{1g} state of Y-Ba-Cu-O is almost the same as those of La-Sr-Cu-O.

From comparison of Fig. 2 with Fig. 3, the spectral shape of the satellite is found to be different between the calculated and experimental results. As shown by van der Laan et al./18/ from an atomic calculation, the experimentally observed structure of the satellite originates mainly from the atomic multiplet structure of the Cu $2p^5 3d^9$ final state. As found from Fig. 3, however, the satellite structure changes systematically in Cu

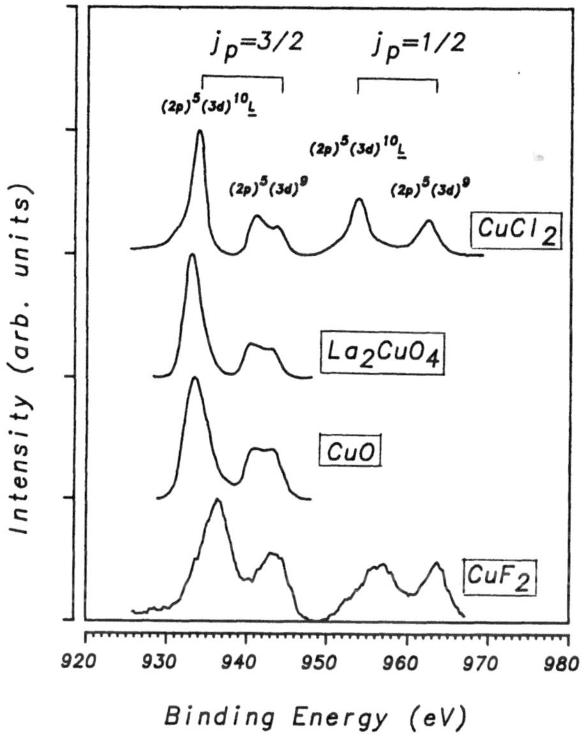

Fig. 3. Experimental results of Cu 2p-XPS in $CuCl_2$, CuF_2 CuO and La_2CuO_4, taken from refs. 4 and 18.

compounds, $CuCl_2$, La_2CuO_4, CuO and CuF_2/4,8/. This suggests that the solid state effect originating from the hybridization is also important. It is an interesting problem to study how the interplay between the atomic multiplet coupling and the solid state hybridization affects the spectral shape, as recently studied for the core-level spectroscopy of Ce compounds/19,20/. For this purpose, we incorporate in our Hamiltonian (1) the interactions described by the Slater integrals F^2, G^1 and G^3 between 2p and 3d states, and the spin-orbit interactions in the 2p and 3d states, and calculate the Cu 2p-XPS. We take the Slater integrals as F^2 = 7.47 eV, G^1 = 5.62 eV and G^3 = 10.95 eV, and the spin-orbit coupling parameters as ζ_d = 0.13 eV and ζ_p = 13.6 eV. We put N = 1 and $\Delta(\Gamma)$ = 2.0 eV for all Γ, since we have checked that the effects of the valence band width and the Γ dependence of Δ is not very important in determining the spectral shape of the satellite. On the anisotropy of the hybridization, we assume $V(a_{1g}) = V(b_{1g})/\sqrt{3}$, $V(b_{2g}) = V(e_g)/\sqrt{2}$, and $V(b_{2g}) = V(b_{1g})/2$, corresponding to the local D_{4h} symmetry of the square planer CuO_4 cluster/21/. The calculated result is shown in Fig. 4, where we take U_{dc} = 8.0 eV and change the value of $V(b_{1g})$ as a parameter/14/. It is found that the spectral shape of the satellite strongly depends on the hybridization strength. When V is small (as in the case of $V(b_{1g})$ = 1.5 eV), the satellite of $2p_{3/2}$ component has a two-peak structure with a stronger peak on the lower energy side, which is similar to the atomic calculation and also similar to the experimental result for $CuCl_2$. When V is increased (as in the case of $V(b_{1g})$ = 2.0 eV), the intensity of the lower energy peak decreases and the satellite becomes of rectangular shape, similarly to the experimental results for La_2CuO_4 and CuO. When V is increased further (as in the case of $V(b_{1g})$ = 3.0 eV), the intensity of the lower energy peak is very much weakened so that the satellite becomes of triangular shape, similarly to the experimental result

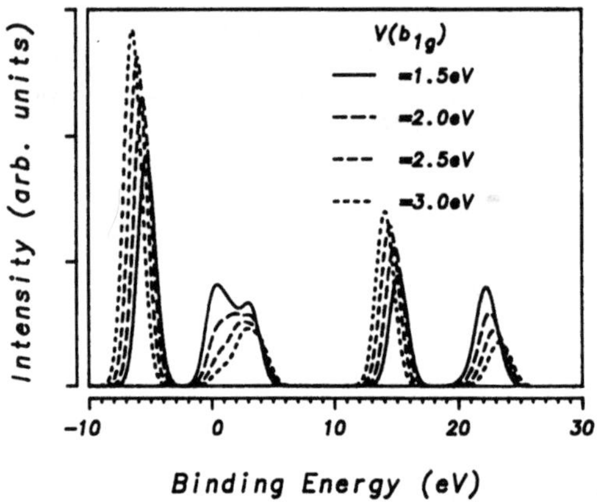

Fig. 4 Cu 2p-XPS in the D_{4h} symmetry as a function of $V(b_{1g})$. We take N = 1 and Δ = 2.0 eV.

for CuF_2. A more detailed analysis of the satellite spectral shape for these Cu halides is possible to be made by using more appropriate values of $V(b_{1g})$, Δ and U_{dc}. It can also be shown that the spectral shape of the satellite depends strongly on the anisotropy of the hybridization; for instance, when we change the ratio of $V(b_{2g})$ and $V(b_{1g})$, the satellite structure changes sensitively/14/. In this way, we have made the first systematic study on the interplay between the atomic multiplet coupling and the solid state hybridization in the Cu 2p-XPS of high T_c materials and the related Cu compounds, and have shown that the multiplet structure in the observed spectra provides us with the information on the hybridization strength and its anisotropy.

4. X-Ray Absorption and Its Polarization Dependence

Let us consider the Cu 2p-XAS in undoped high T_c systems. As shown in Fig. 1, we have only one XAS final state with the $\underline{2p}3d^{10}$ configuration. Therefore the Cu 2p-XAS has a single peak which is called the white line. According to experimental data for $YBa_2Cu_3O_{7-\delta}$, the Cu 2p-XAS has only the white line for δ = 0.5, which corresponds to the undoped system. However, when holes are doped with decreasing δ, a weak satellite peak appears at the photon energy about 1.5 eV higher than the white line/7,9/. The relative intensity of the satellite is proportional to the concentration of the doped hole. A similar satellite is also observed in the Cu 2p-XAS of the Bi-Sr-Ca-Cu-O system/11/. In order to get the information on the symmetry of the doped hole, as mentioned in Sec.1, the polarization dependence of the satellite has been measured. As a result, it is found that the satellite of Y-Ba-Cu-O is mainly allowed for $\varepsilon \parallel c$/9/, where ε and c represent the directions of the X-ray electric field and the crystalline c-axis, respectively, and that the satellite of Bi-Sr-Ca-Cu-O is mainly allowed for $\varepsilon \perp c$/11/. In Fig. 5 we show experimental data/9/ for Y-Ba-Cu-O with δ = 0.2, where the intensities of XAS for α = 15° and 85° (90°-α being the angle between ε and c directions as shown in the inset of Fig. 5) are normalized so that their maximum intensitis are the same. A similar result was also obtained by EELS measurements/10/.

Now we calculate the polarization dependence of the Cu 2p-XAS both in the one-hole and two-hole cases of our impurity Anderson model/13/. We put $V(b_{1g})$ = 2.5 eV, $\Delta(b_{1g})$ = 2.0 eV and U_{dc} = 8.0 eV, and disregard the effect of the multiplet coupling for simplicity. In the one-hole case, the hole is in the b_{1g} state, so that the white line is allowed only for $\varepsilon \perp c$, as shown with the dotted curve in Fig. 6. This corresponds to the dipole

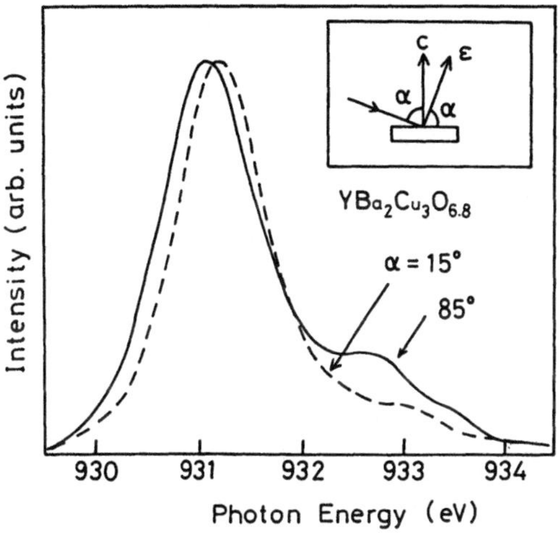

Fig. 5 Experimental results of the polarized Cu 2p-XAS for $YBa_2Cu_3O_{6.8}$ taken from ref. 9.

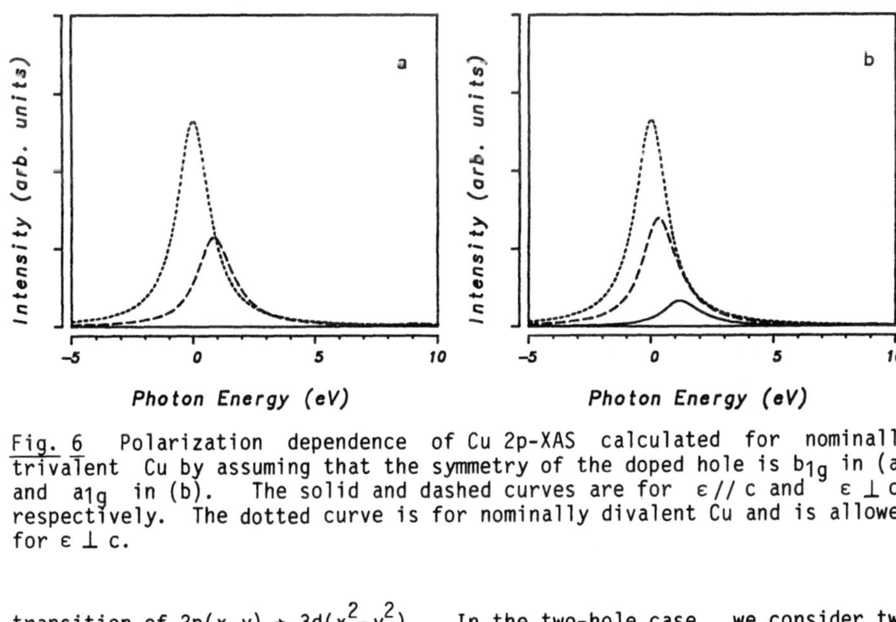

Fig. 6 Polarization dependence of Cu 2p-XAS calculated for nominally trivalent Cu by assuming that the symmetry of the doped hole is b_{1g} in (a) and a_{1g} in (b). The solid and dashed curves are for $\varepsilon // c$ and $\varepsilon \perp c$, respectively. The dotted curve is for nominally divalent Cu and is allowed for $\varepsilon \perp c$.

transition of $2p(x,y) \rightarrow 3d(x^2-y^2)$. In the two-hole case, we consider two possible cases where the symmetry of the doped hole is b_{1g} and a_{1g}, i.e., the two-hole states of (b_{1g},b_{1g}) and (b_{1g},a_{1g}). From energetical consideration, the b_{1g} hole is most naturally expected to be doped because of the strong hybridization $V(b_{1g})$, but the a_{1g} hole might also be possible if $\Delta(a_{1g})$ is small enough, where $\Delta(a_{1g})$ is mainly determined by the charge transfer energy from the 2p(z) state of the apical oxygen to the Cu $3d(3z^2-r^2)$ state. In this calculation we assume $V(a_{1g})$ = 1.8 eV and $\Delta(a_{1g})$ = 0.5 eV, then the energy of the two-hole ground state (b_{1g},a_{1g}) is almost the same as that of (b_{1g},b_{1g}). As shown with the solid curve in Fig. 6(a), the two-hole state (b_{1g},b_{1g}) gives a satellite on the higher energy side of the white line. This satellite corresponds to the $\underline{2p}3d^{10}\underline{L}(b_{1g})$ final state, and is allowed only for $\varepsilon \perp c$. On the other hand, the two-hole state (b_{1g},a_{1g}) gives two XAS peaks corresponding to the final states

$\underline{2p}3d^{10}\underline{L}(a_{1g})$ and $\underline{2p}3d^{10}\underline{L}(b_{1g})$, as shown with the dashed and solid curves, respectively, in Fig. 6(b). The former is allowed for $\varepsilon \perp c$ because of the dipole transition $2p(x,y) \rightarrow 3d(x^2-y^2)$, but its energy position is almost the same as that of the white line. The latter is allowed mainly for $\varepsilon // c$ because of the dipole transition $2p(z) \rightarrow 3d(3z^2-r^2)$, and this becomes a satellite on the higher energy side of the white line. Summarizing the present result, the satellite is allowed for $\varepsilon \perp c$, if the doped hole is in the b_{1g} state. In this case the hole is doped mainly in the O 2p(x,y) states on the CuO_2 plane. On the other hand, the satellite is allowed for $\varepsilon // c$, if the doped hole is in the a_{1g} state. In this case the hole is doped mainly in the 2p(z) state on the apical O site.

Comparing the calculated result with experiments, we find that in the Bi-Sr-Ca-Cu-O system the extra-holes are mainly in the O 2p(x,y) states on the CuO_2 plane. This is also consistent with recent experimental results for O 1s-XAS/10,12/, where the XAS near the absorption edge is allowed only for $\varepsilon \perp c$. In the Y-Ba-Cu-O system, on the other hand, the doped hole seems to enter mainly the 2p(z) state of the apical oxygen, i.e., the oxygen in the BaO plane. However, we should be careful in discussing the Cu 2p-XAS of Y-Ba-Cu-O, because there are two inequivalent Cu sites: the Cu(1) site on the CuO chain in addition to the Cu(2) site in the CuO_2 plane. It is also to be noted that in the calculated result of Fig. 6(b) only the satellite is allowed for $\varepsilon // c$, whereas both of the main line and the satellite are observed for $\varepsilon // c$ in the experimental result of Fig. 5. In order to take account of the contribution from the Cu(1) site, we combine an impurity Anderson model including the Cu(2) site with a cluster model including the Cu(1) site. The impurity Anderson model is essentially the same as the model which we have treated so far (the parameter values are somewhat changed). The cluster model consists of the Cu(1) and the neighboring four O atoms for $YBa_2Cu_3O_{7-\delta}$ with $\delta = 0$, while it consists of the Cu(1) and three O atoms for $\delta = 0.5$. The apical O atom is closer to Cu(1) than to Cu(2), so that it is included in the cluster model. The calculation of the Cu 2p-XAS is made for $\delta = 0$ and 0.5. For a general value of δ, the results for $\delta = 0$ and 0.5 are superposed with the statistical weight. A preliminary result/22/ for $\delta = 0.2$ is shown in Fig. 7. This is obtained by assuming that a half of the doped holes enter the CuO_2 plane with the symmerty of x^2-y^2 around Cu(2) and the remaining half enter the CuO chain and the apical O atom with z^2-x^2 around Cu(1). The white line for $\varepsilon \perp c$ comes mainly from the undoped Cu(2) site, while

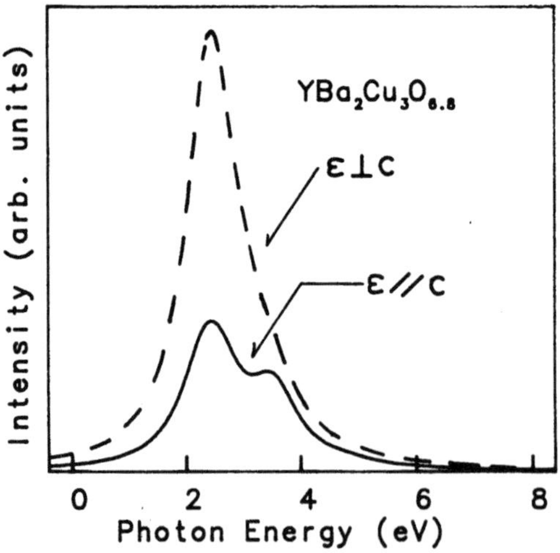

Fig. 7 Polarization dependence of Cu 2p-XAS in $YBa_2Cu_3O_{6.8}$ calculated by taking account of two inequivalent Cu sites.

the main line and the satellite for $\varepsilon // c$ originate from the undoped Cu(1) site and the Cu(1) site with the doped hole on the neighboring O atoms, respectively. The effect of the holes doped in the CuO_2 plane also gives, for $\varepsilon \perp c$, a shoulder on the higher energy side of the white line, but it is not clearly separated as a satellite peak. These features agree rather well with those of experimental results in Fig. 5. Therefore, we interpret that in the Y-Ba-Cu-O system a considerable amount of holes are doped not only in the BaO plane (i.e. the apical O site) but also in the CuO chain and the CuO_2 plane. This is consistent with recent experiments of O 1s-XAS/10/, which has near-edge features both for $\varepsilon // c$ and $\varepsilon \perp c$. Our calculation can also be applied to the O 1s-XAS, and the details will be published elsehwere.

5. X-Ray Emission Spectra

We calculate the Cu 2p-XES for the undoped Cu_2LaO_4 by using the same parameter values as those used in the analysis of Fig. 2 and by assuming the additional parameter values of $V(b_{2g})$ = 1.5 eV, $V(e_g)$ = 1.0 eV and $\Delta(b_{2g}) = \Delta(e_g)$ = 3.0 eV. The multiplet coupling effect is disregarded. As shown in Fig. 1, the intermediate states of the XES are the same as those of the Cu 2p-XPS (i.e. the one-hole system), and the final states are linear combinations of $3d^8$, $3d^9\underline{L}$ and $3d^{10}\underline{L}^2$ configurations (i.e. the two-hole system). There are many transitions which can contribute to the XES as ahown in Fig. 1 and also there are many two-hole final states (b_{1g},Γ) with various symmetries Γ. However, from our calculation it is found that the transitions $\underline{2p}3d^{10}\underline{L}(b_{1g}) \rightarrow 3d^9(\Gamma)\underline{L}(b_{1g})$ with $\Gamma = b_{2g}$ and e_g give the main contribution to the XES. The calculated result is plotted in Fig. 8 together with the results of Cu 2p-XPS and Cu 2p-XAS/15/, and compared with the corresponding experimental data/3,9,23/ (the inset of Fig. 8). The calculated XES is found to be in fairly good agreement with the experimental result/23/, in its spectral shape and the relative peak position to the XAS and XPS peaks. The energy separation between the peaks

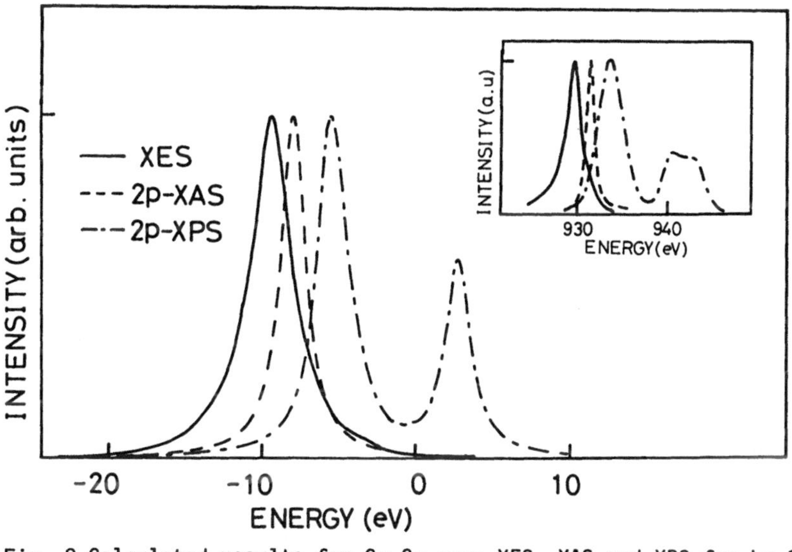

Fig. 8 Calculated results for Cu 2p-core XES, XAS and XPS for La_2CuO_4. Experimental results are shown in the inset.

of XES and XAS is explained mainly by the fact that the hybridization strength between $3d^9\underline{L}$ and $3d^{10}\underline{L}^2$ (in the final state of XES) is differnt from that between $\underline{3}d^9$ and $\underline{3}d^{10}\underline{L}$ (in the initial state of XAS). Furthermore, the XES peak position depends to some extent on the value of U_{dd} through the contribution of the transition $\underline{2p}3d^9 \to 3d^8$, and the result with U_{dd} = 7 eV is consistent with the experimental data/15/. Therefore, the Cu 2p-XES can be well explained by the mechanism which is consistent with the mechanism of Cu 2p-XPS and Cu 2p-XAS.

6. Concluding Remarks

In this paper, we have shown that the core-level spectroscopy gives us important information on the electronic states of High T_c superconductors and related Cu compounds. From the analysis of Cu 2p-XPS with the use of the impurity Anderson model, we can obtain the information on the hybridization between Cu 3d and O 2p states. The structure of the 2p-XPS satellite is found to depend strongly on the interplay between the atomic multiplet coupling and the solid state hybridization. The polarization dependence of the Cu 2p-XAS and O 1s-XAS gives us the most direct information on the character of the doped holes. In the analysis of Cu 2p-XAS in the Y-Ba-Cu-O system, it is necessary to take account of the contribution from two inequivalent Cu sites. The Cu 2p-XES can also be made with the impurity Anderson model, in a manner consistent with the analysis of Cu 2p-XPS and Cu 2p-XAS.

The authors would like to thank Professor A. Bianconi, Professor G.A. Sawatzky, Dr. J.C. Parlebas and Dr. H. Katayama-Yoshida for valuable discussions in the course of the present work.

Rererences

1 Core-Level Spectroscopy in Condensed Systems, ed. J. Kanamori and A. Kotani (Springer-Verlag, Berlin, 1988).
2 A. Kotani, T. Jo and J. C. Parlebas, Adv. Phys. 37, 37 (1988).
3 P. Steiner, J. Albers, V. Kinsinger, I. Sander, B. Siegwart, Hüfner and C. Politis, Z. Phys. B66, 275 (1987).
4 N. Nücker, J. Fink, B. Renker, D. Ewert, C. Politis, P. J. W. Weijs and J. C. Fuggle, Z. Phys. B67, 9 (1987).
5 A. Fujimori, E. Takayama-Muromachi, Y. Uchida and B. Okai, Phys. Rev. B35, 8814 (1987).
6 T. Gourieux, G. Krill, M. Maurer, M. F. Ravet, A. Menny, H. Tolentino and A. Fontaine, Phys. Rev. 37, 7516 (1988).
7 A. Bianconi, A. Clozza, A. Congiu Castellano, S. Della Longa, M. De Santis, P. Delogu, A. Gargano, R. Giorgi, P. Lagarde, A. M. Flank and A. Marcelli, Proceedings of the Adriatico Research Conference on High Temperature Superconductors, ed. Y. Lu, M. Tosi and E. Tosatti (World Science, Singapore, 1987) p. 853.
8 P. Kuiper, G. Kruizinga, J. Ghijsen, M. Grioni, P. J. Weijs, F. M. F. de Groot, G. A. Sawatzky, H. Verweij, L. F. Feiner and H. Peterson, Phys. Rev. B38, 6483 (1988).
9 A. Bianconi, M. De Santis, A. Di Cicco, A. M. Flank, A. Fontaine, P. Lagarde, H. Katayama-Yoshida, A. Kotani and A. Marcelli, Phys. Rev. B38, 7196 (1988).
10 N. Nücker, H. Romberg, X. X. Xi, J. Fink, B. Gegenheimer and Z. X. Zhao, Phys. Rev. B39, 6619 (1989).
11 A. Bianconi, P. Castrucci, M. De Santis, A. Di Cicco, A. Fabrizi, A. M. Flank, P. Lagarde, H. Katayama-Yoshida, A. Kotani, A. Marcelli, Z. X. Zhao and C. Politis, Modern Phys. Letters B2, 1313 (1988).

12 F. J. Himpsel, G. V. Chandrashekkar, A. B. McLean and M. W. Shafer, Phys. Rev. B38, 11946 (1988); P. Kuiper, M. Grioni, G. A. Sawatzky, D. B. Mitzi, A. Kapitulnik, A. Santaniello, P. de Padova and P. Thiry, to be published in Physica C.
13 K. Okada and A. Kotani, J. Phys. Soc. Jpn. 58, 1095 (1989).
14 K. Okada and A. Kotani, to be published in J. Phys. Soc. Jpn.
15 S. Tanaka, K. Okada and A. Kotani, J. Phys. Soc. Jpn. 58, 813 (1989).
16 J. Zaanen, G. A. Sawatzky and J. W. Allen, Phys. Rev. Lett. 35, 418 (1985).
17 P. Steiner, V. Kinsinger, I. Sander, B. Siegwart, S. Hüfner, C. Politis, R. Hoppe and H. P. Müller, Z. Phys. B67, 497 (1987).
18 G. van der Laan, C. Westra, C. Haas and G. A. Sawatzky, Phys. Rev. B23, 4369 (1981).
19 T. Jo and A. Kotani, Phys. Rev. B38, 830 (1988).
20 A. Kotani, H. Ogasawara, K. Okada, B. T. Thole and G. A. Sawatzky, to be published in Phys. Rev. B.
21 H. Eskes and G. A. Sawatzky, Phys. Rev. Lett. 61, 1415 (1988).
22 Y. Seino, K. Okada and A. Kotani, unpublished.
23 T. Takahashi, private communication.

Dispersive Single-Particle Excitations in Strongly Correlated Two-Dimensional Systems

*Y. Kuramoto and H.J. Schmidt**

Department of Applied Physics, Tohoku University, Sendai 980, Japan
*On leave of absence from Institut für Theoretische Physik, Universität zu Köln, D-5000 Köln, Fed. Rep. of Germany

Abstract. The momentum dependence of single-particle excitations is investigated for a Cu_4O_8 cluster model with the periodic boundary condition by means of an exact numerical method. The electron removal as well as addition spectra up to a few eV exhibit considerable dispersion which can partly be interpreted by a band model. In contrast to the band model, however, an energy gap appears in the undoped ground state. At high excitation energies (≃ 10 eV), the electron removal spectrum shows almost dispersionless features corresponding to $3d^8$ (Cu) and $2p^4$ (O) final states. The low energy excitations are further analyzed with use of the t-J model which is solved analytically on the 2×2 lattice.

1. Introduction

The high transition temperature T_c in cuprate superconductors has stimulated strong interest in their electronic structure. The photoemission spectroscopy (PES) and the Bremsstrahlung isochromat spectroscopy (BIS) are the most direct experimental methods for investigating the electronic structure. Recent angle-resolved experiments have reported the presence of dispersive excitations near the Fermi level [1-4]. On the other hand, the momentum dependence in PES is very small at high excitation energies. The coexistence of itinerant (dispersive) and localized characters seems essential for the high T_c superconductivity, since the superconductivity disappears when the itinerant degrees of freedom are increased too much by doping [5].

The coexistence of itinerant and localized characters is symptomatic of the presence of strong electron correlations, and its understanding is beyond the scope of the one-electron theory. Thus detailed theoretical study on the spectral property seems highly relevant as a step toward understanding the superconductivity. Theoretical studies on the electronic spectrum have been performed for impurity models [6-9]. The dispersive character, however, can be studied only if the periodicity of the lattice is taken into account. This paper reports on the momentum dependence of single-particle excitations for two-dimensinal models with strong correlations. We deal with a small-sized cluster with the periodic boundary condition but introduce no approximations to derive the spectrum.

2. Model and Method for Calculation

The crystalline structures of the cuprate high T_c superconductors are rather complex and vary in each compound. The basic structure common to them is the two-dimensional CuO_2 square lattice. We take into account the $3d(x^2-y^2)$ orbital of Cu at each lattice point and the 2pσ orbital of O which is sandwitched by Cu atoms on both sides. The strong correlation between holes in these orbitals seems relevant to the occurrence of superconductivity. We include the on-site Coulomb repulsion U_d for 3d

Springer Series in Solid-State Sciences, Vol. 89 Strong Correlation and Superconductivity
Editors: H. Fukuyama · S. Maekawa · A.P. Malozemoff

holes and U_p for 2p holes. The transfer integral between the nearestneighbor 3d and 2pσ orbitals is denoted by t_{dp} and that between the two nearest 2pσ orbitals by t_p. Then the model Hamiltonian in terms of hole operators is given by

$$H = \varepsilon_d \sum_{i\sigma} d^\dagger_{i\sigma} d_{i\sigma} + \varepsilon_p \sum_{i\sigma} p^\dagger_{j\sigma} p_{j\sigma} - t_{dp} \sum_{<ij>} \sum_{\sigma} (d^\dagger_{i\sigma} p_{j\sigma} + p^\dagger_{j\sigma} d_{i\sigma})$$

$$-t_p \sum_{<jm>} \sum_{\sigma} p^\dagger_{j\sigma} p_{m\sigma} + U_d \sum_i d^\dagger_{i\uparrow} d_{i\uparrow} d^\dagger_{i\downarrow} d_{i\downarrow} + U_p \sum_j p^\dagger_{j\uparrow} p_{j\uparrow} p^\dagger_{j\downarrow} p_{j\downarrow}, \quad (1)$$

where ε_d and ε_p denote the energy levels of holes.

It has turned out that the model (1) involves rich but elusive physics which escapes from simple approximations like the mean-field theory. We deal with the ground state and the single-particle excitation of the model in two different ways which are complementary to each other: First, a numerical method is employed to derive the spectrum for a Cu_4O_8 cluster without approximations; secondly, the model is transformed to the t-J model and is solved analytically on the 2×2 lattice. In both cases we impose the periodic boundary condition. The size of the cluster is the smallest possible one that still gives information on the momentum dependence. We can deal with momenta such as (0,0), (0,π/a), (π/a,0), and (π/a,π/a) in the square Brillouin zone for the lattice parameter a. To simplify the notation we take the units such as π/a = 1. In eq.(1) the phase of the wave function has been shifted so that the transfer integrals have the same sign for all bonds. In order to make correspondence with experimental results and the band theory, we express the momentum with respect to the unshifted phase of the wave function.

The ground state of the system is derived with use of the modified Lanczos algorithm [10]. The basis function is specified by location of holes for each spin direction. The number of basis functins for 4 holes per cluster for example is 4356 with the total spin zero. Using the ground state $|\psi_n\rangle$ with n holes we define the 3d hole-addition (electron-removal) Green function $D_{n\to n+1}(\mathbf{k},z)$ by

$$D_{n\to n+1}(\mathbf{k},z) = \langle\psi_n|d_\sigma(\mathbf{k})(z-H)^{-1}d^\dagger_\sigma(\mathbf{k})|\psi_n\rangle,$$

where $d_\sigma(\mathbf{k})$ is the annihilation operator of the 3d hole with momentum $\mathbf{k}$ and spin σ. In the case of $\mathbf{k}$ = (1,1), for example, it is given by

$$d_\sigma(1,1) = \frac{1}{2}\sum_i d_{i\sigma},$$

where i runs over four sites in the cluster. Here we have compensated the shift of phase in $d_{i\sigma}$. The 3d electron-removal spectrum $\rho^{(d)}_{PES}(\mathbf{k},\varepsilon)$, where PES indicates the relevance to the angle-resolved phtoemission spectroscopy, is obtained from the Green function by

$$\rho^{(d)}_{PES}(\mathbf{k},\varepsilon) = \frac{1}{\pi}\,\mathrm{Im}\, D_{n\to n+1}(\mathbf{k},\varepsilon+E_n-i\gamma),$$

where E_n is the ground-state energy associated with $|\psi_n\rangle$. Thus the excitation energy ε is measured from the chemical potential. We include a finite imaginary part γ = 0.1 eV to smooth the spectrum which otherwise consists of delta-function peaks in the finite cluster. The Green function is calculated with use of the continued-fraction representation:

$$D_{n\to n+1}(\mathbf{k},z) = \frac{\langle 0|0\rangle}{z-a_0-}\,\frac{b_1}{z-a_1-}\,\frac{b_2}{z-a_2-}\cdots$$

The coefficients a_i and b_i are derived iteratively [10] by

$$|0\rangle = d_\sigma^\dagger(\mathbf{k})|\psi_n\rangle,$$

$$|i+1\rangle = H|i\rangle - |i\rangle\langle i|H|i\rangle/\langle i|i\rangle - |i-1\rangle\langle i-1|H|i\rangle/\langle i-1|i-1\rangle$$

$$= H|i\rangle - a_i|i\rangle - b_i|i-1\rangle. \quad (b_0 = 0)$$

We notice that each state $|i\rangle$ can be obtained as a superposition of basis functions for n+1 holes. In the practical calculation we truncate the continued fraction at i = 100 since the spectrum hardly changes after about 80 iterations.

Similar procedure is available for the 3d electron addition spectrum $\rho_{BIS}^{(d)}(\mathbf{k},\varepsilon)$ where BIS indicates the relevance to the angle-resolved Bremsstrahlung isochromat spectrum. Namely we calculate

$$\rho_{BIS}^{(d)}(\mathbf{k},\varepsilon) = \frac{1}{\pi}\,\mathrm{Im}\, D_{n\to n-1}(\mathbf{k},\varepsilon+E_n-i\gamma),$$

$$D_{n\to n-1}(\mathbf{k},z) = \langle\psi_n|d_\sigma^\dagger(\mathbf{k})(z-H)^{-1}d_\sigma(\mathbf{k})|\psi_n\rangle,$$

by using the continued-fraction representation. The 2p component of the spectrum $\rho_{PES}^{(p)}(\mathbf{k},\varepsilon)$ and $\rho_{BIS}^{(p)}(\mathbf{k},\varepsilon)$ can be calculated in the same way. Since there are two O sites per unit cell, the annihilation operator $p_\sigma^{(\alpha)}(\mathbf{k})$ in the momentum representation has two components with α = 1 and 2. One of these components does not hybridize with 3d holes at the **k** points we consider. Without hybridization the many-body effect on the 2p spectrum is less interesting. Therefore in the following we mainly discuss such 2p component that hybridizes with 3d holes.

3. Numerical Results

The parameters used in the numerical calculation are summarized in Table 1. These values are based on band theoretical estimates [8]. However t_p and t_{dp} have been reduced in view of the result of detailed fit for Ce pnictides [11] where the band theoretical value turns out to be too large. We do not claim that these parameters are the optimum ones for fitting the experimental spectra but rather regard the values as reasonable but tentative.

Table 1: Numerical values of parameters (eV).

$\Delta = \varepsilon_p - \varepsilon_d$	t_{dp}	t_p	U_d	U_p
1.2	0.8	0.325	8.5	4.1

We derive the ground state of the cluster with the number n of the holes being 3, 4, and 5. These correspond respectively to the electron-doped case (n = 3), the undoped case (n = 4), and the hole-doped case (n = 5). The undoped ground state has the total spin zero and the wave function corresponds to the resonating valence bond (RVB) state [12]. The first excited state is the superposition of two equivalent Néel states with opposite directions of sublattice magnetization. The stability of the RVB state against the Néel state is lost if both t_p and t_{dp} are doubled, but is restored if in addition U_p is set to zero [13]. Thus the RVB state and the Néel state are competing for the delicate stability which should also depend on the size of the system. On the other hand, both doped cases have degenerate ground states with the total spin 1/2. They have finite momentum (0,1) or (1,0) each of which corresponds to the Fermi momentum in the noninteracting system. This aspect is further discussed in Sect.4 using analytic solution of the t-J model.

In deriving the spectrum we choose ε_d in such a way that the ground state is stable against both PES and BIS excitations. The charge transfer energy $\Delta = \varepsilon_p - \varepsilon_d$ is kept constant for all n. Let us first discuss the hole-doped case (n = 5) where ε_d = -0.1 eV is taken. Figure 1 shows the PES spectra from the ground state with **k** = (0,1). The spin of the added hole is directed opposite from that of the ground state. The results show considerable **k** dependence near the Fermi level, while at higher energies the **k** dependence is small. The exception is the 2p spectrum with **k** = (0,0) in Fig.1(b). Here hybridization with 3d holes vanishes for both 2p components with α = 1 and 2 by reason of symmetry. The large peak at $\varepsilon \simeq$ 10 eV in Fig.1(a) corresponds to the $3d^8$ final state where two holes come to the same Cu site at the cost of U_d. The final state is almost localized as seen from the lack of the **k** dependence. On the other hand, the large value of U_p leads to similar quasi-localized final states which correspond to the $2p^4$ configuration at an O site. These states appear in Fig.1(b) at energies from 7 to 10 eV. They constitute multiple peaks since the energy depends on occupation of nearby 3d orbitals. The intensity is smaller than the $3d^8$ peaks because of smaller population of 2p holes in the ground state. We have confirmed [13] that the $2p^4$ peaks disappear when U_p is neglected. Thus the appearance of the $2p^4$ peaks signifies the strong correlation among 2p holes.

The spectra near the Fermi level can partly be interpreted in terms of a band model. Namely, in the uppermost electron band that intersects the Fermi level, **k** = (1,1) is empty, (0,1) is closest to the Fermi level, and (0,0) is filled by electrons. Thus in Fig.1(a) the band model is consistent with the relative position of the first (0,1) and (0,0) peaks counted from the Fermi level, as well as the low intensity of (1,1) spectra in this energy region. The first 2p (0,0) peak in Fig.1(b) is shifted from the 3d (0,0) peak in Fig.1(a) because of absence of hybridization. The

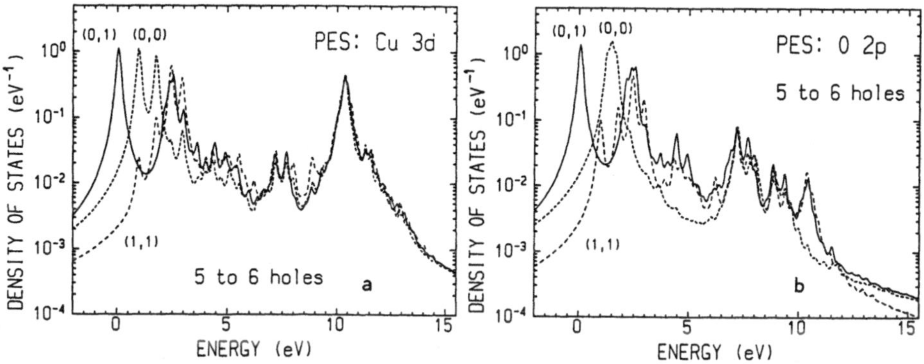

Fig.1: The hole addition spectra from the ground state with 5 holes and with momentum (0,1). The 3d component (a) and the 2p component (b) are shown separately. The momentum attached to each curve indicates that of the added hole.

band picture ceases to be valid for excitations above 2 eV. We remark that the **k** = (1,0) spectrum [13], which is not shown in Fig.1, is different from the (0,1) spectrum because the ground state has the momentum (0,1). This difference is specific to the small size of the system.

The BIS spectra from the n = 5 ground state is shown in Fig.2. The spin direction of the hole removed is taken to be the same as that of the ground state. Localized final states do not appear because of low probability of having two holes in the same site as a result of the hole removal. The presence of multiple peaks for each **k** makes the interpretation in terms of the band model difficult even at low energies. The difficulty is related to the nature of the n = 4 final state which in fact is a Mott insulating state. Nevertheless the band picture is consistent with the low intensity of the (0,0) peaks, particularly in Fig.2(b), and relative position of (0,1) and (1,1) peaks.

We have performed similar computation starting from the n = 4 ground state. The choice ε_d = 0.6 eV satisfies the stability against the change of the hole number. Figure 3 shows the results for the 3d component. It appears that the band model applies well in the energy range within 2 eV from the Fermi level. However, there is a large energy gap (0.84 eV) between the first (0,1) peak in Fig.3(a) and that in Fig.3(b). This energy gap shows the insulating nature of the RVB ground state. We notice that

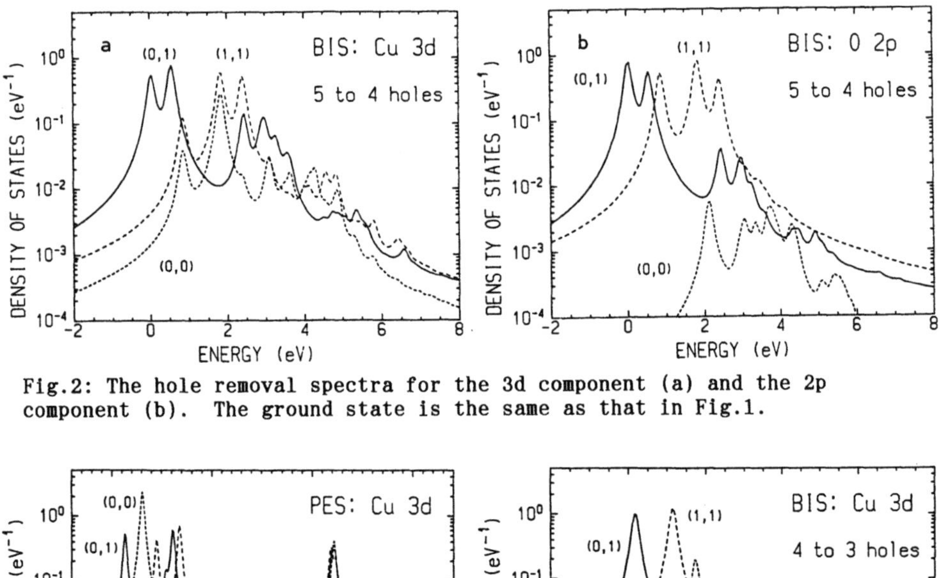

Fig.2: The hole removal spectra for the 3d component (a) and the 2p component (b). The ground state is the same as that in Fig.1.

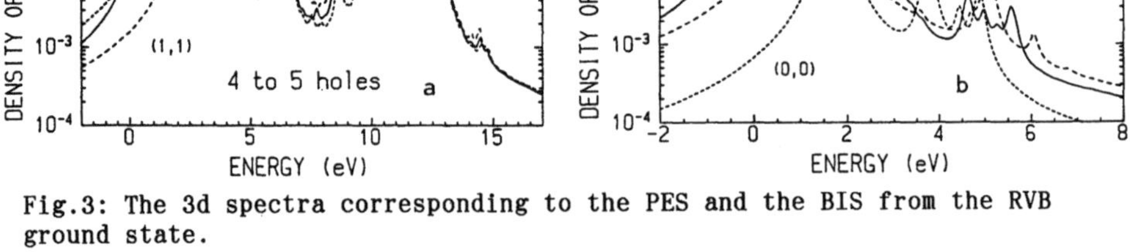

Fig.3: The 3d spectra corresponding to the PES and the BIS from the RVB ground state.

the energy gap for the n = 5 ground state is much smaller (0.1 eV), and becomes slightly negative (-0.008 eV) with the neglect of U_p [13]. The negative value was first obtained in [14] and was ascribed to the pairing of two holes.

We examine how much the many-body effect influences the band width. Setting $U_d = U_p = 0$ and using the same tight-binding parameters as those in Table 1, we can calculate the band width in the one-electron theory. On the other hand the band width in the presence of the many-body effect can be estimated from the separation of peaks in Figs.2 and 3. Table 2 summarizes the comparison for the n = 4 ground state. The overall reduction of the band width is about 20 %.

Table 2: Comparison of the band width (eV) for the ground state with n = 4. The single-particle excitation energy with momentum $\mathbf{k}$ is denoted by $\varepsilon(\mathbf{k})$.

	$\varepsilon(0,0)-\varepsilon(0,1)$	$\varepsilon(1,1)-\varepsilon(0,1)$
band theory	1.1	1.2
present theory	0.87	0.97

4. Analytic Results

The model (1) includes many dynamical processes which span a wide energy range as we have seen in the previous section. In order to have more insight into the dispersive excitations at low energy, it is instructive to work with a simpler effective Hamiltonian where high-energy states enter only implicitly as renormalization of parameters in the Hamiltonian. We start from the limiting case where Δ and U_d are much larger than t_{dp} and t_p, and make a canonical transformation to eliminate charge-transfer excitations. Then the undoped system is well described by the Heisenberg model with spin 1/2 for each Cu site. Zhang and Rice have pointed out [15] that a doped 2p hole in a B_{1g} type molecular orbital on the square of oxygens makes a stable singlet state with a 3d hole surrounded by them. If one neglects the spin polarization of this singlet, the low energy excitations are described by transfer of the singlets and spin excitations. The resultant single-band model is called the t-J model. In this paper we include the effect of t_p in the lowest order and express the effective transfer t of the t-J model in terms of parameters in eq.(1). Then we solve the model analytically on the 2×2 lattice and aim at better understanding of the numerical results in Sect.3.

The annihilation operator $a_{i\sigma}$ of a 2p hole in the B_{1g} molecular orbital around a Cu site $\mathbf{R}_i$ is expressed as

$$a_{i\sigma} = \frac{1}{2} \sum_{j(i)} p_{j\sigma},$$

where the sum over j runs over four O sites around $\mathbf{R}_i$. Note that the shift in the phase of $p_{j\sigma}$ has been compensated. Since the B_{1g} orbital is not orthogonal to another B_{1g} orbital at adjacent sites, it is convenient to introduce the Wannier orbital by making orthogonalization. The annihilation operator $b_{j\sigma}$ corresponding to the Wannier orbital at $\mathbf{R}_j$ is related to $a_{i\sigma}$ by

$$a_{i\sigma} = \sum_j \lambda(\mathbf{R}_i - \mathbf{R}_j) b_{j\sigma},$$

$$\lambda(\mathbf{R}) = \frac{1}{4} \int_{-1}^{1} dk_x \int_{-1}^{1} dk_y \; \{1 + \frac{1}{2}[\cos(\pi k_x) + \cos(\pi k_y)]\}^{1/2} \exp(i\mathbf{k}\cdot\mathbf{R}).$$

The value of $\lambda(\mathbf{R})$ for the same site is $\lambda_0 = 0.96$, for the nearest neighbor $\lambda_1 = 0.14$ and for the next nearest one $\lambda_2 = -0.02$. In the hole-doped case we introduce a projection operator P_i such as

$$P_i = 1 - \frac{3}{4}\sum_{\alpha} b_{i\alpha}^{\dagger} b_{i\alpha} - \frac{1}{2}\sum_{\alpha\beta} \mathbf{S}_i \cdot b_{i\alpha}^{\dagger} \boldsymbol{\sigma}_{\alpha\beta} b_{i\beta} ,$$

where $\mathbf{S}_i$ is the spin operator for the 3d hole at $\mathbf{R}_i$, and where $\boldsymbol{\sigma}$ is the vector composed of the Pauli matrices. The operator P_i can be expressed as $P_i = 1-Q_i$ where Q_i is the projection operator to the 3d-2p triplet state. We operate the product of P_i over all lattice points $\mathbf{R}_i$ on both sides of the Hamiltonian which has been transformed so as to eliminate the 3d charge fluctuation. The projection is equivalent to neglecting the spin polarization of the 3d-2p singlets. Then the electronic states at each site consist of the spin doublet $|^2B_{1g}(\mathbf{R}_i);\sigma\rangle$ and the 3d-2p singlet $|^1A_{1g}(\mathbf{R}_i)\rangle$. To describe the motion of the singlet we introduce the operator $X_{0\sigma}^{(i)} = |^1A_{1g}(\mathbf{R}_i)\rangle\langle^2B_{1g}(\mathbf{R}_i);\sigma|$ and its hermitian conjugate $X_{\sigma 0}^{(i)}$. Then we obtain after some algebra the effective Hamiltonian

$$H = t\sum_{\langle ij\rangle} X_{0\sigma}^{(i)} X_{\sigma 0}^{(j)} + J\sum_{\langle ij\rangle} \mathbf{S}_i \cdot \mathbf{S}_j , \tag{2}$$

where $\langle ij\rangle$ denotes a nearest-neighbor pair in the square lattice and

$$t = 6\lambda_0\lambda_1 t_{dp}^{\;2}\left(\frac{1}{\Delta} + \frac{1}{U_d-\Delta}\right) - \frac{1}{4} t_{dp}^{\;2}\left(\frac{1}{\Delta} - \frac{1}{U_d-\Delta}\right) + \mu_1 t_p . \tag{3}$$

Here we have restored the original phase back from that of eq.(1). Terms with λ_2, $\lambda_1^{\,2}$, and even smaller factors have been neglected. These terms include the next-nearest-neighbor transfer which does not disrupt the Néel order [16]. This point is interesting but is not discussed here. The factor $\mu_1 = 0.55$ takes care of the 2p-2p transfer. The superexchange interaction J is given by

$$J = 4\left(\frac{V^2}{\Delta}\right)^2 \left(\frac{1}{U_d} + \frac{2}{U_p+2\Delta}\right). \tag{4}$$

We notice that in the order of $(t_p/\Delta)^2$ a next-nearest-neighbor antiferromagnetic exchange appears and causes frustration in the spin ordering.

The result given by eq.(3) is a generalization of the one by Zhang and Rice [15] where $\Delta = U_d/2$ and $t_p = 0$ have been assumed. We emphasize, however, that these results are accurate only in the limiting case of large U_d and Δ. For the parameters in Table 1 we find that eqs.(2) - (4) poorly fit the numerical data. In order to obtain more accurate parameters in the effective Hamiltonian one needs partial summation of infinite-order perturbation series. The actual procedure as well as the effects of next-nearest-neighbor interactions will be discussed elsewhere.

In the case of the electron doping the state $|^1A_{1g}(\mathbf{R}_i)\rangle$ is reinterpreted as the hole vacuum at $\mathbf{R}_i$. In this case one obtains again the t-J model but t is now given by $t = -t_{dp}^{\;2}/\Delta$ as discussed in [17]. Note that the sign of t is opposite from that for the hole doping.

By imposing the periodic boundary condition on the 2×2 lattice, one can classify the many-hole wave function according to its momentum. Then the Hamiltonian matrix is decomposed into small blocks and is solved analytically. One can also calculate the matrix element of creation and annihilation operators between the eigen states. We summarize the results of the calculation in Table 3. The results for n = 3 is obtained from that for n = 5 by taking the negative value of t.

Table 3: Properties of eigen states $\psi_{\mathbf{k}}$ of the t-J model on the 2×2 lattice. In (b) $X_{0\sigma}(\mathbf{k})$ denotes the linear combination of $X_{0\sigma}^{(1)}$ to create the momentum $\mathbf{k}$.

(a) 4 holes

momentum	energy	spin
(0,0)	-4J 0 2J	0 (RVB) 0 2
(1,1)	-2J	1
(1,0), (0,1)	0	1

(b) 5 holes

momentum	energy	spin	$\langle\psi_{\mathbf{k}}\vert X_{0\sigma}(\mathbf{k})\vert RVB\rangle$
(0,0)	4t+J 2t -2t-2J	3/2 1/2 1/2	0 0 $1/\sqrt{2}$
(1,1)	2t-2J -2t -4t+J	1/2 1/2 3/2	$1/\sqrt{2}$ 0 0
(1,0), (0,1)	$-J-\sqrt{(J^2+12t^2)} \equiv E_-$ J $-J+\sqrt{(J^2+12t^2)} \equiv E_+$	1/2 3/2 1/2	$E_-/\sqrt{(2E_-^2+24t^2)}$ 0 $E_+/\sqrt{(2E_+^2+24t^2)}$

Let us examine to what extent the low-energy transition from the undoped ground state can be interpreted by the t-J model. The n = 5 states with $\mathbf{k}$ = (0,0) and (0,1) given in Table 3 find corresponding peaks near the Fermi level in Fig.3(a). Similar correspondence is found for the n = 3 states with $\mathbf{k}$ = (1,1) and (0,1) in Table 3 and Fig.3(b). However, in Fig.3 the first PES peak with $\mathbf{k}$ = (1,1) has much smaller intensity than the (0,0) peak, which is not explained from the result in Table 3. The same difficulty appears in the first BIS peak with $\mathbf{k}$ = (0,0). The small intensity of these peaks is better understood as a many-body correction to the band picture where the peaks are forbidden.

The results for 5 holes in Table 3 make clear the competition for the stability between the spin 3/2 state with $\mathbf{k}$ = (1,1) and the spin 1/2 state with $\mathbf{k}$ = (1,0) or (0,1) and energy E_-. The spin 3/2 state gains the kinetic energy as discussed by Nagaoka [18], while the spin 1/2 state takes advantage of the combined effect of the kinetic energy and the exchange energy. In the present size of the cluster the Nagaoka polarization occurs for $J/t < 0.26$. In larger systems the critical ratio of J/t becomes smaller [14,19].

It can be shown that the transition from the n = 5 ground state to each of the n = 4 states in Table 3 has finite matrix element except for the fully polarized state with spin 2. Thus extra peaks in Fig.2, which cannot be interpreted by the band picture, can partly be identified. Namely the second peak with $\mathbf{k}$ = (0,1) corresponds to the (0,0) final state with energy

0 and spin 0 in Table 3. The spin 1 final states with the total momentum (0,1) and (1,0) appear as the first peaks of the removed hole with **k** = (0,0) and (1,1), respectively, in Fig.2(a). Their energy should be the same as that of the second (0,1) peak if the t-J model applies accurately. Thus the difference in the peak positions represents a deviation from the t-J model. The Néel-like final state with the total momentum (1,1) appears in removal of the (1,0) hole from the n = 5 ground state [13], but it is not shown in Fig.2. We have found that the resultant peak is very close to the first (0,1) peak. Thus the energy difference between the RVB state and the Néel state in the model (1) is much smaller than 2J which is given in the t-J model.

5. Discussion and Concluding Remarks

The small cluster calculation distinguishes the undoped case from the doped one by the large difference in the energy gap. On the other hand, the band width is not much different between these cases. The first conclusion of the present study is that the single-particle excitations from the undoped insulating state have considerable dependence on the momentum. We have also calculated the spectrum from the Néel-like ground state and have found similar low-energy dispersive excitations [13]. However in the small cluster it is not possible to study the momentum dependence near the spectral edge. It is conceivable in the doped case that the dispersion is much suppressed by the correlation effect in the vicinity of the Fermi level. What has become clear in our study is that the overall bandwidth is not much reduced. Since the energy scale involved is of the order of t_{dp}, a larger size of the cluster will not modify the overall width. On the other hand approximate treatments [20-23] of the specral property in the t-J model predict a narrow quasi-particle bandwidth of the order of J. In order to compare with these results it is necessary for the exact numerical study to perform calculation with a much larger size of the system.

The second conclusion is that the dispersive excitations near the Fermi level have strong hybridization of 3d and 2p components. This is seen by comparison of the 3d and 2p spectra in Figs.1 and 2. The strong hybridization indicates that the system is far from the limiting case where charge fluctuations of 3d holes are negligible. It is not yet resolved how the strong hybridization picture, supported by the present calculation and also by the core level XPS [6] is reconciled with the resonant PES [1] and the magnetic form factor [23] measurements which suggest the alternative picture [24] where the moments are almost on Cu sites and the carriers are almost 2p like.

We make two remarks on comparison with experiment. The first concerns the quasi-localized excitations at high energy. The peak position of the $3d^8$ final states in the present result shows a good correspondence with the experimentally observed one [25]. On the other hand, experimental identification of the $2p^4$ final states is not yet confirmed. Our result shows that the intensity is by an order of magnitude smaller than the features near the Fermi level. We encourage, however, further experimental effort since the $2p^4$ peaks are the direct signature of the presence of large U_p.

The final remark concerns the simplification in the model. If one compares the spectral shape between the present results and the experimental PES and BIS results, one should be careful about contribution of those electronic states which are neglected in the model (1). Especially important are the other 2p states of O atoms which are present not only in the CuO_2 layer but in other places in the crystal. Thus they make dominant contribution to the PES spectrum in the energy range from about 2 to 6 eV.

We thank Prof. T. Watanabe for encouragement and Mr. Y. Honda for his help in constructing the computer program.

References

[1] T. Takahashi et al., Phys. Rev. B**39**, 6636 (1989).
[2] Y. Sakisaka et al., Phys. Rev. B**39**, 9080 (1989).
[3] F. Minami, T. Kimura and S. Takekawa, Phys. Rev. B**39**, 4788 (1989).
[4] R. Claessen et al., Phys. Rev. B**39**, 7316 (1989).
[5] J.B. Torrance et al., Phys. Rev. Lett. **61**, 1127 (1988).
[6] A. Fujimori et al., Phys. Rev. B**35**, 8874 (1987).
[7] H. Eskes and G.A. Sawatzky, Phys. Rev. Lett. **61**, 1415 (1988).
[8] A.K. McMahan, R.M. Martin and S. Satpathy, Phys. Rev. B**38**, 6650 (1988).
[9] K. Okada and A. Kotani, J. Phys. Soc. Jpn. **58**, 1095 (1989).
[10] E.R. Gagliano and C.A. Balseiro, Phys. Rev. Lett. **59**, 2999 (1987).
[11] M. Takeshige, O. Sakai and T. Kasuya, in: Theory of Heavy Fermions and Valence Fluctuations, edited by T. Kasuya and T. Saso (Springer, Berlin, 1985) p.120.
[12] P.W. Anderson, Mat. Res. Bull. **8**, 153 (1973).
[13] H.J. Schmidt and Y. Kuramoto, to be published.
[14] M. Ogata and H. Shiba, J. Phys. Soc. Jpn. **57**, 3074 (1988).
[15] F.C. Zhang and T.M. Rice, Phys. Rev. B**37**, 3759 (1988).
[16] Y. Kuramoto and T. Watanabe, in: Research Report on Mechanism of Superconductivity (Science Research on Priority Area, No.031, Ministry of Education, Science and Culture of Japan, March 1989) p.223.
[17] H. Matsukawa and H. Fukuyama, to be published in J. Phys. Soc. Jpn.
[18] Y. Nagaoka, Phys. Rev. **147**, 392 (1966).
[19] Y. Hasegawa and D. Poilblanc, preprint.
[20] S. Maekawa, J. Inoue and M. Miyazaki, in: Mechanisms of High Temperature Superconductivity, edited by H. Kamimura and A. Oshiyama (Springer, Berlin, 1989) p.68.
[21] S. Schmitt-Rink, C.M. Varma and A.E. Ruckenstein, Phys. Rev. Lett. **60**, 2793 (1988).
[22] C.L. Kane, P.A. Lee and N. Read, Phys. Rev. B**39**, 6880 (1989).
[23] C. Stassis et al., Phys. Rev. B**38**, 9291 (1988).
[24] V.J. Emery, Phys. Rev. Lett. **58**, 2794 (1987).
[25] T. Takahashi et al., Phys. Rev, B**36**, 5686 (1987).

Magnetism and Superconductivity in Strongly Correlated Systems in One and Two Dimensions

M. Imada

Department of Physics, College of Liberal Arts, University of Saitama, Shimo-Okubo, Urawa 338, Japan

Abstract. Several magnetic and superconducting correlations are examined by numerical analyses to have insights on the nature of strongly correlated systems. In the spin-1/2 Heisenberg antiferromagnets, exact diagonalization study suggests that chirality correlation remains short-ranged in the ground state on the triangular lattice as well as on the square lattice. The probable picture of the ground state of the triangular lattice is singlet liquid. A characteristic feature of the magnetic correlation away from the half-filling is the short-ranged incommensurate structure. The incommensurate peak seen in the two-dimensional Hubbard model is in qualitative agreement with recent neutron scattering data in high-Tc oxides. No support of the superconductivity has been found in Monte Carlo data of the Hubbard model in the ground state. A hint of superconductivity is found in the coupled spin-fermion model as well as in the $t-J$ model with a relatively large exchange coupling J. The formation of an extended singlet around a doped fermion is common phenomena in a wide class of effective hamiltonians of two-band CuO_2 models.

1. Introduction

High temperature superconductivity is still far from satisfactory understanding. Because the strong electron correlation plays a major role in the high temperature superconductivity, there occurs a difficult problem in discriminating various possible mechanisms of superconductivity from purely experimental data. Purely theoretical approach also seems to be a hopeless way if one wants to establish the existence of superconductivity in various theoretical models. The present stage of understanding strongly urges us to collect various experimental data and numerical analyses of theoretical models to combine in inductive reasoning. From this motivation, I have started accumulating information obtained from the numerical analyses making use of every possible techniques such as Monte Carlo, diagonalization, cluster analysis...... I will discuss and summarize the present status of numerical efforts done in several relevant models of high-Tc oxides. In the first part, recent proposal on the possible time reversal and parity symmetry breaking is examined in the half-filled case. In the second part, the ground state of the two-dimensional Hubbard model is discussed. Effective hamiltonians are the third topics. The coupled spin-fermion model and the $t-J$ model as its strong coupling limit are main issues of discussion. I will emphasize that

Springer Series in Solid-State Sciences, Vol. 89 **Strong Correlation and Superconductivity**
Editors: H. Fukuyama · S. Maekawa · A.P. Malozemoff

a satisfactory framework of theoretical formulation has to contain the charge-spin composite excitation induced around an (extended) singlet with charge.

2. Half-Filled System

It seems to be always true that the half-filled case scales to the infinite interaction. The spin-1/2 antiferromagnet in two-dimension defined by

$$H = -2J \sum_{<i,j>} \mathbf{S}_i \cdot \mathbf{S}_j \tag{2.1}$$

is then the starting point in this section. In non-frustrated lattices such as the square lattice, the existence of the antiferromagnetic order predicted by the spin wave theory [1,2] seems to be true. On the triangular lattice, although the spin wave theory predicts the existence of long range order of three-sublattice structure[3], numerical studies seem to support the absence of such a classical order[4-9]. Comparison of the ground state energy with variational estimates[10] does not provide any reliable argument, because quite different states may have very similar energies in this problem. Since the energy is insensitive to the choice of states, we need the accuracy beyond the level of present variational argument.

Recent argument on the possibility of fractional statistics of the elementary excitation has proposed another type of symmetry breaking[11]. As the time reversal and parity symmetry broken state, the long-range order of the chirality defined by

$$c_{123} = \mathbf{S}_1 \cdot (\mathbf{S}_2 \times \mathbf{S}_3) \tag{2.2}$$

on a unit triangle has been the subject of mean field study[12]. However, the exact diagonalization of finite $L_x \times L_y$ systems with the railroad trestle extrapolation in the free boundary condition strongly suggests that the chirality correlation and hence the flux correlation is short-ranged on the triangular lattice as well as on the square lattice[13]. Figures 1 and 2 illustrate the railroad trestle extrapolation of squared chirality

$$\Gamma_+ = < C_+^* C_+ > /N$$

defined from

$$C_+ = \sum_{<i,j,k>} c_{i,j,k},$$

where the summation with respect to $< i, j, k >$ is over the unit triangles with the sites i, j and k in the clock-wise direction. It suggests that regular frustration is not sufficient to break the chiral symmetry in the half-filled systems. The possible time reversal and/or parity symmetry breaking in non-half-filled systems remains to be examined.

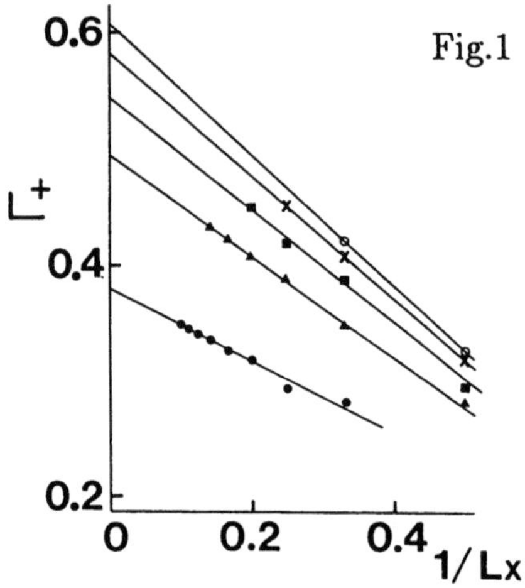

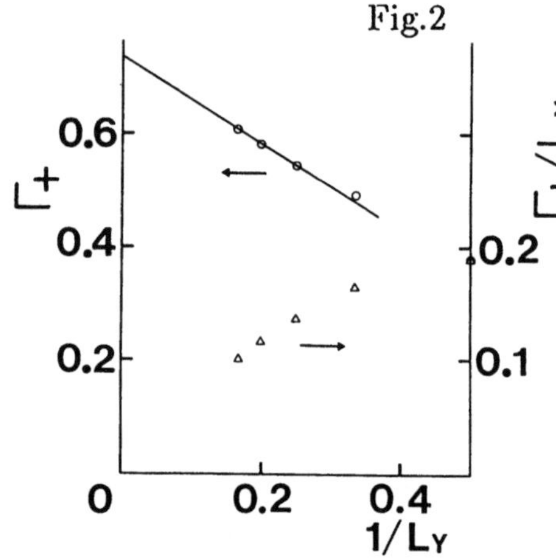

Fig.1: The railroad extrapolation of Γ_+ to $L_x \to \infty$ on the triangular lattice. The widths(L_y) are 2(solid circles), 3(solid triangles), 4(solid squares), 5(crosses) and 6(open circles).

Fig.2: The railroad extrapolation of Γ_+ to $L_y \to \infty$ on the triangular lattice is shown as circles. The triangles show Γ_+/L_y.

3. Non-Half-Filled System

i) Hubbard Model

The Hubbard model is defined from the hamiltonian

$$H = -t \sum_{<ij>,\sigma} (c^\dagger_{i\sigma} c_{j\sigma} + c^\dagger_{j\sigma} c_{i\sigma}) + U \sum_i n_{i\uparrow} n_{i\downarrow}, \tag{3.1}$$

where $c^\dagger_{i\sigma}$ is the fermion creation operator with spin σ at the i-th site. The summation with respect to $< ij >$ is over the nearest neighbor pairs. The numerical data have been obtained from the ground state Monte Carlo method[14] proposed by Sorella et al.[15]. In the magnetic correlation, short-ranged incommensurate peak is a characteristic feature of the system away from half-filling[14] in the ground state. A typical example of the equal time magnetic correlation defined by

$$S(\mathbf{q}) = \frac{1}{N} \sum_{ij} \langle S^z_i S^z_j \rangle \exp(i\mathbf{q} \cdot (\mathbf{r}_i - \mathbf{r}_j)) \tag{3.2}$$

is seen in Fig.3. The deviation of the peak position from the commensurate point (π, π) is proportional to the doping concentration and in qualitative agreement with the so-called "$2k_F$" anomaly. The $short-range$

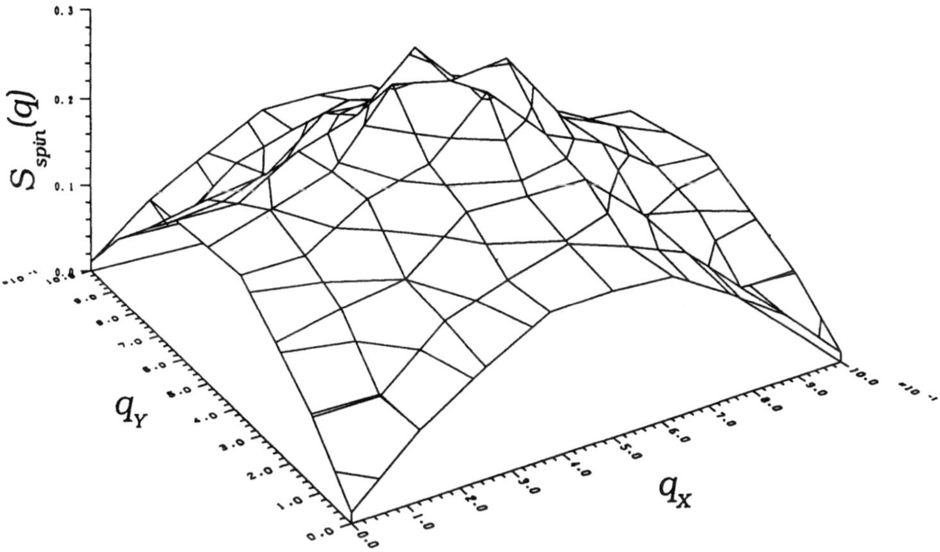

Fig.3: Equal time spin correlation in the momentum space for the non-half-filled Hubbard model at t=1.0 and U=3.0.

correlation of incommensurability, however, indicates that various arguments leading to the incommensurate long range order are certainly incorrect. We note qualitative agreement of our result with the neutron scattering experiment for high-Tc oxides[16-18]. The short-ranged incommensurate correlation itself seems to be a general feature of non-half-filled systems.

As for the superconducting correlation measured in the ground state, no support of the superconductivity has been found[14]. The result is consistent with the suppression of superconducting correlation seen at finite temperatures[19,20]. It should also be noted that the ground state Monte Carlo at this moment has difficulty in getting reliable data at the fillingness around 0.9.

ii) Effective hamiltonians of two-band systems[14,21-25]
The coupled spin-fermion system as a model for the CuO_2 plane[1-5] is defined from the hamiltonian

$$H = -t \sum_{<i,j>\sigma} (c_{i\sigma}^{\dagger} c_{j\sigma} + c_{j\sigma}^{\dagger} c_{i\sigma}) + U_h \sum_i n_{i\uparrow} n_{i\downarrow} - 2J_K \sum_{<i,l>} \mathbf{S}_l \cdot \sigma_i - 2J_S \sum_{<l,m>} \mathbf{S}_l \cdot \mathbf{S}_m, \tag{3.3}$$

where

$$\vec{\sigma}_i = \frac{1}{2} c_{i\sigma}^{\dagger} (\sigma)_{\sigma\sigma'} c_{i\sigma'} \quad , \quad n_{i\sigma} = c_{i\sigma}^{\dagger} c_{i\sigma}$$

The $t-J$ model is the strong coupling limit $J_K \to -\infty$ of the coupled spin-fermion model:

$$H = -t \sum_{<ij>} ((1-n_{i,-\sigma})c^{\dagger}_{i\sigma}c_{j\sigma}(1-n_{j,-\sigma}) + h.c.) \\ -2J \sum_{<ij>} \mathbf{S}_i \cdot \mathbf{S}_j, \tag{3.4}$$

The extended singlet around a doped fermion in the coupled spin-fermion model is reduced to a local singlet in the $t-J$ model. Because of the formation of an extended singlet between "localized" spins and the fermion, perturbative treatment with respect to J_K may not be justified. It, however, does not justify the reduction of the CuO_2 model to "one-band" models in two senses. The first point is concerned with the ratio of J to t, if the $t-J$ model is employed. Several numerical data support that the attractive interaction of two-holes in the $t-J$ model arises only when the ratio is relatively large. It is quite possible that the $t-J$ model with small and large J/t belong to different classes of ground states. So many data suggest that the originally one-band model i.e. Hubbard model is not connected to the large J/t region. The second point is the role of spatial extension of singlet seen in the coupled spin-fermion model. In the realistic situation of the high-Tc oxides, J_K and t are supposed to have comparable values, which necessarily causes the spatial extension of the singlet. Other considerations not taken account of in the coupled spin-fermion model such as the "three-site term"[25–27] also seem to contribute to the spatial extension of the singlet[28,29]. I, therefore, assume that it is dangerous to restrict the relevant model to the local-singlet case as in the $t-J$ model without any firm analysis on the scaling behavior in the parameter space. The difference between spatially extended singlet and local singlet in the role toward the superconductivity has to be clarified.

The formation of extended singlet by the doping means the appearance of a state in the so-called charge excitation gap. This state is split off from the oxygen band edge in the energy scale of J_K. The appearance of such states seems to be consistent with recent photoemission and soft-X-ray absorption experiments[30–32]. The connection to the heavy fermion problem has also been discussed[33]. Although the "fermi level" lies in this newly formed "singlet level", it is absolutely not clear whether the excitation in the states follows fermi liquid behavior.

The attractive interaction of two fermions(holes) seems to be the consequence of spin distortion caused around the (extended) singlet. It implies that the coupling between spin and charge degrees of freedoms is crucially important for the formation of bound two fermions. The numerical data seem to support not the separation of charge and spin but their inherent coupling. It has to be incorporated in the anticipated formalism of the low energy excitation of the effective hamiltonians. At this moment, we have no numerical evidence to support the fractional nature of the statistics of the elementary excitation. It is however, a quite interesting problem to clarify the statistics of the excitation at extended singlet object surrounded by the spin distortion. This problem may also be clarified by examining the role of dimensionality.

References

1) P. W. Anderson: Phys. Rev. 86 (1952) 694.
2) R. Kubo: Phys. Rev. 87 (1952) 568.
3) T. Oguchi: J. Phys. Soc. Jpn. 52, (1983) Suppl. 183
S. Miyake and H. Nishimori: Prog. Theor. Phys. 73 (1985) 18.
4) T. Oguchi, H. Nishimori and H. Taguchi: J. Phys. Soc. Jpn. 55, (1986) 323.
5) M. Imada: J. Phys. Soc. Jpn. 56, (1987) 311.
6) M. Takasu, S. Miyashita and M. Suzuki: Prog. Theor. Phys. 75 (1986) 1254.
7) H. Nishimori and H. Nakanishi: J. Phys. Soc. Jpn. 57, (1988) 626.
8) S. Fujiki: Can. J. Phys. 65, (1987) 489.
9) M. Imada: "*Quantum Monte Carlo Methods in Equilibrium and Nonequilibrium Systems*" ed. by M.Suzuki (Springer Verlag, 1987)p.125.
10) D. A. Huse and V. Elser: Phys. Rev. Lett. 60 (1988) 2531.
11) V. Kalmeyer and R. B. Laughlin: Phys. Rev. Lett. 59 (1987) 2095.
12) W. G. Wen, F. Wilczek and A. Zee: preprint(1988).
13) M. Imada: "*Chirality and Flux in Two−Dimensional Spin−1/2 Antiferromagnets −Exact Diagonalization Study*" submitted to J. Phys. Soc. Jpn.
14) M. Imada and Y. Hatsugai: "*Numeriacl Studies on the Hubbard Model and the t − J Model in One − and Two − Dimensions*" submitted to J. Phys. Soc. Jpn.
15) S. Sorella, E. Tosatti, S. Baroni, R. Car and M. Parrinello: Int. J. Mod. Phys. B 1 (1988) 993.
16) R. J. Birgeneau et al.: Phys. Rev. B 39, (1988) 2868.
17) T. R. Thurston, et al.: preprint(1989).
18) G. Shirane: this proceedings.
19) M. Imada: J. Phys. Soc. Jpn. 57, 3128 (1988).
20) J .E. Hirsch H. Q. Lin: Phys. Rev. B 37 (1988) 5070.
21) M. Imada, N. Nagaosa and Y. Hatsugai: J. Phys. Soc. Jpn. 57 (1988) 2901 .
22) M. Imada, Y. Hatsugai and N. Nagaosa: Int. J. Mod. Phys. B 1 (1988) 959 .
23) Y. Hatsugai, M. Imada and N. Nagaosa: J. Phys. Soc. Jpn. 58 (1989) 1347.
24) N. Nagaosa, Y. Hatsugai and M. Imada: J. Phys. Soc. Jpn. 58 (1989) 978.

25) M. Imada: Proceedings of 2nd.NEC Symposium "*Mechanisms of High Temperature Superconductivity*" ed. by H. Kamimura and A. Oshiyama (Springer Verlag, 1989)p.53.
26) H. Eskes, G. A. Sawatzky: Phys. Rev. Lett. 61 (1988) 1415.
27) H. Fukuyama, H. Matsukawa and Y. Hasegawa: J. Phys. Soc. Jpn. 58 (1989) 367.
28) H. Matsukawa and H. Fukuyama: Technical Report of ISSP No.2141
29) H. Fukuyama: this proceedings.
30) A. Fujimori, Y. Tokura, H. Eisaki, H. Takagi, S. Uchida and M. Sato: "*Origin of the Electronic States near the Fermi Level in High − Tc Superconductors*" and this proceedings.
31) T. Takahashi et al.: this proceedings
32) G. A. Sawatzky: this proceedings
33) M. Tachiki: this proceedings.

Characteristics of the Charged Hole in the One-Dimensional Mott Insulator

N. Kawakami[1] *and A. Okiji*[2]

[1]Research Institute for Fundamental Physics,
Kyoto University, Kyoto 606, Japan
[2]Department of Applied Physics,
Osaka University, Suita, Osaka 565, Japan

Abstract. Charge excitations in the one-dimensional Hubbard model are investigated, with the use of the Bethe-Ansatz method, by inserting the charged hole into the Mott insulator. The density of states and the momentum-dependent susceptibility for the charged hole are shown to have a large enhancement due to the formation of the Hubbard gap. The logarithmic anomaly near $4k_F$ for the latter quantity is investigated in the vicinity of the Mott insulator. The compressibility and the specific heat calculated in the whole temperature region are discussed in connection with the characteristics of the charged holes.

1. Introduction

Highly correlated electron system in low dimension has attracted much interest since the recent discovery of the high T_c superconducting copper oxides [1,2]. In particular, the electronic properties in the metallic system close to the Mott insulator [3] have been studied extensively. The one-dimensional Hubbard Hamiltonian is considered as a simplified model to treat the essential properties of the highly correlated electron system including the metal-insulator transition. In this model precise arguments are possible for various physical quantities by means of the Bethe Ansatz solution [4-13]. In this article, on the basis of the Bethe Ansatz solution, we investigate the characteristic properties of the charge excitations in the one-dimensional Hubbard model by inserting the charged hole (definition is given in section 3) into the Mott insulator [13]. First, we calculate the density of states for the excitation spectrum of the charged hole and show that it has a divergent term in the lower energy region according to formation of the Hubbard gap. Then we introduce the momentum-dependent susceptibility for the charged hole and discuss the $4k_F$ anomaly (k_F: Fermi momentum in non-interacting case) of this quantity in the metallic system close to the Mott insulator. In this connection, we show the results for the compressibility and the specific heat calculated in the whole temperature region [12].

2. Basic Equation in the Bethe Ansatz Solution

We consider the Hubbard model in one dimension described by the Hamiltonian

$$H = -t \sum_{<ij>,\sigma} C_{i\sigma}^{\dagger} C_{j\sigma} + 4U \sum_{i} C_{i\uparrow}^{\dagger} C_{i\uparrow} C_{i\downarrow}^{\dagger} C_{i\downarrow}, \tag{1}$$

where $C_{i\sigma}$ is the annihilation operator of the electron state with the σ-spin at the ith site and the transfer of the electron is assumed to occur between the nearest neighbor sites. The Bethe Ansatz solution of (1) was obtained by Lieb and Wu in following two steps [4]. First one seeks for the solution of the Schrodinger equation as a superposition of wave functions for the noninteracting case ($U = 0$). The imposition of the periodic boundary condition on this Bethe wave function results in the additional problem in the spin space, which can be diagonalized by the generalized Bethe Ansatz developed by Yang [14]. Consequently, the Bethe-Yang equation is obtained for two kinds of the quasi-momenta k_j and Λ_α which specify the elementary excitations concerning the charge and spin fluctuations [4],

$$\exp(ik_j N_a) = \prod_{\beta=1}^{M} \frac{\sin(k_j) - \Lambda_\beta + iU/t}{\sin(k_j) - \Lambda_\beta - iU/t}, \qquad j = 1 \sim N, \tag{2a}$$

$$-\prod_{j=1}^{N} \frac{\sin(k_j) - \Lambda_\alpha + iU/t}{\sin(k_j) - \Lambda_\alpha - iU/t} = \prod_{\beta=1}^{M} \frac{\Lambda_\alpha - \Lambda_\beta - 2iU/t}{\Lambda_\alpha - \Lambda_\beta + 2iU/t}, \qquad \alpha = 1 \sim M, \tag{2b}$$

where N_a is the number of the lattice point and $N(M)$ is the total (up-spin) electron number. The total energy and the total momentum are given by $E = -2t\sum_j \cos k_j$ and $K = \sum_j k_j$, respectively. At zero temperature the distribution of the above quasi-momenta is expressed in terms of the density functions $\rho(k)$ and $\sigma(\Lambda)$ which are obtained by taking the logarithm of (2) and differentiating them with respect to k and Λ. In the ground state for the half filled case, $\rho(k)$ and $\sigma(\Lambda)$ have continuous values within the regions $[-\pi,\pi]$ and $[-\infty, \infty]$, respectively [4].

3. Charged Hole Inserted into the Mott Insulator

3.1 Density of States

In order to specify the charge excitations, we necessarily examine the properties of the quasi-momentum k [7,10,13]. The most simple excitation including the change of the electron charge is obtained by inserting a *hole* into the half filled band at a given value of the quasi-momentum $k_0(-\pi \leq k_0 \leq \pi)$. This kind of excitation describes that the charged hole is doped into the Mott insulator with total spin of the system kept in singlet state [15]. The quasi-particle specified by the above excitation is referred to as the *charged hole* in the following discussion. The change of the density function $\rho(k)$ from the ground state is described by [13],

$$\rho_1(k) = -\delta(k - k_0) + \cos k \int_{-\pi}^{\pi} dk' R(\sin k - \sin k')\rho_1(k'), \tag{3}$$

where the integral kernel $R(x)$ is given by the Fourier transform of $[1 + \exp(U|\omega|/t)]^{-1}$. Integral equation (3) is easily solved, resulting in the expression for the change of the energy defined by $\epsilon = -2t\int_{-\pi}^{\pi} \cos k\rho_1(k)dk$ [7,13],

$$\begin{aligned}\epsilon = &-2U + 4t\cos k_0\theta(\pi/2 - |k_0|) \\ &+ t^2 \int_1^\infty \frac{dz}{U}(z^2 - 1)^{1/2}\{\operatorname{cosech}\frac{\pi t}{2U}(z - \sin k_0) + \operatorname{cosech}\frac{\pi t}{2U}(z + \sin k_0)\},\end{aligned} \tag{4}$$

where $\theta(x)$ is the step function. The change of the total momentum is expressed in a similar way as,

$$q = -2\pi \int_0^{k_0} \rho_0(k) dk, \tag{5}$$

where $\rho_0(k)$ is the density function for the ground state [4,13],

$$\rho_0(k) = (1/\pi)\theta(\pi/2 - |k|) - \frac{t\cos k}{2\pi U} \int_1^{\infty} dz (z^2 - 1)^{-1/2} \{\mathrm{cosech} \frac{\pi t}{2U}(z - \sin k) + \mathrm{cosech} \frac{\pi t}{2U}(z + \sin k)\}. \tag{6}$$

With the use of above formulae, the density of states for the charged hole can be obtained by

$$D_c(\varepsilon) = 2\rho_0(k_0)/[d\varepsilon/dk_0], \tag{7}$$

with $k_0(> 0)$ given by (4) as a function of ε.

The results calculated for the density of states $D_c(\epsilon)$ are shown in Fig.1 for several choices of the Coulomb interaction [13]. It is seen that $D_c(\epsilon)$ exhibits a divergent behaviour in the lower energy region in the presence of the Coulomb interaction, which shows the clear contrast with the non-interacting case. This fact directly reflects the formation of the Hubbard gap accompanied with the Mott insulating phase (half filled model is a Mott insulator as far as U exists [4]). In order to understand this behaviour more clearly, we derive the expression for $D_c(\epsilon)$ in a closed form in the energy region close to the gap-edge. First, the quantities ϵ and q are expanded up to the lowest order of $\delta k_0 (= \pi - k_0)$,

$$\varepsilon \sim \alpha_0 (k_0)^2, \quad q \sim 2\pi\rho_0(\pi)k_0, \tag{8}$$

where

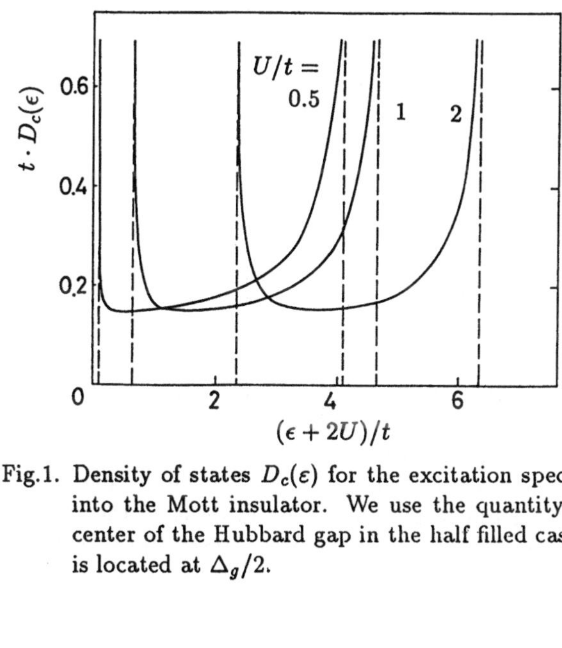

Fig.1. Density of states $D_c(\varepsilon)$ for the excitation spectrum of the charged hole inserted into the Mott insulator. We use the quantity $\varepsilon + 2U$ as the energy ($2U$ is the center of the Hubbard gap in the half filled case) so that the lower edge of $D_c(\varepsilon)$ is located at $\Delta_g/2$.

$$\alpha_0 = \frac{\pi^2 t^3}{4U^3} \int_1^\infty dz(z^2-1)^{1/2}[2\coth^2(\pi t z/2U) - 1]\mathrm{cosech}(\pi t z/2U). \tag{9}$$

Then the density of states in the lower energy region is written as,

$$D_c(\varepsilon) \sim \rho_0(\pi)(t\alpha_0)^{-1/2}[\varepsilon - (-2U + \Delta_g/2)]^{-1/2}, \tag{10}$$

where the magnitude of the Hubbard gap Δ_g, defined by $4U + 2\varepsilon(k_0 = \pi)$ [4,7], takes the form,

$$\Delta_g = (4t^2/U) \int_1^\infty dz(z^2-1)^{1/2}\mathrm{cosech}(\pi t z/2U), \tag{11}$$

In the case of the weak interaction ($U \sim 0$), expressions (10) and (11) are reduced to

$$D_c(\epsilon) \sim \frac{1}{2\pi}\sqrt{\Delta_g/t^2}[\epsilon - (-2U + \Delta_g/2)]^{-1/2}, \tag{12a}$$

$$\Delta_g \sim \frac{16}{\pi}\sqrt{tU}\exp(-\pi t/2U). \tag{12b}$$

This formula explicitly suggests that the divergent behaviour is accompanied with the formation of the Hubbard gap. In contrast with the low-energy region, the behaviour near the band edge is dominated chiefly by the density of states for the non-interacting case as far as U is not so strong. It is worthwhile to note that the band width of the energy spectrum is not renormalized and takes the value of $4t$ irrespective of the strength of the Coulomb interaction.

If the renormalized particle picture is adopted, expression (12) is rewritten as $D_c(\varepsilon) \sim (2\pi\sqrt{t^*})^{-1}[\varepsilon - (-2U + \Delta_g/2)]^{-1/2}$, where the renormalized transfer integral is given by $t^* = t^2/\Delta_g$. Therefore the effective transfer integral t^* becomes exponentially large with the decrease of the Coulomb interaction, while it leads to the bare value t in the large U region (spinless fermion). It is to be emphasized that the excitations near the gap-edge in the weak interaction case show the universal behaviour with the one scaling parameter $t^* = t^2/\Delta_g$. This is mentioned again in the section for the momentum dependent susceptibility and also for the attractive case. It is clarified in the following discussion that the density of states for the charged hole introduced above plays the essential role for the various physical quantities concerning the charge excitations near the Mott insulator.

3.2 Momentum-Dependent Susceptibility

The formulation of the correlation functions is difficult within the framework of the Bethe Ansatz solution, so that such quantities have not been investigated so far. Here we should notice a simple fact that in the Bethe Ansatz formalism, two kinds of the quasi-particles concerning the charge and spin fluctuations behave as renormalized *free particles*. For instance, two charged holes inserted into the Mott insulator (half filled case) do not interact with each other, as easily seen from (3). This fact enables us to introduce the susceptibility concerning the charge fluctuation in terms of the charged particle. On such a standpoint of view, we define the *momentum-dependent susceptibil-*

ity for the charged hole (or particle) in a manner somewhat similar to the non-interacting case [13],

$$\chi_c(Q) = \sum_{k_0,k_1} \frac{P(k_0,k_1)}{E_c(k_0,k_1) - E_g}, \tag{13}$$

where E_g is the ground state energy and the $E_c(k_0, k_1)$ is the energy of the particle-hole excited states with quasi-momentum $k_0(k_1)$ for hole (particle). The function $P(k_0, k_1)$ restricts the summation for the quasi-momenta to the possible region of the excited states with real momentum-transfer Q. The susceptibility defined by (13) is considered to give useful informations for the charge excitations with the momentum transfer Q.

Henceforth, we deal with the metallic case close to the Mott insulator. The deviation from the half filled band can be described by changing the range of k from $[-\pi,\pi]$ to $[-\nu_F,\nu_F]$ with $\nu_F \leq \pi$. The *Fermi momentum* for the charged particle, ν_F, is related with the electron concentration n as, $\nu_F \sim \pi - (1-n)/(2\rho_0(\pi))$, for nearly half filled case $n \sim 1$. It is seen from (3) that in the metallic case close to the Mott insulator the low energy excitations are possible for $Q \sim 0$ and $Q \sim 2\pi$. In this case the energy for the particle-hole pair excitation is expressed as

$$E_c(k_0,k_1) = \varepsilon(k_0) + \varepsilon(k_1), \tag{14}$$

where the real momentum-transfer Q is related to the quasi-momenta through $Q = q(k_0) - q(k_1)$ with $|k_0| < \nu_F$ and $|k_1| > \nu_F$ in (5). By completing the summation over k_0 and k_1 in (13) with the momentum transfer fixed as $Q \sim 0$, we obtain the contribution to the susceptibility in the limiting case of the zero-momentum transfer [13],

$$\begin{aligned} \chi_c(Q \sim 0) &\sim \rho_0(\pi)/(2\alpha_0) \lim_{z\to 0} \log \frac{2(\pi-\nu_F)+z}{2(\pi-\nu_F)-z} \\ &= [2\rho_0(\pi)^2/\alpha_0](1-n)^{-1}. \end{aligned} \tag{15}$$

Note that this result is same as the ordinary charge susceptibility (compressibility) derived directly from the differentiation of the electron number with respect to the chemical potential (see next section) [11,12]. This fact is considered to suggest that the present approach in terms of the quasi-particle may be relevant one for the investigation of the momentum-dependent susceptibility. It should be noted that the quantity $\chi_c(0)$ is considerably enhanced by the factor $(1-n)^{-1}$ near the Mott insulator according to the formation of the Hubbard gap. In the limiting case $U \sim 0$, $\chi_c(0)$ is reduced to

$$\chi_c(0) \sim \frac{1}{2\pi^2}(\Delta_g/t^2)(1-n)^{-1}. \tag{16}$$

Therefore in the small U region we have again a close relationship between the divergent behaviour of the charge susceptibility and the formation of the Hubbard gap. With the use of the renormalized transfer integral t^* introduced in 3-1, (16) is rewritten as $[2\pi^2 t^*(1-n)]^{-1}$ like the non-interacting case in the vicinity of the band edge. Above simple results deduced from (12) and (16) lead us to the conclusion that the charged hole in the one-dimensional Mott insulator behaves as if it is a band-edge hole with the effective transfer integral t^*, which becomes exponentially large with the decrease of the Coulomb interaction.

The most interesting aspect for $\chi_c(Q)$ appears at $Q \sim 4k_F$ when the electron concentration deviates from the half filled case [13]. In order to see this explicitly, we complete the summation for the quasi-momenta in (13) with the real momentum-transfer fixed as $Q \sim 2\pi$. Consequently, we have the susceptibility $\chi_c(Q \sim 2\pi)$ as,

$$\chi_c(Q) \sim \frac{1}{2}[\rho_0(\pi)^2/\alpha_0](1-n)^{-1}\log\frac{2\pi - Q}{2\pi n - Q}, \tag{17}$$

for $Q \leq 2\pi n$. As is seen from (15) the factor $\rho_0(\pi)^2/\alpha_0(1-n)^{-1}$ appearing in (17) is proportional to the compressibility, $\chi_c(0)$, and the Fermi momentum k_F in non-interacting case is related to the electron concentration through $k_F = (\pi/2)n$. Hence, formula (17) is rewritten into the following simple one,

$$\chi_c(Q \sim 4k_F) \sim \frac{1}{4}\chi_c(0)\log\frac{2\pi - Q}{4k_F - Q}. \tag{18}$$

It is seen that the logarithmic $4k_F$ anomaly should appear as soon as the electron concentration deviates from the half filled case and that it is considerably enhanced by the factor $(1-n)^{-1}$ due to the same mechanism for the compressibility $\chi_c(0)$ (see (15)). The logarithmic divergent behaviour reflects the fact that in the metallic case there exist zero-energy charge excitations at $Q = 4k_F$ as well as at $Q = 0$ [10]. Even in the metallic case far from the Mott insulator, we have an expression for $\chi_c(Q \sim 4k_F)$ similar to (18). It is to be noted, however, the absolute value of $\chi_c(Q)$ has a tendency to be suppressed in a ordinary way according to the increase of the Coulomb interaction, except for the case $n \sim 1$ discussed above. Note that the logarithmic divergence at $2k_F$ does not appear in the charged hole as far as the Coulomb interaction takes the finite value.

The susceptibility for the spin particle $\chi_s(Q)$ is formulated in a quite similar way employed for the susceptibility $\chi_c(Q)$ [13]. As a reference, we summarize the characteristic properties for $\chi_s(Q)$ below. The spin wave spectrum is hardly affected by the formation of the Hubbard gap [6], so that the $2k_F$ anomaly always exists irrespective of the band filling. As a matter of course, the anomalous enhancement $(1-n)^{-1}$ does not appear in this quantity. In the half filled case, the analytic calculation is possible for $Q \sim 0$ and $Q \sim \pi$. As for the homogeneous case, $\chi_s(0)$ calculated within the present framework agrees with the exact one already obtained by Takahashi [5]. On the other hand, for $Q \sim \pi(2k_F)$, the logarithmic divergent behaviour is derived as $\chi_s(Q \sim 2k_F) \sim \chi_s(0)\log(2k_F - Q)$ for $Q \leq 2k_F$. Since $\chi_s(0)$ is increased with increasing U, the $2k_F$ anomaly is enhanced as well, as expected [16].

3.3 Comparison with Attractive Case

In order to see the characteristic properties for the charged hole more clearly, it is of much help to compare them with those for the *spin* excitation spectrum in the attractive case ($U < 0$). As is well known, the ground state in the attractive case consists of the bound-state solution of the charge complex k and the spin excitation spectrum can be described by breaking the bound state [5]. The pair-breaking energy E_b and the corresponding density of states $D_b(\epsilon)$ are obtained by the similar procedure outlined above. The results are given by the same expressions as (12) for the charge excitations in the repulsive case, although the underlying physics is completely different from each other. In the weak coupling limit ($U \sim 0$), we rewrite them in a more popular form as,

$$D_b(\epsilon) \sim N(0)\sqrt{E_b}[\varepsilon - E_b]^{-1/2}, \tag{19a}$$

$$E_b \sim \frac{2}{\pi} D_0 \sqrt{2\pi\mu^*} \exp(-1/\mu^*) \tag{19b},$$

where $D_0 (= 4t)$ is the band width, $N(0)$ the density of states at the Fermi level and $\mu^* (= -4U/2\pi t)$ the dimensionless coupling constant. The above expressions are quite similar to those expected for pair breaking excitations in BCS theory in the weak coupling limit (prefactor $\sqrt{\mu^*}$ in (19b) may be due to the effects of the one-dimensional fluctuation). Therefore it is naturally understood in the light of the attractive case that the divergent behaviour derived for the charged hole is closely related to the formation of the Hubbard gap and that it shows the universal behaviour in the small U region. Furthermore, this comparison may suggest that the divergent behaviour appearing in the small U region may be slightly different from the one expected for the spinless fermion in the large U region. It is considered that the crossover occurs from the former to the latter as the Coulomb interaction is increased.

4. Thermodynamic Properties

The characteristic properties discussed in the previous section appear in the thermodynamic quantities concerning the charge fluctuations. In this section, we investigate the compressibility and the specific heat in the whole temperature region. The formalism for the Bethe Ansatz solution at an arbitrary temperature $T(k_B = 1)$ has been derived by Takahashi [8]. The basic equations describing the equilibrium-thermodynamics are written by an infinite set of the nonlinear integral equations for the so-called pseudo-energy [8]. Following the method employed in the Anderson model [17], we have solved these equations numerically so as to calculate the thermodynamic quantities [12]. Here, we quote from ref.12 some of the numerical results for the the compressibility and the specific heat calculated at arbitrary temperatures.

In Fig.2 the compressibility, defined by $\chi_c = dn/d\mu$ (μ: chemical potential), is plotted as a function of the electron concentration n with the temperature being fixed. While the compressibility is not so sensitive to the electron concentration at higher temperatures, it is strongly dependent on the electron concentration as the temperature is decreased. The hump developed in the lower concentration regime reflects the shape of the density of states for the non-interacting case. Just at the half filled case ($n = 1$, Mott insulating phase), the compressibility is seen to be suppressed with the decrease of the temperature and leads to the vanishing value at $T = 0$. On the other hand, the compressibility is considerably enhanced with decreasing temperature in the metallic regime close to the Mott insulator ($n \sim 1$). As discussed in previous section, the compressibility diverges as $(1-n)^{-1}$ when the system approaches the Mott insulator from the metallic side at zero temperature. The characteristic properties for χ_c summarized above are due to divergent term in the density of states for the charged hole which is accompanied with the formation of the Hubbard gap.

The specific heat C has the similar anomaly as the compressibility near the half filled case. In order to see it more clearly, we plot the quantity C/T as a function of the

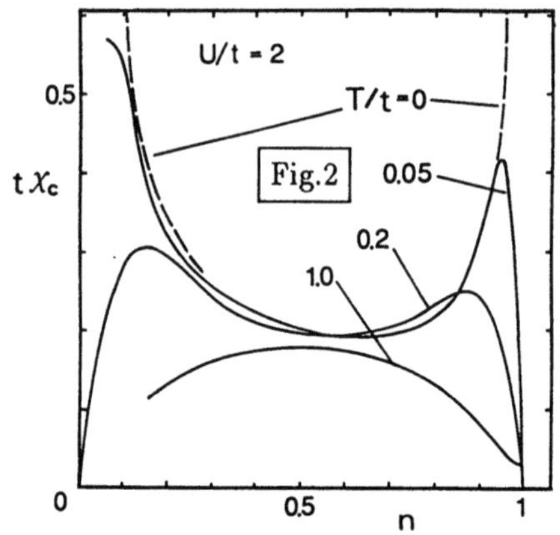

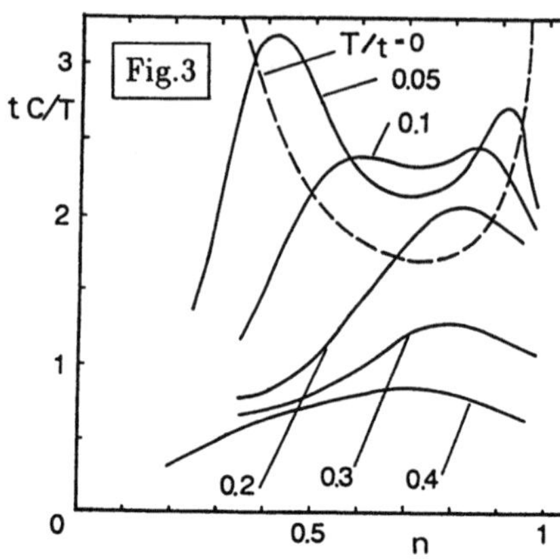

Fig.2. Compressibility χ_c as a function of n : $U/t = 2$. It is seen that χ_c near the half filled case is considerably enhanced with decreasing temperature.

Fig.3. Specific heat coefficient C/T as a function of n : $U/t = 2$. The enhancement due to the electron correlation exists near the half filled case, as in the compressibility.

concentration n in Fig.3. It is seen that as the temperature is decreased C/T is enhanced near the half filled case. At low temperatures, we can express the specific-heat coefficient in terms of the compressibility and the spin susceptibility, $C/T = (\pi/6)(\beta\tilde{\chi}_c + \tilde{\chi}_s)$ with $1 \leq \beta \leq 2$, where the quantities $\tilde{\chi}_c$ and $\tilde{\chi}_s$ are normalized such that $\tilde{\chi}_c = \tilde{\chi}_s = 1$ at $U = 0$ and $n = 1$. In the vicinity of the half filled case the specific heat coefficient is proportional to the compressibility, $C/T \sim (\pi/3)\tilde{\chi}_c$, because $\beta \sim 2$ and $\chi_c \gg \chi_s$ in this region. It is easily understood from this fact that the large enhancement of quantity C/T has the same origin of that for the compressibility (note that the spin susceptibility does not have any anomaly mentioned above [11,12]).

5. Summary

We investigated the charge excitations in the one-dimensional Hubbard model by means of the Bethe Ansatz solution. It was shown that the density of states for the charged hole is considerably renormalized near the Fermi level in the presence of the Hubbard gap and it has a divergent term in the vicinity of the gap edge. We pointed out analytically that the charged hole in the one-dimensional Mott insulator behaves like a band-edge hole with the characteristic energy scale $t^* = t^2/\Delta_g$ in the small U region while it behaves like the spinless fermion for the large U region. Furthermore, the comparison with the attractive case suggests more clearly that the divergent behaviour near the gap edge is a direct evidence for the formation of the Hubbard gap. The universal behaviour in the small U region (one characteristic energy scale $t^* = t^2/\Delta_g$) is understood rather naturally in the light of the pair breaking excitations in the weak coupling case with the attractive interaction.

In terms of the quasi-particle picture in the Bethe Ansatz method, the logarithmic $4k_F$ anomaly is caused by the nesting effect of the charged particle introduced here. In this sense, the one-dimensional Mott insulator may be referred to as the state in which the $4k_F$ nesting anomaly of the charged particle is stabilized by the commensurability condition $n = 1$, with the total spin of the system kept in the singlet state. The Hubbard gap accompanied with the insulating phase gives rise to the dispersion anomaly of the charged particle, which results in the somewhat peculiar aspects of the charge excitations: divergent behaviour of $D_c(\varepsilon)$ and large enhancement of the susceptibility $\chi_c(Q)$ by the factor $(1-n)^{-1}$ for $Q \sim 4k_F$ as well as for $Q \sim 0$ in the vicinity of the Mott insulator.

The present approach based on the quasi-particle picture in the the Bethe Ansatz method is very useful to understand the characteristics of the one-dimensional Hubbard model [15]. For instance, we have shown here that the $4k_F$ anomaly of the charged hole plays a crucial role for the charge excitations. We think that the enhancement expected for the ordinary charge susceptibility at $4k_F$ reflects the one discussed here for the charged hole [16,18,19]. The formulation of the dynamics for the charged hole is possible by extending the procedure proposed here, which may give useful informations about the dynamical properties concerning the charge excitations.

Acknowledgments

The authors express their sincere thanks to T. Usuki for useful discussions.

References

[1] J. G. Bednorz and K. A. Muller: Z. Phys. **B64** (1986) 189.

[2] C. W. Chu, P. H. Hor, P. L. Meng, L.Gao, Z. J. Huang and Q. Wang: Phys. Rev. Lett. **58** (1987) 405.

[3] P. W. Anderson: Science **235**(1987) 1196

[4] L. H. Lieb and F. Y. Wu: Phys. Rev. Lett. **20** (1968) 1445.

[5] M. Takahashi: Prog. Theor. Phys. **42** (1969) 1098.

[6] A. A. Ovchinnikov: Soviet Physics JETP **30** (1970) 1160.

[7] I. A. Musurkin and A. A. Ovchinnikov: Soviet Physics Solid State **12** (1971) 2031.

[8] M. Takahashi: Prog Theor. Phys. **47** (1972) 69.

[9] H. Shiba: Phys. Rev. **B6** (1972) 930.

[10] C. F. Coll III: Phys. Rev. **B9** (1974) 2150.

[11] T. Usuki, N. Kawakami and A. Okiji: Phys. Lett. **135A** (1989) 476.

[12] N. Kawakami, T. Usuki and A. Okiji: Phys. Lett.(1989) in press.

[13] N. Kawakami and A. Okiji: submitted to Phys. Rev. **B** (1989).

[14] C. N. Yang: Phys. Rev. Lett. **19** (1967) 1312.

[15] This specification of excitations is similar to the one discussed in the RVB theory, see [3] and also S. Kivelson, D. Rokhsar and J. Sethna, Phys. Rev. **B35** (1987) 8865.

[16] J. E. Hirsch and D. J. Scalapino: Phys. Rev. **B27** (1983) 7169.

[17] See for the thermodynamics of the Anderson model, A. Okiji and N. Kawakami, J. Appl. Phys. **55** (1984) 1931.

[18] V. Emery, Phys. Rev. Lett. **37** (1976) 107.

[19] J. B. Torrance, Phys. Rev. **B17** (1978) 3099.

On the Application of Projection Techniques to the Electron Correlation Problem

K.W. Becker and P. Fulde

Max-Planck-Institut für Festkörperforschung,
D-7000 Stuttgart 80, Fed. Rep. of Germany

Expressions for the ground-state energy and for correlation functions at zero temperature are presented which are suitable for the application of projection techniques. They can be used for weak as well as strong correlations and allow for a unified approch to the electronic correlation problem. Three examples are presented. Two of them concern the ground-state energy of a two-band Hubbard Hamiltonian and of a two-dimensional Heisenberg antiferromagnet, respectively. As a third example the motions of a hole in a one-dimensional quantum antiferromagnet is treated.

1 Introduction

Considerable experience has been gained with respect to treating and understanding electron correlations when the latter are weak. This holds true for molecules [1] as well as solids [1,2]. With respect to strong electron correlations the situation is quite different. Here we are still at the very beginning of understanding them and most of the calculations which have been done are based on uncontrolled approximations. Because the new high-temperature superconductors are materials in which electron correlations are strong, and because it is difficult to conceive that an understanding of these materials is possible without being able to handle them, strong electron correlations are presently under intense investigations.

The origin of the difficulty of treating strong correlations is easily traced. Assume a Hamiltonian of the form $H = H_0 + H_1$ with H_1 small as compared with H_0. Then one is interested in the effects of the operator

$$U = exp[-\lambda H] \tag{1}$$

on different states. When $\lambda = it$ this is the time evolution operator while for $\lambda = (k_B T)^{-1}$ this is the statistical operator. Traditionally U is decomposed into the form [3]

$$U = e^{-\lambda H_0} T_\tau e^{-\int_0^\lambda d\tau H_1(\tau)} \tag{2}$$

where T_τ is the "time" ordering operator and

$$H_1(\tau) = e^{\tau H_0} H_1 e^{-\tau H_0}. \tag{3}$$

When the exponential containing $H_1(\tau)$ is expanded, one ends up with expressions of the form

$$\int_0^\lambda d\tau_1 ... \int_0^\lambda d\tau_n \langle T_\tau H_1(\tau_1) ... H_1(\tau_n) \rangle . \quad (4)$$

Here $\langle .... \rangle$ stands for an average with respect to the ground state of H_0 or, at finite temperatures, for a thermodynamic ensemble average with respect to H_0. Averages of the form of (4) can be linked to Feynman diagrams and are easily evaluated provided Wick's theorem holds. This is the case when H_0 is a single electron operator. This is not the case when the correlations are strong. Then H_0 contains the strong Coulomb repulsions [4] or spin operators, which do not fulfill simple commutation relations.

One way to reestablish Wick's theorem is by introducing auxiliary fields. A well-known example is the slave boson field [5,6], which is used in the theory of heavy fermion systems. After it is introduced, the fermion operators fulfill again the usual anticommutation relations, but the prize is that one must handle in addition the auxiliary boson field, which interacts with the fermion fields.

In order to avoid the above mentioned difficulties, a different approach is chosen here from the very beginning. Instead of the decomposition (2) the following one is used as a starting point [7]

$$U = e^{-\lambda(H_1+L_0)} e^{-\lambda H_0}. \quad (5)$$

Here L_0 is a Liouville - or superoperator, which acts on other operators A according to

$$L_0 A = [H_0, A]_- . \quad (6)$$

Based on Eq. (5) an expression for the ground-state energy can be derived, which is in a form appropriate for applying projection techniques. The same holds true for the one-particle Green's function or any other correlation function.

Three applications of the above formalism will be presented. Two refer to the ground-state energy of a two-band Hubbard Hamiltonian at half-filling and of a two-dimensional Heisenberg antiferromagnet, respectively. In a third example the spectral function of a hole in a one-dimensional antiferromagnet is computed.

2 Formalism

Starting point is a Hamiltonian of the form

$$H = H_0 + H_1. \quad (7)$$

The eigenstates and eigenvalues of H_0 are supposed to be known. The ground state and its energy are denoted by $| \phi_0 \rangle$ and ϵ_0, respectively. It is assumed that the ground state is nondegenerate, but the theory can be also formulated for degenerate ground states.

By starting from the decomposition (5) it was shown in Ref. [7, 8] that the shift δE_0 in the ground-state energy due to H_1 is given by

$$\delta E_0 = \langle \phi_0 \mid H_1 - H_1 \frac{1}{H_1 + L_0} H_1 \mid \phi_0 \rangle^c \tag{8}$$

Here L_0 is again the Liouville operator belonging to H_0. The index c in Eq. (8) implies that the cumulant of that expression must be taken. The concept of cumulants was applied by Kubo [9] and we refer to that paper for details.

Eq. (8) can be rewritten by introducing the following metric in operator - or Liouville space

$$(A \mid B) = \langle \phi_0 \mid A^+ B \mid \phi_0 \rangle^c. \tag{9}$$

Because of the cumulants this is not a scalar product, i.e. it is not necessarily non-negative.

This way one finds

$$\delta E_0 = (H_1 \mid \Omega) \tag{10}$$

where

$$\mid \Omega) = \mid (1 - \frac{1}{H_1 + L_0} H_1)) \tag{11}$$

is analogous to the wave - or Möller operator and may be used to characterize the exact ground state $\mid \psi_0 \rangle$.

In deriving Eq.(10) nothing has been required about the division (7) of H into H_0 and H_1. The equation holds equally well when the Coulomb interactions are contained in H_0 or in H_1, respectively. No time dependent operators nor Wick's theorem need to be considered. Therefore Eq. (10) can be used for weakly as well as strongly correlated systems, which is a major advantage. For weakly correlated systems, H_0 is the self-consistent field Hamiltonian H_{SCF}, while H_1 contains the residual interactions. Then δE_0 coincides with the correlation energy and standard quantum chemical methods like the Local Ansatz or the Coupled Electron Pair Approximation ($CEPA$) can be rederived [10]. When Eq. (12) is expanded in powers of H_1 one obtains

$$\delta E_0 = \sum_{n=0}^{\infty} (-1)^n (H_1 \mid (\frac{1}{L_0} H_1)^n). \tag{12}$$

This is a generalization of Goldstone's linked cluster expansion [11] for arbitrary Hamiltonians H_0, H_1, i.e. it also applies when H_0 is a two-particle operator.

The above formalism can be generalized to correlation functions of the form

$$G_{AB}(\tau) = \langle \psi_0 \mid A^+ e^{-\tau(H - E_0)} B \mid \psi_0 \rangle - \\ - \langle \psi_0 \mid A^+ \mid \psi_0 \rangle \langle \psi_0 \mid B \mid \psi_0 \rangle \tag{13}$$

where E_0 is the ground-state energy and τ is an imaginary time. For the Laplace transform $G_{AB}(z)$, the form

$$G_{AB}(z) = (A\Omega \mid \frac{1}{z - L_0 - H_1} B\Omega) \tag{14}$$

can be derived [12].

Eqs. (10) and (14) are of a form which is suitable for applying projection techniques [13,14]. In the following this method will be used repeatedly.

3 Applications: ground-state energies

As a first example the ground-state energy of a two-band Hubbard Hamiltonian at half-filling is calculated [7]. It is supposed to describe the $(CuO_2)_n$ planes in La_2CuO_4. For H_0 the following form is chosen

$$\begin{aligned} H_0 &= \epsilon_d \sum_{i\sigma} n^d_{i\sigma} + \epsilon_p \sum_{j\sigma} n^p_{j\sigma} + H' \\ H' &= -4 \sum_{j,i\neq\ell} \frac{t^2_{ij} t^2_{j\ell}}{(U_p + \Delta)\Delta^2} \{ \frac{n^d_i n^d_\ell}{4} - \underline{S}_{id} \cdot \underline{S}_{\ell d} \}. \end{aligned} \tag{15}$$

Here ϵ_d, ϵ_p are the orbital energies of the Cu $3d$ and O $2p$ holes, and $\Delta = \epsilon_p - \epsilon_d$. The hybridization matrix element of the two orbitals is t_{ij}, where i refers to a copper and j to an oxygen site. Furthermore $n^d_{i\sigma}(n^p_{j\sigma})$ and $\underline{S}_{id}(\underline{S}_{jp})$ are the density (for spin σ) and spin operator of the $d(p)$ holes, respectively. It is $n^d_i = \sum_\sigma n^d_{i\sigma}$. The Coulomb interaction of the p holes is U_p, while the one for the d holes has been assumed to be infinitely large. The Hamiltonian H_1 is of the form

$$H_1 = H_t + H_p - H' \tag{16}$$

where

$$\begin{aligned} H_t &= \sum_{ij\sigma} t_{ij} \{ \hat{c}^+_{j\sigma} \hat{d}_{i\sigma} + \hat{d}^+_{i\sigma} \hat{c}_{j\sigma} \} \\ H_p &= -2 \sum_{ij} \frac{t^2_{ij}}{U_p + \Delta} \{ \frac{n^d_i n^p_j}{4} - \underline{S}_{id} \cdot \underline{S}_{jp} \} \\ &- \sum_{i\neq m, j\sigma} \frac{t_{ij} t_{jm}}{U_p + \Delta} \{ \hat{d}^+_{i\sigma}\, n^p_{j-\sigma}\, \hat{d}_{m\sigma} - \hat{d}^+_{i\sigma} S^{-\sigma}_{jp} \hat{d}_{m-\sigma} \} \end{aligned} \tag{17}$$

The Hubbard operators

$$\begin{aligned} \hat{d}_{i\sigma} &= d_{i\sigma}(1 - n^d_{i-\sigma}) \\ \hat{c}_{j\sigma} &= c_{j\sigma}(1 - n^p_{j-\sigma}) \end{aligned} \tag{18}$$

ensure that the number of d and p holes on a site is not larger than one (the $d_{i\sigma}, c_{j\sigma}$ are destruction operators for d and p holes, respectively).

The superexchange Hamiltonian H' has been added to H_0 and subtracted again in H_1 in order to have a good unperturbed starting Hamiltonian.

The ground state of H_0 contains one hole per Cu site which are interacting via H'. The corresponding energy is $\epsilon_0 = N(\epsilon_d - \delta\epsilon_0)$, where N is the number of sites and $\delta\epsilon_0$ is the contribution due to H'. The computation of the latter is a separate problem (see below). When the projection technique is applied to Eq. (10) one obtains δE_0 in form of a continued fraction [7]

$$\delta E_0 = N(\delta\epsilon_0 - \cfrac{4x^2\Delta}{1 - \cfrac{\delta\epsilon_0}{4x^2\Delta} + \cfrac{5x^2}{1 + \cfrac{5.93x^2}{1+\delta}}}) \tag{19}$$

where $x = t/\Delta$ and it has been assumed that t is the nearest neighbor hopping matrix element. The continued fraction has been terminated by writing an undetermined term δ in its place. The radius of convergency of the continued fraction is $x \to \infty$, but the smaller x is, the more rapidly convergency takes place. Eq. (19) contains all terms to order $(t/\Delta)^6$. By differentiating Eq. (19) with respect to the chemical potential one can determine $\langle n_j^P \rangle$, i.e. the number of p holes. While for $t = 0$ it is $\langle n_j^P \rangle = 0$, one finds for $t/\Delta = 0.3$ that $\langle n_j^P \rangle \simeq 0.13$.

The second example concerns the ground state energy of a $2D$ Heisenberg antiferromagnet. In that case it is convenient to choose for H_0 the Ising part, i.e.

$$H_0 = -J\sum_{\langle ij \rangle} S_i^z S_j^z + h\sum_i S_i^z \tag{20}$$

while

$$\begin{aligned} H_1 &= \frac{J}{2}\sum_{\langle ij \rangle}(S_i^+ S_j^+ + S_i^- S_j^-) \\ &= H_I + H_I^+. \end{aligned} \tag{21}$$

The spins on sublattice B have been rotated so that in the ground state the spins on both sublattices point in the downward direction. Furthermore, h is an external field corresponding to the staggered field in the unrotated frame. The function

$$\phi(z) = (H_1 \mid \frac{1}{z - (L_0 + H_1)} H_1) \tag{22}$$

is evaluated by projection techniques. According to Eq. (10) it is $\delta E_0 = \phi(0)$. One finds [8]

$$\phi(z) = \frac{2N(J/2)^2}{z - (3J + 2h) - M(z)/(H_I \mid H_I)} \tag{23}$$

where the memory function $M(z)$ is given by

$$M(z) = (H_I^2 \mid \frac{1}{z - Q(L_0 + H_1)Q} H_I^2) \tag{24}$$

$$Q = 1 - \frac{\mid H_I)(H_I \mid}{(H_I \mid H_I \mid}. \tag{25}$$

When H_1 in the denominator of $M(z)$ is neglected, i.e. $M(z) \to M_0(z)$, one finds

$$M_0(z) = 2N(J/2)^4\{\frac{8}{z - 4J - 4h} + \frac{28}{z - 5J - 4h} - \frac{46}{z - 6J - 4h}\} \tag{26}$$

This shows that one does not obtain a simple continued fraction for δE_0 like in the previous case (see Eq. (19)), but instead one, which splits into several different continued fractions.

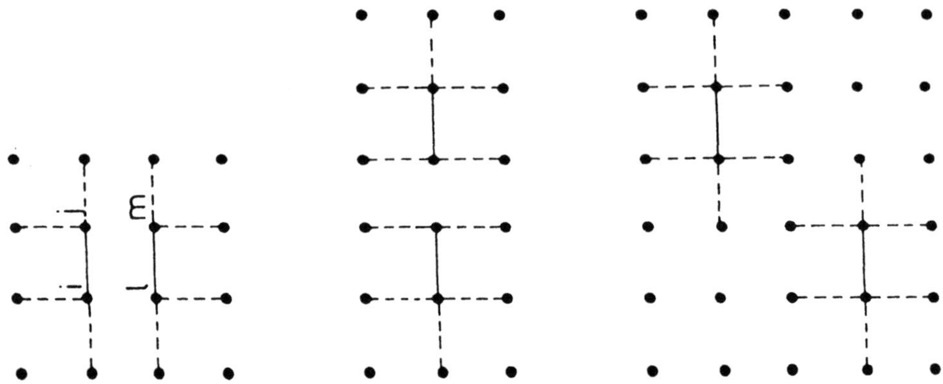

Fig. 1:
Different classes of configurations generated by the action of H_I^2 on the ground state. Solid lines indicate bonds with spin flips. Dashed lines mark bonds with energies affected by the spin flips.

This is so, because $H_I^2 \mid \phi_0\rangle$ is the sum of three types of configurations, which vary with respect to the number of affected bonds due to the spin flips generated by H_I^2. They are shown in Fig. 1. One can continue by evaluating again $M(z)$ by projection techniques. This yields a new memory function in each of the three fractions of $M_0(z)$. The results for the ground-state energy can be summarized as follows.

When one sets $M(z) = 0$, then

$$E_0 = \epsilon_0 + \delta E_0 = \frac{-2NJ}{3} \tag{27}$$

where $\epsilon_0 = -2NJ/4$ has been used. When the approximation $M(0) = M_0(0)$ is made, one obtains

$$E_0 = -2NJ \cdot 0.33287. \tag{28}$$

If the next step is made, i.e. when the new memory functions in each of the fractions in $M_0(z)$ are evaluated to lowest order in H_1, one obtains

$$E_0 = -2NJ \cdot 0.33372. \tag{29}$$

This can be compared with the result of 6th order perturbation theory [15], i.e. $E_0^{(6)} = -2NJ \cdot 0.33366$ and that of the most detailed Monte Carlo calculations [16], i.e. $E_0^{MC} = -2NJ \cdot 0.3367$. By differentiating $E_0(h)$ with respect to h one can determine the ground-state sublattice magnetization m. One finds $m = 0.36$ in units of μ_B, as compared with $m = 1/2$ in the absence of H_1. The series for m seems to be rapidly convergent. Our result is therefore at variance with that of other investigations, based on Monte Carlo calculations in which a value close to $m = 0.31$ was suggested. (Note, that the value obtained within spin-wave theory is m=0.303). However, in that approach it remains unclear how one has to relate m to the numerically computed quantity

$$C_L = \frac{1}{N}\sum_i \langle S_i^z S_{i+L}^z \rangle . \tag{30}$$

L is the linear dimension of the N-site system, i.e. $L = N^{1/2}$. Often it is assumed that $m = (3C_L)^{1/2}$, but there is no justification for that. For a penetrating analysis of the problem see Ref. [18].

4 Application: motion of a hole in a 1D antiferromagnet

In the following the spectral density of a hole in a one-dimensional, anisotropic antiferromagnet is discussed by applying Eq.(14). The Hamiltonian is separated into

$$H_0 = J_\parallel \sum_{\langle ij \rangle} [S_i^z S_j^z + \frac{1}{4}(1 - n_i n_j)] \tag{31}$$

and

$$\begin{aligned} H_1 &= H_\perp + H_t \\ H_\perp &= \frac{J_\perp}{2} \sum_{\langle ij \rangle} [S_i^+ S_j^- + S_i^- S_j^+] \\ H_t &= -t \sum_{\langle ij \rangle, \sigma} [\hat{c}_{i\sigma}^+ \hat{c}_{j\sigma} + \hat{c}_{j\sigma}^+ \hat{c}_{i\sigma}]. \end{aligned} \tag{32}$$

The ratio $J_\parallel / J_\perp$ depends on the degree of anisotropy. As before, $\hat{c}_{i\sigma} = c_{i\sigma}(1 - n_{i-\sigma})$, i.e. double occupancies of sites are excluded. The ground state of H_0 is the $1D$ Néel state.

In order to determine the spectral density of a moving hole, the following correlation function has to be evaluated [19]

$$R(\underline{k}, z) = (\hat{c}_{\underline{k}\uparrow}\Omega \mid \frac{1}{z - L_0 - H_1} \hat{c}_{\underline{k}\uparrow}\Omega), \tag{33}$$

with $z = \omega + i\eta$. Hereby the approximation is made that $\mid \Omega)$ is limited to that part of the Liouville- (or operator-) space, which is spanned by the elements $\mid 1)$ and $\mid H_\perp)$. Thus the ansatz is made

$$\mid \Omega) = \mid 1) - (\frac{2}{J_\perp})\alpha \mid H_\perp). \tag{34}$$

The parameter α is defined through

$$\alpha = \frac{J_\perp}{2} \frac{1}{(H_\perp \mid H_\perp)} \cdot (H_\perp \mid \frac{1}{L_0 + H_1} H_\perp). \tag{35}$$

According to Eq. (10), α determines the energy shift δE_0. For the isotropic case one obtains a value of $\alpha = \ell n 4 - 1 \simeq 0.39$ from the exact Bethe result. When in addition $L_0 + H_1$ in Eq. (32) is replaced by the dominant part H_t, one can calculate $R(\underline{k}, z)$ analytically by using again the projection method. It is found [19] that

$$-ImR(k,x) = \frac{1}{2t}\frac{1}{\sqrt{4-x^2}}\{1 - \alpha[2x\cos(ka) + 2(x^2-2)\cos(2ka)] \tag{36}$$

$$+\alpha^2[2x\cos(ka) + (x^2-2)\cos(2ka) - 2(3x - x^3)\cos(3ka) +$$

$$+(x^4 - 4x^2 + 2)\cos(4ka)]\} \quad ; \quad |x| \langle 2$$

Here $x = \omega/t$. For $|x| \rangle 2$ the spectral function vanishes. One notices that for $\alpha = 0$ the result reduces to that of Ref. [20]. But when the effect of $H_\perp$ on the ground state is taken into account the spectral function is drastically modified (see Fig. 2). In distinction to the case of $\alpha = 0$ the hole is delocalized. The spectral function in then k dependent and changes its form. Due to the approximation (33) it suffers from the defect though, that for larger values of the anisotropy parameter it may become negative in some frequency range. Presumably this shortcoming will be removed in the next higher approximations in $H_\perp$. Related work is under way.

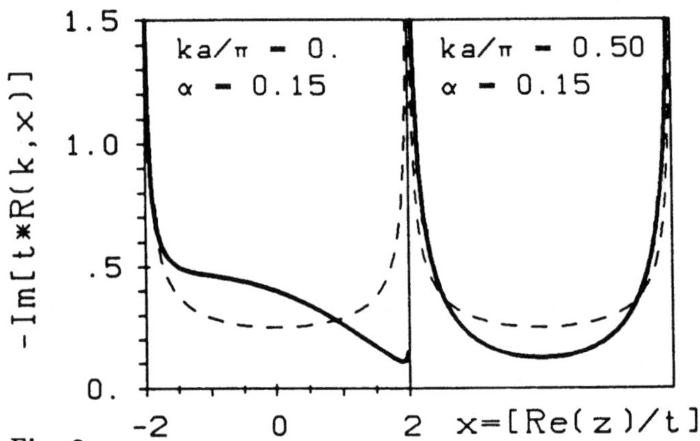

Fig. 2:
Thick solid lines: One-hole spectral function as given by Eq. (36) for spin-fluctuation parameter $\alpha = 0.15$ and two different k-vectors. Dashed lines: k-independent result for α=0 which coincides with that of Ref. [20].

Acknowledgement

We would like to thank Dr. W. Brenig and Dr. H. Won for many valuable discussions.

References

[1] see e.g. "Methods of Electronic Structure Theory" edited by H.F. Schaefer III, Plenum Press, New York, 1977

[2] M.V. Gandulia Pirovano, G. Stollhoff, P. Fulde and K. Bohnen, Phys. Rev. B 39, 5156 (1989)

[3]R.P. Feynmann "Statistical Mechanics" W. A. Benjamin, Inc. (Reading, Mass.) 1972
[4] J. Hubbard, Proc. Royal Soc.A 276, 238 (1963)
[5] S. E. Barnes, J. Phys. F 6, 1375 (1976), ibid. F 7, 2637 (1977)
[6] P. Coleman, Phys. Rev. B 29, 3035 (1984)
[7] K.W. Becker and P. Fulde, Z. Phys. B 72, 423 (1988)
[8] K.W. Becker, H. Won and P. Fulde, Z. Phys. B (in print)
[9] R. Kubo, J. Phys. Soc. Japan 17. 1100 (1962)
[10] K.W. Becker and P. Fulde, submitted for publication
[11] J. Goldstone, Proc. Roy. Soc. (London) A 239, 267 (1957)
[12] K.W. Becker and W. Brenig (to be published)
[13] E. Fick and G. Sauermann "Quantenstatistik dynamischer Prozesse" H. Deutsch Verlag, Frankfurt/M., 1986
[14] D. Forster "Hydrodynamic Fluctuations, Broken Symmetry, and Correlation Functions" Benjamin Publ. (Reading, Mass.) 1975
[15] M. Parinello and T. Arai, Phys. Rev. 10, 265 (1974)
[16] J. D. Reger and A.P. Young, Phys. Rev. B 37, 5978 (1988)
[17] S. Liang, B. Doucot and P.W. Anderson, Phys. Rev. Lett. 61, 365 (1988)
[18] T. Kaplan, P. Horsch and W. von der Linden (submitted for publication)
[19] W. Brenig and K.W. Becker, Z. Phys. B (in print)
[20] W. F. Brinkmann and T. M. Rice, Phys. Rev. B 2, 1324 (1970)

Slave-Fermion Mean Field Theory of the Hubbard Model

D. Yoshioka

Department of Physics, Kyushu University,
4-2-1 Ropponmatsu, Chuoh-ku, Fukuoka 810, Japan

Abstract. Square lattice Hubbard model is investigated near half-filled in the limit of large U. A slave-fermion scheme is used, where electron operators are replaced by bose operators with auxiliary fermi operators. The ground state by a mean field approximation with RVB order parameter for spin shows long-range antiferromagnetic order, which is incommensurate except for half-filled case.

1. Introduction

Since Anderson's proposal that the high-T_c superconductivity can be explained by resonating valence bond (RVB) state of the two-dimensional Hubbard model [1], there have been a lot of investigations on this model. Originally single electron state is considered for each unit cell. This is the single band Hubbard model. It is known that this model reduces to so called *t-J* model when the on-site Coulomb repulsion U is large [2]. On the other hand Emery pointed out the importance of the oxygen sites besides copper sites [3]. This means that we must consider two-band Hubbard model. However, it has been clarified that this model also reduces to the *t-J* model in certain situations [4]. Thus the *t-J* model deserves extensive investigation.

The *t-J* model is a quite simple model, but it is quite hard to draw a reliable conclusion from it. Here we show results of an investigation on this model by a simple mean field theory [5,6]. There have been many mean field theories on this model [5-9]. In most of these, slave-boson scheme has been used [8,9]. In this scheme auxiliary bose particles which keep track of vacant sites are introduced. However, what we consider in this paper is not such a scheme. Instead, we show results by an investigation where slave-fermion scheme is used. In this scheme spin freedom of an electron is expressed by bose operators and charge freedom is expressed by (slave-)fermion operators.

In section 2. the model and the method are briefly explained. In section 3. we show the results of this scheme when applied to the half-filled

Springer Series in Solid-State Sciences, Vol. 89 **Strong Correlation and Superconductivity**
Editors: H. Fukuyama · S. Maekawa · A.P. Malozemoff

case [5]. The application of the present scheme to this case is intended as a test of this method. Here the t-J model is equivalent to the antiferromagnetic Heisenberg model. Comparison with established results shows that the present slave-fermion scheme gives reasonable results in this case. This is in contrast with the fact that the slave-boson scheme gives results which do not describe the system even qualitatively. Based on this success of the slave-fermion mean field theory at half-filled case, we have done the mean field theory at general fillings [6]. The results are shown in section 4. Near half-filled the ground state shows long-range antiferromagnetic spin-spin correlation. The period of the correlation deviates from that of the commensurate Néel order to an incommensurate one as the system deviates from the half-filled case. We discuss this result and superconductivity in the final section.

2. Method

The Hubbard model reduces to the t-J model, when the on-site Coulomb repulsion U is large. We consider the t-J model near half-filled case. The Hamiltonian is given as follows:

$$\begin{aligned}\mathcal{H} = & -t \sum_{<i,j>} \sum_{\sigma} [e_i e_j^\dagger s_{i,\sigma}^\dagger s_{j,\sigma} + \text{h.c.}] \\ & - \frac{J}{2} \sum_{<i,j>} \sum_{\sigma} [s_{i,\sigma}^\dagger s_{j,-\sigma}^\dagger s_{j,-\sigma} s_{i,\sigma} - s_{i,\sigma}^\dagger s_{j,-\sigma}^\dagger s_{j,\sigma} s_{i,-\sigma}] \\ & - \mu_e \sum_i e_i^\dagger e_i - \mu_s \sum_i \sum_\sigma s_{i,\sigma}^\dagger s_{i,\sigma}.\end{aligned}$$

Here i and j are site indices, $\sum_{<i,j>}$ means summation over nearest neighbor bonds, $\sigma =\uparrow$ or $\downarrow$, and $J = 4t^2/U$. The electron annihilation operator $c_{i,\sigma}$ of the original Hubbard model has been replaced by bose operators $s_{i,\sigma}$ and by fermi operators e_i and d_i as $c_{i,\sigma} = e_i^\dagger s_{i,\sigma} + \text{sgn}(\sigma) d_i s_{i,-\sigma}^\dagger$ where sgn($\uparrow$)=1 and sgn($\downarrow$)=-1. The operator $d_i^\dagger$ makes the i-th site doubly occupied, and $e_i^\dagger$ makes it empty. Since we consider cases where the system is half-filled or less than half-filled, the operators d_i and $d_i^\dagger$ have dropped out of the Hamiltonian. The replacement of the electron operator by these operators is valid only when the condition $e_i^\dagger e_i + \sum_\sigma s_{i,\sigma}^\dagger s_{i,\sigma} + d_i^\dagger d_i = 1$ is satisfied at each site. However, in the mean field theory this condition is relaxed to be satisfied on the average. Namely we require that $\langle e_i^\dagger e_i \rangle = \delta$ and $\langle \sum_\sigma s_{i,\sigma}^\dagger s_{i,\sigma} \rangle = (1 - \delta)$, where δ is the hole concentration. The last two terms of the Hamiltonian with the chemical potentials μ_e and μ_s are added to enforce these conditions.

Three kinds of order parameters are introduced to decouple this Hamiltonian. They are

$$\Delta_z = \frac{1}{2}\langle s_{i+\hat{z},\uparrow}s_{i,\downarrow} - s_{i+\hat{z},\downarrow}s_{i,\uparrow}\rangle,$$

$$\kappa_{z,\sigma} = \langle s_{i,\sigma}^{\dagger}s_{i+\hat{z},\sigma}\rangle,$$

and

$$\epsilon_z = \langle e_{i+\hat{z}}^{\dagger}e_i\rangle.$$

Here and hereafter z stands for x or y, and $i+\hat{x}(i+\hat{y})$ indicates a site next to the site i in the $x(y)$ direction. After the docoupling, the Hamiltonian is easily diagonalized through the consecutive fourier and Bogoliubov transformations. Then these order parameters are determined self-consistently at a given hole concentration δ.

3. Half-filled case

For the half-filled case ($\delta = 0$) the Hamiltonian becomes equivalent to that of the antiferromagnetic Heisenberg model. In this case the free energy is minimized by the following choice of the order parameters: $\Delta_x = \Delta_y \equiv \Delta$, $\kappa_{z,\sigma} = 0$, and $\epsilon_z = 0$. The order parameter Δ becomes finite below $T_{\rm RVB} = 0.91081J$. Of course, this transition is a consequence of the mean-field theory, and it is not real. Fluctuation in the phase of Δ will destroy the order. $T_{\rm RVB}$ merely gives a temperature below which the amplitude of Δ becomes appreciable. However, since Δ is an expectation value of the destruction operator of nearest neighbor singlet pair, we could consider the system below $T_{\rm RVB}$ to be in the RVB state.

The Hamiltonian has been diagonalized by the Bogoliubov transformation of the fourier transform of $s_{i,\sigma}$. The new bosons which appear in the diagonalized Hamiltonian Bose-condense at $T = 0$. The condensation occurs at momentum $\mathbf{k} = \mathbf{K}_{\pm} = \pm(\pi/2, \pi/2)$. The fraction of the Bose-condensate is determined by a definite integral as follows:

$$\begin{aligned} n_{\rm B} &= 2 - \frac{1}{4\pi^2}\int_{-\pi}^{\pi} {\rm d}k_x \int_{-\pi}^{\pi} {\rm d}k_y [1 - \frac{1}{4}(\sin k_x + \sin k_y)^2]^{-1/2} \\ &= 0.6068. \end{aligned}$$

The order parameter at $T = 0$ is also given by a definite integral as follows:

$$\begin{aligned}\Delta &= \frac{1}{16\pi^2}\int_{-\pi}^{\pi} dk_x \int_{-\pi}^{\pi} dk_y \frac{(\sin k_x + \sin k_y)^2}{\sqrt{4-(\sin k_x + \sin k_y)^2}} + \frac{1}{2}n_{\rm B} \\ &= 0.5790.\end{aligned}$$

The order parameter Δ gives the ground state energy, and $n_{\rm B}$ gives the long-range antiferromagnetic spin correlation as follows:

$$\langle \mathbf{S}_i \cdot \mathbf{S}_{i+\hat{x}} \rangle = -\frac{3}{2}|\Delta|^2 = -1.5 \times 0.3352,$$

$$\langle \mathbf{S}_i \cdot \mathbf{S}_{i+\hat{R}} \rangle = (-1)^{X+Y}\frac{3}{8}n_{\rm B}^2 = -1.5 \times (-1)^{X+Y} \times 0.0921.$$

Here $\mathbf{S}_i$ is the spin operator at site i in the Heisenberg model, and site i and site $i+\hat{R}$ are separated by $\mathbf{R} = (Xa, Ya)$, $|\mathbf{R}| >> a$, where a is the lattice constant.

These values are about 1.5 times larger than results of other investigations [10-12]. However, the uniform susceptibility χ and specific heat C seem to be given correctly at low temperatures: These are given as follows:

$$\begin{aligned}\chi(T) &= \frac{\mu_{\rm s}^2}{4J}\frac{n_{\rm B}}{4\Delta + n_{\rm B}}\left[1 + \frac{1}{\pi}\frac{n_{\rm B}}{4\Delta + n_{\rm B}}T + O(T^2)\right] \\ &\simeq \frac{\mu_{\rm s}^2}{J}(0.05190 + 0.03726\frac{T}{J}),\end{aligned}$$

and

$$\frac{C}{N} = \frac{3}{2\pi}\zeta(3)(\frac{T}{J\Delta})^2 + O(T^3) \simeq 1.712(\frac{T}{J})^2,$$

where $\mu_{\rm s}/2$ is the magnetic moment of an electron, N is the total number of the lattice sites, and $\zeta(3) = 1.2021$. The susceptibility is compared with Monte Carlo data [13] in Fig.1. The agreement is quite good below $T = 0.6J$. The T^2 dependence of the specific heat also agrees with Monte Carlo data: the prefactor 1.712 is within the error bar of the Monte Carlo data 1.96 ± 0.36 [14].

At finite temperatures, antiferromagnetic spin correlation decays exponentially. The correlation length ξ is given by

$$\xi = \sqrt{2\Delta}\frac{J}{T}\exp(\frac{\pi J \Delta}{T}n_{\rm B}).$$

This temperature dependence coincides with a result of one-loop approximation of the non-linear-σ model [15], and is consistent with an experiment [16]. Thus the present method gives quite reasonable results at half-filled case.

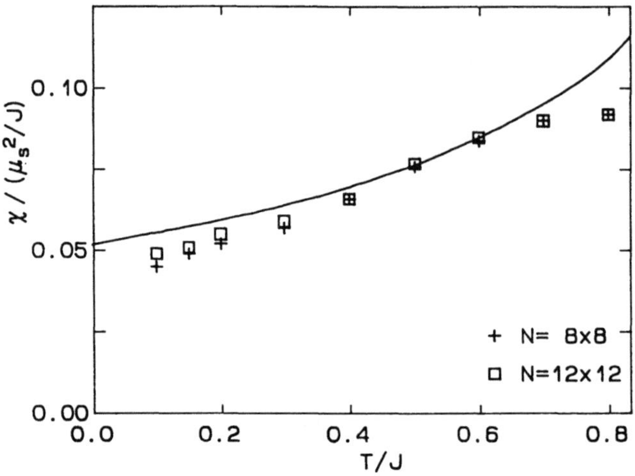

Fig.1 Uniform magnetic susceptibility at half-filled case by the present method is shown by a full line. Crosses and squares show the results of Monte Carlo calculation [14] for 8×8 and 12×12 sites systems, respectively.

4. Less than half-filled case

Based on the success of the present method at half-filled, the method is applied to less than half-filled case [6]. In this case the order parameters $\kappa_{z,\sigma}$ and ϵ_z can also become finite. We consider the case where all the $\kappa_{z,\sigma}$ are the same (equal to κ) and $\epsilon_x = \epsilon_y \equiv \epsilon$. The fermion part of the mean field Hamiltonian only consists of hopping terms. Thus the fermions are localized when $\kappa = 0$ and have finite fermi surface when κ becomes finite.

On the other hand, the boson part is diagonalized by a Bogoliubov transformation similar to that used at the half-filled case. We define transition temperatures $T_{\rm RVB}$ and T_ϵ as temperatures below which Δ and ϵ become finite, respectively. T_ϵ also gives a transition temperature below which κ becomes finite. These transition temperatures depend on the choice of the ratio t/J. The reduction of the Hubbard model to the t-J model is justified only when $U >> t$, or $t >> J$. Therefore a result for the case of $t/J \equiv U/4t = 10$ is shown in Fig.2. As seen in this figure coexistence of these three order parameters is possible. However, $T_{\rm RVB}$ is suppressed in the presence of ϵ and κ, especially. The behavior for other choices of t/J is qualitatively similar. When a smaller value of t/J is chosen, T_ϵ becomes lower, and the suppression of $T_{\rm RVB}$ becomes smaller.

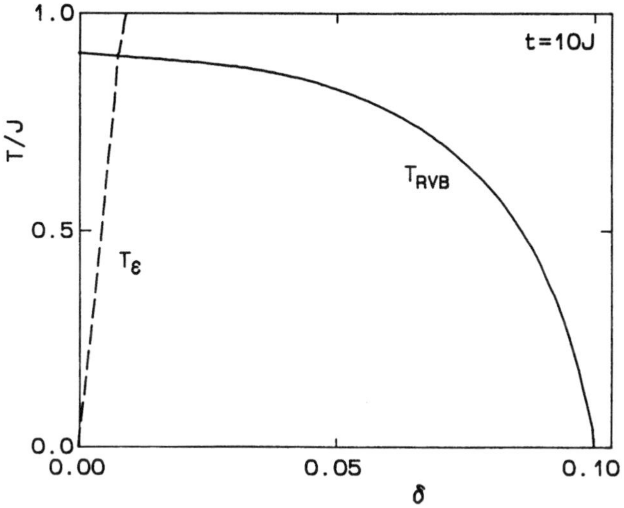

Fig.2 The transition temperatures T_{RVB} and T_ϵ are shown by solid and dashed lines, respectively. Here δ is the hole concentration, and $t = 10J$.

The hole concentration δ_c at which T_{RVB} vanishes depends on t/J. For $t/J = 10$, δ_c is 0.101. For smaller t/J, δ_c is larger. For larger values of t/J, δ_c is given approximately by $0.9317J/t$.

Like the half-filled case there is an antiferromagnetic long-range order at $T = 0$, when Δ is finite. However, in the presence of finite κ and ϵ, the Bose condensation occurs at wave vectors $\mathbf{k} = \mathbf{K}_\pm = \pm(K_0, K_0)$, where $\cos K_0 = -\tilde{t}\epsilon/\sqrt{(\tilde{t}\epsilon)^2 + (J\Delta)^2}$, and $\tilde{t} = t + J\kappa/2\epsilon$. Therefore the long-range spin-spin correlation that is given as

$$\langle \mathbf{S}_i \cdot \mathbf{S}_j \rangle = \frac{3}{8} n_{\mathrm{B}}^2 \cos[2\mathbf{K}_+ \cdot (\mathbf{r}_i - \mathbf{r}_j)],$$

has an incommensurate period. The dependence of K_0 on δ is shown in Fig.3 for $t/J = 10$. When δ is small, K_0 is given as

$$K_0 a \simeq \frac{\pi}{2} + 1.25 \frac{t}{J} \delta.$$

Thus the incommensurability depends both on t/J and δ. It is independent of the size of the Fermi surface of the holes. This result is interesting, since incommensurate antiferromagnetic correlation has been observed in $La_{1.89}Sr_{0.11}CuO_4$ [17].

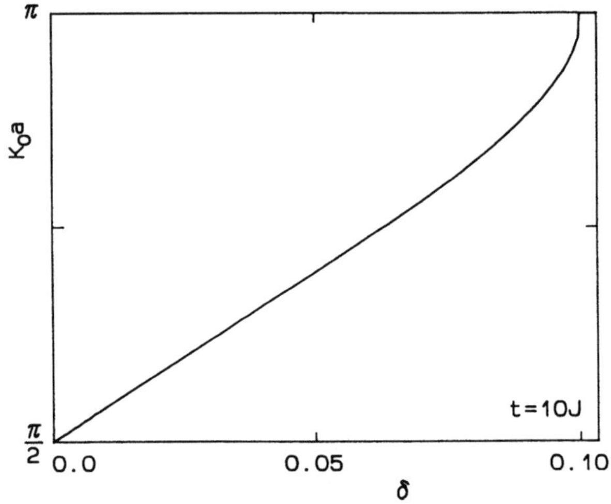

Fig.3 The wave number K_0 at which the Bose condensation occurs is shown. Here a is the lattice constant, and $t = 10J$.

5. Discussion

In this paper results of a slave-fermion mean field theoy for a nearly half-filled Hubbard model are shown. As seen in section 3 the present theory gives a reasonable results when applied to the half-filled case. These results indicate that the slave-fermion scheme should be used for the case of less than half-filled case also, if we treat the Hamiltonian by a mean field theory. The results of such a theory have been shown in section 4. There it has been shown that the ground state shows incommensurate antiferromagnetic order, when the hole concentration is small.

We do not know if this simple t-J model can well approximate the actual system or not. We also do not know appropriate values for the parameters t and J, even if the model turns out to be applicable to the actual system. Therefore comparison with experiments may be meaningless at present. However, if we compare the experimentally obtained incommensurability, 0.05Å^{-1} to 0.08Å^{-1} at $\delta = 0.11$ [17], with that of the present result, $(2K_0 - \pi/a) \simeq 2.5(t/J)\delta/a$, t/J must be around unity. Although our method will not be applicable to too large δ, we remark that when $t/J \leq 2.25$, $T_{\rm RVB}$ is finite for the whole range of δ $(0 \leq \delta \leq 1)$.

Superconductivity has not been discussed in the present slave-fermion scheme. If we consider the order parameter for the Cooper pair, $\langle c_{i,\uparrow} c_{j,\downarrow} \rangle$, in the present scheme, it should be decoupled to $\langle e_i^\dagger e_j^\dagger \rangle \langle s_{i,\uparrow} s_{j,\downarrow} \rangle$. When $T < T_{\rm RVB}$, the latter factor becomes finite, but the former factor never becomes finite in the present scheme. However, it is presumable that the singlet pairs in the RVB state become Cooper pairs for sufficiently high hole concentration [1]. We will need some extension of the present scheme to describe such a state. This is a future problem.

Acknowledgment

The author thanks Y. Okabe for providing him detailed results of their investigation. The numerical computation was done by FACOM M780 of the Computer Center of Kyushu University. This work is financially supported by Priority Areas, New Functional Materials—Design, Preparation and Control(63604014) of Grant-in-Aid for Scientific Research from the Ministry of Education, Science and Culture of Japan.

References

[1] P.W.Anderson, Science **235**, 1196 (1987).
[2] J.E.Hirsch, Phys. Rev. Lett. **54**, 1317 (1985).
[3] V.J.Emery, Phys. Rev. Lett. **58**, 2794 (1987).
[4] F.C.Zhang and T.M.Rice, Phys. Rev. B **37**, 3759 (1988).
[5] D.Yoshioka, J. Phys. Soc. Jpn. **58**, 32 (1989).
[6] D.Yoshioka, J. Phys. Soc. Jpn. **58**, 1516 (1989).
[7] G.Baskaran, Z.Zou, and P.W.Anderson, Solid State Commun. **63**, 973 (1987).
[8] Y.Suzumura, Y.Hasegawa, and H.Fukuyama, J. Phys. Soc. Jpn. **57**, 2768 (1988).
[9] G.Kotliar, Phys. Rev. B **37**, 3664 (1988).
[10] S.Liang, B.Doucot, and P.W.Anderson, Phys. Rev. Lett. **61**, 365 (1988).
[11] P.W.Anderson, Phys. Rev. **86**, 694 (1952).
[12] J.D.Reger and A.P.Young, Phys. Rev. B **37**, 5978 (1988).
[13] Y.Okabe, M.Kikuchi, and A.D.S.Nagi, Phys. Rev. Lett. **61**, 2971 (1988).
[14] Y.Okabe and M.Kikuchi, J. Phys. Soc. Jpn. **57**, 4351 (1988).
[15] S.Chakravarty, B.I.Halperin, and D.R.Nelson, Phys. Rev. Lett. **60**, 1057 (1988).
[16] G.Shirane, Y.Endoh, R.J.Birgeneau, M.A.Kastner, Y.Hidaka, M. Oda, M.Suzuki, and T.Murakami, Phys. Rev. Lett. **59**, 1613 (1987).
[17] R.J.Birgeneau, Y.Endoh, K.Kakurai, Y.Hidaka, T.Murakami, M.A.Kastner, T.R.Thurston, G.Shirane, and K.Yamada, Phys. Rev. B **39**, 2868 (1989).

Disorder and Interactions in the Hubbard Model

G.T. Zimanyi [a] *and E. Abrahams*

Serin Physics Laboratory, Rutgers University,
P.O. Box 849, Piscataway, NJ 08855, USA

Abstract. We investigate the single-band repulsive Hubbard model with random site energies as a model for the disordered electron system in the strong-interaction limit. Previous scaling treatments of this problem, pioneered by Finkel'shtein, are based on a perturbative treatment of the interactions and the scaling eventually fails as the couplings become large. We treat the disordered Hubbard model at large repulsion U by a mean-field slave-boson technique coupled with the matrix non-linear sigma model method to average over the disorder. In first approximation, we find an effective hamiltonian with an attractive coupling inversely proportional to U. A new feature is an enhancement of band narrowing due to the interplay of correlation and disorder. The scaling drives $1/U$ to weak coupling consistent with the apparent scaling of U to strong coupling in the conventional approach, to which our method is complementary. We find that the conductivity decreases monotonically as the temperature decreases, a new result.

1. Introduction

The modern scaling theory of localization, which has its origins in the work of Wegner [1] and Abrahams, Anderson, Licciardello and Ramakrishnan [2] was developed to include effects of electron-electron interaction primarily by Finkel'shtein [3] and by Castellani et al [4]. The scaling theory with interactions is somewhat complicated and we give only a brief summary here. First, in the absence of interactions [2] and in two dimensions, logarithmic singularities appear in the two-particle correlations as modified by the disorder. The conductivity decreases logarithmically as temperature decreases (or length scale increases). At dimensions greater than two, the logarithmic singularities sum to power law behavior which describes the continuous disappearance of the conductivity at a mobility edge determined by a critical value of the disorder. All of these results were formulated as a renormalization-group (RG) treatment of a matrix non-linear σ model field theory of the long wavelength disorder-coupled (particle-particle and particle-hole) diffusion modes of the system [5].

Finkel'shtein [3] extended the RG treatment of the scaling theory to include interaction effects. The disorder was treated in lowest order and all interaction contributions to the leading logarithmic behavior were summed. In $2+\varepsilon$ dimensions, different scenarios are realized depending on the nature of the disorder.

Springer Series in Solid-State Sciences, Vol. 89 Strong Correlation and Superconductivity
Editors: H. Fukuyama · S. Maekawa · A.P. Malozemoff

For non-magnetic impurity scattering, in the region of low disorder, the system scales (as $T \to 0$) to a conductor (non-zero conductivity). However, with stronger disorder, the interaction scales to strong coupling although the conductivity remains non-zero. Therefore, the RG equations break down in a region where the charge diffusion is non-zero but where the spin-diffusion constant is going to zero and the spin susceptibility is diverging, possibly signifying the onset of local-moment behavior.

The general phenomenological picture on the metallic side of the metal-insulator transition which seems to be consistent with experiment is that of itinerant electrons coexisting with local moments which arise from a large spectral weight of low-frequency spin fluctuations. The traditional physical system to which these ideas are applied is the doped semiconductor, in particular, phosphorous-doped silicon in which this description has received experimental support [6]. The failure of earlier approaches to succeed to give a complete picture is due in part to the fact that they are essentially perturbative in that the interactions are assumed to be smaller than the bandwidth.

The failure of the Finkel'shtein approach [3] to extend into the strong-interaction regime makes it natural to investigate this problem starting from the strong-coupling limit at the outset. Some progress describing the situation microscopically from the large-correlation limit has been made recently [7] within a Hartree-Fock treatment of the Hubbard model with off diagonal disorder which is treated exactly by means of a numerical calculation. We expect the strong-correlation point of view to be of interest not only for doped semiconductors near their metal-insulator transition but also, in particular, for heavy-fermion materials and high-T_C superconductors.

In the present work, we begin with a disordered Hubbard model in the large-repulsion (U) limit. The disorder appears as a random distribution of local site energies (diagonal disorder). At large U, we represent the no-double-occupancy constraint by a slave-boson field which we treat in a mean field approximation. This yields a renormalized disordered Fermi liquid with weak attractive interactions to which we apply the Finkel'shtein scaling. The approach complements earlier work [3]; the large repulsion is treated non-perturbatively from the outset and the attractive deviations from this starting point scale to weak coupling. Our results are in part consistent with the conclusions of earlier work [3,4], but there are differences which we describe below.

2. The model

We begin with the standard Hubbard model

$$H = t \sum_{<ij>,\sigma} c_{i\sigma}{}^{+} c_{j\sigma} + U \sum_i n_{i\sigma} n_{j-\sigma} + \sum_{i,\sigma} (\varepsilon_i - \mu) n_{i\sigma} \tag{1}$$

The local site energies ε_i are gaussian random variables with zero mean and rms value = W. To construct the low-temperature effective hamiltonian, we take the large-U limit and perform an expansion in t/U. At leading order we obtain the usual t - J hamiltonian with the constraint that no site have double occupancy. The projection onto the singly-occupied subspace is realized by the introduction of slave boson operators [8] b_i, b_i^+.

$$H = t\,\Sigma_{<ij>,\sigma}\, c_{i\sigma}^+ c_{j\sigma}\, b_i b_j^+ \; + \; J\,\Sigma_{<ij>}(S_i S_j - n_i n_j/4)$$
$$+ \Sigma_{i,\sigma}\,(\varepsilon_i - \mu) n_{i\sigma} - \Sigma_i\, \lambda_i (b_i^+ b_i + \Sigma_\sigma\, c_{i\sigma}^+ c_{i\sigma} - 1), \tag{2}$$

where $J = 4t^2/U$. Minimization of the free energy with respect to the Lagrange multipliers λ_i enforces the local constraints $b_i^+ b_i + n_i = 1$ $(n_i = \Sigma_\sigma c_{i\sigma}^+ c_{i\sigma})$.

The conventional mean field treatment in the slave-boson picture involves assuming, consistent with translational invariance, constant values for the b_i and the λ_i. It is equivalent to the leading term in a large-N expansion. At this level, one obtains [8], in Kondo-type problems, the Kondo scale T_K and a reduced bandwidth $(b^2 t)$. When one minimizes the free energy with.respect to λ, one has the mean-field constraint $b^+ b + n_c = 1$, where n_c is the average electron number per site. One sees then a strong reduction of the effective bandwidth as one nears the half-filled point $(n_c = 1)$ where $b \to 0$. This represents the Mott–Hubbard metal-insulator transition at the mean-field level.

3. Effect of disorder

Disorder destroys the translational invariance of the system and the approximation $b_i \to b$ is inadequate. This situation arose in the heavy-fermion problem where there is evidence of rather large effects of impurity scattering which disrupt the low-temperature coherence. This has been discussed by Fukuyama [9] and Rice and Ueda [10]. The latter authors pointed out that b_i is correlated with ε_i. In our case, we write

$$b_i\,(\varepsilon_i) \;\approx\; b(0) + (\chi/2b\,)\varepsilon_i, \tag{3}$$

where $\chi = \partial b^2/\partial\varepsilon$ is a charge susceptibility. Our approximation is to consider one site imbedded in a Fermi liquid representing the average behavior $(\varepsilon = 0)$ of the other sites. This one-site problem resembles that of a single Anderson impurity for which the charge susceptibility is known [8]. It is simply $\chi \approx b^2/t$.

We use Eq.(3) in Eq.(2) and obtain

$$H = tb^2 \Sigma_{<ij>,\sigma}\,[1 + (\chi/2b^2)\varepsilon_i]\;[1 + (\chi/2b^2)\varepsilon_j] c_{i\sigma}^+ c_{j\sigma}$$
$$+ \Sigma_{i,\sigma}(\varepsilon_i - \mu) n_{i\sigma} + H_c \tag{4}$$

where H_c is the part of the hamiltonian Eq. (2) which does not contain the random site energies ε_i. The approximation in Eq. (3) may be improved by including the influence of nearest-neighbor site energy variations by means of a non-local charge susceptibility which may be obtained from the treatment of the two-Anderson impurity (RKKY) problem [8].

We perform a quenched average over disorder configurations by the standard replica trick [3]. The Green's functions of the hamiltonian may be found from a generating functional integral over replicated (Matsubara) frequency-dependent Grassmann variables. The disorder is integrated out over a Gaussian distribution of width W. Since the terms linear in the ε_i have coefficients bilinear in the c's, the result of this procedure is to produce a translationally invariant action with a mixed replica four-Fermion interaction. This is exactly similar to Ref. [3]. However, the terms of Eq. (4) which contain the charge susceptibility introduce, in addition to the usual on-site terms, terms coupling nearest neighbor sites. As in Ref. [3], a mean-field factorization of the on-site terms defines a lifetime τ. The new feature comes from a similar treatment of the nearest neighbor terms. Due to the disorder, an *enhancement* of the band narrowing is found. In particular, we have, for bare density of states ν,

$$\begin{aligned} &<c_{i\sigma}^{+}\, c_{i\sigma}> \;\propto W = 1/(2\pi\nu\tau) \\ &<c_{i\sigma}^{+} c_{j\sigma}> \;\propto tb^2(1 - W^2/t^2). \end{aligned} \tag{5}$$

4. Scaling equations.

At this point, we have obtained an interacting Fermi liquid with a reduced effective bandwidth $tb^2(1 - W^2/t^2)$ which renormalizes the density of states ν upward and an attractive interaction $-t^2/U$. As is usually the case in mean-field treatments of the Hubbard model, this Fermi liquid picture is not valid too near the point of half filling. Although we began with a square lattice, we shall ignore the k-space structure factors since we are only investigating the long wavelength properties of the system. Thus, we may simply apply the Finkel'shtein short-range interaction scaling equations [3] to our situation. These are

$$\mathrm{d}g/\mathrm{d}\xi = g\{(2-d)/2 + g[4-3(1+1/\gamma_s)\ln(1+\gamma_s)]\}$$

$$\mathrm{d}\gamma_s/\mathrm{d}\xi = g(1+\gamma_s)^2/2$$

$$\gamma_c = (1+\gamma_s)/2, \quad \xi = -\ln\lambda^2, \quad \sigma_\lambda = \lambda^{d-2}/g(\lambda). \tag{6}$$

The quantities in Eq. (6) have the following meanings: g is the dimensionless disorder scaling variable proportional to W (inversely proportional to the diffusion constant). d is the dimension. γ_s, γ_c are the spin and charge channel scattering amplitudes, each divided by the frequency rescaling factor z [4]. λ^{-2} is the length scale in the scaling and is interpreted as the inelastic scattering length. Therefore, λ is proportional to a power of the temperature and ξ may be thought of as the logarithm of the temperature. σ_λ is the conductivity at "temperature" λ.

In contrast to the usual case [3], the initial value of the spin channel scattering amplitude γ_s is negative and therefore, as may be seen from Eq. (6), it scales to weak coupling. Since the coupling is inversely proportional to U, the behavior is consistent with that found previously [3], namely the interaction scales to strong coupling. However, the difficulty in the usual approach in which the scaling fails as the coupling gets too large, is avoided. Therefore, we expect there is an overlap of the validity regions of the two approaches.

The scaling trajectories in the disorder-interaction plane are divided by a separatrix

$$W/t_{eff} = W/[tb^2(1 - W^2/t^2)] = (d-2)t/U, \tag{7}$$

where t_{eff} is the effective bandwidth obtained in Eq. (5). In terms of t_{eff}, this is the same as that found in earlier work [3,4]. In our case, on the metallic side (small W), γ_s scales to a small negative value, while γ_c is of order unity. As expected, charge and spin diffusion and conductivity are all non-zero. In this regime, the system appears to have the properties of a conventional Fermi liquid. However, it must be remembered that there are underlying strong correlations represented by the slave bosons, so far treated in mean-field approximation. It is necessary to investigate the slave-boson fluctuations [8] to expose the effect of correlations and have a complete physical picture of the metallic state.

On the insulating side of the separatrix (large W), the conductivity scales monotonically downward as temperature decreases. This new result is in contrast to the behavior in Refs. [3,4], where a counter-intuitive and experimentally-unobserved rise of the conductivity with decreasing temperature is found before the breakdown of the RG equations.

5. Conclusion.

So far, we have not included the Cooper channel in our analysis. The Cooper coupling (in the particle-particle channel) is attractive and grows stronger in the scaling. This is expected in the non-disordered t - J model where the dominant instability is to d-wave superconductivity. We remarked above that slave-boson fluctuations are essential to bring out the correlated nature of the ground state. In particular, as shown by Kotliar and Liu [11], in the weakly interacting mean-field treatment of the t - J model, the fluctuations drive a d-wave superconducting instability. An investigation in our model of the interplay of superconductivity, disorder and local-moment formation is a subject for further study. However, we expect that in a range of parameters, superconductivity will be suppressed by disorder through pair-breaking and local-moment formation. In this region, the scenario we have outlined will be relevant.

Acknowledgement

This work was supported in part by National Science Foundation (US) grant DMR 85-20190

References

[a] Current address: Physics Department, University of California, Davis CA, 95616, USA.

[1] F.J.Wegner, Z.Phys.B **35**, 207 (1979).

[2] E.Abrahams et al, Phys. Rev. Lett.**42**, 673 (1979).

[3] A.M.Finkel'shtein, Zh. Eksp. Teor. Fiz. **84**, 168 (1983) [Sov. Phys. JETP **57**,97 (1983)]; Z.Phys.B **56**, 189 (1984).

[4] C. Castellani, et al, Phys. Rev. **B 30**, 527, 1596 (1984).

[5] S. Hikami, Phys. Rev. **B 24**, 2671 (1981).

[6] M. Palaanen et al, Phys. Rev. Lett. **61**, 597 (1988).

[7] M.Milovanovic et al, Phys. Rev. Lett. **63**, 82 (1989).

[8] D.M. Newns and N. Read, Adv. in Physics **36**, 799 (1987); P. Coleman, Phys. Rev. **B 35**, 5072 (1987).

[9] H.Fukuyama in *Theory of Heavy Fermions and Valence Fluctuations*, eds. T.Kasuya and T.Saso (Springer. 1985). p. 209

[10] T.M.Rice and K.Ueda, Phys. Rev. **B 34,** 6420 (1986).

[11] G.Kotliar and J.Liu, Phys. Rev. Lett. **61**, 1784 (1988).

Characteristic Features of Electronic States in High-T_c Superconducting Oxides

M. Tachiki

Institute for Materials Research, Tohoku University, Katahira 2-1-1, Aoba-ku, Sendai 980, Japan

Abstract By use of a model of highly correlated electron system, the electronic structure of the sueprconducting oxides is theoretically studied. It is shown that if holes are doped in an insulating sample, a new narrow energy band grows up inside the charge transfer gap and develops to a metallic band. The formation of the new band is quite suitable for explaining various anomalous properties of the oxides in the normal state.

1. Introduction

Since the discovery of the high T_c superconductivity in La-Ba-Cu-O by Bednorz and Müller, 1986, enormous efforts to clarify the mechanism of the superconductivity have been made. The characteristic feature of the superconducting oxides is that the superconducting state appears in the transition region from the semiconducting state to the metallic state when holes are doped in the crystals. In this paper we try to understand what structural change of the electronic state occurs in the transition region.

2. Experimental results of Electronic Structure

The copper oxides which show the high T_c superconductivity have two-dimensional-like CuO_2 layers. When holes are doped, the holes enter into the layers and contribute to the electric conduction. The experimental results of photoemission spectroscopy indicate that the valence of copper ions is approximately divalent, even when holes are doped. The electronic state near the Fermi level E_F was investigated by the resonant photoemission spectroscopy using $Bi_2Sr_2CaCu_2O_8$ [1]. The results indicate that there is an band which shows the Fermi edge and its character is approximately of the 2p electron of oxygen ions. The same results were obtained by the experiments of electron loss spectroscopy using $La_{2-x}Sr_xCuO_4$ [2], $YBa_2Cu_3O_{7-y}$ [2,3], $Bi_2Sr_2CaCu_2O_8$ [3] and $TlBa_2Ca_{n-1}Cu_nO_{2n+3}$ [4]. The spectra in the experiments show that the

Springer Series in Solid-State Sciences, Vol. 89 Strong Correlation and Superconductivity
Editors: H. Fukuyama · S. Maekawa · A.P. Malozemoff

strength of the transition from the 1s level of oxygen to the 2p level of oxygen just above E_F increases as holes are doped. This fact may imply that the doped holes occupy oxygen sites rather than copper sites.

The next problem is whether the energy level near E_F has an energy dispersion, and the dispersion curve cuts the Fermi level, and thus the electron system has the Fermi surface. For this purpose, the experiments of angle resolved photoemission spectroscopy has been done using single crystals of $Bi_2Sr_2CaCu_2O_8$ [1]. The experimental results show that there is a narrow energy band with a dispersion as small as 0.5 eV and the dispersion curve cuts the Fermi level. Similar resutls have been obtained by the experiments using single crystals of $YBa_2Cu_3O_{7-y}$ [5].

The following infrared experiments reveal how the energy band near E_F develops when holes are doped in an insulating sample [6]. The absorption coefficient of infrared light was measured using single crystal films of $La_{2-x}Sr_xCuO_4$. A rapid increase of the absorptin coefficient which corresponds to the charge transfer gap appears at 2 eV in an insulating sample. When holes are doped, in addition to this absorption a new peak of the absorption appears at 1.5 eV and develops with an increase of Sr doping. The infrared experiment implies that a narrowband appears inside the charge transfer energy gap as shown in Fig. 1. The band width of the new band near E_F is very narrow when the number of doped holes is small. Thus, the band may have a localization nature and the electric conduction is expected to be of the variable range hopping type. When the number of holes increase, the band is widen and the electric conduction is expected to become metallic. This feature of the conductivity is consistent with the temperature dependence of the electric resistivity observe in the normal state of the superconducting oxides.

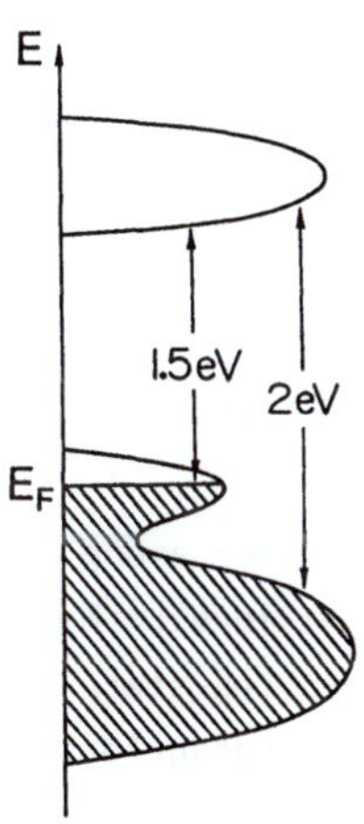

Fig. 1 Schematic illustration of the density of states in the oxide superconductors.

3. Theoretical Model and Formulation

From the experimental facts mentioned above, it may be appropriate to take the following model. We consider a 3d electron orbital $3d_{x^2-y^2}$. When two electron occupy this orbital, the strong Coulomb repulsive interaction acts between them. In this case this interaction opens the so-called Hubbard gap. The energy separation between the occupied lower Hubbard level and the empty upper Hubbard level is about 7 eV. Since the charge transfer energy gap which may correspond to the transition from the oxygen 2p band to the upper Hubbard level is about 2 eV, the oxygen band is considerd to be situated about 2 eV below the upper Hubbard level. Since the lower Hubbard level is situated much below the oxygen 2p band, the p-d mixing interaction mixes the oxygen 2p band mainly with the upper Hubbard level. The intensity of the mixing interaction is aobut 1 eV. The model mentioned above is schematically shown in Fig. 2.

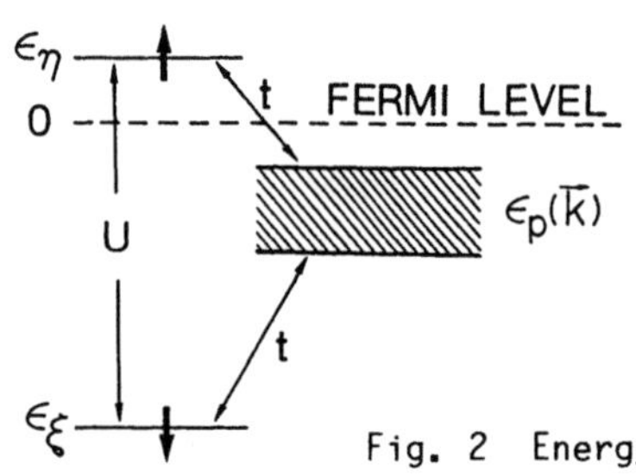

Fig. 2 Energy levels

The present model is expressed by the following Hamiltonian

$$H = H_0 + H_I \tag{1}$$

with

$$H_0 = \sum_{i\sigma} \varepsilon_d d^\dagger_{i\sigma} d_{i\sigma} + \sum_i U n_{i\uparrow} n_{i\downarrow} + \sum_{jj'\sigma} (\varepsilon_p)_{jj'} p^\dagger_{j\sigma} p_{j'\sigma} \quad , \tag{2}$$

$$H_I = \sum_{ij\sigma} t_{ij} (d^\dagger_{i\sigma} p_{j\sigma} + p^\dagger_{j\sigma} d_{i\sigma}) \ , \tag{3}$$

where $d_{i\sigma}$ is the d-electron operator with a spin σ ($\sigma = \uparrow, \downarrow$) at the i-th site, $p_{j\sigma}$ is the p-electron operator at the j-th site, $n_{i\sigma} = d^\dagger_{i\sigma} d_{i\sigma}$. In eq.(2) ε_d is a constant and ε_p contains a possible direct transfer among p electrons. The energies ε_d and ε_p are measured from the Fermi level. We try to evaluate the H_I effect to the one electron- or hole-excited state from the ground state, which can be described by the electron propagators.

The important effect of H_I to electronic states is induced by the process of p-d transition accompanied by spin and charge fluctuations.

Let us assume that H_0 is an unperturbed part and H_I is its perturbation. The d-electron part of H_0 is diagonalizable by introducing operators ξ_i and η_i as [7]

$$\xi_{i\sigma} = d_{i\sigma}(1 - n_{i-\sigma}) \quad , \qquad \eta_{i\sigma} = d_{i\sigma}n_{i-\sigma} \quad , \tag{4}$$

which give

$$[\ \xi_{i\sigma} ,\ H_0\ ] = \varepsilon_\xi \xi_{i\sigma} , \qquad [\ \eta_{i\sigma} ,\ H_0\ ] = \varepsilon_\eta \eta_{i\sigma} \tag{5}$$

with

$$\varepsilon_\xi = \varepsilon_d \quad , \qquad \varepsilon_\eta = \varepsilon_d + U \quad . \tag{6}$$

The operator ξ_i describes the transition between states with $n_i = 0$ and $n_i = 1$ and η_i does between those with $n_i = 1$ and $n_i = 2$. The energies ε_ξ and ε_η are necessary energies to increase one electron. As mentioned above we consider the case where the p electron level is situated between the Hubbard splitted d electron levels.

When the mixing interaction or the charge transfer interaction H_I is considered, it mixes the p electron and the Hubbard splitted d electron levels and forms new mixed bands. However, this interaction also induces the processes of electron hopping from d-sites to p-sites (and vice versa) with emitting (absorbing) spin and charge fluctuations.

In general, there exist four types of fluctuations at each site of the d electron system, which are given by the following operators:

$$n_{i\mu} = \sum_{\sigma\sigma'} d^\dagger_{i\sigma}(\sigma_\mu)_{\sigma\sigma'} d_{i\sigma'} \quad , \qquad (\mu = 0,1,2,3, \quad \sigma_\mu = (1,\vec{\sigma})\ ) \tag{7}$$

$$n_{i4} = n_{i\uparrow}n_{i\downarrow} \quad , \tag{8}$$

$$b_i = d_{i\uparrow}d_{i\downarrow} \quad , \tag{9}$$

The first one describes the charge (μ=0) and spin fluctuations (μ=1,2,3), the second one does the double charge fluctuation and the last one does the Cooperon fluctuation.

As discussed above the ξ level is deep below the Fermi level, the contribution from this level is negligible. In this case we can treat the problem just using the operators $p_{i\sigma}$, $\eta_{i\sigma}$, and $n_{i\mu}$. Hereafter, we use the notations Q_i=$Q(x)$ and $\Sigma\, C_{ij}Q_j$= $C(\partial)Q(x)$ for an operator Q_i and c-number C_{ij}. The equations of motions are

$$[i\frac{\partial}{\partial t} - \varepsilon_p(\partial)]\, p_\sigma(x) = t(\partial)\eta_\sigma(x)\ , \tag{10}$$

$$[i\frac{\partial}{\partial t} - \varepsilon_\eta]\, \eta_\sigma(x) = -\frac{1}{2}\sum_{\sigma'}(\sigma^\mu)_{\sigma\sigma'} n_\mu(x) t(\partial) p_{\sigma'}(x), \tag{11}$$

$$i\frac{\partial}{\partial t} n_\mu(x) = \sum_{\sigma\sigma'}\{\eta^\dagger_\sigma(x)(\sigma_\mu)_{\sigma\sigma'} t(\partial) p_{\sigma'}(x) - p^\dagger_\sigma(x) t(\partial)(\sigma_\mu)_{\sigma\sigma'}\eta_{\sigma'}(x)\}, \tag{12}$$

and

$$n(x) - \langle n(x)\rangle = n_4(x) - \langle n_4(x)\rangle\ , \tag{13}$$

where $\sigma^\mu n_\mu$ is an abbreviation of $(-n + \vec{\sigma}\cdot\vec{n})$. The canonical relations are

$$\{ p_\sigma(\vec{x}),\ p^\dagger_s(\vec{y})\} = \delta(\vec{x} - \vec{y})\delta_{\sigma s}\ , \tag{14}$$

$$\{ \eta_\sigma(\vec{x}),\ \eta^\dagger_s(\vec{y})\} = -\frac{1}{2}(\sigma^\mu)_{\sigma s} n_\mu(\vec{x})\delta(\vec{x} - \vec{y})\ , \tag{15}$$

$$[\eta_\sigma(\vec{x}),\ n_\mu(\vec{y})] = \sum_{\sigma'}(\sigma_\mu)_{\sigma\sigma'}\eta_{\sigma}(\vec{x})\delta(\vec{x} - \vec{y})\ , \tag{16}$$

where $\delta(\vec{x} - \vec{y})$ indicates δ_{ij} for the lattice sites i and j. Note that in the two-dimensional lattice

$$\delta(\vec{x} - \vec{y}) = \frac{\Omega}{(2\pi)^2}\int_{\Omega_B} d^2k\, e^{ik(\vec{x}-\vec{y})}\ , \tag{17}$$

with Ω being the volume of the unit cell and Ω_B indicating the first Brillouin zone. We also have

$$\langle\eta_\sigma(\vec{x})\eta^\dagger_s(\vec{x})\rangle = (1 - \frac{n}{2})\delta_{\sigma s}\ ,\quad \langle\eta^\dagger_s(\vec{x})\eta_\sigma(\vec{x})\rangle = (n - 1)\delta_{s\sigma}\ , \tag{18}$$

$$\langle(n(\vec{x}) - n)^2\rangle = (2 - n)(n - 1)\ , \tag{19}$$

$$\langle n_3(x)^2\rangle = (2 - n)\ . \tag{20}$$

The electronic states are investigated by use of the retarded Green's functions of p and η electrons whose energy-momentum representations are denoted as $S_{pp^\dagger}(\omega,\vec{k})$ and $S_{\eta\eta^\dagger}(\omega,\vec{k})$, which are expressed by use of the self-energy $\Sigma(\omega,\vec{k})$ for the η-level and by the vertex correction $\Gamma(\omega,\vec{k})$ for the mixing interaction as

$$S_{pp^\dagger}(\omega,\vec{k}) = \frac{1}{\omega - \varepsilon_p(k) - t^2(k)\dfrac{n/2 + \Gamma(\omega,\vec{k})}{\omega - \varepsilon_\eta - \Sigma(\omega,\vec{k})}} , \tag{21}$$

$$S_{\eta\eta^\dagger}(\omega,\vec{k}) = \frac{n/2 + \Gamma(\omega,\vec{k})}{\omega - \varepsilon_\eta - \Sigma(\omega,\vec{k}) - t^2(\vec{k})\dfrac{n/2 + \Gamma(\omega,\vec{k})}{\omega - \varepsilon_p(\vec{k})}} . \tag{22}$$

The above equations are roughly interpreted as the mixing between the levels $\varepsilon_p(\vec{k})$ and $\varepsilon_\eta+\Sigma(\omega,\vec{k})$ with the mixing interaction $t^2(\vec{k})(n/2 + \Gamma(\omega,\vec{k}))$. The corrections Σ and Γ are induced from the spin and charge fluctuations due to the charge transfer interaction. Using some approximation, we obtained the self-energy and vertex correction, and calculated the density of states of the system.

4. Numerical results

For simplicity, we consider a two dimensional lattice in which one d electron orbit and p electron orbit are in each unit cell. An example of the calculated density of states is shown in Fig. 3. The bold solid lines are for the density of states (DOS) per spin of the d electron and the fine solid lines are for DOS per spin for the p electron. Fig.3 (a) shows DOS for an insulating state in which 3 electrons are filled in each unit cell. When holes are doped, a new narrow band appears inside the charge transfer gap as shown in Figs. 3 (b)-(d). The number of holes increases in alphabetical order in Fig. 3. The new band grows up to a wider metallic band as the number of holes increases. The state near E_F (the energy zero in Fig.3) is mainly of the p electron character. It should be noted that the distance between the lower p band and the new band is almost constant independent of the hole concentration.

The formation of the new band inside the charge transfer energy gap can be understood in the following way. In the insulating state, the p electron band renormalized by the mixing interaction is below the Fermi level. When holes are doped in the insulating sample, the Fermi level approaches the electron p band. Before the top of the p electron band touches the Fermi level, a formation of new electronic state at the Fermi level may be energetically favored by the following reason. If a new electronic state is created near the Fermi level, the state yields many quantum mechanical processes that the p electrons near the Fermi level are

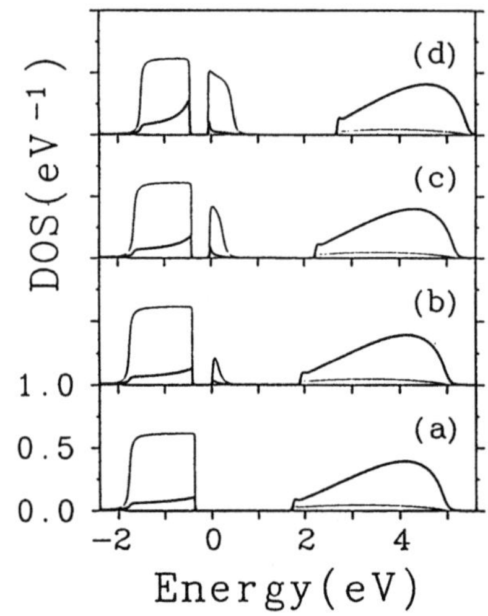

Fig. 3 Electronic density of states. The bold solid lines are for the density of states per spin of the d electron and the fine solid lines are for the density of states per spin for the p electorn. The number of holes increases in alphabetical order in the figure.

virtually excited to the upper Hubbard level and decay back to the contiuum state of unoccupied p electron state. This process is accompanied by spin fluctuations and is enhanced strongly due to the localized nature of the Cu spins and the sharp Fermi edge effect of the p electron state. The formation of the new electronic state at the Fermi level due to this process lowers the energy arising from the mixing interaction. When the transfer energy is much larger than the energy difference between the top of the p electron band and the Fermi level, the energy loss to create an electronic state near the Fermi level is overcome by the gain in the mixing interaction, and the new electronic state may be formed spontaneously at the Fermi level.

The formation of the new narrow band may be quite suitable to explain the features of electric conduction, infrared absorption, optical conductivity, and photoemission experiments. The observed new absorption at 1.5 eV in $La_{2-x}Sr_xCuO_4$ [6] is interpreted as the transition from the new narrow band to the upper Hubbard level. A broad peak observed around 0.5 eV [6] may be identified as the transition from the lower p electron band to the levels above the Fermi level in the new band. This broad absorption peak is in the region same as the energy region identified as the plasma frequency. The frequency usually obtained from the refectivity experiments is almost independent of the hole concentration, contradictorily with the free carrier model [8]. In the optical conductivity, the sharp peak at $\omega \sim 0$ which is strongly depends on temperature and concentration of the hole is followed by a broad peak in the mid-infrared region which has milder dependence on temperature and concentration [9]. We may identify the behavior near $\omega \sim 0$ as the intra-band transition of the new narrow band which is strongly dependent on temperature and concentration of the hole, and the peak in the mid-infrared range as the interband transition from the lower p electron band to the new band.

The quasi particles near the Fermi level in the new band have the nature of Fermi liquid. We consider that the pairing of the quasi particles is responsible for the formation of the superconducting state in the oxides. What is the origin of the attractive force for the pairing is the important problem in future.

The author would like to thank Professor H. Matsumoto and Mr. M. Sasaki for their collaboration on this work and valuable discussions, and Dr. T. Koyama and Dr. S. Takahashi for useful discussions.

References

[1] T.Takahashi, M.Matsuyama, H.Katayama-Yoshida, Y.Okabe, S.Hosoya, K.Seki, H.Fujimoto, M.Sato, and H.Inokuchi: Nature 334, 691 (1988).

[2] N.Nücker, J.Fink, J.C.Fuggle, P.J.Durham, and W.M.Temmerman: Phys. Rev. B37, 5158 (1988).

[3] N.Nücker, H.Romberg, X.X.Xi, J.Fink, B.Gegenheimer, and Z.X.Zhao: Phys. Rev. B39, 6619 (1989).

[4] D.Shindo (private communication)

[5] Y. Sakisaka, T. Komeda, T. Maruyama, M. Onchi, H. Kato, Y. Aiura, H.Yanashima, T.Terashima, Y.Bando, K.Iijima, K.Yamamoto, and K.Hirata: Phys. Rev. B39, 2304 (1989).

[6] M. Suzuki: Phys. Rev. B39, 2312 (1989).

[7] H.Matsumoto and H. Umezawa: Phys. Rev. B31, 4433 (1985).

[8] S.Tajima, T.Nakanishi, S.Uchida, and S.Tanaka: Physica C136, 90 (1988).

[9] G.A.Thomas, J.Orenstein, D.H.Rapkine, M.Capizzi, A.J.Millis, R.N. Bhatt, L.F.Schneemeyer, and J.V.Waszczak: Phys. Rev. Lett. 61, 1313 (1988).

Anderson Lattice Model of High-T_c Superconductors

D.M. Newns and P.C. Pattnaik

IBM Research Division, T.J. Watson Research Center, Box 218,
Yorktown Heights, NY 10598, USA

Abstract: An accurate Anderson lattice model, solved using slave boson techniques, gives a remarkably quantitative picture of the thermodynamic and spectroscopic properties of the normal state. The proposed pairing mechanism, based on the exchange J_{pd} to order $1/U$, gives an order parameter of the correct symmetry and explains several trends in T_c.

1. High-T_c's as Transition Metal Oxides

High temperature superconductors are first and foremost transition metal oxides. In such systems, the Coulomb repulsion, U, which two electrons experience when occupying the same d-orbital is less well screened than it is in a metallic environment. Hence U is large (6-9 eV in high T_c superconductors), leading to numerous strong coupling effects, such as magnetism, metal-insulator transitions and spectral satellites.

In high-T_c's, the well-known generic energy level diagram [1] is shown in Fig. 1. It is convenient to adopt the convention of working with holes, so the direction of the energy axis is the reverse of the usual one. The key orbitals are the Cu $d_{x^2-y^2}$ and the planar oxygen $2p_x$ and $2p_y$ orbitals. A zeroth approximation is obtained by ignoring the coupling matrix elements V between the Cu and oxygen orbitals. In this approximation the lowest level in Fig. 1 is seen to be the single hole level of Cu $d_{x^2-y^2}$. Hence the Cu is essentially d^9. In hole-doped systems the extra holes would, in the $V=0$ approximation, reside on the oxygen $2p_x - 2p_y$ bands, separated from the d^9 state by the oxide gap Δ , of order 1 eV. Additional holes can not be added to the copper without incurring the energy penalty of the two-hole interaction U. In electron doped systems the right picture for the added electrons seems not yet clear.

Charge fluctuations involving the d-orbitals are seen from the energy level scheme in Fig. 1 to be inhibited in high-T_c systems. Hole motion from Cu to oxygen must overcome the energy barrier Δ, while promotion of a hole from oxygen to Cu must overcome the even higher barrier U. These hole

Springer Series in Solid-State Sciences, Vol. 89 **Strong Correlation and Superconductivity**
Editors: H. Fukuyama · S. Maekawa · A.P. Malozemoff

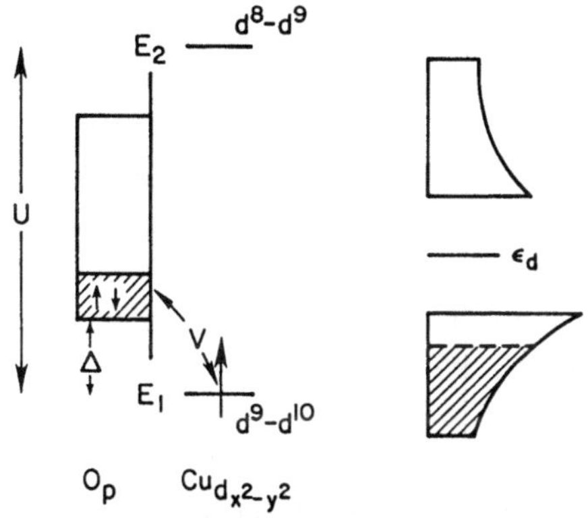

Fig. 1 Energy level diagram for high-T_c oxides (hole notation). Left, energy levels when the p-d hopping matrix element V is zero. Δ is oxide gap, from $d^9 - d^{10}$ level at E_1 to bottom of oxygen band. Right, MFA or heavy fermion solution to the nonzero V problem redefines the Cu d-level to ε_d, and generates an effective p-d hopping, resulting in two new hybrid bands.

promotion energies are manifested as satellites in inverse photoemission and photoemission spectra of the high-T_c materials. The resistance of the system to charge fluctuations is unfavourable for pairing mechanisms involving charge fluctuations in high-T_c superconductors.

Spin fluctuations of the d-electrons, in contrast to charge fluctuations, have no energy barrier in the $V=0$ case. Since the charge on the Cu's is stabilised at one hole, the magnitude of whose spin is 1/2, and the spin is free to rotate, we can speak of Cu spins.

How the system behaves when the perturbation V is switched on is of course the prime issue at the present time. At low doping, correlations between d spins dominate [2]. They are mediated by the antiferromagnetic nearest neighbour d-d exchange due to virtual fluctuations across the oxide gap [3], usually considered to be of order $J_{dd} \sim V^4/\Delta^3$. The holes also see an antiferromagnetic exchange interaction between themselves and the spins, of order $J_K \sim V^2/\Delta$. In this low doping regime, in which there is significant disorder induced by the dopant, the holes are presumably bound to the spin system by J_K and effectively localised. The charge gap Δ is retained, and the system is a Mott insulator.

At doping levels corresponding to the metallic superconducting regime, the nature of the state is at present the subject of lively discussion [4 – 6]. Our position in this paper is based on the strong evidence that the Cu d^9 spins and the added holes *condense* into the heavy fermion state well known in certain rare earth and uranium compounds.

2. Heavy Fermion Condensation

The heavy fermion state is a fermi liquid state obeying the Luttinger Theorem. The Luttinger Theorem, we recall, maps the strongly interacting model ($U = 6$-9 eV) onto the $U = 0$ model. Now the Fermi surface (FS) of the $U = 0$ model, which is just a set of non-interacting holes in a periodic potential, of course contains all valence holes in the system, in this case $1 + x$, where x is doping. The Luttinger theorem states that the FS of the interacting system must then also contain $1 + x$ holes, as illustrated in Fig. 2. The proviso is that there be no qualitative change in the state of the system as the interaction U is turned on, such as a phase transition. The presence of the large U is manifested as a significant mass enhancement of the quasiparticles - hence the term heavy fermion. It is as if the holes cannot get round each other due to their strong interaction, which slows them up, appearing as an enhanced mass. The heavy fermions have a quasiparticle band structure which we shall discuss shortly. Enhanced mass is far from the only many-body effect in heavy fermion systems, some of whose properties resemble the $V = 0$ limit discussed in Sec. 1 - partial memory of the charge gap is retained.

Fig. 2

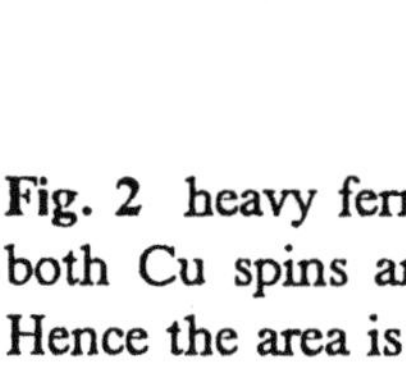

Fig. 3

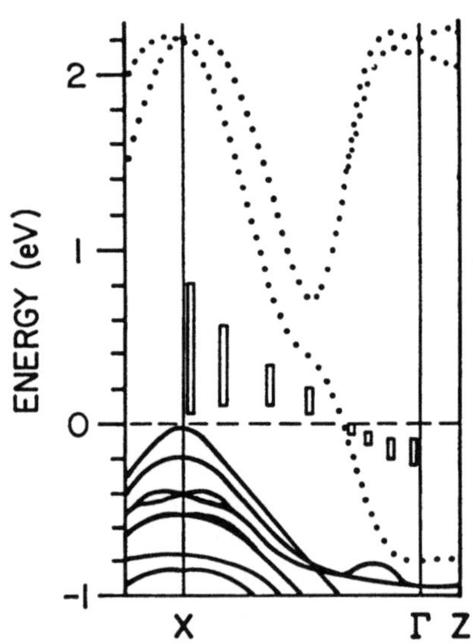

Fig. 2 heavy fermion solution has a Fermi surface area which includes both Cu spins and added holes, with concentration x_h per planar Cu. Hence the area is $1 + x_h$.

Fig. 3 Dispersion of quasiparticle bands along ΓX in Bi 2212 from photoemission plus inverse photoemission data [8,9] (rectangles), compared with band structure calculations (dotted and full curves).

Experimental photoemission data clearly show a sharp Fermi edge characteristic of a Fermi liquid [7]. Angle resolved photoemission is even more interesting, in that it explicitly shows the heavy fermion bands. Examples are the pioneering work of T. Takahashi [8] (see Fig. 6 below and elsewhere in this volume) and the work of Fink et al.[9] (see Fig. 3), which also includes inverse photoemission. Note in Fig. 3 that the band crosses the Fermi level at the same k_F as in the band structure calculation. Since the band structure obeys Luttinger's Theorem (it is essentially the solution to the non-interacting problem), then this is evidence from one direction in k-space that the quasiparticle band structure also obeys Luttinger. Just as obvious as the agreement between the band structure and measured k_F's, is the huge disagreement in the masses. The observed mass is about five times that calculated from the band structure. This is the classic heavy fermion scenario.

Positron annihilation offers in principle another way to measure the Fermi surface, but this is not a spectroscopy. Only the momentum distribution function is measured, and the Fermi discontinuity must be sought on top of a large background. It has indeed been claimed [10] that a Fermi surface in the 123 material like that predicted in band structure is seen in positron annihilation, but the results seem not to be universally accepted at this time.

In the following we shall present more indirect evidence for the heavy fermion state, which depends on a quantitative solution to the heavy fermion problem.

3. Normal State - $U=\infty$ ALM

The model described by Fig. 1 is mathematically expressed in terms of the Anderson Lattice model, which also describes heavy fermion systems such as those containing Ce and Yb, for example. The oxygen band is represented by a set of states $|k>$ of energy ε_k. The Cu d-states are most simply assumed strongly split in the tetragonal crystal field, so that only the $d_{x^2-y^2}$ has any significant occupation of holes and is the only d-state included in the simplest model (but see model B below). The Cu $d_{x^2-y^2}$ state is given a bare energy E_1 . Let us first consider the model in the absence of any electron-electron interaction U. The $U=0$ Hamiltonian for a CuO_2 plane and its environment can be written

$$\mathscr{H}_0 = \sum_{k\sigma}\varepsilon_k C^+_{k\sigma}C_{k\sigma} + E_1\sum_{i\sigma}D^+_{i\sigma}D_{i\sigma} + \sum_{ik\sigma}V_k D^+_{i\sigma}C_{k\sigma}e^{ik.R_i} + \text{H.c.} \quad (1)$$

In (1), $C_{k\sigma}$ and $D_{i\sigma}$ are standard fermion operators for an oxygen band state k and for a d state at the site given by the lattice vector $\mathbf{R}_i$; σ is spin. The

V_k are the Cu-O hopping matrix elements. We use hole operators so that E_l is negative (see Fig. 1). The first two terms in (1) describe respectively the energy of the oxygen and copper subsystems, and the third term describes the simple hopping-type coupling between them.

The non-interacting case is however of less interest than the strongly interacting case $U=\infty$, which should be sufficient to give a reasonable account at least of the normal state, since the oxide gap Δ is found to be smaller than the gap to the d^8 state. Because of its systematic nature, an increasingly widely accepted approach for handling the $U=\infty$ ALM is the slave boson field theory [11 – 13].

The essence of slave boson theory is to describe the d-shell by an overcomplete representation in which the singlet d^{10} state is represented by a boson b, and the doublet d^9 state is represented by the fermion operator d_σ. Working with holes, then the real hole operator D_σ is, when $U=\infty$

$$D_\sigma = b^+ d_\sigma \quad , \tag{2}$$

i.e. annihilating a spin-sigma hole involves *both* annihilating a d^9 state of spin σ *and* creating a d^{10} state. This overcomplete representation only corresponds to reality in the presence of a constraint

$$b^+ b + \sum_\sigma d_\sigma^+ d_\sigma = 1 \quad , \tag{3}$$

implying that the probabilities of $d^9_\uparrow$, $d^9_\downarrow$ and d^{10} add to one. It is seen that the only d-states admitted by (3) are $|\uparrow 0>$, $|\downarrow 0>$, and $|0\,1>$, in an obvious notation; d^8 is excluded. The operator (2) acts on these states to give

$$D_\sigma |\sigma' 0> \; = \; |0\,1> \delta_{\sigma\sigma'} \tag{4}$$

$$D_\sigma^+ |0\,1> \; = \; |\sigma\, 0> \quad .$$

Now with these tricks for projecting out the d^8 state, the $U = \infty$ version of (1) is

$$\mathscr{H}_\infty = \sum_{k\sigma} \varepsilon_k C_{k\sigma}^+ C_{k\sigma} + E_l \sum_{i\sigma} d_{i\sigma}^+ d_{i\sigma} + \sum_{ik\sigma} V_k b_i d_{i\sigma}^+ C_{k\sigma} e^{i\mathbf{k}.\mathbf{R}_i} + \text{H.c.} \tag{5}$$

$$+ \lambda \sum_i [b_i^+ b_i + \sum_\sigma d_{i\sigma}^+ d_{i\sigma} - 1] \quad .$$

Now the simple hopping interaction in (1) has been replaced by the field theoretical vertex illustrated in Fig. 4. The constraint (3), whose application at every site i is essential to the slave boson representation, has been replaced by

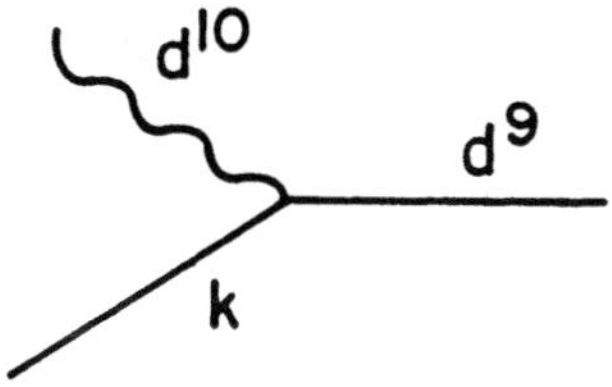

Fig. 4 In slave boson theory for the $U = \infty$ ALM, pd hopping is replaced by a field theoretic vertex, where transfer of a hole from a 2p (k) state to a d^9 site (both full lines) involves emitting or absorbing a slave boson (wavy line).

a global constraint acting only in a mean field sense via a Lagrange multiplier λ. At first it seems very surprising that this global form of the constraint should be satisfactory; we shall see how the matter is resolved later.

4. The 1/N Expansion

A conventional Feynman graph expansion may be developed from (5), formally as a perturbational theory in V_k. Comparing with (1), we notice that there is no simple copper- oxygen hopping term. Let us generate one !

In Fig. 5 we show graphs for a k to d self energy, i.e. for spontaneously generating a Cu-O hopping term. The route to organising such an expansion has been discovered to be via an extension of the spin degeneracy from 2 to a fictitious degeneracy of N, i.e. the symmetry of the model is extended from SU(2) to SU(N).

The order of a graph is assigned by two rules: a) a factor N is introduced for each Fermi loop; b) a factor $1/N^{1/2}$ is introduced for every vertex V. With these rules Fig. 5 displays the beginning of a series in descending powers of N, i.e. a series in 1/N, the 1/N expansion.

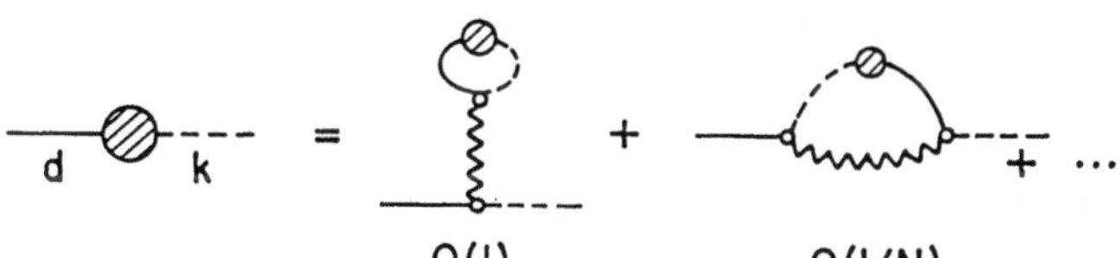

Fig. 5 Generation of an effective k to d hopping self energy replaces the direct p-d hopping matrix element of non-interacting band structure theory. The self energy is expanded as a series in 1/N. Shaded circle, dk self energy, full line, d^9 state, broken line, 2p (k) state, open circle, vertex of Fig.4.

It turns out that quite a lot can be gained from just the leading term in the 1/N expansion ('leading-N'). This is seen to be equivalent to taking an expectation value of the b-boson, so that

$$Vb \rightarrow V<b> \quad , \tag{6}$$

the mean field approximation (MFA).

It is a widely credited myth that the leading-N or MFA is not good because of its cavalier treatment of the constraint (2). But, this has been checked by Read [12] and by Millis and Lee [13] for the ALM. The constraint correlation function

$$<[b_i^+(0)b_i(0) + \sum_\sigma d_{i\sigma}^+(0)d_{i\sigma}(0)]\,[b_j^+(\tau)b_j(\tau) + \sum_\sigma d_{j\sigma}^+(\tau)d_{j\sigma}(\tau)]> \tag{7}$$

has been calculated, also to leading-N approximation, and found to be zero. This should be the case since the constraint operator in (3) commutes with $\mathscr{H}$.

The assumption is that if the 1/N expansion is carried out systematically, involving a recalculation of $<b>$ (and λ) to each order, then the constraint will be satisfied to each order. Ward identities are also automatically satisfied by the systematic procedure. There is no a priori reason for trying to satisfy the constraint exactly, while the theory is perturbational, since this implies paying the heavy price of discarding the Feynman graph technique.

The key step is seen to be the introduction of the boson expectation value $<b>$ (which actually breaks the U(1) symmetry of the model). This immediately 'jumps' the system into a state with Fermi liquid symmetry, enabling the FL state to be reached by a V-expansion. Without this trick, it is difficult to reach the FL state by a V-expansion, because the unperturbed state is a Mott insulator, containing free spins and thus with a different symmetry from the sought-for final state. The lurking threat is that the 1/N expansion may turn out to be asymptotic. However it is certainly more serviceable than a U-expansion in this problem, for though the latter is expected to converge, it will be necessary to calculate a large number of terms when U is large.

5. Mean Field Approximation

Inserting (6) into (5), and introducing the renormalised d-level illustrated in Fig. 1 by $\varepsilon_d = E_1 + \lambda$, we get the mean field Hamiltonian

$$\mathscr{H}_{mf} = \sum_{k\sigma} \varepsilon_k C_{k\sigma}^+ C_{k\sigma} + \varepsilon_d \sum_{i\sigma} d_{i\sigma}^+ d_{i\sigma}$$

$$+ \ <b> \sum_{ik\sigma} V_k d^+_{i\sigma} C_{k\sigma} e^{ik.R_i} + \text{H.c.}$$

$$+ \ (\varepsilon_d - E_1) \sum_i (<b>^2 - 1) \ . \tag{8}$$

The prescription is that $<b>$ and ε_d are to be determined by minimising the free energy of (8), leading to certain mean field equations. Eq. (8) is exceedingly similar to the non-interacting model (1), with the d-level E_1 and the hopping matrix element V_k replaced by effective values ε_d and $<b>V_k$. These effective values are uniquely related, via the mean field equations, to the bare parameters E_1 and V_k. Space is unfortunately lacking to describe several other works essentially derived from the approximation (8). Kotliar et al. [14] gave an analytic treatment in the limit of small oxygen bandwidth, while recently Balseiro et al. [15] considered the Mott transition, a topic not discussed below.

We have worked with two models [16 – 18] for the system, both optimised for the 214 material.

The simpler model [16] ('A') is an isotropic 2-band model, with $V_k = V$ and only one k-band per Cu. In this case (8) can be nearly diagonalized at sight by introducing a (2 ×2) one-body Hamiltonian h_k at each k

$$h_k = \varepsilon_k C^+_{k\sigma} C_{k\sigma} + \varepsilon_d d^+_{k\sigma} d_{k\sigma} + [<b> V d^+_{k\sigma} C_{k\sigma} + \text{H.c.}] \ . \tag{9}$$

This model gives two bands on diagonalising (9), a lower (in hole notation) d-like one and a higher k-like one, with the level ε_d in between. These are the renormalised bands illustrated in Fig. 1 (right).

The second model [17,18] ('B') is a generalisation of (8). It is based on the NRL tight binding parametrisation of the 214 band structure including all O 2p, La 5d, Cu 3d and Cu 4s orbitals. We included slave bosons with both the $d_{x^2-y^2}$ and d_{z^2} orbitals, since there turns out to be a significant number of holes in the latter.

6. Results of Model B for the 214 Compound in the MFA

A good feature of our MFA treatment of the 214 material [17,18] using model B is that it is almost parameter-free. Only one parameter, E_1 , in model B could not be safely taken from the NRL parameterisation, as this would give not the required bare value of E_1 , but an LDA mean field value. So E_1 was treated as an adjustable parameter, derived from a fit to the paramagnetic susceptibility χ. There were no other free parameters. From this $<b>^2$ was deduced to be about 0.2, implying a very significant reduction in the CuO_2 plane bandwidth of order 5.

The MFA band structure derived for the 214 material is compared in Fig. 6 with angle resolved photoemission data [8] for a different material, $Bi_2Sr_2CaCu_2O_8$, or 2212. We see that, despite the differences in detail in the materials, the illustrated bands, which derive essentially from the CuO_2 planes, show strong similarities between experiment and theory. These bands are around five times heavier than expected from band structure, and may be called heavy fermion bands. From comparison with experiment, our effective mass deduced from the 214 is seen to be about 50% too light relative to the bands of the 2212. The effective mass derives from the mean field equations, and is a very sensitive function of E_1 and V_k.

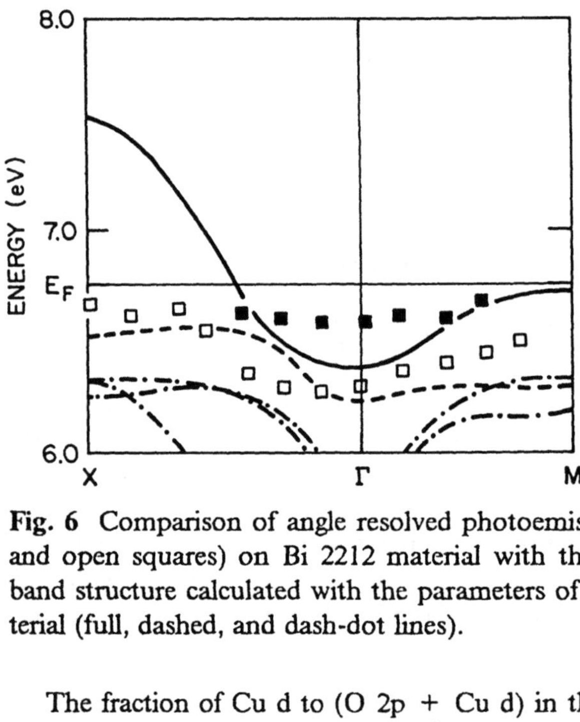

Fig. 6 Comparison of angle resolved photoemission data of Ref. 8 (full and open squares) on Bi 2212 material with theoretical heavy fermion band structure calculated with the parameters of Ref. 17 for the 214 material (full, dashed, and dash-dot lines).

The fraction of Cu d to (O 2p + Cu d) in these bands was estimated to be around 40% by Olson et al. [19]. Theoretically, the fraction is found to be about $0.5 - \langle b \rangle^2/4$, or 45%, in reasonable accord with the data.

We hope that the above quantitative successes will help to lay to rest the frequently heard myth that the carriers are 'oxygen holes', based on a false analogy with a semiconductor. If the carriers were oxygen holes, the Fermi wavevector would have to be quite small, and the d-character much less than 50%. In contrast to this picture, the Fermi wavevector is that expected from the Luttinger theorem and the Cu character is substantial.

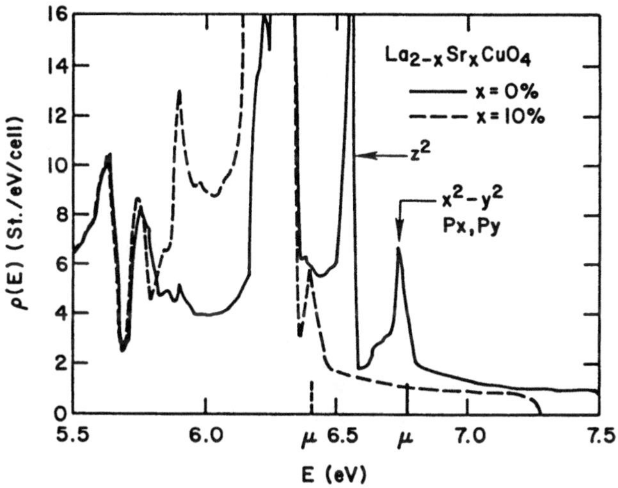

Fig. 7 DOS, including spin, for 214 material at two dopings as calculated in Ref. 17. Features arising from $d_{x^2-y^2}$, $2p_x$, $2p_y$ and from d_{z^2} bands are indicated.

The DOS near ϵ_F is dominated by the heavy fermion bands. In Fig. 7 we illustrate (in electron notation) the DOS at two dopings (in fact 0% doping would be unstable relative to the insulating state). The highest peak and its foot are part of the CuO_2 x^2-y^2 /2p heavy fermion band, which contains the Fermi level. The narrow peak 0.4 eV below the above-mentioned peak is from the occupied d_{z^2} band. The other structures are O_{2p} and 3d t_{2g} derived. Note that the lowest peak in the band structure (way below Fig. 7) is actually an oxygen 2p - Cu 4s derived feature.

The key result is then the presence of a heavy fermion band about 1 eV wide, derived from $3d_{x^2-y^2}$ and $2p_x$ and $2p_y$, which is evident in the band structure of Fig. 6 and in the DOS of Fig. 7. This band is where most of the action is in these materials and has been very widely discussed from different points of view, such as the Hubbard model [20].

The calculated heavy fermion DOS should be approximately consistent with the paramagnetic susceptibility and the specific heat. In Fig. 8a we illustrate the density of states ρ_F (including spin) at the Fermi level as a function of doping x in the 214 material. It is seen that the calculated DOS is in fact in fair agreement with the DOS deduced from the paramagnetic susceptibility [21] χ and specific heat coefficient γ [22], using the MFA equations

$$\chi = \mu_B^2 \rho_F \quad , \tag{10a}$$

$$\gamma = \frac{\pi^2}{3} k_B^2 \rho_F \quad . \tag{10b}$$

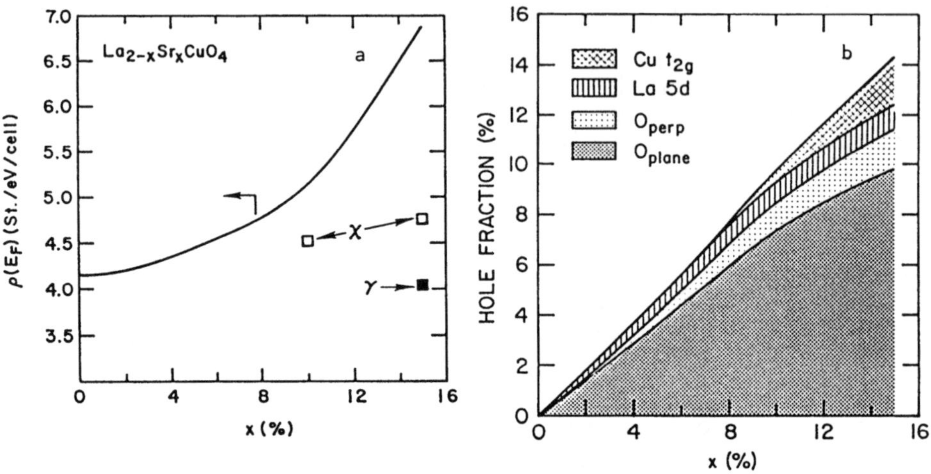

Fig. 8a DOS, including spin, for 214 material plotted vs. doping (calculation of Ref. 17). Also data from paramagnetic susceptibility [21] χ and specific heat [22] γ are included via Eq's 10.

Fig. 8b Plot of fraction of holes in various orbitals vs. added holes x in 214 material, in calculation of Ref. 17.

Note that our value of E_1 has been selected to fit the 214 data at $x = 10\%$, so whereas the fit of χ to theory is not significant, it is found that the value of E_1 required to achieve the fit is realistic in the light of other calculations [23], which is significant.

An interesting experimental trend is that χ increases with doping, as has been clearly established in the work of Torrance et al. reported elsewhere in this volume. This trend is clearly established in the theoretical curve of Fig. 8a.

Should we be using the Anderson lattice or Hubbard model to treat high-T_c systems? An interesting distinction can be made using this χ trend between the two models. This is because although in MFA both models involve a similar picture of a narrow band, the MFA equations are quite different. In particular, the DOS decreases roughly as $1/x$ in the Hubbard model, whereas the DOS tends rather to increase slightly with doping in the ALM. Hence in the Hubbard model χ should decrease rapidly with doping, in the ALM χ should increase with doping. It is the latter behaviour which is found in the experimental data, which then provides support for the ALM rather than the Hubbard model. It should be mentioned in this context that the spectroscopic fraction of 60% oxygen seems too high for the Hubbard model according to its derivation by the Zhang-Rice transformation, in which the oxygen states enter only as dopant holes bound to the Cu spins, and should be present to an amount of order the doping.

Where do the added holes go ? As a final application of the MFA to model B, we illustrate in Fig. 8b the decomposition of added hole density among various orbitals. Most holes go into the p_x and p_y planar orbitals, as has been found in various experimental tests.

7. Satellites

There are numerous properties of the normal state which depend on effects beyond the MFA. Since in this paper we are concentrating on thermodynamic and one-body properties (i.e. the simplest ones), we consider just one example, the Inverse Photoemission spectrum. In a Cu $d^8 - d^9$ material the Inverse Photoemission spectrum is analogous to the photoemission spectrum of Ce $f^1 - f^0$ valence fluctuation systems, where the comparison of theoretical and experimental spectra has proved very informative.

In Fig. 9 we illustrate the angle-averaged spectrum [24] and notice essentially two structures, a bump in the region 0-1 eV, and a peak at 2.9 eV. The structure at 0-1 eV arises from electrons going into the empty part of the uppermost band in Fig. 7, which is seen to have roughly this width. Incidentally the width itself is again evidence of a Fermi liquid state obeying the Luttinger theorem - the large empty state width corresponds to the large hole Fermi Surface area (Fig. 2). A hole gas consisting purely of holes of concentration x would have a very narrow unoccupied energy window of order 0.2 eV.

The 2.9 eV peak is interpreted as a satellite of the single particle bump. In order to calculate the satellite, we assume it is dominated by d-spectral weight,

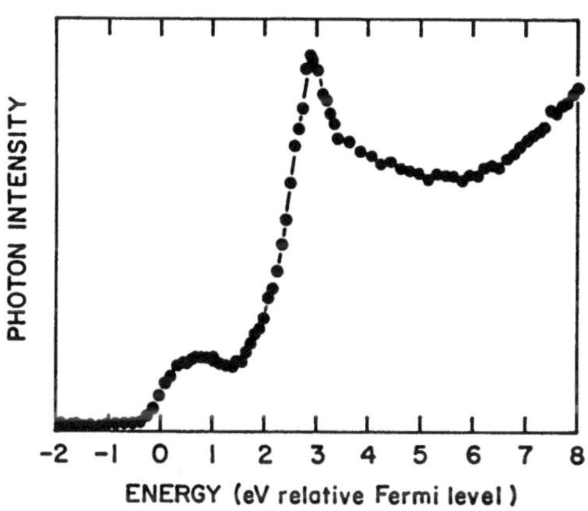

Fig. 9 Experimental angle-averaged inverse photoemission spectrum for Bi 2212 material, according to Ref. 24.

and that it will be sufficient to calculate the spectral density of the d-electron Green's function

$$- < TD(0)D^{+}(\tau) > \quad , \tag{11}$$

for which the diagrams up to order 1/N are illustrated in Fig. 10. The diagrams are complicated because, since $D_\sigma = b^+ d_\sigma$, the vertices emit and absorb bosons as well as fermions.

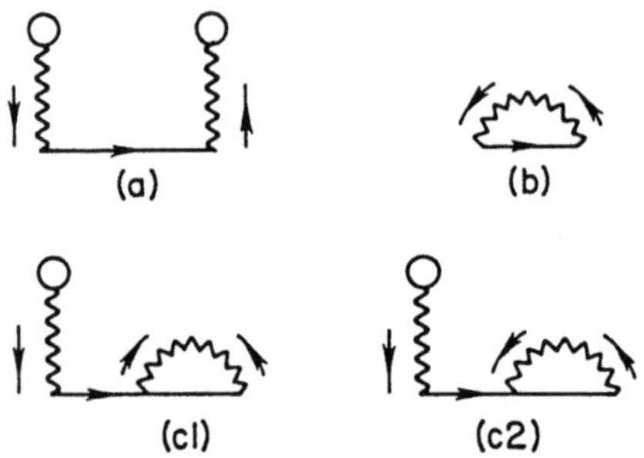

Fig. 10 Diagrams for dd Green's function to O(1/N). Full line, complete heavy fermion propagator; wavy line, slave boson.

The leading-N expression for the b-boson is derived by summing the RPA diagrams (bubble sum). The calculation is complicated enough that we have only done it for the simpler model A of Sec. 5. The b-boson has an extremely rich structure involving an electron-hole continuum, an interband feature, and a collective zero-sound mode which turns into the $d^9 - d^{10}$ satellite.

When the b-boson is inserted into the diagrams of Fig. 10, the inverse photoemission spectrum obtained [18] is illustrated in Fig. 11. It is seen to have the essence of the two-peak structure found in the data, but the physics deserves comment. First the quasiparticle structure is rather peaked, due to an unrealistic feature of model A. Secondly the weight of the satellite peak is not, as we would have expected in the $d^9 - d^{10}$ satellite, but in the interband transition. The transition involves excitation of a hole between the two solutions of Eq. (9), accompanying the absorption of an electron into the empty part of the band.

This is the first time that the calculation of a spectral satellite has been done for the coherent heavy fermion state. The fact that the satellite is found to be predominantly interband, and thus intimately related to the coherence of the HF state, is interesting and is further evidence for the existence of the heavy fermion state in the high-T_c materials. Note that in conventional heavy fermion materials, which have very high masses, the interband splitting is too small to be resolvable; it is the much larger $f^1 - f^0$ splitting which is resolved.

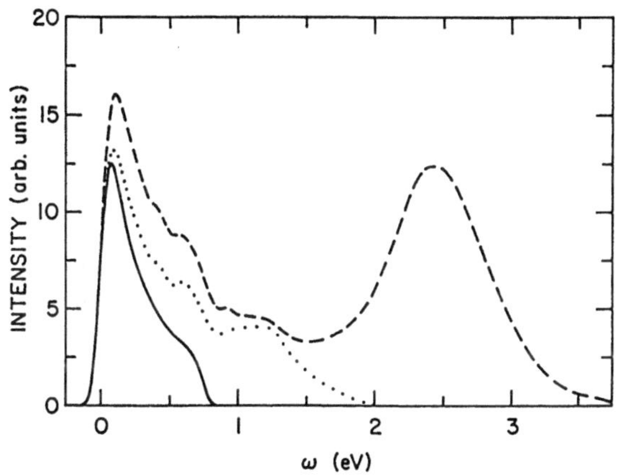

Fig. 11 Inverse photoemission spectrum calculated from Fig. 10. Full line is quasiparticle contribution; dotted line sum of quasiparticle and $d^9 - d^{10}$ contributions; dashed line is sum of these two plus the interband contribution (involving transitions between the two bands schematised at Fig. 1, right).

8. High-T_c Superconductivity

We have only brief space here to discuss the essentials of the experimental data on superconductivity. The nature of the gap parameter can be deduced from two critical inputs: a) Josephson tunneling between conventional and high-T_c superconductors [25] and b) direct [26] and other measurements of the London penetration depth. Experiment (a) suggests that the order parameter is a singlet, while experiment (b) indicates a nearly isotropic gap in the plane with no zeros. The combination of these results indicates conventional s-wave pairing as in BCS superconductivity.

However, the exceedingly small isotope shift, particularly in the 123 material [27], does not suggest phonon-mediated pairing. The search then, is for a rather strong attractive interaction, for the interaction must overcome the intrinsic repulsion between quasiparticles. The investigation needs to take into account the intrinsic screening out of charge fluctuations in the heavy fermion fluid, assuming that to be the correct description of the normal state.

We shall discuss here only one pairing interaction, in which the finite value of U generates an attractive interaction [16,17,28,29]. One way to see this is to introduce, in addition to the d^{10} , or b-boson, an additional d^8, or a-boson [16,17]. Then pairing proceeds via the diagram of Fig. 12a, which gives an interaction $\sim - V^2/\tilde{E}_2$, where $\tilde{E}_2$ is the d^8 energy with some corrections (the

Fig. 12 Two versions of finite-U pairing mechanism. a) a d^8 - or A-boson (zig-zag line) is introduced; b) performing a lowest-order canonical transformation on the finite-U ALM, the infinite-U ALM is retrieved but with the added exchange interaction J_{pd}.

bare d^8 energy is E_2 , obeying $U = E_2 - 2E_1$). The procedure can be formalised under SU(2)×O(N/2) symmetry.

Perhaps more directly, Spalek [29] has made a canonical transformation to eliminate U from the finite-U Anderson lattice model, to order 1/U. The result is an infinite-U model containing an extra exchange term J_{pd} (Fig. 12b), which is essentially identical to the interaction derived from the a-boson. It is found that $J_{pd} \sim -V^2/\tilde{E}_2$.

In either case, an approximate expression (not in practice used for numerical evaluation) for T_c in model A is

$$T_c = 1.14\omega_c \exp[-1/(\frac{4V^2\rho_0}{\tilde{E}_2} - 0.5)] \ . \tag{12}$$

Eq. (12) has a similar structure to the usual BCS formula, where the attractive term is essentially 4 $J_{pd}\rho_0$, with ρ_0 the bare oxygen DOS in the ε_k band. The repulsive term, replacing the usual empirical constant μ^*, comes from the single b-boson exchange [30] illustrated in Fig. 13. It is *non-empirical.* The cut-off ω_c is essentially a mean half band width of the heavy fermion band, and is of order 0.5 eV.

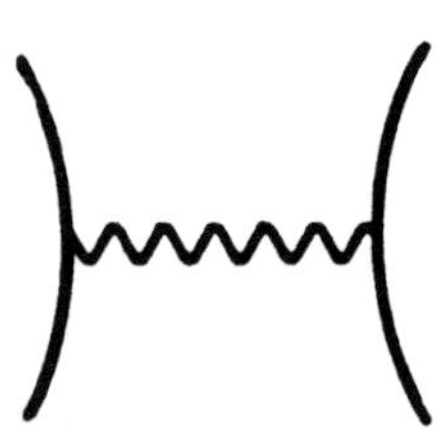

Fig. 13 Direct exchange of a single slave boson (B-boson) generates the quasiparticle-quasiparticle repulsion (which in BCS theory is modelled by μ^*) in a non-empirical way.

Many of the properties of high-T_c superconductivity calculated in the present leading-N approximation are BCS-like, e.g. a gap $2\Delta = 3.5k_BT_c$ is predicted.

There is one well-defined but almost undetectable effect which would act as a 'smoking gun' experiment for this coupling mechanism. To notice it we must return to the a-boson formulation of Fig. 12a. In this formulation, an expectation value $< a > \neq 0$, for $T < T_c$, appears. Unlike the b-boson expectation value, the appearance of nonzero $< a >$ signals a real phase transition. In terms of measurable quantities, the kind of thing that would be observed is an increase in the d^8 fraction from, say, 0.04 above T_c to 0.041 at $T = 0$ - not an easy effect to detect !

9. Trends in T_c

One approach to learning about the mechanism of high temperature superconductivity is to look at trends in experimental data. The same procedure resulted in the discovery of the isotope shift in phonon-mediated superconductivity. Here we consider the following trends: a) the saturation of T_c at a certain doping, b) the pressure effect on T_c, and c) the effect on T_c of substituting Ba for Sr.

The saturation of T_c in our theory comes from a normal state effect: as doping is increased in a given high-T_c material, a point is reached where added holes no longer reside in the planes, but start to go into other orbitals. This is illustrated in Fig. 14, where it is seen that soon after 10% doping in the 214

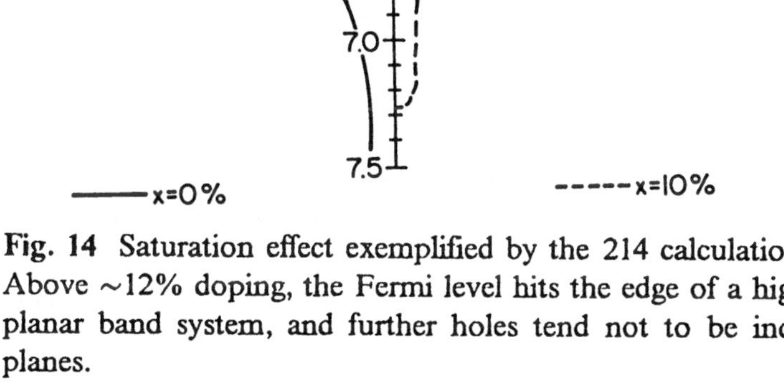

Fig. 14 Saturation effect exemplified by the 214 calculation of Ref. 17. Above ~12% doping, the Fermi level hits the edge of a high DOS non-planar band system, and further holes tend not to be included in the planes.

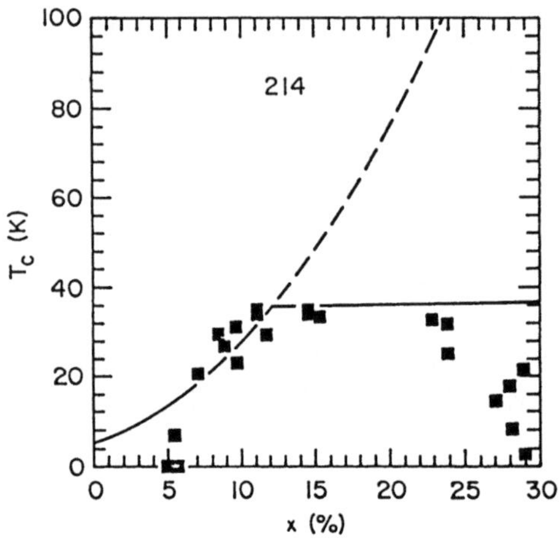

Fig. 15 T_c vs. doping in the 214 material. Full curve, calculation of Ref. 16, including saturation effect; dashed curve, continuation with no saturation effect; squares, data of Torrance et al.

we expect further holes to go into the high DOS non-planar band rather than into the planar band which has a relatively low DOS. The planar doping is then expected to be pegged, leading to a saturation of T_c which is indeed seen experimentally (Fig. 15). Recently Uemura et al. [31] have published work showing evidence for higher T_c plateaus in the 123, 2212 and 2223 materials. These could be explained along the same lines as the 214 plateau.

This argument suggests that an important consideration in achieving high T_c is the location of the 'non-planar' bands, which are seen to control the maximum planar doping.

Our approach to the pressure coefficient of T_c in the 214 material (it is small in many other materials) is based on the size effect well known in rare earth intermediate valence materials which are in some ways the closest analogue to the high-T_c materials. It is assumed that the size of the d-shell decreases in the order $d^{10} > d^9 > d^8$, so there will be a shift in the energy levels of the atomic configurations under pressure like that illustrated in Fig. 16. The lowering of E_2 under pressure is expected to increase T_c.

It might be thought that these shifts will be dominated by shifts in the hopping matrix elements V. However the example of the Intermediate Valence compound YbCuAl, whose characteristic energy scale T_K decreases under pressure, suggests that the contrary is indeed the case. If then we ignore the V-shifts under pressure, we can deduce from Fig. 16 the quantity [16]

$$\frac{d \ln T_c/dp}{d \ln \chi/dp} = 5.8 \quad . \tag{13}$$

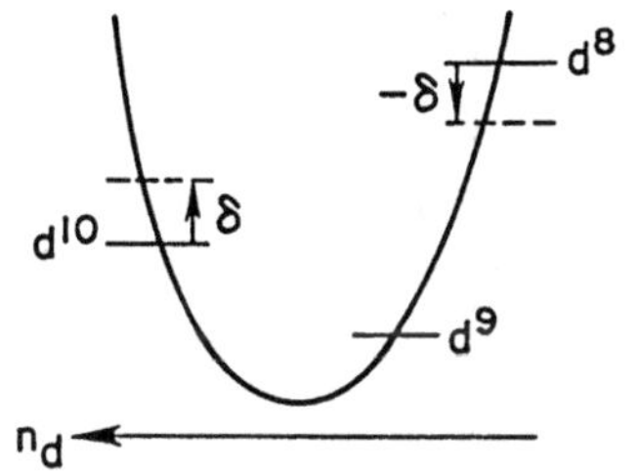

Fig. 16 Energy level of Cu $d_{x^2-y^2}$ shell showing assumed pressure effect by shift δ in d^8 and d^{10} levels relative to d^9.

This is quite close to the experimental result, 3.9, of Schilling [32], and is brought even closer by considering the effect of the residual electron-phonon scattering.

Finally, we turn to a more invasive probe, the substitution of Ba for Sr in a range of materials. We exclude the 214 material, with its high pressure coefficient, so as to exclude the coupling of T_c purely to the size difference between these two ions. It is found by Ronay and Newns [33] that the substitution of Ba for Sr systematically leads to a ~ 15% increase in T_c in non-214 materials. We interpret the effect as a reduction in U (and hence in E_2) in the Ba materials due to screening - essentially a different situation to that in Fig. 16 in that *both* the d^{10} and the d^8 states move down as screening increases.

The reduction in U in better screened environments is a well known effect in Surface Physics and is responsible for the smaller values of U in metals like Ni relative to their oxides. The more efficient screening by Ba than by Sr was originally attributed [33] to the higher polarizability of the fractionally charged positive ions. A more precise description might be in terms of screening by virtual transitions from the filled oxygen-p like valence band to the unoccupied alkaline earth like states in the conduction band. The lowest lying of these empty conduction band states is 4d-like in Sr and 5d-like in Ba substituted materials. The 5d of Ba is particularly low-lying, and it may be this more easily accessible acceptor state for virtual interband transitions which is responsible for the more efficient screening, and hence higher T_c's, in the Ba-substituted materials.

As the data base increases with further and increasingly more reliable experimental work, we expect more definitive examination of the foregoing and other trends of T_c with various parameters and substitutions to become possible.

10. Conclusion

Taking into account the strong electron-electron interaction U is surely the most important task in the theory of high temperature superconductivity. That this problem has been defined almost immediately following the discovery of this new class of materials is the product of the genius of one individual and has been a tremendous bonus to the field. In the detailed approach to the U-problem two schools of thought can be distinguished. One school, associated with RVB, has its roots in magnetic problems in oxides involving the Heisenberg and Hubbard models. Another school developed spectroscopy for illuminating the energy level scheme in oxides, has links with heavy fermion materials, and employs the Anderson model.

Our approach has closely followed the heavy fermion approach, being based on the Fermi liquid solution to the Anderson lattice model. This approach has we think been useful, in suggesting numerous correlations between data many of which are upheld, as we have shown in this article. In particular it is hard to escape from the conclusion that the 'smoking gun' of angle resolved photoemission supports a heavy fermion state or something close to it. If the outstanding problems (linear resistivity, Hall coefficient), which have not so far received a convincing solution in the heavy fermion framework, indeed require some modification to the Fermi liquid state, as RVB proponents argue, we believe that the changes must be in some sense minor.

Higher energy spectroscopic information (e.g. satellites in photoemission and inverse photoemission) also plays an essential role, for the high and low energy scale phenomena are intimately linked in problems involving scaling, of which this is probably an example. In high energy physics this characteristic has made it an expensive discipline.

If the widely accepted view that the pairing is singlet s-wave is correct, phonons or other low energy charge excitations seem unlikely to be responsible for pairing. This is because, in a heavy fermion state, the coupling of quasiparticles near the Fermi level to charge excitations is weak. A high frequency interaction, perhaps involving extended s character, seems likely. We have proposed the exchange interaction J_{pd}, which lives on a copper -oxygen bond, but it is impossible to be conclusive at this stage.

References

1. V. J. Emery, Phys. Rev. Lett. **58**, 2794 (1987); H. Eskes and G. A. Sawatzky, Phys. Rev. Lett. **61**, 1415 (1988).

2. G. Shirane, Y. Endoh, R. J. Birgenau, M. A. Kastner, Y. Hidaka, M. Oda, M. Suzuki, and T. Murakami, Phys. Rev. Lett. **59**, 1613 (1987).

3. F. C. Zhang and T. M. Rice, Phys. Rev. B **37**, 3759 (1988).

4. P. W. Anderson, G. Baskaran, Z. Zou, and T. Hsu, Phys. Rev. Lett **58**, 2790 (1987).

5. C. L. Kane, P. A. Lee, and N. Read, Phys. Rev. B **39**, 6880, (1989).

6. R. B. Laughlin, Phys. Rev. Lett. **60**, 2677 (1988).

7. R. S. List, A. J. Arko, Z. Fisk, S.-W. Cheong, S. D. Conradson, J. D. Thompson, C. B. Pierce, D. E. Peterson, R. J. Bartlett, J. A. O'Rourke, N. D. Shinn, J. E. Schirber, C. G. Olson, A.-B. Yang, T.-W. Pi, B. W. Veal, A. P. Paulikas, and J. C. Campuzano, J. Vac. Sci. Tech., in press.

8. T. Takahashi, H. Matsuyama, H. Katayama-Yoshida, Y. Okabe, S. Hosoya, K. Seki, H. Fujimoto, M. Sato, and H. Inokuchi, Phys. Rev. B **39**, 6636 (1989).

9. J. Fink, N. Nucker, H. Romberg, M. Alexander, S. Nakai, R. Manzke, T. Buslaps, and R. Claessen, Proc. of Int. Winterschool on Electronic Properties of Polymers and Related Compounds, Kirchberg, Austria, to be published in Springer Series in Solid State Science.

10. L. Hoffman, A.A. Manuel, M. Peter, E. Walker, and M. A. Damento, Physica C **153**, 129 (1988); A. Bansil, R. Pankaluoto, R. S. Rao, P. E. Mijnarends, W. Dlugosz, R. Prasad, and L. C. Smedskjaer, Phys. Rev. Lett. **61**, 2480 (1988).

11. P. Coleman, Phys. Rev. B **25**, 5255 (1987); Phys. Rev. B **29**, 3035 (1987); Phys. Rev. B **35**, 5072 (1987).

12. N. Read and D. M. Newns, J. Phys. C. **16**, 3273 (1983); N. Read, J. Phys. C. **18**, 2651 (1983); N. Read and D. M. Newns, Adv. Phys. **36**, 800 (1989).

13. A. J. Millis and P. A. Lee, Phys. Rev. B **35**, 3394 (1987).

14. G. Kotliar, P. A. Lee, and N. Read, Physica C **153-155**, 538 (1988).

15. C. A. Balseiro, M. Avignon, A. G. Rojo, and B. Alascio, Phys. Rev. Lett. **62**, 2624 (1989).

16. D. M. Newns, Phys. Rev. B **36**, 5595 (1987); D. M. Newns. M. Rasolt, and P. C. Pattnaik, Phys. Rev. B **38**, 6513 (1988).

17. D. M. Newns, P. C. Pattnaik, M. Rasolt, and D. A. Papaconstantopoulos, Phys. Rev. B **38**, 7033 (1988).

18. P. C. Pattnaik and D. M. Newns, submitted to Phys. Rev. B.

19. C. G. Olson, R. Liu, A.- B. Yang, D. W. Lynch, A. J. Arko, R. S. List, B. W. Veal, Y. C. Chang, P. Z. Jiang, and A. P. Paulikas, preprint.

20. A. E .Ruckenstein, P. J. Hirschfeld, and J. Appel, Phys. Rev. B **36**, 857 (1987).

21. R. L. Greene, H. Maletta, T.-S. Plaskett, J. G. Bednorz, and A. Muller, Solid State Commun. **63**, 379 (1987).

22. A. Junod, A. Bezinge, D. Caltani, J. Cors, M. Decroux, O. Fischer, P. Genoud, L. Hoffman, J.-L. Jorda, J. Muller, and E. Walker, Jpn. J. Appl. Phys. **26**, Suppl 26-3, 1119 (1987).

23. M. S. Hybertson, M. Schluter, and N. E. Christenson, Phys. Rev. B **39**, 9028 (1989); A. K. McMahan, R. M. Martin, and S. Satpathy, Phys. Rev. B **38**, 6650-66 (1988).

24. W. Drube, F. J. Himpsel, G. V. Chandrasekhar, and M. W. Shafer, submitted to Phys. Rev..

25. J. S. Tsai, K. Kubo, and J. Tabuchi, Phys. Rev. Lett. **58**, 1979 (1987), H. Akoh, F. Shinoki, M. Takashi, and S. Takada, Ja. J. Appl. Phys. **27**, L519 (1988).

26. L. Krusin-Elbaum, R. L. Greene, F. Holtzberg, A. P. Malozemoff, and Y. Yeshurun, Phys. Rev. Lett. **62**, 217 (1989).

27. L. C. Bourne, A. Zettl, T. W. Barbee, and M. L. Cohen, Phys. Rev. B **36**, 3990 (1987).

28. B. H. Brandow, Solid State Commun. **69**, 915 (1989).

29. J. Spalek, Phys. Rev. B **37**, 533 (1988).

30. M. Lavagna, P. A. Lee, and A. J. Millis, Phys. Rev. Lett. **58**, 266 (1987).

31. Y. J. Uemura, G. M. Luke, B. J. Sternlieb, J. H. Brewer, J. F. Carolan, W. N. Hardy, R. Kadono, J. R. Kempton, R. F. Kiefl, S. R. Kreitzman, P. Mulhearn, T. M. Risemann, D. L. Williams, B. X. Yang, S. Uchida, H. Takagi, J. Gopalakrishnan, A. W. Sleight, M. A. Subramanian, C. L. Chien, M. Z. Cieplak, G. Xiao, V. Y. Lee, B. W. Statt, C. E. Stronach, W. J. Kossler, and X. H. Yu, Phys. Rev. Lett. **23**, 2317 (1989).

32. C. Allgeier, J. S. Schilling, and K. Andres, International Conference on High Temperature Superconductors and Materials and Mechanisms of Superconductivity, Interlaken 1988, Abstract.

33. M. Ronay and D. M. Newns, Phys. Rev. **39**, 819 (1989).

Quasi-Particles in the Degenerate Lattice Anderson Model

Y. Kuroda [1], *Y. Ōno* [1], *T. Matsuura* [1], *and H. Jichu* [2]

[1]Physics Department, Nagoya University, Chikusa-ku, Nagoya, Japan
[2]Aichi College of Technology, Gamagori, Aichi, Japan

Abstract. We present a modified bosonic version of the infinite-U lattice Anderson model with large orbital degeneracy N, which yields finite intersite coherence effects even in the limit $N\to\infty$ in contrast to the traditional models. Using the new model, we study the renormalized energies and life-times of the conduction electrons, the effective electron-electron interaction and the effects of the strong electron correlation on the electron-phonon interaction rigorously and explicitly in the limit N>>1. These results show that the system is in the Fermi liquid state at $T\lesssim T_0$, where T_0 is the coherence temperature, while it behaves just like a collection of single Kondo impurities at $T\gg T_0$.

1. Introduction

The lattice Anderson model has been used extensively by many physicists during last decade to study the heavy electron systems and more recently to study the oxide superconductors. However, unfortunately, no well-established method has been developed to treat the model rigorously. Therefore any theoretical arguements based on the model must have suffered from more or less inevitable ambiguities arising from their approximation procedures. This is a significant contrast to the case in the impurity Anderson model which can be treated rigorously to a considerable extent, say by the Bethe Ansatz method etc. Now it is highly desired to have methods to treat the model somehow rigorously, which enable us to extract the essential features of the problems.

Among the various methods taken so far the boson technique originally introduced by P. Coleman[1] is one of the most extensively used and, we believe, the most promising to be developed to the end mentioned above. The method has been extensively used by Jin and Kuroda[2] to treat the impurity Anderson model as well as the lattice Anderson model with an infinite intra-atomic Coulomb repulsion U rigorously in the large limit of the orbital degeneracy N and to obtain a unified view describing the high temperature Kondo regime as well as the low temperature Fermi liquid regime for each model. However, in such traditional model as used in [1] and [2], the intersite coherence effects in the lattice model appear always only as higher order corrections in power of (1/N), which is not convenient to be used for studying the heavy electron systems nor the oxide superconductors.

In the present study we modified the traditional model so as to avoid such difficulties and to obtain the finite intersite coherence effects even in the limit $N\to\infty$. Using such model we study the renormalized energies and the life times of the conduction electrons, properties of the effective electron-electron interaction and the effects of the strong electron correlation on the electron-phonon interaction and show that the system is in the Fermi liquid state at $T \lesssim T_0$ where T_0 is the coherence temperature while it behaves just like a collection of the single Kondo impurities at $T \gg T_0$.

In Section 2, we describe our new model and the formulation. In Section 3, we calculate the single particle Green's functions. In Section 4, we study the effective interactions among the renormalized electrons. Finally concluding remarks are given in Section 5.

2. Model and Formulation

The present study is a straightforward extension of the study given in [2]. We begin with the same model as taken there, i.e. the bosonic version of the infinite-U lattice Anderson model given by

$$H = H_f + H_s + H_{sf} \quad , \tag{1}$$

with

$$H_f \equiv \sum_{i=1}^{N_L} \sum_{m=1}^{N} \varepsilon_f \, f_{im}^{+} f_{im} \quad , \tag{2}$$

$$H_s \equiv \sum_{k} \sum_{\sigma} \varepsilon_k \, C_{k\sigma}^{+} C_{k\sigma} \quad , \tag{3}$$

$$H_{sf} \equiv N_L^{-1/2} \sum_{k} \sum_{\sigma} \sum_{i} \sum_{m} [V(k\sigma;m) \, C_{k\sigma}^{+} f_{im} b_i^{+} e^{-ik\cdot R_i} + h.c.] \quad , \tag{4}$$

where f_{im}^{+} is for the f-electron localized at lattice site i and in the m-th spin-orbit state, $C_{k\sigma}^{+}$ is for the itinerant s-electron with wave vector k and spin σ and b_i^{+} is for the slave boson denoting the empty state at lattice i. This model is equivalent to the infinite-U Anderson model if treated under the local constraints,

$$Q_i \equiv \sum_{m=1}^{N} f_{im}^{+} f_{im} + b_i^{+} b_i = 1, \ (i=1,2,\cdots, N_L) \ . \tag{5}$$

In order to enforce the local constraints strictly while calculating physical quantities, we introduce a grand canonical ensemble as

$$H_\lambda \equiv H + \sum_{i=1}^{N_L} \lambda_i \, Q_i \quad , \tag{6}$$

and calculate an expectation value of an operator $\hat{O}$ in such a way as

$$\langle \hat{O} \rangle = \lim_{\{\lambda_i\}\to\infty} [\langle \hat{O} \prod_{i=1}^{N_L} Q_i \rangle_\lambda / \langle \prod_{i=1}^{N_L} Q_i \rangle_\lambda] \tag{7}$$

where $\langle \hat{O} \rangle_\lambda$ is defined by

$$\langle \hat{O} \rangle_\lambda \equiv \mathrm{Tr}\, [e^{-\beta H_\lambda} \hat{O}] / \mathrm{Tr}\, [e^{-\beta H_\lambda}] . \tag{8}$$

In actual calculations, we apply the conventional Feynman diagram expansion method, where we divide H_λ into two parts, the diagonal part

$$H_\lambda^{(0)} \equiv H_f + H_s + \sum_{i=1}^{N_L} \lambda_i Q_i , \tag{9}$$

and the off-diagonal part H_{sf}, and treat H_{sf} as a perturbation.

Now we modify our model slightly so that we might make some rigorous investigations and elucidate some of the essential features in the model. First we divide the total degrees of freedom for the s-electrons into N subspaces somehow so that we might rewrite H_s and H_{sf} as

$$H_s \equiv \sum_{m=1}^{N} \sum_{k_m} \varepsilon_{k_m} C_{k_m}^{+} C_{k_m} , \tag{10}$$

$$H_{sf} \equiv N_L^{-1/2} \sum_{m=1}^{N} \sum_{k_m} \sum_{i} V_{sf} [C_{k_m}^{+} f_{im} b_i^{+} e^{-ikR_i} + h.c.] . \tag{11}$$

Secondly, we assume the large limit of the orbital degeneracy, i.e. $N\to\infty$, but with keeping the total degrees of freedom for the s-electrons unchanged from those of the original model, i.e.

$$\sum_{m=1}^{N} \sum_{k_m} 1 = \sum_{k} \sum_{\sigma} 1 . \tag{12}$$

This means that the summation over yields a factor of O(N) while the summation over k_m yields a factor of O(1/N) compared to the ordinary summation over k. This limiting procedure is a contrast to the traditional one, where the limit $N\to\infty$ is taken under the condition, NV_{sf}^2=const. These two types of limiting procedures make no difference for studying the impurity system but do some for the lattice system as will be shown later.

3. Single Particle Green's Functions

Now we introduce the unperturbed component of the single particle Green's function for the f-electrons, s-electrons and slave-bosons defined in an ordinary way[2] as

$$G_{im}^{(0)}(i\omega_n) \equiv (i\omega_n - \lambda_i - \varepsilon_f)^{-1} , \tag{13}$$

$$G^{(0)}_{k_m}(i\omega_n) \equiv (i\omega_n - \varepsilon_{k_m})^{-1} \ , \qquad (14)$$

$$D^{(0)}_{i}(i\nu_n) \equiv (i\nu_n - \lambda_i)^{-1} \ , \qquad (15)$$

where $\omega_n \equiv (2n+1)\pi T$ and $\nu_n \equiv 2n\pi T$ with an integer n. Then, Feynman diagrams are drawn in terms of these single particle Green's functions and the vertex V_{sf}, and are rearranged into groups according to their power with respect to (1/N).

3.1. Most Dominant Terms

The most dominant terms in power of (1/N) contributing to the perturbed single particle Green's functions are described by the Dyson equations illustrated in Fig. 1. Their analytic forms are given by

$$G_{im}(i\omega_n) \simeq G^{(0)}_{i}(i\omega_n) \ , \qquad (16)$$

$$G_{k_m}(i\omega_n) \simeq [i\omega_n - \varepsilon_{k_m} - \Sigma^{(0)}_{k_m}(i\omega_n)]^{-1} \ , \qquad (17)$$

$$D_i(i\nu_n) \simeq [i\nu_m - \lambda_i - \Pi^{(0)}_{i}(i\nu_m)]^{-1} \ , \qquad (18)$$

where the self-energies are defined as

$$\Sigma^{(0)}_{k_m}(i\omega_n) \equiv \lim_{\lambda_i \to \infty} [V^2_{sf}\, T \sum_{\ell} G^{(0)}_{i}(i\omega_n + i\nu_\ell)\, D_i(i\nu_\ell)/\langle Q_i \rangle_\lambda] \ , \qquad (19)$$

$$\Pi^{(0)}_{i}(i\nu_\ell) \equiv N_L^{-1}\, V^2_{sf}\, T \sum_{n} \sum_{m} \sum_{k_m} G_{k_m}(i\omega_n) G^{(0)}_{i}(i\omega_n + i\nu_\ell) \ . \qquad (20)$$

Although equations (17)-(20) form a set of coupled equations, we can solve them analytically. First we assume that $\Sigma^{(0)}_{km}(i\omega_n)$ is given by

$$\Sigma^{(0)}_{k_m}(i\omega_n) \simeq \alpha\, V^2_{sf}(i\omega_n - T_K)^{-1} \ , \qquad (21)$$

where α and T_K are the parameters to be determined self-consistently. In fact, it is straightforward to see that using equations (21), (17) (20) and

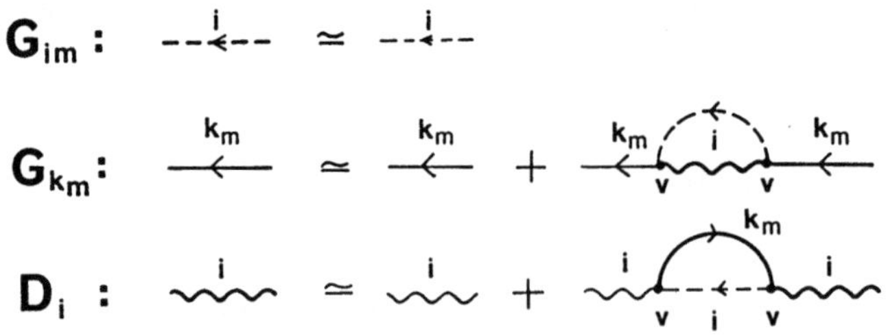

Fig. 1 Dyson equations for the single particle Green's functions. The broken, solid and wavy lines denote the f-, s-, and slave boson Green's functions, respectively.

(18) we can calculate $\Sigma^{(0)}_{km}(i\omega_n)$ in (19) and obtain a set of coupled equations to determine α and T_K, which simplifies in the sd-limit, i.e. $V_{sf}^2/2D \ll |\varepsilon_f|$, as

$$T_K - \varepsilon_f - (V_{sf}^2/D)\ln(D/T_K) = 0 , \tag{22}$$

$$\alpha = (T_K D/V_{sf}^2)[1 + N\exp(-\beta T_K)]^{-1} . \tag{23}$$

In deriving (22) and (23), we have replaced the summations over k_m by integrations over ε_k with a constant density of state (1/2ND) for $-D \le \varepsilon_{km} \le D$. Finally we obtain the s-electron Green's function given by

$$G_{k_m}(i\omega_n) = \sum_{\gamma=\pm} A^{(\gamma)}_{k_m}[i\omega_n - E^{(\gamma)}_{k_m}]^{-1} , \tag{24}$$

with

$$E^{(\gamma)}_{k_m} \equiv \frac{1}{2}[(\varepsilon_{k_m} + T_K) + \gamma\{(\varepsilon_{k_m} - T_K)^2 + 4\alpha V_{sf}^2\}^{1/2}] , \tag{25}$$

$$A^{(\gamma)}_{k_m} \equiv (E^{(\gamma)}_{k_m} - T_K) / (E^{(\gamma)}_{k_m} - E^{(-\gamma)}_{k_m}) . \tag{26}$$

Thus we obtain the renormalized bands for the s-electrons. This is essentially due to the intersite coherence effect. We note that $E^{(\gamma)}_{km}$ describes heavy electrons near the Fermi level in the sd-limit where $\alpha \ll 1$. We also note that such intersite coherence effect diminishes at $T \gg T_0$ ($T_0 \equiv T_K/\ln N$) as can be seen from (23). So we may call T_0 the coherence temperature. If we would take the conventional limiting procedure, $N \to \infty$ with NV_{sf}^2=const, the intersite coherence effect, i.e. $\Sigma^{(0)}_{km}$ in (17) drops out as a higher order correction of O(1/N) in the present stage.

3.2. Higher Order Corrections

Using the results obtained in Sec.3.1, we calculate higher order self-energy corrections of O(1/N) to the s-electron Green's function G_{km}, whose diagrammatic expressions are illustrated in Fig. 2. The diagram in

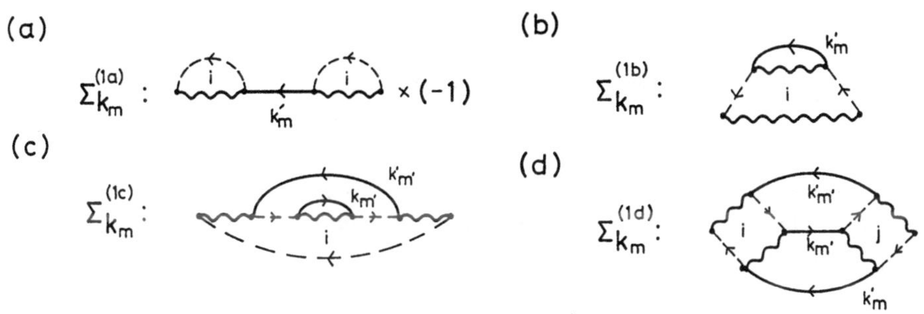

Fig. 2 Self-energy corrections of O(1/N) to the s-electron Green's function G_{k_m}.

Fig. 2(a) violates the local constraint (5), but has been included in writing (17) together with (19), because it simplifies the resultant expressions while its contribution is smaller by O(1/N) compared to the most dominant terms. But when we study the corrections of O(1/N) such term should be subtracted. So $\Sigma_{km}^{(1a)}$ carries the factor -1. This term gives a finite life time to the renormalized electrons, which is given in the sd-limit by

$$\mathrm{Im}\ \Sigma_{k_m}^{(1a)}(0+i\delta) = -(2\pi D/N). \tag{27}$$

Similarly we can calculate the contributions $\Sigma_{km}^{(1b)}$ and $\Sigma_{km}^{(1c)}$ from the diagrams in Figs. 2(b) and 2(c) to find

$$\mathrm{Im}\ \Sigma_{k_m}^{(1b)}(0+i\delta) = -\mathrm{Im}\ \Sigma_{k_m}^{(1a)}(0+i\delta)\ , \tag{28}$$

$$\mathrm{Im}\ \Sigma_{k_m}^{(1c)}(0+i\delta) = 0\ . \tag{29}$$

Here we note that in the impurity problem the term corresponding to the one in Fig. 2(b) leads to the residual resistance, while in the lattice problem the term is canceled exactly by the term ensuring the local constraint strictly. In this sense, the local constraint plays a crucial role to take account of the intersite coherence effect correctly. Finally, the contribution $\Sigma_{km}^{(1d)}$ from the diagram in Fig. 2(d) gives Fermi liquid type life times to the renormalized electrons. In the sd-limit we find

$$\mathrm{Im}\ \Sigma_{k_m}^{(1d)}(\omega+i\delta) \simeq (\pi D/12N\ T_K^2)\{(\pi T)^2 + \omega^2\}\ ,$$
$$(\text{at } T,|\omega|\ll T_0)\ . \tag{30}$$

In fact, the vertex parts in Fig. 2(d) should have been replaced by the full vertex Γ defined by the BS equation shown in Fig. 3 in order to study the corrections of O(1/N) correctly. For that sake we have to take very careful account of the analytic structure in Γ, which is still in progress and will be reported separately.

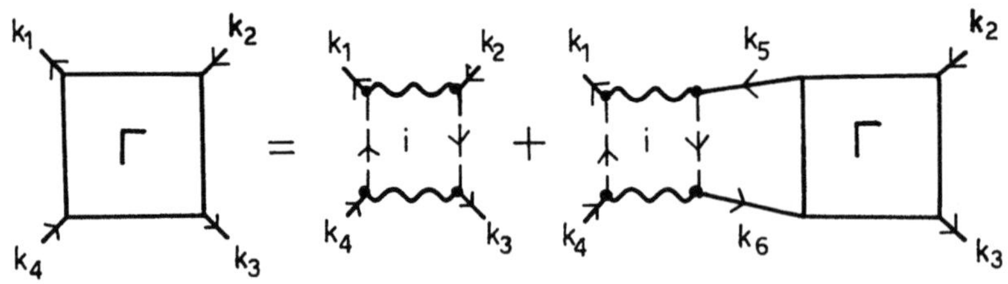

Fig. 3 Bethe-Salpeter equation for the s-electron - s-electron vertices of $O(1/N^0)$. Conventional four-dimensional momentum notation such as $k_1 \equiv (k_{m1}, i\varepsilon_1)$ is used.

4. Pairing Interactions

The present model and formulation enable us to investigate the interactions among the renormalized s-electrons systematically. In the present study we briefly discuss the electron-electron interaction and the phonon mediated interaction to form the Cooper pairs.

4.1 Electron-Electron Interaction

From the vertex function $\Gamma(k_1,k_2,k_3,k_4)$ we can derive the effective interaction $\tilde{\Gamma}(k_1,k_2,k_3,k_4)$ among the renormalized electrons with energies $E_{k1}^{(\gamma 1)}$, $E_{k2}^{(\gamma 2)}$, $E_{k3}^{(\gamma 3)}$, and $E_{k4}^{(\gamma 4)}$ by the relation,

$$\tilde{\Gamma}(k_1,k_2,k_3,k_4) \equiv \Gamma(k_1,k_2,k_3,k_4)[A_{k_1}^{(\gamma_1)} A_{k_2}^{(\gamma_2)} A_{k_3}^{(\gamma_3)} A_{k_4}^{(\gamma_4)}]^{1/2} . \tag{31}$$

Then the Cooper pairing interaction $\tilde{\Gamma}(k_1,k_2) \equiv \tilde{\Gamma}(k_1,-k_2,-k_1,k_2)$ is given in the sd-limit in the weak-coupling limit by

$$\tilde{\Gamma}(k_1,k_2) = T_K \, [1 + T_K \chi^{(0)}(k_1 - k_2)]^{-1} , \tag{32}$$

with

$$\chi^{(0)}(q) = (ND^2/T_K) T \sum_n \sum_k G_k(i\omega_n) G_{k+q}(i\omega_n) . \tag{33}$$

In order to determine the q-dependence of $\chi^{(0)}(q)$ we need to use an explicit form of the band structure ε_k. Such study will be reported separately. Here we only note that $\tilde{\Gamma}$ given in (32) is entirely due to charge fluctuation and is positive for arbitrary k_1 and k_2. Therefore this interaction is unfavorable to any S-like pairing state.

4.2 Phonon-Mediated Interaction

Using the present model and formulation we can investigate the effects of the strong electronic correlation on the electron-phonon interaction. To this end, we introduce a model for the electron-phonon coupling as

$$\begin{aligned} H_{e\ell\text{-}ph} \equiv & \frac{1}{2} \sum_m \sum_{k_m} \sum_{q_m} \frac{\partial}{\partial \Omega} (\varepsilon_{k_m} + \varepsilon_{k_m+q_m}) C_{k_m}^{+} C_{k_m+q_m} \sum_i \delta\hat{\Omega}_i e^{iq_m \cdot R_i} \\ & + \sum_i \sum_m \frac{\partial \varepsilon_f}{\partial \Omega} f_{im}^{+} f_{im} \delta\hat{\Omega}_i \\ & + \frac{1}{\sqrt{N_L}} \sum_m \sum_i \sum_{k_m} \frac{\partial V_{sf}}{\partial \Omega} [C_{k_m}^{+} f_{im} b_i^{+} \delta\hat{\Omega}_i e^{-ik_m R_i} + h.c.] , \end{aligned} \tag{34}$$

$$H_{ph} \equiv \sum_m \sum_{q_m} \omega_{q_m} (a_{q_m}^{+} a_{q_m} + \tfrac{1}{2}) , \tag{35}$$

where Ω is the molar volume of the system and the local change of the molar volume, $\hat{\delta\Omega}$ is expressed in terms of the phonon operator a_{q_m} as

$$\delta\hat{\Omega}_i \equiv \Omega \sum_m \sum_{q_m} (2N_L M \omega_{q_m})^{-1/2} |q_m| (a_{q_m} + a^+_{-q_m}) e^{iq_m R_i} . \qquad (36)$$

Some typical diagrams contributing to the renormalized electron-phonon vertex $\Lambda_i(k_1,k_2;q)$ in the approximation to the most dominant terms in power of (1/N) are illustrated in Fig. 4. We have 4 more similar terms contributing to Λ_i. By introducing the effective electron-phonon interaction $\tilde{\Lambda}_i$ defined by

$$\tilde{\Lambda}_i(k_1,k_2;q) \equiv \Lambda_i(k_1,k_2;q)(A_{k_1}^{(\gamma_1)} A_{k_2}^{(\gamma_2)})^{1/2} , \qquad (37)$$

we find in the sd-limit

$$\tilde{\Lambda}_i(k_1,k_2;q) \simeq (|q|/(8M\omega_q)^{1/2}) \exp(iq\cdot R_i)\cdot\eta(i\varepsilon_1+i\varepsilon_2) ,$$
$$(\text{for } |\varepsilon_1|,|\varepsilon_2|<<T_K) . \qquad (38)$$

where $\eta\equiv(\Omega/T_K)(\partial T_K/\partial\Omega)$. Using this result we obtain the effective pairing interaction $\hat{V}(k_1,k_2)$ as

$$\hat{V}(k_1,k_2) \sim - \eta^2(\varepsilon_1+\varepsilon_2)^2 / D . \qquad (39)$$

If η is as large as ~100 (in fact, $\eta\simeq80$ for $CeCu_2Si_2$), $\hat{V}$ may dominate Γ in (32) and we may find a S-like pairing state. In such a case, no isotope effect in the ordinary sense will not be obserbed, because $\hat{V}$ diminished at $|\varepsilon_1|,|\varepsilon_2|\gtrsim T_K$ instead of ω_D, Debye frequency.

The results obtained in this section recover explicitly the results obtained in a more general Fermi liquid theoretical study of the original Anderson model by the present authors[3].

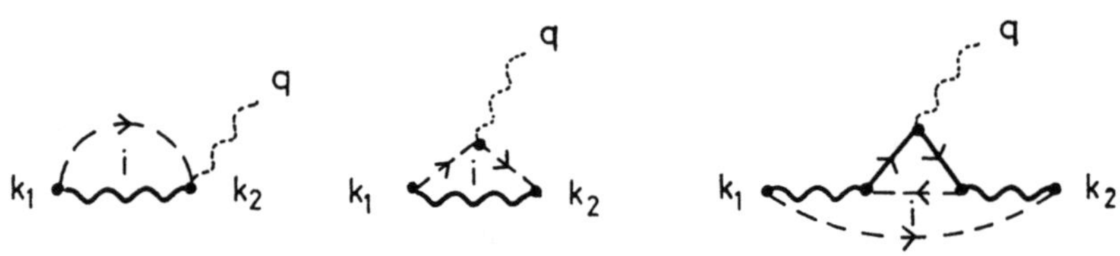

Fig. 4 Typical examples of the renormalized s-electron -phonon vertices of $O(1/N^0)$.

5. Concluding Remarks

We have presented a modified bosonic version of the infinite-U lattice Anderson model with large orbital degeneracy N, which yields finite intersite coherence effects even in the limit $N\to\infty$. Using such model we have calculated the single particle Green's functions, the electron-electron interaction and the electron-phonon interaction rigorously in the limit N>>1 and found in the sd-limit that: 1. At $T\lesssim T_0$ ($T_0\equiv T_K/\ln N$ where T_K

is the Kondo temperature) the electrons form bands with band width of $O(T_K)$ and have the Fermi liquid type life time as higher order corrections of $O(1/N)$, while at $T \gg T_0$ the system behaves just like a collection of the single Kondo impurities. 2. The effective interaction Γ among the renormalized electrons is repulsive and of $O(T_K)$. 3. Phonon-mediated pairing interaction is significantly modified by the strong electronic correlation to become of $O(\eta^2 T_K^2/D)$ (η is the Grüneisen constant of the system and D is the band width of the conduction electrons), but still can be comparable with Γ if η is as large as ~ 100.

In the present paper, we have presented the explicit results obtained only in the sd-limit because our original results are rather lengthy. However, the latter results can be applied to discuss more general cases other than the sd-limit, including oxide superconductors, though for that sake we need to take account of more explicit information about the band structure ε_k. Such study will be reported near in future.

References

[1] P. Coleman: Phys. Rev. B29, 3035 (1984).

[2] B. Jin and Y. Kuroda: J. Phys. Soc. Japan 57, 1687 (1988).

[3] H. Jichu, A.D.S. Nagi, B. Jin, T. Matsuura and Y. Kuroda: Phys. Rev. B35, 1692 (1987).

Experimental Constraints and Theory of Layered High-T_c Superconductors

T. Schneider and M. Frick

IBM Research Division, Zürich Research Laboratory,
CH-8803 Rüschlikon, Switzerland

Abstract. Recent experiments have revealed several key features of the normal and superconducting states of high-temperature superconductors. They impose strong constraints on theoretical models of the phenomenon. We developed a tight-binding BCS-type model. The carriers form a narrow and anisotropic band, and are subject to on-site and interlayer singlet pairing. The resulting properties agree remarkably well with experimental results, resolve conflicting interpretations of experimental results, offer an easily understandable physical picture, and point to the nature of the pairing mechanism.

1. Experimental Constraints

Recent experiments have revealed several key features of the normal and superconducting states of the Bi-Ca-Ba-Cu-O and Tl-Ca-Ba-Cu-O families of high-temperature superconductors. Measurements of the penetration depth [1,2], the coherence length ξ as derived from the upper critical fields [3,4], the conductivity σ [5-7], and infrared reflectivity [8-10] clearly point to the presence of strong anisotropy, as expressed by the ratios

$$\frac{\lambda_{ab}}{\lambda_c}, \quad \frac{\xi_\perp}{\xi_\parallel}, \quad \sqrt{\frac{\sigma_\perp}{\sigma_\parallel}}, \quad \sqrt{\frac{m_\parallel}{m_\perp}} < 1, \tag{1}$$

where m is the effective mass, $\parallel$ denotes directions parallel and $\perp$ perpendicular to the layers. Compared to the conventional superconductors, the zero-temperature correlation length is very short and the penetration depth much longer. Current estimates are

$$\frac{\lambda_{ab}(0)}{\lambda_c(0)} \simeq \frac{1400\ \text{Å}}{7000\ \text{Å}}, \quad \frac{\xi_\perp(0)}{\xi_\parallel(0)} \simeq \frac{5\ \text{Å}}{25\ \text{Å}}. \tag{2}$$

Because $\xi_\perp$ turns out to be much smaller than the mean spacing, the interlayer interaction is small and a description in terms of a system of weakly interacting layers appears to be valid.

Springer Series in Solid-State Sciences, Vol. 89 Strong Correlation and Superconductivity
Editors: H. Fukuyama · S. Maekawa · A.P. Malozemoff

The temperature dependence of the penetration depth [1,2,11] closely follows BCS-type behavior, consistent with singlet pairing, weak coupling and a nodeless gap. In contrast to conventional BCS superconductors, there is considerable evidence for an anisotropic gap from tunneling [12-14] and infrared reflectivity measurements [8-10]. As far as the normal state is concerned, Hall measurements for H parallel to the c-axis and I parallel to the a,b-plane yield a positive value for the appropriate element of the Hall tensor, i.e. hole-like, while an element for H parallel to the a,b-plane is negative [15] Finally, photoemission [16-18] and positron annihilation [19-21] experiments provide clear evidence for anisotropic Fermi liquid states, forming a narrow band with dominant O-2p nature and low carrier density.

These experimental facts place strong constraints on theoretical models.

2. Model

We adopt the tight-binding Hamiltonian

$$\mathscr{H} = \sum_{i,j,\sigma} t_{ij}\, c_{i\sigma}^{+} c_{j\sigma} - \sum_{i,j} g_{ij}\, c_{i\uparrow}^{+} c_{j\downarrow}^{+} c_{j\downarrow} c_{i\uparrow} \, . \tag{3}$$

For simplicity, we concentrate on a system with one Cu-O layer per unit cell. The first term describes the band structure of the carriers, while the second term represents the unretarded pairing interaction. To match the band structure with the experimental facts, in particular with photoemission [16] and Hall effect [7,13] measurements, we introduce nearest and next-nearest neighbors hopping within the layers and between adjacent sheets. Diagonalization of the one-particle term then yields

$$\varepsilon(\vec{k}) = A\Big\{ -2\big(\cos k_x a + \cos k_y a\big) + 4B \cos k_x a \cos k_y a - 2C \cos k_z s - E_F \Big\} \tag{4}$$

A, AB and AC correspond to the values of the hopping matrix elements, and E_F denotes the Fermi energy given in terms of the band filling

$$\rho = \frac{1}{N} \sum_{\vec{k}} \Big(\exp(\beta \varepsilon(\vec{k})) + 1 \Big)^{-1} \, . \tag{5}$$

N is the number of sites; a and s denote lattice constants of the square lattice within and perpendicular to the layers, respectively. Guided by the photoemission results [16], band structure calculations [22] and the anisotropy of the coherence length, we fixed the parameters entering Eq. (4) as follows

$$A = 0.153 \text{ eV}, \quad B = 0.45, \quad C = 0.1. \tag{6}$$

Assuming BCS-type pairing, the gap equation is then given by

$$\Delta(\vec{k}) = \sum_{\vec{q}} V(\vec{k} - \vec{q}) \frac{\Delta(\vec{q})}{2E(\vec{q})} \tanh \frac{\beta E(\vec{q})}{2}, \tag{7}$$

where

$$E(\vec{q}) = \left(E^2(\vec{q}) + \Delta^2(\vec{q})\right)^{1/2} \tag{8}$$

and

$$V(\vec{q}) = \sum_{l} g_{l0} e^{i\vec{q} \cdot \vec{R}_l}, \tag{9}$$

where l labels the sites. Considering singlet pairing, s-wave behavior within the layers, but allowing for gap anisotropy along the z-axis, we are left with on-site pairing within the sheets and interlayer nearest-neighbor pairing. Thus,

$$V(\vec{q} - \vec{k}) = g_0 + g_1 \cos q_z s \cos k_z s \tag{10}$$

and

$$\Delta(\vec{k}) = \Delta_0 + 2\Delta_2 \cos k_z s. \tag{11}$$

The gap thus becomes anisotropic and varies between

$$\Delta_0 - 2|\Delta_1| = \Delta_{min} \le \Delta(k_z) \le \Delta_{max} = \Delta_0 + 2|\Delta_1|. \tag{12}$$

According to Eqs. (7), (10) and (11), the gap parameters and T_c follow from [23]

$$\Delta_0 = \frac{g_0}{2}(\Delta_0 f_0 + 2\Delta_1 f_1) \quad \text{and} \quad \Delta_1 = \frac{g_1}{2}(\Delta_0 f_1 + 2\Delta_1 f_2), \tag{13}$$

where

$$f_n = \sum_{\vec{k}} (\cos k_z s)^n \frac{\tanh\left(\beta E(\vec{k})/2\right)}{2E(\vec{k})}. \tag{14}$$

For $g_1 = 0$, Δ_1 vanishes. Thus the gap anisotropy is due to the interlayer pairing interaction of strength g_1.

In solving Eq. (13), we assume unretarded pairing. In this case, the k-summation extends over the full Brillouin zone and the limits of integration with respect to energy correspond to bottom and top of the band. For $|g_1| < g_0$, the weak-coupling solution for T_c reads

$$k_B T_c = 1.14\, k_B \theta \ \exp(-1/N_N(0) g_0), \tag{15}$$

where

$$k_B \theta = \left(\tilde{E}_F (W - \tilde{E}_F)\right)^{1/2}. \tag{16}$$

$\tilde{E}_F = |\varepsilon(\vec{k} = 0)|$ is the relative Fermi energy, W the band width and $N_N(0)$ the density of states in the normal state at the Fermi level. Weak coupling holds for $g_0 > |g_1|$ to guarantee a nodeless gap, and for $k_B T_c \ll k_B \theta$. In contrast to standard BCS, where the retarded nature of the phonon-mediated pairing interaction leads to a cutoff given by the Debye energy, unretarded pairing yields a prefactor fixed by top and bottom of the band. Thus, all carriers are subject to pairing and modest values for $N_N(0)\, g_0$ lead to high T_c values, determined by the band width W and the Fermi energy $\tilde{E}_F$.

In summary, the model we have used is of BCS-type, assumes a tight-binding band for the carriers and unretarted pairing, on-site within the layers and nearest-neighbor between the sheets. In the weak-coupling limit, T_c is related to the band width and the Fermi energy, yielding high T_c's for modest coupling strengths. The tight-binding band of the carriers includes nearest and next-nearest neighbors hopping within the layers, and nearest-neighbor hopping between the sheets.

3. Model Properties

In this section, we describe some of the key properties of the model for comparison with the experimental constraints. These properties also offer a physical explanation of key features, such as the short coherence length, long penetration depth, and pronounced anisotropy. In doing so, we consider normal state properties at zero temperature next. The elements of the Hall tensor are given by

$$R_{xyz} = \frac{\sigma_{xyz}}{\sigma_{xx}\sigma_{yy}} = \frac{E_y}{J_x H_z}, \quad R_{yzx} = \frac{\sigma_{yzx}}{\sigma_{yy}\sigma_{zz}} = \frac{E_y}{J_z H_x},$$
$$R_{zxy} = \frac{\sigma_{zxy}}{\sigma_{zz}\sigma_{yy}} = \frac{E_z}{J_y H_x}. \tag{17}$$

Assuming an isotropic relaxation time, the transport coefficients read [24]

$$\sigma_{\alpha\alpha} = e^2 \tau n \frac{1}{m_{\alpha\alpha}}, \quad n = \frac{2\rho}{V_0} \tag{18}$$

and

$$\sigma_{xyz} = -2 \frac{e^3 \tau^2}{\hbar^4 V_0 N} \sum_{\vec{k}} \frac{\partial \varepsilon}{\partial k_x} \left(\frac{\partial \varepsilon}{\partial \vec{k}} \times \nabla_k \right)_z \frac{\partial \varepsilon}{\partial k_y} \delta(\varepsilon) \tag{19}$$

where

$$\frac{1}{m_{\alpha\alpha}} = \frac{1}{\rho N \hbar^2} \sum_{\vec{k}} \left(\frac{\partial \varepsilon}{\partial k_\alpha} \right)^2 \delta(\varepsilon) \tag{20}$$

is the effective mass, $V_0 = a^2 s$, s denotes the mean spacing of the layers, and the z-direction is perpendicular to the sheets. In this approximation, the R_{xyz} are independent of the relaxation time. Moreover, it is important to observe that the free electron formula $R = -1/ne$ becomes meaningless, except for a parabolic band, which is not applicable here. In fact, the Fermi surface corresponds to a corrugated column with a cross section depending on the filling [23].

In the London approximation, the zero-temperature value of the penetration depth is also fully determined in terms of normal-state properties by

$$\lambda_{\alpha\alpha}^{-2} = \frac{4\pi n e^2}{c^2} \frac{1}{m_{\alpha\alpha}} . \tag{21}$$

The numerical results listed in Table I agree remarkably well with available experimental data. R_{xyz} is positive, i.e. hole-like, and has the correct order of magnitude as does the carrier density n [7]. R_{zxy} is negative, as observed in Y-Ba-Cu-O [15]. To obtain these magnitudes and signs, it is necessary to include the nearest-neighbor hopping term, Eq. (4), which also strongly affects the shape of the Fermi surface parallel to the sheets. While the R-values are independent of the width of the band, i.e. A in Eq. (4), this quantity enters the effective mass and the penetration depth. In fact, $m \sim A^{-1}$ and $\lambda \sim A^{1/2}$. Nevertheless, the estimates listed in Table I are in the range of the current experimental data. Moreover, the anisotropy and long penetration depth can be understood in terms of the low carrier density and the large effective mass, Eq. (21). Accordingly, our simple tight-binding parameterization leads to remarkable agreement with the available experimental facts. The narrow anisotropic band implies carriers with both hole- and electron-like character

Table I. Normal state properties and zero temperature penetration depth for several band fillings and the parameters listed in Eq. (6); $\tilde{E}_F = |\varepsilon(\vec{k} = \vec{0})$ is the relative Fermi energy, $m_{xx} = m_{yy}$ and m_{zz} are the elements of the effective mass tensor, while R_{xyz}, R_{yzx} and R_{zxy} denote independent elements of the Hall tensor in units of $10^{-3}\,cm^3C^{-1}$. The other units are: energy in A [Eq. (4)] and the carrier density in $10^{21}\,cm^{-3}$ $(n = 2\rho/V_0)$. As characteristic lattice constants we used $a = 3.85$ and $s = 7.76$ Å.

ρ	$\tilde{E}_F$	$N_N(0)$	$\frac{m_{xx}}{m_e}$	$\frac{m_{zz}}{m_e}$	R_{xyz}	R_{yzx}	R_{zxy}	λ_{xx}	λ_{zz}	n
0.6	1.62	0.091	2.70	140	0.86	-0.25	-0.25	940	6738	10.43
0.5	1.51	0.175	3.64	119	0.68	-0.59	-0.59	1088	6214	8.69
0.4	1.04	0.264	5.64	100	0.58	-0.98	-0.98	1356	5698	6.96
0.3	0.75	0.461	10.84	77	0.57	-1.68	-1.68	1878	4992	5.22

and low density, a column-like Fermi surface, and rather large effective masses.

Next, we turn to the superconducting state by assuming unretarded on-site and next-nearest-neighbor interlayer pairing. In Table II, we summarized estimates for the gap parameters and T_c for $\rho = 1/2$ as a function of g_1, the strength of the interlayer pairing interaction. These estimates, obtained from a numerical solution of Eq. (13), reveal that for $g_0 > |g_1|$ the gap becomes anisotropic but remains nodeless; in this coupling regime T_c is rather insensitive to the value of g_1, while interlayer pairing fully determines the gap anisotropy; the standard BCS formula $2\Delta/k_BT_c \simeq 3.52$ is no longer applicable because the gap varies between Δ_{min} and Δ_{max}.

The transition temperatures are in the range of 23 to 33 K and are found to increase initially for fixed g_0 and g_1 by reducing ρ, owing to an increase in the density of states. Even higher T_c's are obtained by increasing g_0.

To clarify the coupling regime and the validity of the weak-coupling solution, Eq. (15), we also studied $T_c(g_0)$ for fixed ρ and $T_c(\rho)$ for fixed g_0 by solving Eq. (13) numerically. In the range $g_0 \lesssim 2$ and for $\rho \simeq 1/2$, the weak-coupling solution, Eq. (15), turns out to be reasonable.

Next, we turn to the implications of the anisotropic gap, the coherence length, and some thermodynamic properties. Tunneling conduc-

Table II. Numerical estimates for Δ_0, Δ_1 and T_c as obtained from Eq. (13) for $\rho = 1/2$. Energies are in units of $A = 0.153$ eV. Δ_{min} denotes the minimum, Δ_{max} the maximum value of the gap.

g_0	g_1	Δ_0	Δ_1	$k_B T_c$	$\frac{2\Delta_{min}}{k_B T_c}$	$\frac{2\Delta_{max}}{k_B T_c}$	$\frac{\Delta_{min}}{\Delta_{max}}$
1	-0.99	0.0232	0.00067	0.0131	3.34	3.75	0.89
1	-0.9	0.0232	0.00063	0.0131	3.35	3.73	0.90
1	-0.8	0.0232	0.00059	0.0131	3.36	3.72	0.91
1	-0.6	0.0233	0.00049	0.0132	3.38	3.68	0.92
1	-0.4	0.0233	0.00036	0.0132	3.42	3.64	0.94
1	-0.2	0.0235	0.00020	0.0132	3.50	3.62	0.97
1	0	0.0235	0	0.0133	3.53	3.53	0
1	0.2	0.0236	-0.0003	0.0134	3.43	3.61	0.96
1	0.4	0.0237	-0.0006	0.0135	3.33	3.69	0.90
1	0.6	0.0239	-0.0013	0.0138	3.09	3.84	0.80
1	0.8	0.0243	-0.0026	0.0146	2.62	4.04	0.65
1	0.9	0.0244	-0.0039	0.0158	2.10	4.08	0.51
1	0.99	0.0244	-0.0061	0.0186	1.32	3.92	0.33

tance [12-14] and infrared reflectivity [8-10] measurements yield increasing evidence for a nodeless anisotropic gap. To illustrate a fingerprint of this anisotropy, we depicted the density of states in the superconducting phase in Fig. 1. The standard BCS behavior, a square-root singularity at the gap energy, is removed. The density of states is zero up to the minimum gap Δ_{min}. For larger energies, up to Δ_{max}, the structure is a characteristic of the specific gap anisotropy. In the present case, the anisotropy is uniaxial and a peak occurs at Δ_{max}, corresponding to a Van Hove singularity. In any case, there no longer is a unique gap, its energy varies between Δ_{min} and Δ_{max}. Clearly, for small anisotropy, the modifications of the standard BCS behavior will shrink to a small energy interval.

In principle, information on the gap anisotropy can be obtained from tunneling and infrared reflectivity measurements. Unfortunately, these techniques are extremely surface-sensitive. Moreover, substantial anisotropy invalidates interpretations based on standard BCS behavior [23]. To illustrate this point, we consider here specular and diffuse tunneling. For specular tunneling, requiring a perfect surface, the tunnel current is given by [23]

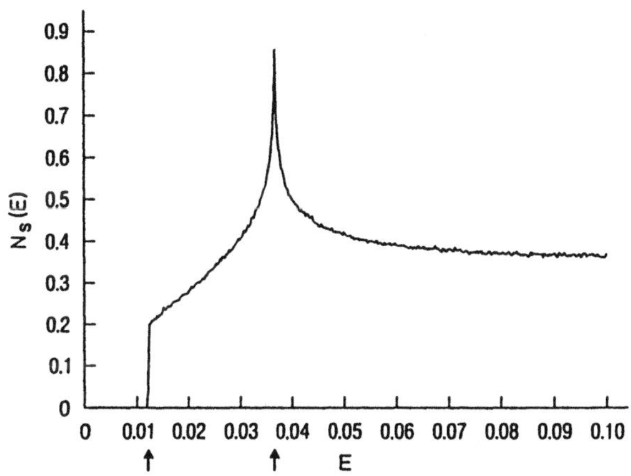

Figure 1. Density of states $N(E)$ in the superconducting state for $\Delta_0 = 0.0244$, $\Delta_1 = -0.0061$, $g_0 = 1$, $g_1 = 0.99$ and $\rho = 1/2$ (see Table II). The arrows mark the minimum ($\Delta_{min} = 0.0124$) and the maximum ($\Delta_{max} = 0.0364$) value of the gap. Energies are in units of $A = 0.153$ eV.

$$I \sim 2\pi e \sum_{\vec{k}} \left| \frac{\partial \varepsilon}{\partial k_L} \right| \left(1 \mp \frac{\varepsilon(\vec{k})}{E(\vec{k})} \right) \theta\left(|eV| - E(\vec{k}) \right) \exp\left(-\frac{k_T^2}{\kappa^2} \right) \tag{22}$$

where

$$\kappa^{-2} = \frac{\hbar d}{\sqrt{2m^* U}} . \tag{23}$$

U denotes the barrier height, m^* the effective mass in the barrier and d the barrier thickness. The index L denotes the direction parallel to the tunnel current, T that perpendicular to it. In the numerical estimates, we used $m^* = m$, $U = 1$ eV and $d = 10$ Å. Figure 2 shows the voltage dependence of the conductance for tunneling currents parallel and perpendicular to the layers and $C = 0.1$ [Eq. (4)]. In actual measurements, the peak position is taken as the gap. Thus, a smaller gap would be observed for a tunnel current parallel to the layers. This seems to contradict tunneling experiments on cryogenically cleaved YBCO surfaces [13,14]. In fact, the larger gap appeared for I parallel to the layers. However, if we change the sign of C [Eq. (4)], where C characterizes the matrix element for next-nearest-neighbor interlayer hopping, Δ_1 is found to change sign as well. This change of sign has no effect on most properties, including the density of states depicted in Fig. 1, but it modifies the specular tunneling conductance with the current parallel to the

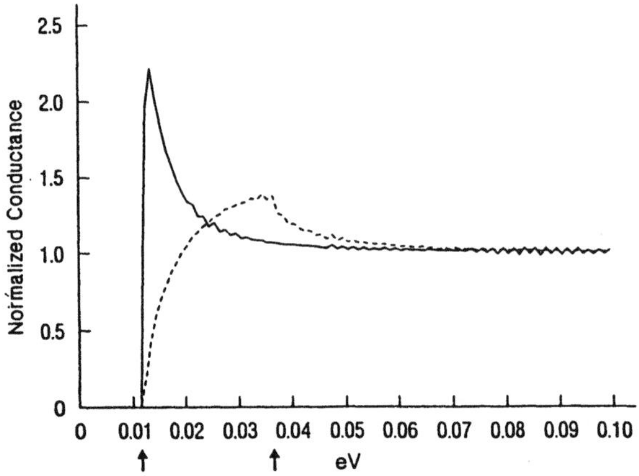

Figure 2. Specular tunneling conductance versus voltage for $C = 0.1$, $\Delta_0 = 0.0244$, $\Delta_1 = -0.0061$, $g_0 = 1$ and $g_1 = 0.99$. Solid (dashed) line: tunneling current parallel (perpendicular) to the layers. The arrows mark Δ_{min} and Δ_{max}.

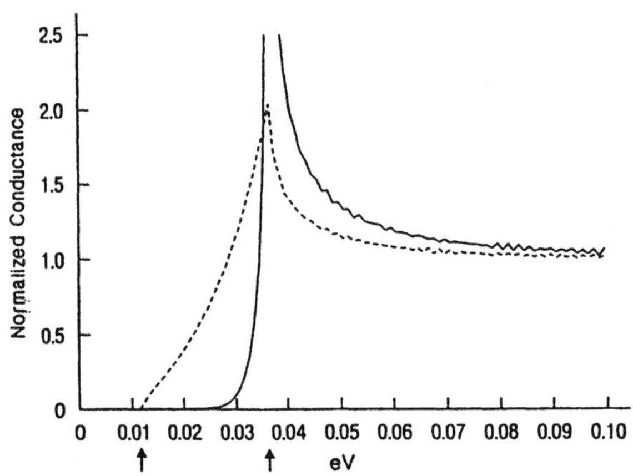

Figure 3. Specular tunneling conductance versus voltage for $C = -0.1$, $\Delta_0 = 0.0244$, $\Delta_1 = 0.0061$, $g_0 = 1$ and $g_1 = 0.99$. Solid (dashed) line: tunneling current parallel (perpendicular) to the layers. The arrows mark Δ_{min} and Δ_{max}.

layers, as seen in Fig. 3. This difference is due to the exponential term in Eq. (23), which for a tunneling current parallel to the layers favors small wave vectors in the z-direction. Thus for $C > 0$, where $\Delta_1 < 0$, the minimum gap is favored, while for $C < 0$ the situation is reversed. These examples clearly illustrate the sensitivity of specular tunneling to the specific gap anisotropy. It is also worth noting that the behavior shown in Fig. 3 is closer to the experimental findings [13,14]. Specular tunneling requires, however, a perfect junction. In view of this, one expects that diffuse tunneling is closer to reality.

For diffuse tunneling, the current is given by [23]

$$I \sim \pm 2\pi e\, N_N(0) \sum_{\vec{h}} \left| \frac{\partial \varepsilon}{\partial k_L} \right| \left(1 \mp \frac{\varepsilon(\vec{k})}{E(\vec{k})} \right) \theta\big(|eV| - E(\vec{k})\big) . \qquad (24)$$

Owing to the absence of the exponential term appearing in the specular case, see Eq. (22), the current depends much less on the orientation of the junction with respect to the layers. In fact, neglecting the gradient and the ε/E-terms, the conductance becomes proportional to the density of states, which reflects the gap anisotropy as shown in Fig. 1. Thus, the standard BCS square-root singularity is removed, and by taking the position of the maximum in the conductance one determines the maximum gap value [23]. In absolute infrared reflectivity measurements the situation is reversed: $R = 1$ up to photon energies corresponding to twice the minimum gap value. Accordingly, the ratio of gaps evaluated from infrared reflectivity and diffuse tunneling measurements then provides a measure of the gap anisotropy in terms of $\Delta_{min}/\Delta_{max}$. In view of this, it is not surprising that for YBCO the infrared reflectivity estimate $2\Delta/k_BT_c = 3.5$ [10] and the tunneling data, converging to $2\Delta/k_BT_c = 6$, differ markedly. In fact, this discrepancy should be taken as clear evidence for an anisotropic gap with $\Delta_{min}/\Delta_{max} = 3.5/6 = 0.6$.

Gap anisotropy will of course affect also other properties, such as the critical field and the temperature dependence of the specific heat, nuclear magnetic relaxation rate and penetration depth.

The critical field at zero temperature is obtained from the energy difference between the superconducting and normal states in terms of

$$E_s - E_N = \sum_{\vec{k}} \left(|\varepsilon(\vec{k})| - E(\vec{k}) + \frac{\Delta^2(\vec{k})}{2E(\vec{k})} \right) = - \frac{H_c^2(0)}{8\pi} . \qquad (25)$$

Numerical results for $H_c^2(0)/8\pi$ and the ratio $\gamma T_c^2/H_c^2(0)$, yielding $1/4\pi = 0.084$ in standard BCS, are given in Table III. As gap anisotropy increases with g_1 (Table II), the critical field is seen to increase with anisotropy and the ratio becomes larger and progressively deviates from the standard BCS value. γ is the Sommerfeld constant

Table III. Zero temperature values of the critical field and the ratio $(\gamma T_c^2)/H_c^2(0)$ for several g_1 values for the parameters given in Eq. (6), $\rho = 1/2$ and $g_0 = 1$.

g_1	$\frac{\Delta_{min}}{\Delta_{max}}$	$k_B T_c$	$\frac{H_c^2(0)}{8\pi}\,10^5$	$\frac{\gamma T_c^2}{H_c^2(0)}$
0	1	0.0133	9.34	0.087
0.4	0.9	0.0136	10.16	0.083
0.6	0.8	0.0140	10.47	0.086
0.8	0.65	0.0147	11.14	0.089
0.9	0.51	0.0160	11.40	0.103
0.99	0.33	0.0186	12.41	0.128

$$\gamma = \frac{2\pi^2}{3} N_N(0)\, k_B^2 , \tag{26}$$

where $N_N(0)$ is 0.175.

In Table IV, we listed the normalized jump of the specific heat, $\Delta C/\gamma T_c$, for several g_1 values. The jump is seen to decrease with increasing anisotropy and for $g_1 = 0$, one regains the standard BCS ratio 1.45. Anisotropy also affects the asymptotic low-temperature behavior, because $C \sim \exp(-\beta\Delta_{min})$.

To calculate the penetration depth at finite temperature, local electrodynamics is appropriate ($\lambda(0) \gg \xi(0)$). The finite temperature extension of Eq. (21) then reads

$$\lambda_{\alpha\alpha}^{-2} = \frac{4\pi n e^2}{c^2} \sum_{\vec{k}} v_\alpha^2(\vec{k}) \left[\left(-\frac{\partial f}{\partial \varepsilon} \right) - \left(-\frac{\partial f}{\partial E} \right) \right] , \tag{27}$$

Table IV. Specific heat jump for the parameters listed in Eq. (6), $\rho = 1/2$ and $g_0 = 1$.

g_1	0	0.2	0.4	0.6	0.8	0.9	0.99
$\frac{\Delta_{min}}{\Delta_{max}}$	1	0.96	0.9	0.8	0.85	0.51	0.33
$\frac{\Delta C}{\gamma T_c}$	1.45	1.47	1.42	1.39	1.14	0.85	0.63

where

$$f(x) = (\exp \beta x + 1)^{-1}, \quad V_\alpha(\vec{k}) = \frac{1}{\hbar}\frac{\partial \varepsilon}{\partial k_\alpha}. \tag{28}$$

The effect of the anisotropy of the temperature dependence is shown in Fig. 4. Above $T/T_c > 0.1$, anisotropy appears to modify the temperature dependence markedly. In particular, the amplitude of the asymptotic behavior $(\lambda_{\alpha\alpha}(0)/\lambda_{\alpha\alpha}(T))^2 = A_{\alpha\alpha}(1 - T/T_c)$ decreases from $A_{\alpha\alpha} = 2$ with increasing anisotropy. It is important to note, however, that substantial deviations from the isotropic case require close proximity to nodes in the temperature-dependent gap. In the present case, nodes appear for $g_1 = 1$, corresponding to $\Delta_{min}/\Delta_{max} \lesssim 1/3$ (Table II). Thus, the anisotropy effects seen in Fig. 4 ($g_1 = 0.9$) just signal the close proximity to the nodes appearing for $g_1 \geq 1$. For $g_1 < 0.6$, where $\Delta_{min}/\Delta_{max} \gtrsim 0.65$, the deviations become much smaller. Experiments on YBCO clearly point to BCS behavior [1,2,11], justifying first of all a BCS treatment and revealing the absence of nodes. However, as pointed out above, consistency with isotropic BCS behavior does not contradict the presence of gap anisotropy for $\Delta_{min}/\Delta_{max} \gtrsim 0.65$, as required to resolve the conflict between the infrared and tunneling estimates for the gap.

Finally, we turn to the nuclear spin relaxation rate given by [25,26]

$$\frac{1}{T_1} \simeq \frac{1}{2}\sum_{\vec{k},\vec{k}'}\left(1 - \frac{\varepsilon(k)\varepsilon(k')}{E(k)E(k')} + \frac{\Delta(k)\Delta(k')}{E(k)E(k')}\right) \times f(E(k'))\,(1 - f(E(k))\delta(E(k) - E(k'))). \tag{29}$$

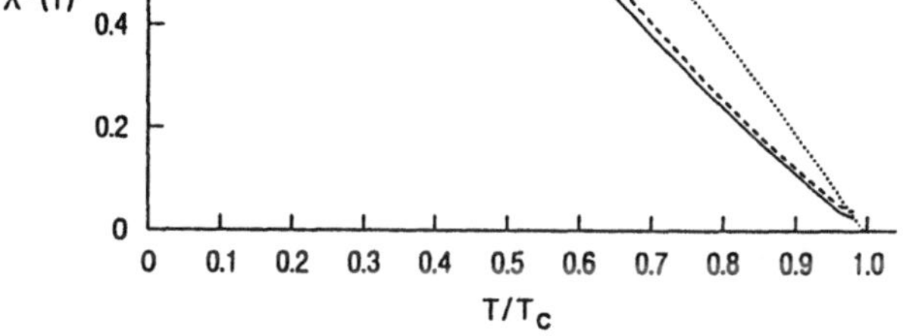

Figure 4. Temperature dependence of the penetration depth for the parameters listed in Eq. (6), and dotted line: $g_1 = 0$, λ_{xx}, λ_{zz}; dashed line: $g_1 = 0.9$, λ_{zz}; solid line: $g_1 = 0.9$, λ_{xx}.

For an isotropic gap, where the density of states exhibits a square-root singularity, the integral is know to diverge [25,26]. This problem can be resolved by taking into account the lifetime of the quasiparticles and the anisotropy of the gap. Here we neglect the lifetime effect. Figure 5 shows a comparison between the temperature dependences resulting from a small and a large anisotropy. For $T/T_c > 1$, Korringa behavior appears, while the broad peak occurring just below T_c, reminiscent of conventional superconductors, is seen to become weaker with increasing anisotropy. Recently, the NMR relaxation of O^{17} has been measured in YBCO [27], revealing a very small enhancement just below T_c. This reduction clearly points to the presence of gap anisotropy and a finite lifetime of quasiparticles. Taking $\Delta_{min}/\Delta_{max} \simeq 0.6$, consistent with the infrared, tunneling and penetration depth measurements, an enhancement still remains (Fig. 5), becoming further reduced by taking the finite lifetime of the quasiparticles into account. In fact, the enhancement appears close to T_c, which is high so that lifetime effects play a role. Further work is needed to disentangle the lifetime and anisotropy induced reduction of the $1/T_1$ enhancement just below T_c.

Finally, we turn to the coherence length in the ground state. It can be obtained from the exponential decay of the correlation function of two quasiparticles, one in the origin, the other at $\vec{r}$, whose spins are antiparallel. For a spherical Fermi surface and an isotropic gap, it is given by $\xi_0 = \hbar^2 k_F/m\Delta$. The numerical estimates listed in Table V reflect the strong anisotropy determined by the small interlayer hopping matrix element. In particular, $\xi_\perp$ is found to be smaller than the spacing of the

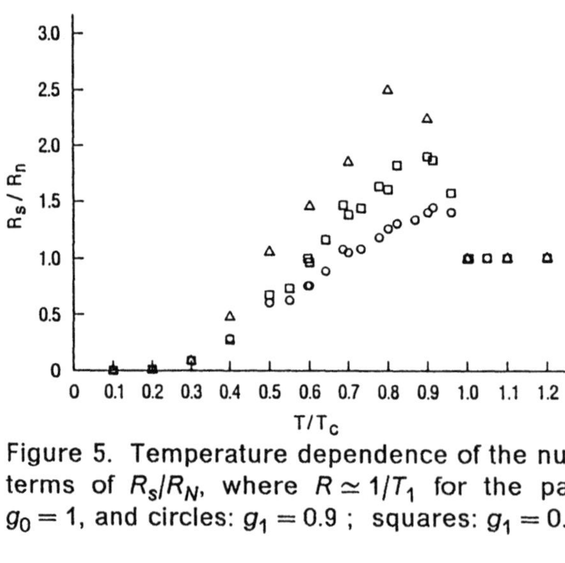

Figure 5. Temperature dependence of the nuclear spin relaxation time in terms of R_s/R_N, where $R \simeq 1/T_1$ for the parameters listed in Eq. (6), $g_0 = 1$, and circles: $g_1 = 0.9$; squares: $g_1 = 0.6$; triangles $g_1 = 0.2$.

Table V. Numerical estimates of the coherence length for the parameters cited in Eq. (6).

ρ	g_0	g_1	Δ_0	Δ_1	ξ_{xx}/a	ξ_{zz}/s	$\frac{\xi_{zz}}{\xi_{xx}}\frac{a}{s}$
0.5	1	0	0.0244	0	7.6	0.39	0.05
0.5	1	0.99	0.0244	-0.0061	7.9	0.47	0.06

layers and the ratio $\xi_\perp/\xi_\parallel$ is in the range of values estimated from the upper critical fields [3,4]. Because the gap anisotropy does not affect the magnitude of the coherence length, a slight extension of the free electron expression

$$\xi_{\perp,\parallel} = \frac{\hbar}{<\Delta>}\left(\frac{2\tilde{E}_F}{m_{\perp,\parallel}}\right)^{1/2} \tag{30}$$

appears to be appropriate. Thus the anisotropy comes from the effective mass, and the small magnitude is due to the large average gap and mass, and the small relative Fermi energy.

4. Concluding Remarks

We sketched a BCS-type tight-binding model, consistent with experimental constraints. The carriers form a narrow and anisotropic band. Normal state properties, including the elements of the Hall tensor, the effective mass, the carrier density and the penetration depth, agree remarkably well with experiment. The carriers, having hole- and electron-like character, are subject to unretarded pairing of on-site and interlayer nearest-neighbor origin. Thus, T_c is proportional to the Fermi energy and can adopt rather large values even for a weak pairing interaction. T_c turned out to be rather insensitive to the strength of interlayer pairing, giving rise to an uniaxial gap anisotropy which was found to modify the standard BCS behavior. The gap energy varies between Δ_{min} and Δ_{max}, and interpretations based on a unique ratio $2\Delta/k_BT_c$ become meaningless. Δ_{min} is accessible in infrared reflectivity measurements, while diffuse tunneling probes Δ_{max}. This interpretation offers an explanation for the apparent discrepancy of $2\Delta/k_BT_c$ values, as determined from infrared reflectivity and tunneling experiments. Moreover, we identified the effects of gap anisotropy on the temperature dependence of the nuclear spin relaxation rate T_1 and of the penetration depth. For $\Delta_{min}/\Delta_{max} \gtrsim 0.6$, the penetration depth was found to be rather insensitive to gap anisotropy. Thus, the conflict between the penetration depth measurements, which are consistent with isotropic BCS behavior, and the infrared reflectivity, as well as tunneling measurements, which point to an anisotropic gap, has been resolved. A consistent picture requires that $\Delta_{min}/\Delta_{max} \gtrsim 0.6$. The missing enhancement in $1/T_1$ then is a natural consequence of gap anisotropy and the lifetime of quasiparticles.

The remarkable agreement with the experimental findings points to an unretarded pairing interaction, on-site within the layers and of nearest-neighbor nature between the sheets. The origin is of electronic nature, yielding a cutoff proportional to the Fermi energy.

Acknowledgements

The authors thank A. Aharony, A. Baratoff, A.P. Malozemoff, K.A. Müller and N. Schopohl for stimulating discussions and pertinent suggestions.

References

1. Y. J. Uemura *et al.*, Phys. Rev. B **38**, 909 (1988).
2. L. Krusin-Elbaum *et al.*, Phys. Rev. Lett. **62**, 217 (1989).
3. T. T. M. Palstra *et al.*, Phys. Rev. B **39**, 5102 (1988).
4. J. H. Kang *et al.*, Appl. Phys. Lett. **53**, 2560 (1988).
5. S. Martin *et al.*, Phys. Rev. Lett. **60**, 3194 (1988).
6. A. Zettl *et al.*, in *Mechanisms of High Temperature Superconductivity*, edited by H. Kamimura and A. Oshiyama (Springer-Verlag, Berlin Heidelberg, 1988), p. 249.
7. H. Takagi *et al.*, Nature **332**, 236 (1988).
8. U. Hofman *et al.*, preprint.
9. R. T. Collins *et al.*, preprint.
10. G. A. Thomas *et al.*, Phys. Rev. Lett. **61**, 1313 (1988).
11. A. T. Fiory *et al.*, Phys. Rev. Lett. **61**, 1419 (1988).
12. M. Lee *et al.*, in Ref. 6, p. 220.
13. J. S. Tsai *et al.*, in Ref. 6, p. 229.
14. J. S. Tsai *et al.*, Physica C **153-155**, 1385 (1988).
15. Y. Iye, in Ref. 6, p. 263.
16. T. Takahashi *et al.*, Nature **334**, 691 (1988).
17. J. M. Imer *et al.*, Phys. Rev. Lett. **62**, 336 (1989).
18. F. J. Himpsel *et al.*, Phys. Rev. B **38**, 11946 (1988).

19. L. Hoffmann *et al.*, Physica C **153-155**, 129 (1988).
20. L. C. Smedskkjaer *et al.*, Physica C **156**, 269 (1988).
21. A. Bansil *et al.*, Phys. Rev. Lett. **61**, 2480 (1988).
22. J. Ju *et al.*, Physica C **152**, 273 (1988).
23. T. Schneider, H. de Raedt, and M. Frick, Z. Phys. B (to be published).
24. P. B. Allen and W. E. Pickett, Phys. Rev. B **36**, 3926 (1987).
25. L. C. Hebel and C. P. Schlichter, Phys. Rev. **113**, 1504 (1959).
26. M. Fibich, Phys. Rev. Lett. **14**, 561 (1965).
27. Y. Kitaoka *et al.*, in Ref. 6, p. 148.

Part III

Experiments

Physical Properties of High-T_c Oxide Superconductors – Effect of Doping on the Electronic State

S. Uchida[1], *H. Takagi*[1], *Y. Tokura*[2], *N. Koshihara*[2], *and T. Arima*[2]

[1]Engineering Research Institute, University of Tokyo Yayoi, Tokyo 113, Japan

[2]Department of Physics, University of Tokyo, Hongo, Tokyo 113, Japan

Abstract : We discuss the physical meaning of the doping high T_c oxides, P-type $La_{2-x}Sr_xCuO_4$ and recently discovered N-type $Nd_{2-x}Ce_xCuO_{4-y}$, based on the experimental results of the Hall effect and the plasma reflectivity spectrum. It is suggested that for light doping the Mott-Hubbard picture gives proper description of the transport properties whereas for heavy doping or for higher frequencies the Fermi lquid picture seems more appropriate for both N- and P-type materials.

1. Introduction

So far discovered high T_c cuprates can be expressed by the following general chemical formula

$$(L_{1-x}A_x)_{m+1}Cu_mO_{3m+1-\delta},\ m = 1,\ 2 \text{ and } 3,$$

In this formula L represents either rare earth or alkaline earth ions with relatively large, ionic radius and the integer m the multiplicity of the Cu-O layers in the structure unit.

The high-T_c superconductivity is realized by substituting the element by other element A with different valence charge or by changing the oxygen content. We call this substitution or the compositional change as doping, because so small change in the composition makes the material superconducting. For example, in the case of La_2CuO_4 ($m = 1$) only 3 % substitution of La with Sr is sufficient to make this compound a high T_c superconductor[1] and in the case of yttrium 1-2-3 compound, ($m = 2$ but strictly speaking $m = \infty$) the base material being probably $YBa_2Cu_3O_6$, a superconducting phase is realized when the oxygen content is increased only to 6.3 or 6.4 [2].

Recently quite a new class of superconducting cuprates has been discovered by our group at University of Tokyo [3]. The new compounds have the formula $Ln_{2-x}Ce_xCuO_{4-y}$ where Ln is one of the lanthanides Pr, Nd, Sm or Eu. In contrast to the previous cuprate superconductors where the charge carriers are holes, the carriers in the new compounds are doped electrons. The parent compounds Ln_2CuO_4 has so-called T' structure where apical oxygen atoms in the K_2NiF_4 structure are displaced so as to makes an isolated Cu-O square plane. The electron doping is made by substituting trivalent Ln with tetravalent Ce and or by reducing oxygens. Superconductivity with T_c max = 25 K is observed in a narrow composition range $0.14 < x < 0.18$ only after the sample is annealed under reducing atmosphere.

In the chemical picture such doping processes changes the average valence of Cu. In the parent material all the Cu ions in the Cu-O planes have the valence of 2+ and the hole doping appears to make Cu^{3+} and the electron doping, Cu^{1+} at the sites in the neighborhood of the dopant elements. The chemical analysis, such as the iodometric titration, give support for this view. This paper address the question if such chemical viewpoint of doping is correct also from the physical viewpoint and how the electronic state varies with doping.

Springer Series in Solid-State Sciences, Vol. 89 Strong Correlation and Superconductivity
Editors: H. Fukuyama · S. Maekawa · A.P. Malozemoff

2. P-Type $La_{2-x}Sr_xCuO_4$

La_2CuO_4 is probably the primary parent material of all the P-type high-T_c oxides discovered to date. This material is an insulator in the stoichiometric compound and exhibits an antiferromagnetic order below 250 K [4]. Various experiments, such as NQR [5] and neutron diffraction [6], gave an evidence for the presence of magnetic moment on Cu ions. Therefore, the spins are localized on Cu site in La_2CuO_4 due to the strong Coulomb repulsion between two d-electrons on one Cu site. From this, La_2CuO_4 may be regarded as a Mott insulator.

The doping, the substitution of La with Sr, has the following notable effects on the above mentioned features of La_2CuO_4. The antiferromagnetic order is rapidly suppressed and disappears at 1 % doping or less and instead the spin-glass like state appears in the region $0.02 \leq x \leq 0.05$ [7]. For doping more than 2.5 % the system undergoes an insulator to metal transition and becomes a superconductor. A question arises whether or not the localized spins on Cu sites persist into the metallic phase. The answer would be crucial for the majority of theoretical models which emphasize close connection between magnetism and high T_c superconductivity. However, definite experimental evidences have not been presented, although the photoemission [8] and Raman scattering experiments [9] are suggestive of the localized spins.

The metal-insulator transition at such low level of Sr doping cannot be hardly observed for other transition metal oxides. For example, in isotructural La_2NiO_4 and La_2CoO_4 which are also considered to be Mott-insulators the doping makes localized states inside the insulating gap. The insulating state persist up to 40 $\sim$ 50 % of doping [10].

In contrast, it is clear from the resistivity data (Fig. 2) that the doping of Sr less than 1 % makes mobile carriers. Even for 0.25 % substitution which corresponds to the hole density of $5 \times 10^{19} cm^{-3}$ the metallic T-linear dependence of the resistivity becomes

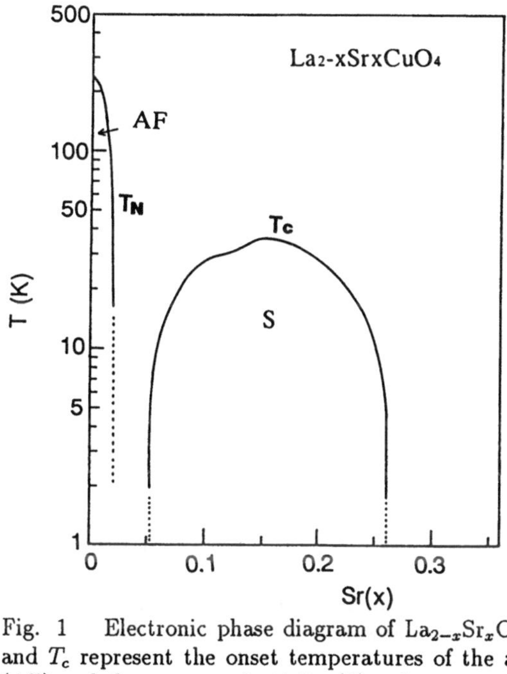

Fig. 1 Electronic phase diagram of $La_{2-x}Sr_xCuO_4$ system. T_N and T_c represent the onset temperatures of the antiferromagnetic (AF) and the superconductivity (S) order, respectively.

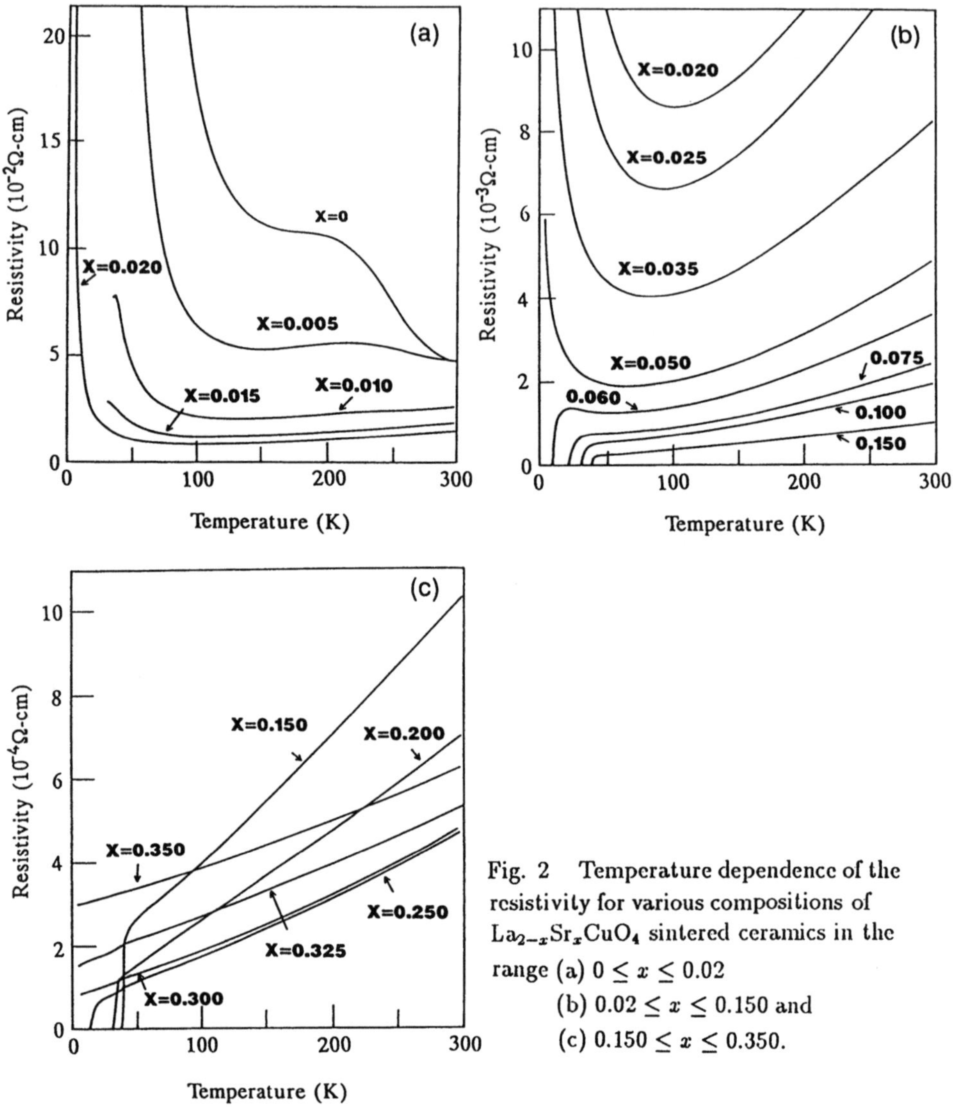

Fig. 2 Temperature dependence of the resistivity for various compositions of $La_{2-x}Sr_xCuO_4$ sintered ceramics in the range (a) $0 \le x \le 0.02$
(b) $0.02 \le x \le 0.150$ and
(c) $0.150 \le x \le 0.350$.

pronounced at higher temperatures, although these carries are localized at low temperatures. The temperature at which carriers shows the trends for localization decreases as the doping proceeds. The localization finally gives way to superconductivity when doping exceeds 2.5%.

The superconducting transition temperature T_c increases rapidly for $x \ge 0.06$ and reaches a maximum $T_c = 38$ K at $x = 0.15$. For larger x, T_c decreases and the superconducting transition cannot be observed in the resistivity measurement when x exceeds 0.25. From the Meissner measurement the bulk superconductivity is seen in narrower range $0.08 < x < 0.22$. Although superconducting disappears in the heavily doped region, the material remains metallic and the resistivity continues to decrease with doping at least up to $x \sim 0.30$.

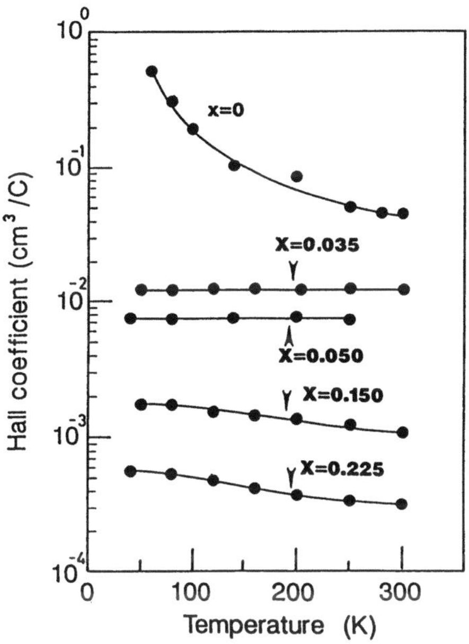

Fig. 3 Temperature dependence of the Hall coefficient for several composition of Sr doped La_2CuO_4 ceramics.

When we want to go into the details of the physical picture of the doping, the Hall effect gives useful information on the type of carriers, electron or hole, as well asthe density of carriers. Figure 3 shows the temperature dependence of the Hall coefficient for various values of x in $La_{2-x}Sr_xCuO_4$. If we apply a simple relationship between R_H and carrier density ($R_H = 1/ne$), the result indicates that the doping makes holes in La_2CuO_4 and the hole density increases with increasing doping. In Fig. 4 the value of the Hall coefficient at 80 K is plotted as a function of Sr composition.

As is seen in the Fig. 4 the value of R_H is inversely proportional to the Sr-composition as far as the Sr-doping less than 5 %, so that the hole density seems to increase proportionally to the Sr-doping amount. Then, if the hole are createson Cu sites by doping, the experimental result of the Hall coefficient is in agreement with the chemical point of view. That is, in terms of the Mott-Hubbar energy scheme, the holes are created in the lower Hubbard band of the host Mott insulator corresponding to the formation of Cu^{3+} ionic state in the host Cu^{2+} network.

However, the apparent agreement between physical and chemical picture is up to the doping of about 5 %. When the doping is larger than 7.5 % which is the optimum doping for the superconductivity, the data significantly deviate from the straight line. At the earlier stage of the investigation, the deviation was ascribed to the oxygen vacancies [11]. The vacancies act to compensate the holes created by the Sr-doping which would decrease the value of R_H. However, the possibility of oxygen vacancies can be ruled out by the chemical titration analysis. According to the chemical analysis, oxygen content is close to 4 at lease up $x = 0.35$ and the deviation from 4 is, if any, much less than the Sr doping amount. It is found that for heavy doping the Hall coefficient charges sign from positive to negative at the composition near the superconductor to "normal" metal transition.

Another fact which does not allow for simple interpretation of the Hall effect is its temperature dependence. The Hall coefficient of the sample of 7.5 % Sr which should be good metal shows anomalous temperature dependence, the magnitude increasing with decreasing temperature. Similar temperature dependence is also observed for YBC and Bi-based samples showing good superconducting transition [12].

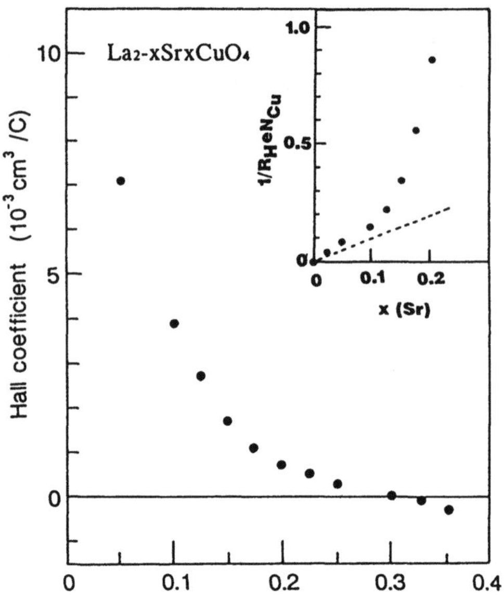

Fig. 4 The value of Hall coefficient (80 K) plotted as a function of the Sr composition for $(La_{1-x}Sr_x)_2CuO_4$. The inset shows the reciprocal values in the small compositional region.

So the result implies that the "rigid" Mott-Hubband picture is no longer valid when the doping level increases and the material becomes a good superconductor.

We can also investigate the free carrier behavior by use of the optical reflectivity spectrum. Free carriers make a collective plasma oscillation. The frequency of the plasma oscillation ω_p corresponds to the reflectivity edge position in the lowest frequency region, usually called plasma edge.

A reflectivity edge is observed in the infrared frequency region in this system [13]. We noticed that the edge frequency did not shift appreciably with changing dopant concentration. This fact has been confirmed by the experiment of our group made on series of single crystalline samples with a-b face (Fig. 5).

In the sense that ω_p^2 is proportional to the carrier density or to x in this case, the obscrbed reflectivity edge is not a plasma edge due to doped holes. As a matter of fact, the damping frequency which is estimated from the drude-fit to the observed reflectivity spectrum near the edge is temperature independent and much higher than that associated with the dc resistivity which shows linear-T dependence. Although for lightly doped samples the spectrum does not even fit to the Drude formula, the spectra for moderately and heavily doped samples can fit well to the Drude formula down to the frequenct $10^3 cm^{-1}$ or lower. The integrated oscillator strength below the edge frequency, i.e. the effective number of electrons involved in this spectral region is found to be almost one per Cu atom in a unit cell. Thus, it seems that the holes or electrons originally localized on Cu site ($3d_{x^2-y^2}$ level) contribute to the (dynamical) conductivity. This situation is similar to what the band picture claims where the conduction band is the antibonding band of Cu $3d_{x^2-y^2}$ oribital hybridized with O $2p$ orbital and it is nearly half-filled.

It should be noted that the reflectivity edge is seen even for semiconducting samples with small values of x—including the case of $x = 0$ which might not be pure because of non-stoichiometry in oxygen content or cation ratio which is unavoidable in the ordinary

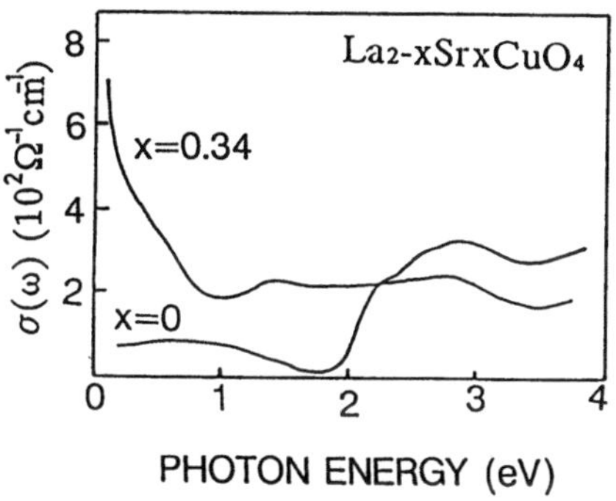

Fig. 5 Optical conductivity spectra for undoped (x =0) and doped (x =0.34) $La_{2-x}Sr_xCuO_4$ single crystals. The spectra were Kramers-Kronig transformed from the reflectiviy spectra measured at room temperature for **E** within a-b plane.

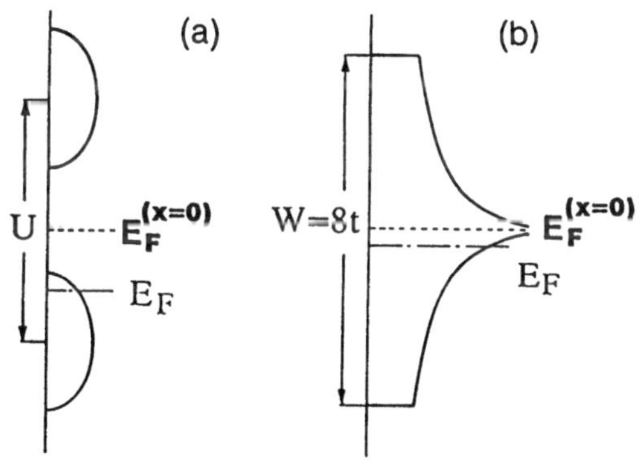

Fig. 6 (a) Schematic energy diagram of the Mott-Hubbard model. (b) State density of the two-dimensional tight-binding band near half-filling.

synthesis process. This indicates that the conductivity is substantially large in the frequency region below the edge, but eventually diminishes at lower frequencies owing to some circumstance which makes the material semiconducting. In this context, the observed reflectivity edge may be looked upon a dormant plasma edge of the material if it were a metal with much weaken Coulomb correlation energy or one might say that we are observing a crossover from the Mott-Hubbard to the band picture on increasing frequency.

3. N-Type $Nd_{2-x}Ce_xCuO_{4-y}$

A similar phase diagram (Fig. 7) has been established for reduced $Nd_{2-x}Ce_xCuO_{4-y}$ compounds [14]. The parent compound $Nd_{2-x}Ce_xCuO_4$ is again an antiferromagnetic insulator. The AF order has been observed recently by the μSR and neutron experiments [15]. Superconductivity is observed in a limited range of Ce composition. The system remains semiconducting up to $x = 0.14$ and the superconductivity abruptly appears at $x =$

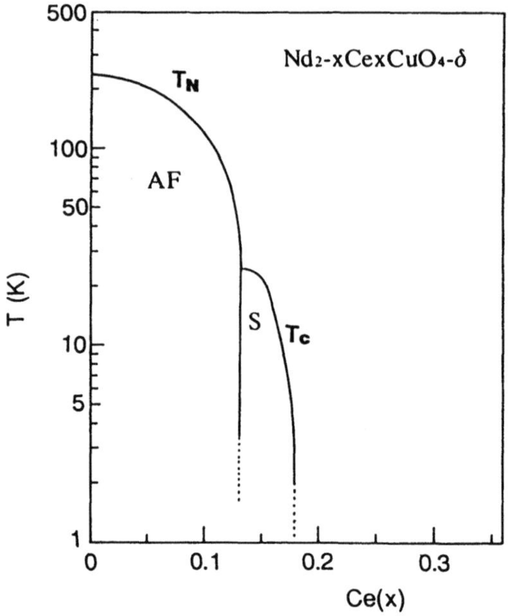

Fig. 7 Electronic phase diagram of reduced $Nd_{2-x}Ce_xCuO_{4-y}$. The temperature T_N of the antiferromagnetic order was roughly estimated from the μSR experiment.

0.14. From 0.14 to 0.15 the magnetude of the Meissner signal rapidly increases without appreciable change of the onset T_c. With further increase of x over 0.15, T_c decreases and the superconductivity disappears at $x = 0.18$. For $x > 0.18$ the compound remains metallic, so that the superconductor-to-nonsuperconducting metal transition takes place at $x = 0.18$ as in the case of heavily doped P-type $La_{2-x}Sr_xCuO_4$. Almost the same diagram is drawn for reduced $Pr_{2-x}Ce_xCuO_{4-y}$.

It should be noted that no superconducting nor "normal" metallic region is observed for as grown, i.e. non-reduced, compounds. They never show truly metallic behavior up to $x = 0.20$ which is the solubility limit of Ce into T'- Nd_2CuO_4. Recent μSR measurement on the as-grown compound with $x = 0.15$ has given an evidence for AF order around 30 K [16]. The AF order was not seen when the same sample was annealed under reducing atmosphere.

The Hall coefficient is plotted in Fig. 8 against the Ce composition for reduced samples. It is negative and the magnitude decrecreases following roughly $1/x$ dependence in the semiconducting phase as is supposed from the Mott-Hubbard picture. In the superconducting phase the sign remains negative but the magnitude decreases steeply with x and then the sign reverses when x exceeds 0.18 corresponding to the transition to the "normal" metallic phase. This is quite the same behavior as is observed in hole-doped $La_{2-x}Sr_xCuO_4$ system [17], so that one can imagine a similar change of the electronic state, from the Mott-Hubbard to Fermi liquid state, with electron-doping.

The optical spectrum of undoped Nd_2CuO_4 is dominated by a sharp exciton-like peak at 1.5 eV without any spectral feature in the lower energy region. The peak is observed only for light polarized along Cu-O planes, so that it corresponds to an optical energy gap in the Cu-O plane, probably to the threshold of the charge-transfer (CT) exitations between

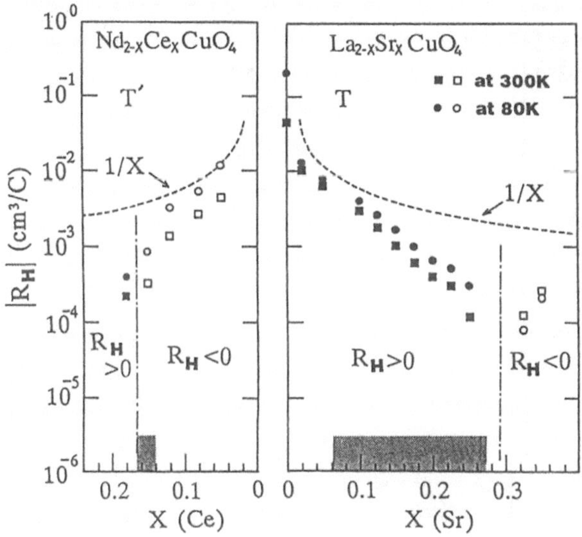

Fig. 8 The absolute value of the Hall coefficient as a function of Ce composition for reduced $Nd_{2-x}Ce_xCuO_{4-y}$. The same ploted for $La_{2-x}Sr_xCuO_4$ are show for composition. The shaded are a indicates composition region where superconductivity is observed.

Cu and O states. The corresponding CT exitation is seen at 2.0 eV in La_2CuO_4 but it is much weaker because the material is slightly doped with holes due to nonstoichiometry introduced during the crystal growth process [18].

From the Kramer-Kroning analysis, the (effective) numberr of electrons involved in this excitation is estimated to be nearly one, indicating that one electron per Cu atom takes part in this excitation. It is clearly demonstrated in Fig. 9 that the doping transfers the effective number of electrons from the CT excitation to the lower energy, perhaps infraband, excitations which makes a reflectivity edge at 1.1 eV. The transfer is completed at $x \sim 0.15$ above which the reflectivity spectrum almost fits the Drude formula down to 0.1 eV. Again, the electron doping has very similar effect on the optical spectrum to the hole doping.

4. Summary

We have discussed the change of the electronic state against the doping, either electrons or holes, into the parent compound, Nd_2CuO_4 of La_2CuO_4 of the high-T_c materials based on the data of the Hall effect and the optical reflectivity spectrum. For light doping, both transport and optical properties recoincile with very strongly correlated electronic state, Mott-Hubbard or charge-transfer insulating state. On the other hand, for heavy doping the Fermi liquid picture seems more appropriate.

Doping creates mobile charges and the cuprates become metallic by small amount of doping. This is quite unique to the cuprates. Similar doping into other transiton-metal oxides cannot easily create charge carriers. To question, why the doping makes the many theorists have stressed on the role of the holes doped onto oxygen atoms in the Cu-O planes. Because the Coulomb repulsion is very large on Cu site, the holes

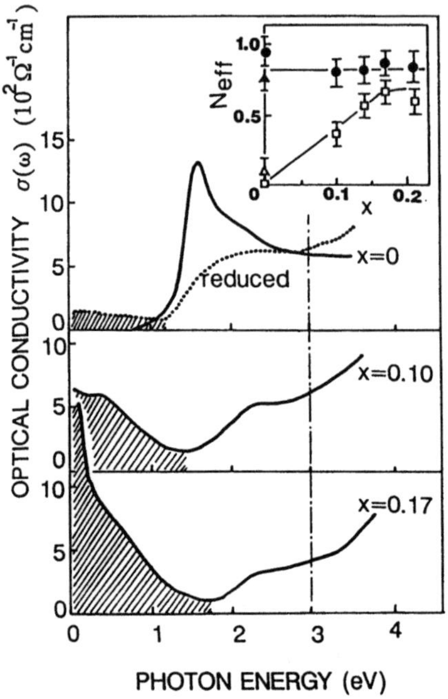

Fig. 9 Optical conductivity spectra for three composition of $Nd_{2-x}Ce_xCuO_{4-y}$. The dotted curve is the spectrum of undoped sample after reducing process. The data were obtained by the Kramers-Kronig analysis on the reflectivity spectra at room temperature for **E** within plane. The integrated value up to 3eV are plotted in the inset. This corresponds roughly to the effective electron number performula unit which is involved in the low energy excitations.

are preferentially created on oxygen sites. The recent discoverry of the electron-doped superconductors has shown that the electrons are equally mobile in the cuprates, even producing superconductivity at moderate doping level. The electrons are supposed to be doped into the Cu-O planes as the holes are in the P-type materials, although there is a debate concerning the location of doped electrons : where they occupy half-vacant Cu 3 $d_{x^2-y^2}$ orbital or vacant Cu 4 s orbital. It is unlikely that the doped electrons move the space out of the Cu-O planes, i.e. Nd/O layers, since the Nd/O layers are highly disordered due to random substitution of Nd with Ce, and since the prperties are uncharged for Pr compound and for Th substitution.

The experimental results presented here demonstorate that the doped electrons affect the electronic state and possibily the spin state of the Cu-O plane in a similar or roughly the same manner as the doped holes do. This is suggestive of an approximate symmetry of the electronic state with respect to hole and electron doping which produces high temperature superconductivity in the Cu-O plane.

References

1. T. Fujita, Y. Aoki, Y.Maeno, J. Sakura H. Fukuba and H. Fujii ; Jpn. J. Appl. Phys. 26, L368 (1987)

2. J.M. Tarascon, W.R. McKinnon, L.H.Greene, G.W. Hull, B.G. Bagley, E.M. Vogel and Y. Lepage, Proc. of Symposium S, 1987 Spring Meeting of the Material Research Society, Edited by D.V. Gubsean and M. Schluter, p. 65.

3. Y. Tokura, H. Takagi, and S. Uchida ; Nature 337, 345 (1989).

4. S. Uchida, H. Takagi, H. Yanagisawa, K. Kishio, K. Kitazawa, K. Fueki and S. Tanaka ; Jpn. J. Appl. Phys. 26, L445(1987).

5. Y. Kitaoka, S. Hiramatsu, T. Kohara, K. Asayama, K. Ohishi, M. Kikuchi and N. Kobayashi ; Jpn. J. Appl. Phys. 26, L397 (1987).
Y. Kitaoka, S. Hiramatsu, K. Ishida, T. Kohara and Asayama ; J. Phys. Soc.Jpn. 56, 3024 (1987).

6. D. Vaknin, S.K. Sinha, D.E. Moncton, D.C. Johnston, J.M. Newson, C.R. Safinya and H.E. King, Jr. Phys. Rev. B58, 2802 (1987)

7. I. Watanabe, K. Kumagai, Y. Nakamura, T. Kimura, Y. Nakamichi and H. Nakajima, J. Phys. Soc. Jpn. 56, 3028 (1987).

8. A. Fujimori, E. Takayama-Muromachi, Y. Uchida and B. Okai, Phys. Rev. B35, 8814 (1987).

9. K.B. Lyons, P.A. Fleury, J.P. Remeika, A.S. Cooper and T.J. Negran ; Phys. Rev. B37, 2353 (1987).
S. Sugai, S. Shamoto and M. Sato ; Phys. Rev. B38, 6436 (1988).

10. J. Gopalakrishnan, G. Colsmann and B. Reuter ; J. Solid State Chem. 22, 145 (1987).
T. Matsuura, Doctor Thesis ; University of Tokyo, 1985.

11. N.P. Ong, Z.Z. Wang, J. Clayhold, J.M. Tarascon, L.H. Green and W.R. McKinnon ; Phys. Rev. B35, 8807 (1987).

12. N.P. Ong, Z.Z. Wang, J. Clayhold, J.M. Tarascon, L.H. Green and W.R. McKinnon ; Novel Superconductivity (Plenum Press, 1987). edited by S.A. Wolf and V.Z. Kresin, p. 1061.
H. Takagi, H. Eisaki, S. Uchida, A. Maeda, S. Tajima, K. Uchinokura and S. Tanaka ; Nature 332, 236 (1988).

13. S. Tajima, S. Uchida, S. Kanbe, K. Kitazawa, K. Fueki and S. Tanaka ; Jpn. J. Appl. Phys. 26, L432 (1987).

14. H. Takagi, S. Uchida and Y. Tokura ; Phys. Rev. Lett. 62, 1197 (1989).

15. G.M. Luke, B.J. Sternlieb, Y.J. Uemura, J.H. Brewer, R.Kadono, R.F.Kiefl, S.R. Kreitzman, T.M. Riseman, J. Gopalakrishnan, A.W. Sleight, M.A. Subramanian, S. Uchida, H. Takagi and Y. Tokura ; Nature 338, 49 (1989).

16. Y.J. Uemura ; private communication

17. H. Takagi, T. Ido, S. Ishibashi, M. Uota, S. Uchida and T. Tokura ; Phys. Rev. (in press)

18. S. Tajima, H. Ishii, T. Nakahashi, H. Takagi, S. Uchida, M. Seki, S. Suga, Y. Hidaka, M. Suzuki, T. Murakami, K. Oka and H. Unoki ; J. Opt. Soc. Am. B6, 475 (1989).

Andreev Reflection, Thermal Conductivity, Torque Magnetometry, and Hall Effect Studies on High-T_c Systems

N.P. Ong, T.W. Jing, Z.Z. Wang, J. Clayhold, S.J. Hagen, and T.R. Chien

Joseph Henry Laboratories of Physics, Princeton University,
Princeton, NJ 08544, USA

Abstract. We survey recent experiments which probe the interface between the high-T_c oxides and Au (Andreev reflection, and proximity induced superconductivity). We also discuss experiments which measure the anisotropy of *123* (thermal conductivity, torque magnetometry) and the Hall effect in the Bi- and Tl-based superconductors.

1. The general problem of the normal metal-*123* interface

In the past, much has been learnt about superconductivity by the technique of current injection into well-characterized junctions[1]. These include strong barrier junctions (Giaever and Josephson tunneling) and weak-barrier junctions (Andreev reflection[2], and weak-links). Major obstacles in applying these tools to high-T_c superconductors are presented by the extreme sensitivity of the oxides to chemical damage and the very short coherence length. The problems in strong-barrier (tunneling) junctions have been widely discussed. We consider here the opposite limit of weak-barrier junctions.

Most "metallic" contacts to $YBa_2Cu_3O_7$ ("*123* ") crystals are in fact semiconducting because of an intervening inert layer adjacent to the deposited metal. In the last few months we have made progress towards lowering the contact resistance[3]. By cleaving *123* crystals in a high vacuum (10^{-6} torr) and evaporating Au *in situ*, we find that the contact resistance becomes comparable to the best reported in ceramics. Direct soldering with In results in contacts 6 orders of magnitude more resistive than the cleaved crystal junctions. If Pb is evaporated onto cleaved crystals, we find that specific contact resistivity $\rho_{\square}$ is surprisingly low (10^{-6} Ωcm^2) even though a PbO layer forms at the interface[3]. (Below 7.2 K the Pb gap is clearly observed in these junctions.) The best junctions are obtained by evaporating Au onto cleaved crystals. These junctions behave as superconducting contacts (see below). The best value obtained in ceramics (10^{-10} Ωcm^2, Ekin et al [4]) are comparable to our Au contacts.

2. Andreev Reflection

We have performed Andreev reflection measurements in the Au-*123* junctions, using the Sharvin contact technique of Hoevers et al[5] (who studied the energy gap Δ in

Springer Series in Solid-State Sciences, Vol. 89 **Strong Correlation and Superconductivity**
Editors: H. Fukuyama · S. Maekawa · A.P. Malozemoff

ceramic *123* capped by a Ag film). When an electron in state **k** is incident onto the *N-S* interface from *N*, it has some amplitude to enter the condensate as part of a Cooper pair and some amplitude to enter as a quasi-particle. To conserve charge, the former process requires a hole to be reflected back into *N* in the state **-k**. If we call the probability flux of the reflected hole and the transmitted quasi-particle *A* and *C*, respectively, then for an incident unit flux, probability flux conservation implies that $1 = A + C$ [2]. The electric current incident from the *N* side is $J = 1 + A$ while on the *S* side the outgoing current equals $2A + C$, where the supercurrent component is $2A$.

It turns out that *A* and *C* are strongly dependent on the energy of the injected electron reckoned from the chemical potential, $\xi_k = \varepsilon_k - \mu$. If $\xi_k < \Delta$, *A* is close to 1 whereas *C* tends to zero. On the other hand, as ξ_k becomes progressively larger than Δ, *A* vanishes while *C* approaches 1. In the former case, $J = 1 + A \rightarrow 2$, while in the latter case, $J \rightarrow 1$. Thus, as the junction bias *V* decreases below Δ/e, the resistance $R(V) = dV/dI$ should be halved. In **Fig. 1** we show *R(V) vs. V* for one of our Au-*123* junctions at several temperatures *T* [6]. At 4.2 K, a sharp decrease in *R* is observed at $V = \pm 9$ mV, which we identify with the superconducting gap. The film thickness is 1,000 Å. In two thicker films (3,000 Å) Δ is also observed at 9 mV. Both the magnitude of $\Delta R/R$ and the observed gap Δ are quite similar to Hoevers et al. At higher *T*, the "gap" is less well-defined. The gap ratio $2\Delta/k_BT \sim 2.3$ is smaller than the gap value inferred from "structures" observed in other tunneling experiments. At present we are not sure if our gap value is intrinsic to bulk *123*, or an artifact of the Au-*123* interface. The sensitivity to finite *T* effects is also not understood. We speculate that it may be caused by increased inelastic scattering of the injected electrons or reflected holes in the Au film.

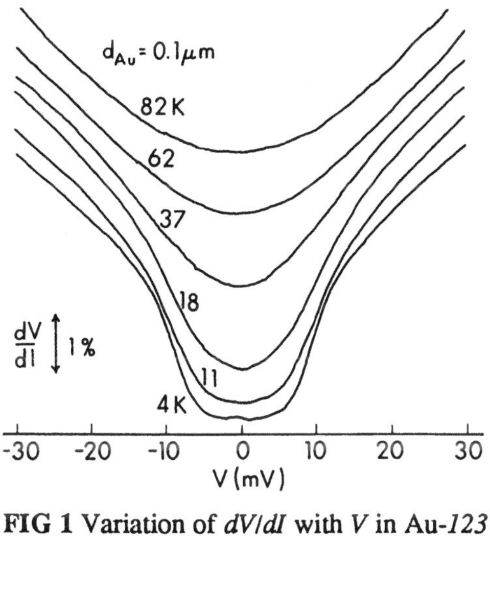

FIG 1 Variation of *dV/dI* with *V* in Au-*123* junction at several temperatures [6].

3. Proximity effect

The question whether a normal metal in proximity with a high-T_c crystal can become superconducting will now be discussed. In the classical superconductors the proximity effect[7] is easily demonstrated because excellent metallic contacts (high transmission probability α for electrons incident on the interface) are readily formed between, say, Pb and Sn[8]. In the high-T_c systems, however, the inert surface layer and the short coherence length ξ conspire to make α very small, so that in most junctions the proximity effect is non-existent. The question is whether the proximity effect is absent because α is simply too small in these junctions, or because a fundamental difference in symmetry prevents the high-T_c pairing potential from infecting conventional metals.

From our work[3] on the Au-*123* interface we believe that the former is in fact the case, although we have not proved it completely to our own satisfaction. In the Au-*123* junctions, we find that the zero-bias contact resistance R_c drops by a factor > 10^4 at T_c (below our resolution of 1 μΩ). The *I-V* characteristics are vertical lines with critical currents exceeding 0.1 mA (our highest dc current). Very near T_c, we observe weak-link behavior with periodic voltage structures reminiscent of Fiske steps. These results suggest that the Au becomes superconducting below the T_c of the crystal, with the whole junction acting as a superconducting contact.

In a second type of junction, Pb is evaporated onto the cleaved surface instead of Au [3]. R_c is fairly high (10-100 mΩ) because of oxidation of the Pb film. However, when this junction is subjected to a large current pulse (0.3 A for 5 μs) at 77 K, R_c drops below 1 μΩ, and *I-V* characteristics similar to *SNS* weak links are observed (**Fig. 2**). At each *T*, the voltage increases abruptly when the critical curent I_c is exceeded. (I_c is linear in T_c -*T* above 60 K.) We speculate that the high current causes partial melting of the Pb which extrudes as fingers into the bulk of the crystal.

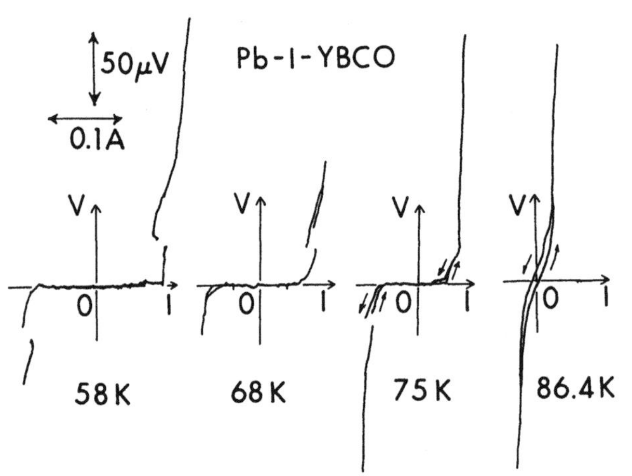

FIG 2 V-I characteristics of a Pb-I-123 junction showing supercurrent behavior. Breaks in the curves are interpreted as caused by flux flow in the junction [3].

Both junctions prove that superconducting contacts with very high I_c can operate at T as high as 88 K. The results also suggest that both Au and Pb can be made superconducting by proximity with *123* if α is made large enough. However, to be entirely convincing one has to demonstrate (by tunneling into the Au or Pb) that a gap is induced in N these high T. These experiments are now in progress.

4. Thermal conductivity

Recently, we have measured the thermal conductivity both in the in-plane (κ_{ab}) and out-of-plane (κ_c) directions in "90K" *123* crystals. κ_{ab} has also been measured in insulating *123* crystals[9]. From **Fig. 3**, it may be seen that there is a strong resemblance of κ_{ab} to the ceramic data[10]. The large anomaly below T_c is apparent in the in-plane conductivity, although its magnitude is sample dependent. Above T_c,we measure reproducibly the value 8 W/mK for κ_{ab}, which is twice the ceramic value. One sample shows a slight increase in κ_{ab} with T above 150 K, which is also observed in some ceramic data. The electronic contribution to κ_{ab} above T_c, is estimated (from ρ_{ab} = 150 $\mu\Omega$cm at 290 K) to be 4 W/mK, which is about half of the total conductance. (The 10% estimate from ceramic samples is incorrect).

In the out-of-plane direction κ_c shows behavior typical of insulating crystals. At high T, κ_c decreases with increasing T, while at low T it attains a broad maximum. However, there is no evidence of an anomaly below T_c. From our data scatter, we estimate that an anomaly in κ_c, if it is present, must be < 1/40 the size of the anomaly in κ_{ab}. In addition to the 90 K crystals we have also measured κ_{ab} in crystals which have been reduced in a flow of Ar gas to the insulating phase. The profile of κ_{ab} vs. T is then observed to be very similar to κ_c vs. T, i.e. no anomaly is seen in κ_{ab}. However, the magnitude of κ_{ab} is around 8 - 9 W/mK near 100 K. When the carriers are removed, κ_{ab}^{ph} is found to double in size above 100 K, so that

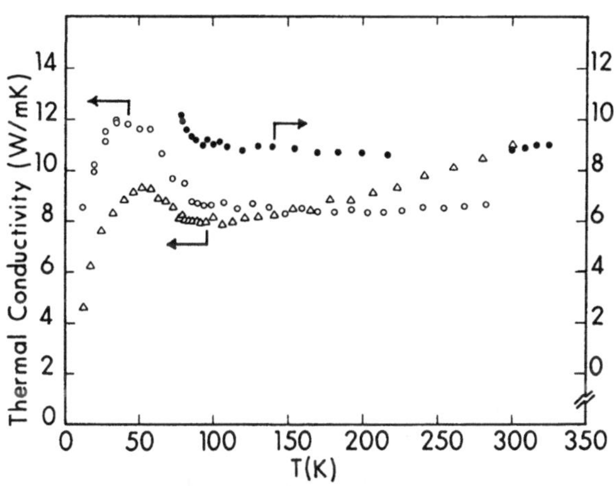

FIG 3 The in-plane thermal conductivity of *123* vs. temperature in three crystals[9].

the total conductance remains the same order of magnitude. This implies that in the 90 K crystal above T_c, the in-plane electron phonon scattering must be rather large. From the large in-plane phonon component found in insulating crystals it is entirely possible that the anomaly in κ_{ab} in the 90 K crystal arises mainly from an increase in the phonon mean free path. It is unnecessary to invoke more sophisticated mechanisms such as the quasi particle backflow contribution [11] (although this may also be present). In contrast, the electron-phonon coupling is entirely negligible in the out-of-plane direction. This very anisotropic electron phonon coupling may be quite ineffective in producing phonon-mediated pairing for the superconductivity, as found in calculations based on the Eliashberg equations[12]. Thus, phonon-scattering in the CuO_2 plane is strong, but far too anisotropic to support a phonon-mediated pairing with a T_c ~100 K. Further information on the anisotropy is provided by the torque magnetometry which we discuss next.

5. Torque magnetometry

Kogan and coworkers[13] have calculated the variation of the torque τ exerted on a crystal when the applied magnetic field $\boldsymbol{H}$ is tilted at an angle θ to the c-axis. Farrell et al [14] have performed some measurements on aligned ceramics and confirmed the validity of the effective mass assumption used by Kogan et al. We have recently extended these measurements to single crystals of *123* in the temperature range near T_c [15]. We have compared our data (**Fig. 4**) with the equation[13]

$$\tau(\theta) = CH \ln \{\gamma\eta H_{c2}^{//}/HD\} (\sin 2\theta)/D \qquad (1)$$

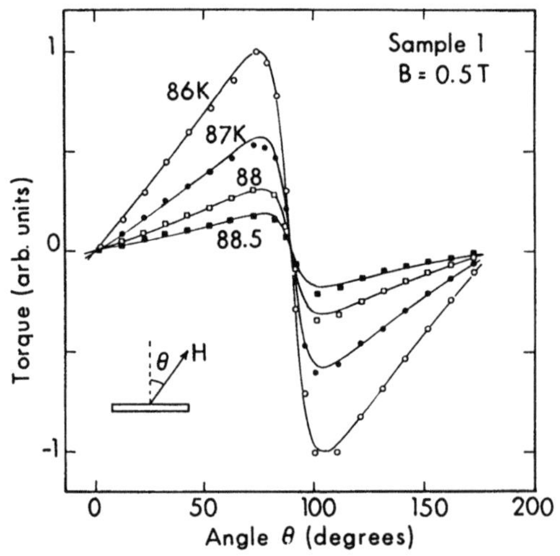

FIG 4 Variation of the torque with angle θ of field orientation in single-crystal *123* near T_c. Solid lines are fits to Eq. 1 [15].

where $D = \sqrt{(\sin^2\theta + \gamma^2 \cos^2\theta)}$, η a constant of the order of unity, and C is proportional to $(m_3-m_1)/\lambda^2$, and λ is the average penetration depth. By fitting the data to Eq. 1, we have extracted the value of $\gamma = 5.8 \pm 0.3$ which implies that an effective mass ratio $m_3/m_1 \sim 34$. Most recent measurements of the anisotropy of H_{c2} in *123* appear to cluster around the value of 6, in agreement with the torque value. If we assume that the conductivity in both directions can be written in the usual Drude form $\sigma = ne^2\tau_{in}/m^*$ where τ_{in} is the inelastic scattering time and m* the effective mass, then we may derive a value for the anisotropy of τ_{in} from the resistivity data and the value of γ. Using the value $\rho_c/\rho_{ab} = 300$ at 100 K obtained by Hagen et al [16], we find that the in-plane scattering *rate* $1/\tau_{in}$ is 9 times slower than the out-of-plane scattering rate. If we further assume that phonon scattering determines most of the scattering in both directions, we arrive at a conclusion in sharp conflict with the thermal conductivity anisotropy measurements which found that the in-plane scattering is much stronger than the out-of-plane. The resolution of this difficulty may be that the transport anisotropy in the normal state cannot be described by an anisotropic mass model. The out-of-plane transport is not Drude like. Despite a very weak electron phonon coupling, carrier mobility perpendicular to the plane is severly hindered by an unknown mechanism. The poor mobility cannot be simply ascribed to a large effective mass.

For T below 83 K, and fields under 1 T, strong pinning effects become dominant, and Eq. 1 is no longer valid. Instead of the smooth variation of τ with θ, we observe a discontinuous jump in τ near $\theta = 70^o$[15]. At fixed T (83K), the magnitude of the discontinuity $\Delta\tau$ decreases with increasing field, and vanishes at a field of 0.8 T. At slightly higher T, the field at which $\Delta\tau$ vanishes is smaller. These discontinuous changes indicate that, below a certain line in the *H vs. T* phase diagram, it is not possible to rotate the vortex lattice smoothly from the in-plane to out-of-plane orientation. Since a number of flux lattice configurations are very close in energy when θ is non-zero, we speculate that the sharp jump corresponds to a rearrangement of the flux lattice, possibly involving spontaneous melting of the lattice structure. These findings are intimately related to the current controversy regarding what happens to the flux lattice close to T_c, and more detailed investigation of this point is underway.

6. Hall effect

One of the basic questions in this field is how do changes in the carrier density n affect T_c. We have studied Bi *2212* doped with trivalent Tm, $Bi_4Sr_3Ca_{3-x}Tm_xCu_4O_{16+y}$[17]. As in previous work on the Y-doped series, both T_c and the resistivity are strongly affected by the Tm content. We measured the resistivity and Hall effect in a series of samples with Tm content x = 0, 0.1, 0.2, 0.5, 1.0, and 1.5. The Hall coefficient is weakly T-dependent or almost T-independent, so we have identified n_H with n. As shown in **Fig. 5**, the Hall density $n_H = 1/(eR_H)$ decreases linearly with increasing x, vanishing at $x = 1.4$. There is a corresponding 100-fold increase in ρ between $x = 0.5$ and 1.4. Our results are similar to results published by Prof. Iye's group on Y-doped Bi *2212* [18].

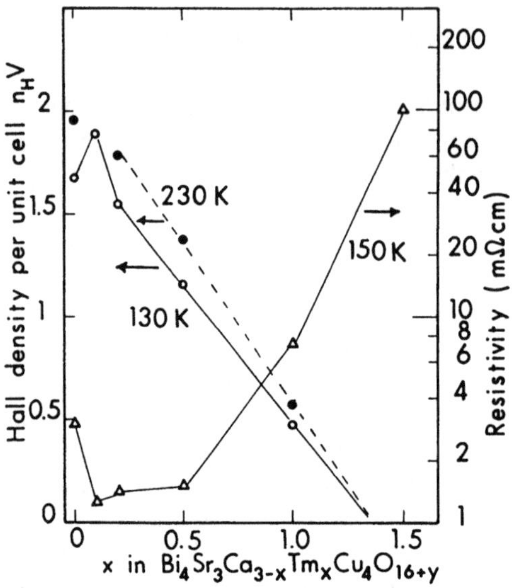

FIG 5 Variation of the Hall density and resistivity in ceramic samples of Bi *2212* doped with Tm [17].

We have interpreted these results in terms of the Mott-Hubbard picture. In the x=0 sample, there is a fairly large carrier density (filling factor $\delta = 0.42$ per Cu ion) which, we argue, arises from cationic deficiency (fewer Bi and Sr ions than given by the nominal *2212* formula) combined with a slight excess in O. By doping with a trivalent ion we raise the chemical potential μ into the Mott Hubbard gap when x = 1.4, thereby driving the system insulating. The variation of n_H with x is very reasonable in the strongly-correlated (large U) scenario, but is in sharp disagreement with the published band structure results[19] which show no gap in the density of states for a range up to 4 eV above μ. An interesting point is that T_c does not scale with n_H. T_c is clamped at the value 83±2 K from $x = 0$ to 0.5. Beyond 0.5, T_c plummets to zero very rapidly (<4.2 K in the $x = 1.5$ sample). We interpret this to indicate that T_c is sensitive to the amount of cation disorder in the system. The Hall mobility is much lower at $x = 0$, than at x =0.5 in our samples.

A previous study[20] of Tl *2223* ($Tl_2Ba_2Ca_2Cu_3O_{10+x}$) showed that the Hall number n_H is quite small ($\delta = 0.15$) compared with *123* ($\delta = 0.5$) and Bi 2212 (δ = 0.42). This runs counter to the orthordox view that n should increase steadily as T_c goes up. We have now performed further measurements on new samples of Tl *2223* and the 2-plane Tl *2212*, which were determined to be single-phase material (by x-ray diffraction) with a large Meissner fraction. The new results confirm our earlier work. n_H at 100 K corresponds to only 0.15 and 0.1 in the 3-plane and 2-plane systems respectively. It is useful to have an overview of the Hall results for the high-T_c oxides based on CuO_2 planes (**Fig. 6**). Both *123* (Co-doping, open circles) and Bi *2212* (Tm doped, open squares) appear to follow trajectories which are close to each other, when n is tuned by chemical doping. The *214* system (open triangles) also

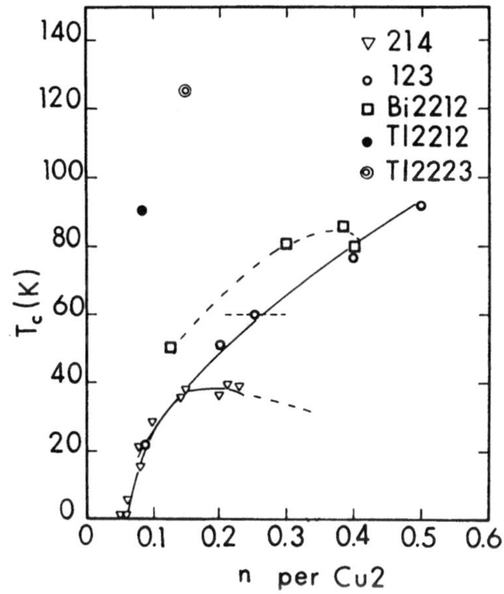

FIG 6 Plot of T_c of the high-T_c oxides vs. the filling factor (carrier density n per Cu ion) as determined by Hall measurements (See text).

follows this trajectory until the break at 40 K. The Tl systems (solid and double circles) appear to lie on a much higher trajectory. The Hall results for the Tl systems are in direct conflict with the μSR results [21] which imply that the carrier density scales linear with T_c. We are inclined to interpret the low carrier density n as intrinsic, rather than a sample-related problem. The increase in T_c observed in the Tl system may be due to the much lower density of defects (especially cation vacancies) in these systems.

We thank P.W. Anderson for guidance and valuable insights. This research is supported by the U.S. Office of Naval Research (Contract N00014-88-K-0283). The Hall effect studies were performed in collaboration with J.M. Tarascon and P. Barboux (Bellcore) and with P.H. Hor and C.W. Chu (Univ. Houston). Some of the studies on the Tl and Bi systems were supported by a subcontract derived from a main DARPA contract (MDA 972-88-J-1002) to the Texas Center for Superconductivity at the University of Houston.

1. For a survey see B.S. Deaver Jr. and D.A. Vincent, in *Methods of Experimental Physics* Vol. 11, edited by R.V. Coleman, (Academic Press) 1974, p. 199.
2. G.E. Blonder, M. Tinkham, and T.M. Klapwijk, Phys. Rev. B **25**, 4515 (1982).
3. T.W. Jing, N.P. Ong, and Z.Z. Wang, to be published.
4. J.W. Ekin et al, Appl. Phys. Lett. 52, 1819 (1988).
5. H.F.C. Hoevers, P.J.M. van Bentum, L.E.C. van de Leemput, H.van Kempen, A.J.G. Schellingerhout, and D.van der Marel, Physica C **152**, 105 (1988).

6. T.W. Jing, Z.Z. Wang, N.P. Ong, and P.W. Anderson, to be published.
7. W.L. McMillan, Phys. Rev. 175, 537 (1968).
8. J. Vrba and S.B. Woods, Phys. Rev. B3, 2243 (1971).
9. S.J. Hagen, Z.Z. Wang, and N.P. Ong, to be published.
10. C. Uher and A.B. Kaiser, Phys. Rev. B 36, 5680 (1987).
11. V. Ginzburg, preprint 1988.
12. M. Prohammer, Physica C **157**, 4 (1989).
13. L. J. Campbell, M.M. Doria, and V. G. Kogan, Phys. Rev. B 38, 2439 (1988).
14. D.E. Farrell, C.M. Williams, S.A. Wolf, N.P. Bansal, and V.G. Kogan, Phys. Rev. Lett. **61**, 2805 (1988).
15. T.R. Chien, Z.Z. Wang, and N.P. Ong, unpublished.
16. S.J. Hagen, T.W. Jing, Z.Z. Wang, J. Horvath, and N.P. Ong, Phys. Rev. B **37**, 7928 (1988).
17. J. Clayhold, S.J. Hagen, N.P. Ong, J.M. Tarascon, and P. Barboux, Phys. Rev. B 39, 7320 (1989).
18. T. Tamegai, K. Koga, K. Suzuki, M. Ichihara, F. Sakai, and Y. Iye, Jpn. Jnl. Appl. Phys. **28**, L112 (1989).
19. L.F. Matthiess and D.R. Harmann, Phys. Rev. B **38**, 5012 (1988).
20. J. Clayhold, N.P. Ong, P.H. Hor, and C. W. Chu, Phys. Rev. B **38**, 7016 (1988).
21. Y.J. Uemura et al, Phys. Rev. Lett. **62**, 2317 (1989).

Resistive State and Hall Effect in High-T_c Superconductors

Y. Iye

The Institute for Solid State Physics, The University of Tokyo,
Roppongi, Minato-ku, Tokyo 106, Japan

Abstract. Magnetotransport studies were carried out on thin film samples of $ErBa_2Cu_3O_{7-y}$ and $Bi_2Sr_2CaCu_2O_{8+y}$ in the superconducting transition region. The fluctuation contribution to the Hall conductivity, $\Delta\sigma_{xy}$, is negative in sign and shows a stronger singularity than $\Delta\sigma_{xx}$, as T_c is approached from above. The magnetic-field-induced broadening of the resistive transition is found to be completely independent of the relative orientation between current and magnetic field, posing a serious question to the conventional picture of Lorentz-force-driven flux creep. The Hall effect in the resistive state exhibits a complicated temperature and field dependence, including threshold behavior and sign reversal.

1. Introduction

The advent of high temperature superconductors (HTSCs) not only sets a challenging problem of the novel mechanism of superconductivity, but also offers a unique opportunity for experimental studies of certain aspects of superconductivity inaccessible by the conventional low temperature superconductors. Examples are found in the distinct manifestations of superconducting fluctuation effect and thermally activated flux motion[1]. These are not only of academic interest but also of great importance for the evaluation of the HTSCs as materials for practical use. We present, in this paper, some results of our recent magnetotransport measurements related to these unique aspects of HTSCs. The first is an angular dependence study of resistive transition, which concerns with the anisotropy of superconductivity and the nature of the resistive state in magnetic field. The second is a study of Hall effect both above and below T_c. The former region is discussed in the context of superconducting fluctuation and the latter is compared with the flux flow Hall effect in conventional superconductors. More details of the present studies than contained in this paper are given in Refs.2 and 3.

2. Experimental Methods

Thin film samples of $ErBa_2Cu_3O_{7-y}$ and $Bi_2Sr_2CaCu_2O_{8+y}$ fabricated by a magnetron sputtering technique on the (001) surface of MgO substrates were used in the present study. The $Bi_2Sr_2CaCu_2O_{8+y}$ samples prepared in our group are highly oriented polycrystalline films with thickness 2 μm having $T_c^{zero} = 76\ K$ and $\Delta T_c \sim 3\ K$. The $ErBa_2Cu_3O_{7-y}$ samples supplied by Dr.Wasa's group at Matsushita Electric Industrial Co. are epitaxially grown 0.25 μm thick films having $T_c^{zero} = 81\ K$ and $\Delta T_c \sim 2\ K$. The samples were patterned into a standard Hall bar shape to facilitate simultaneous measurements of diagonal and Hall conductivities.

Transport measurements were done by a standard *d.c.* method. Probe current in the range of 1 to 10 mA was used, depending on the signal level. At such current levels we did not observe any non-Ohmicity due to Joule heating or other causes. Precise angular dependence measurements with a double axis rotation capability were carried out using a rotating sample holder operated in a 7 *tesla* superconducting Helmholtz coil. Details of this measurement system are described in Ref.2. The system was also conveniently used for automated measurements of temperature dependent resistivity and Hall effect. Measurements of magnetic field dependence at fixed temperatures were done in a 9 *tesla* superconducting solenoid by use of a cryostat with a built-in gas thermometer[4] which enabled us to achieve field-independent temperature regulation during the field sweep.

3. Anisotropy of Superconductivity in $Bi_2Sr_2CaCu_2O_{8+y}$

Figure 1 shows the angular dependence of resistivity in $Bi_2Sr_2CaCu_2O_{8+y}$ at $T = 74.36\ K$ for different magnetic fields. Here, the field angle is swept from $H \parallel c$-axis to $H \perp c$-axis. By taking the 10%, 30% and 50% criteria, the angular dependence of 'H_{c2}' is plotted in Fig.2. The angular dependence near $H \perp c$-axis shows an extremely sharp cusp-like feature, which can be fitted to Tinkham's formula for a thin film superconductor. This is to be contrasted with the case of $YBa_2Cu_3O_{7-y}$, whose anisotropy is well fitted to the anisotropic Ginzburg-Landau model for a layered superconductor.[5] The extreme sharpness of the feature near $H \perp c$-axis proves the well-orientedness of our sample, which is an important experimental requirement for investigation of the issues discussed in this paper.

Figure 3 is an Arrhenius plot of the resistivity for $H \parallel c$-axis and $H \perp c$-axis, demonstrating the thermally activated behavior first recognized by Palstra *et al.*[6] Our result is in qualitative agreement with Palstra *et al.'s*, but shows differences in that (1)the activation energy

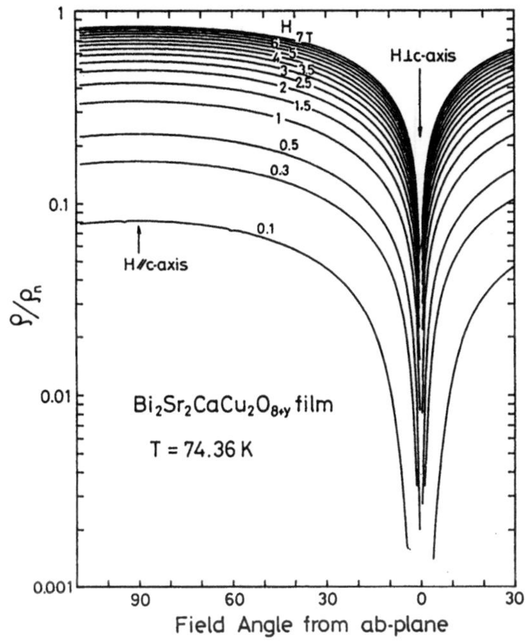

Fig.1 Angular dependence of resistivity of the $Bi_2Sr_2CaCu_2O_{8+y}$ film at $T = 74.36\ K$ for different magnetic fields.

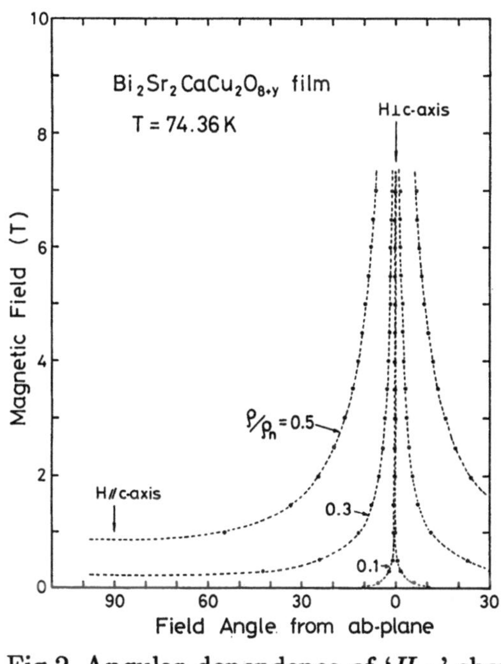

Fig.2 Angular dependence of 'H_{c2}' showing an extremely sharp feature near $H \perp c$-axis.

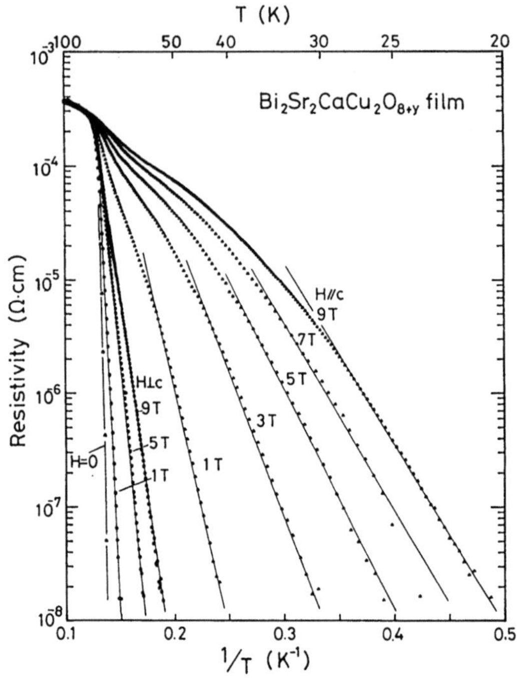

Fig.3 Arrhenius plot of the resistivity of the $Bi_2Sr_2CaCu_2O_{8+y}$ film for different values of magnetic field applied || and ⊥ to the c-axis.

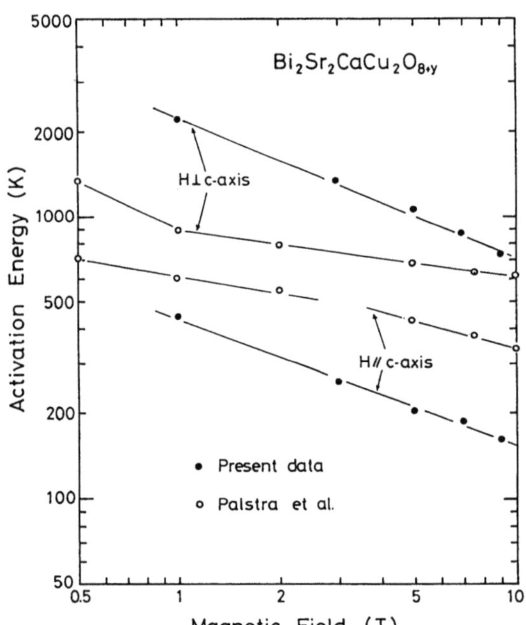

Fig.4 Magnetic field dependence of the activation energy extracted from the data in Fig.3, in comparison with those of Palstra *et al.*[5]

we obtained shows somewhat larger anisotropy as shown in Fig.4, and (2)the pre-exponentaial factor is *not* so universal as Palstra *et al.* report.

4. Current Direction Dependence of Resistive Transition

The thermally activated resistivity in magnetic field has been interpreted by Palstra *et al.*[6] in terms of flux creep. A critical experimental test for the model is to investigate the dependence on the angle between current and magnetic field, because the configurations $H \parallel J$ and $H \perp J$ are fundamentally different in terms of the Lorentz force which is the driving force for the flux lines. We have already pointed out[7] that the resistive transition of $YBa_2Cu_3O_{7-y}$ single crystals under $H \parallel c$-axis is quite similar for $J \parallel c$-axis and for $J \parallel ab$-plane. Similar observation is reported by Kitazawa *et al.*[8] for a $(La_{1-x}Sr_x)CuO_4$ single crystal.

The present experiment investigates this point in a much wider range of resistive transition and to a much higher precision. With the current direction fixed in the *ab*-plane (*i.e.* the plane of the film), an idealized experiment would be to scan the field direction strictly within the *ab*-plane to vary the angle ϕ between the current and the field. However, inevitable angular misalignment coupled with the extremely sharp angular dependence shown in Fig.3 makes the actual traces of ϕ-scans completely dominated by small deflection of the field direction off the *ab*-plane. (Note that what concerns us here is deviation by less than a degree of arc.)

We can still extract the needed information by making use of the fact that any ϕ-scan is graranteed to hit the direction parallel to the *ab*-plane, which is signaled by a minimum of $\rho(\phi)$. At what value of ϕ this occurs depends on the angular misalignment of the sample mounting. Therefore, by deliberately varying the misalignment, we can make the $H \parallel ab$-plane condition occur at different values of the angle between J and H. Figure 5 is an example of a series of ϕ-scans, taken at a temperature where the resistivity is down by about two orders of magnitudes from the normal resistivity. Different traces correspond to minutely different settings of the sample misalignment. It is clearly seen that the minimum value of resistivity is constant irrespective of where it occurs. That the same is true at any point of resistive transition, can be seen in Fig.6, which demonstrates the complete absence of current direction dependence over many orders of magnitude of resistivity.

From the above result, it is clear that the Lorentz-force-driven flux creep picture cannot be true at least in its simplest form, although we *do* believe that the flux motion which is central to the macroscopic magnetic behavior of HTSCs is also an essential ingredient in the resistive state in magnetic field. What should be the alternative is not clear at

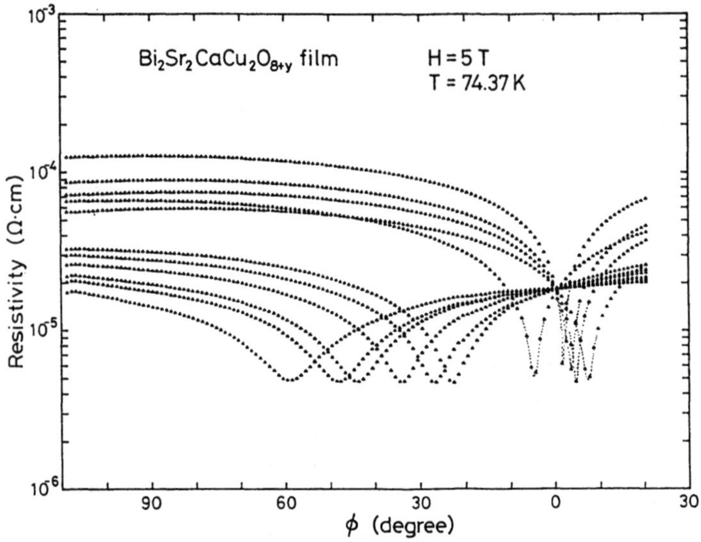

Fig.5 Traces of resistivity in the ϕ-scan. The minimum in each trace corresponds to $H \parallel ab$-plane, which occurs at different values of the angle ϕ between J and H, depending on the angular misalignment of sample setting.

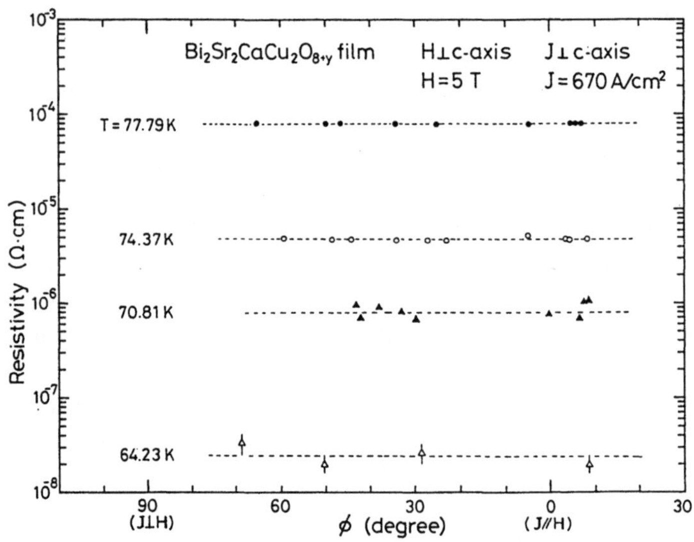

Fig.6 Resistivity as a function of the angle ϕ between J and H, demonstrating complete ϕ-independence of resistivity at any part of the resistive transtion.

the moment. With respect to the point raised by the present result, Tinkham's model[9] based on an analogy with the single overdamped Josephson junction apparently stands safe, as it does not contain any explicit current direction dependence. The fluctuation approach pursued by Ikeda *et al.*[10] also seems attractive in that it gives a result consistent with the current direction independence. If, on the other hand, reconcilation of the flux creep picture with the current direction independence is to be pursued, a key point may lie in such novel concepts as entangled vortex liquid[11] or vortex glass phase[12] as opposed to the conventional Abrikosov lattice. When we remind ourselves that phase-slip and flux motion are, in a general sense, two different ways of describing an essentially same physical process, we like to believe these different pictures to be ultimately united at a deeper level.

5. Fluctuation Effect on the Hall Conductivity above T_c

Figure 7 shows overall temperature dependences of resistivity and Hall coeffiecient for the $ErBa_2Cu_3O_{7-y}$. In the normal state, the resistivity and the reciprocal of Hall coefficient show linear temperature dependences as noted by many authors. The downward deviation of ρ near and above T_c from the linear T-dependence is interpreted as a manifestation of su-

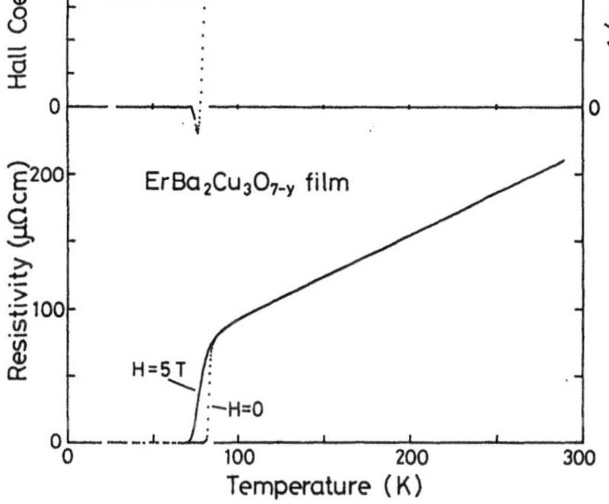

Fig.7 Overall temperature dependences of resistivity and Hall coefficient for the $ErBa_2Cu_3O_{7-y}$ film.

perconducting fluctuation effect. By the same token, the behavior of R_H near and above T_c can be attributed to the fluctuation effect.

Unfortunately, a quantitative analysis of the fluctuation contributions is hampered by the lack of proper understanding of the 'normal' transport properties. As a tentative analysis, we use extrapolations from the high temperature region and assume the 'normal' behavior to be $\rho(T)$ $[\mu\Omega cm] = 37.0 + 0.618\ T\ [K]$ and $R_H^{-1}(T)$ $[\times 10^3\ C/cm^3] = 0.30 + 0.00757\ T\ [K]$, respectively. By subtracting the 'normal' components, the excess conductivity, $\Delta\sigma_{xx}$, and the excess Hall conductivity, $\Delta\sigma_{xy}$, are evaluated and plotted in Fig.8 as functions of $ln(\frac{T}{T_c^*}) \sim \frac{(T-T_c^*)}{T_c^*}$, where $T_c^* = 76.5\ K$ is chosen at the mid-point of the transition. Since this kind of background subtraction is not on a fully justifiable basis, we do not put too much weight on the exact exponents or absolute values of $\Delta\sigma_{xy}$ and $\Delta\sigma_{xx}$. Instead, we make the following qualitative observations for the region $0.05 \leq ln(\frac{T}{T_c^*}) \leq 0.2$, where the data are least susceptible to the ambiguities in the background subtraction and the choice of T_c^*: (1) the sign of $\Delta\sigma_{xy}$ is negative, and (2) the exponent of $\Delta\sigma_{xy}$ is larger than that of $\Delta\sigma_{xx}$. Essentially same features are observed for the $Bi_2Sr_2CaCu_2O_{8+y}$ film.

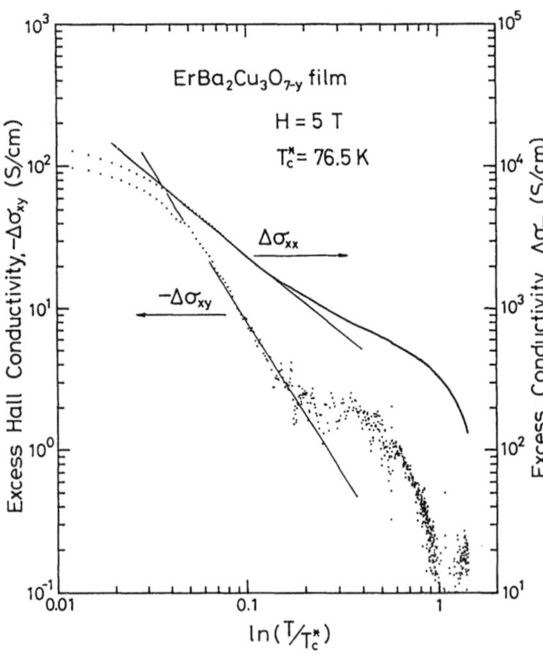

Fig.8 Excess Hall conductivity, $\Delta\sigma_{xy}$, and excess conductivity, $\Delta\sigma_{xx}$, plotted against $ln(\frac{T}{T_c^*})$. Note that $\Delta\sigma_{xy}$ is negative in sign and shows a stronger singularity than $\Delta\sigma_{xx}$.

The effect of fluctuations on the Hall conductivity has been theoretically treated by Fukuyama *et al.*[13] almost two decades ago, but has not been experimentally accessible by conventional superconductors. According to Fukuyama *et al.*, fluctuation-induced supercurrent (Aslamazov-Larkin process) gives a gigantic contribution to the Hall conductivity, and the excess Hall conductivity in the weak field limit is strongly singular, *i.e.* $\Delta\sigma_{xy} \sim (T - T_c)^{\frac{d}{2}-3}$, as compared to $\Delta\sigma_{xx} \sim (T - T_c)^{\frac{d}{2}-2}$, where d denotes the system dimension. The above-mentioned result (2) appears to corroborate this prediction.

Another important conclusion of the theory is that the sign of $\Delta\sigma_{xy}$ is determined by the sign of the energy derivative of the density of states at the Fermi level, $\frac{dN(E)}{dE}$, and hence furnishes different information on the electronic structure than the sign of σ_{xy} itself. The observed negative sign of $\Delta\sigma_{xy}$, indicating a negative $\frac{dN(E)}{dE}$, seems consistent with a picture that the Fermi level lies near the top of the lower Hubbard band (or of the oxygen band in the Hubbard gap), although the applicability of the theory to the present case of correlation-dominated electron system has to be further scrutinized.

6. Hall Effect in the Resistive State below T_c

Figure 9 shows magnetic field dependences of ρ_{xx} and ρ_{xy} at several temperatures near T_c for the $ErBa_2Cu_3O_{7-y}$ film. While ρ_{xy} is proportional to H in the normal phase (*i.e.* Hall coefficient is field independent there), it exhibits a complicated H-dependence below T_c. The Hall voltage stays zero in the low field range, goes negative, and then turns to positive apparently tending to the normal state value. The sign change of ρ_{xy} in the intermediate field range is also manifest in the $R_H(T)$-trace in Fig.7. Similar data for $Bi_2Sr_2CaCu_2O_{8+y}$ are shown in Fig.10.

In a flux flow state of conventional type II superconductors, the hydrodynamical Magnus force on the moving flux lines gives rise to Hall effect. A relevant physical quantity is the Hall angle which corresponds to the direction of the flux line velocity relative to the Lorentz force vector. In the Bardeen-Stephen model[14], the Hall angle is determined by the normal transport in the core region of flux lines, so that it follows the straight extrapolation from the normal behavior above H_{c2}. Microscopic calculations based on the time dependent Ginzburg-Landau equation carried out by Maki[15] and by Ebisawa[16] have shown that in case of a dirty superconductor, the Hall angle just below H_{c2} increases with decreasing H. Experimental studies on many superconducting materials have revealed a much wider range of behavior, including sign change.[17] Such behavior, if it is not dominated by extrinsic causes like guided flux mo-

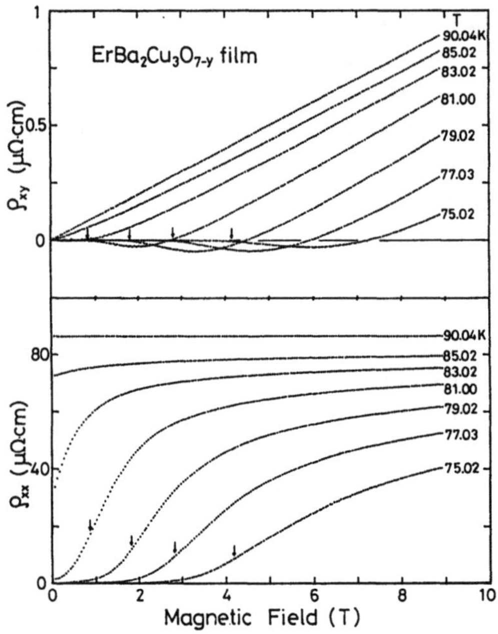

Fig.9 Magnetic field dependences of ρ_{xy} and ρ_{xx} at different temperatures near T_c for $ErBa_2Cu_3O_{7-y}$. The vertical arrows mark the onset points of finite ρ_{xy} and the corresponding points in the ρ_{xx} curves.

tion, must contain useful information on the electronic state on a similar physical basis as the fluctuation contribution to the Hall conductivity. A simplest phenomenological way to explain the sign change may be to incorporate two-carrier conduction in the Bardeen-Stephen model.

One of the most interesting features of the data in Fig.9 is the existence of the finite low field range, in which ρ_{xx} is already finite yet ρ_{xy} remains zero. In the conventional flux flow picture, this implies that there is a threshold flux velocity for the Magnus force to be effective. The longitudinal resistivity values corresponding to the onset points of ρ_{xy} (marked by vertical arrows in Fig.9) lie in the range of 10 to $20\mu\Omega cm$, or ~ 0.1 to $0.2\rho_N$. Malozemoff *et al.*[18] observe that there is a 'knee' in the resistive transition curve around $\rho \sim 10\mu\Omega cm$ for their best crystals of $YBa_2Cu_3O_{7-y}$, which they attribute to a crossover form a flux creep to a flux flow regime. They also note that the crossover region apparently corresponds to an onset of photoresponse found by Zeldov *et al.*[19] It is interesting to note in this connection that the onset of the Hall resistivity apparently occurs at similar values of resistivity. It is tempting to associate these phenomena with the crossover from a flux creep to a flux flow

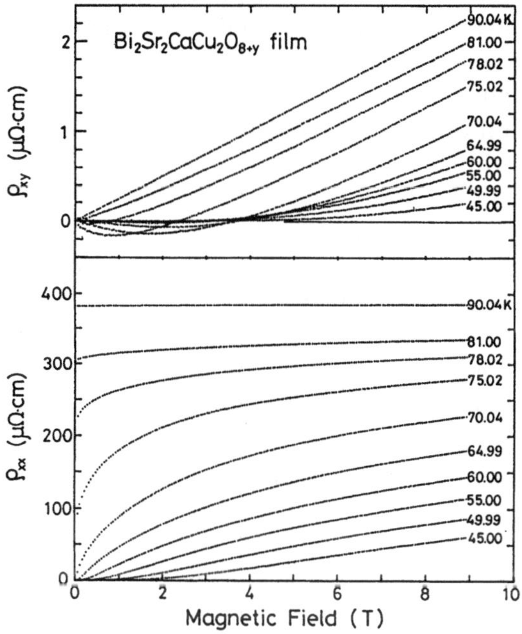

Fig.10 Magnetic field dependences of ρ_{xy} and ρ_{xx} for $Bi_2Sr_2CaCu_2O_{8+y}$. Unlike the $ErBa_2Cu_3O_{7-y}$ case, no clear threshold behavior of ρ_{xy} is seen.

regime, as suggested by Malozemoff, because it gives a rather clear physical picture. Along this line of thought, absence of such threshold behavior of ρ_{xy} in $Bi_2Sr_2CaCu_2O_{8+y}$ may be taken as an indication of different flux dynamics and pinning mechanism between the two materials. We are, however, presently very hesitant to interpret the galvanomagnetic effects in the framework of the conventional flux dynamics, because it is inconsistent with the current direction independence discussed earlier in this paper.

7. Summary

We have studied the superconductivity-related magnetotransport in thin film samples of HTSCs. Fluctuation effect manifests itself in the Hall conductivity as well as in the diagonal conductivity. The former is not easily accessible by conventional superconductors. The qualitative behavior of the fluctuation contribution to σ_{xy} appears to corroborate the theory by Fukuyama *et al.* More quantitative analyses of the fluctuation effect are presently hampered by the lack of proper understanding of the normal transport process. The Hall effect in the magnetic-field-induced resistive state below T_c exhibits complicated temperature- and field-dependent behavior, including threshold behavior and sign change. The complete absence of current direction dependence of the resistive transition calls for either a fundamental renovation of the flux creep model or a different approach. In an attempt to discriminate among theoretical models for the resistive state in magnetic field, it would be interesting to compare what each of those models predicts about the Hall effect.

In conclusion, the Hall effect can furnish important information on the electronic structure and the superconducting state of HTSCs. The Hall effect concerns wit such fundamental issues as polarity and density of carriers, multi-carrier system or not, anisotropy of electronic state, and the nature of resistive state in magnetic field, to name just a few. The Hall effect should also be a key experiment to check Laughlin's intriguing idea[20] of the equivalence of the high temperature superconducting state to the fractional quantum Hall state.

Acknowledgements. This work has been made in collaboration with S.Nakamura and T.Tamegai. We thank Drs. K.Wasa and S.Kohiki of Matsushita Electric Industrial Co. for supply of $ErBa_2Cu_3O_{7-y}$ thin film samples. Helpful conversations with Profs. M.Tinkham, H.Fukuyama, T.Tsuneto and H.Ebisawa and Drs. A.P.Malozemoff and B.Batlogg are appreciated. This work has been supported in part by the Grant-in-Aid for Priority Area of Scientific Research, 'Mechanism of Superconductivity', from the Ministry of Education, Science and Culture, and in part by the Grant for Special Reseach Project from the University of Tokyo.

References

[1] *see, for example*, Y.Iye, *in* 'Studies of High Temperature Superconductors', *ed.* A.V.Narlikar, (Nova Science, New York, 1989) vol.1, p.166.
[2] Y.Iye, S.Nakamura and T.Tamegai, Physica C, *to be published.*
[3] Y.Iye, S.Nakamura and T.Tamegai, Physica C, *to be published.*
[4] Y.Iye, Cryogenics, 28 (1988) 164.
[5] Y.Iye, T.Tamegai, T.Sakakibara, T.Goto, N.Miura, H.Takeya and H.Takei, Physica C153-155, (1988) 26.
[6] T.T.M.Palstra, B.Batlogg, L.F.Schneemeyer, J.V.Waszczak, Phys. Rev. Lett. 60 (1988) 1662.
[7] Y.Iye, *in* 'Mechanisms of High Temperature Superconductivity', *eds.* H.Kamimura and A.Oshiyama, (Springer, Heidelberg, 1989) p.263.
[8] K.Kitazawa, S.Kambe, M.Naito, I.Tanaka and H.Kojima, Jpn. J. Appl. Phys. 28 (1989) L555.
[9] M.Tinkham, Phys. Rev. Lett. 60 (1988) 1658.

[10] R.Ikeda, T.Ohmi and T.Tsuneto, J. Phys. Soc. Jpn. 58 (1989) 1377.
[11] D.R.Nelson, Phys. Rev. Lett. 60 (1988) 1973.
[12] M.P.A.Fischer, Phys. Rev. Lett. 62 (1989) 1415.
[13] H.Fukuyama, H.Ebisawa and T.Tsuzuki, Prog. Theor. Phys. 46 (1971) 1028.
[14] J.Bardeen and M.J.Stephen, Phys. Rev. 140 (1965) A1197.
[15] K.Maki, Prog. Theor. Phys. 41 (1969) 902.
[16] H.Ebisawa, J. Low Temp. Phys. 41 (1972) 11.
[17] *for example,* N.Usui, T.Ogasawara, K.Yasukochi and S.Tomoda, J. Phys. Soc. Jpn. 27 (1969) 574.
[18] A.P.Malozemoff, T.K.Worthington, N.C.Yeh, E.Zeldov, M,W.McElfresh and F.Holtzberg, *in this Proceedings.*
[19] E.Zeldov, N.M.Amer, G.Koren, A.Gupta, R.J.Gambino and M.W.McElsresh, Phys. Rev. B, *to be published.*
[20] R.B.Laughlin, *in.this Proceedings.*

Raman Probing of Electronic Gap Excitations and Their Coupling to Phonons in High-T_c Superconductors

M.V. Klein, S.L. Cooper, F. Slakey, J.P. Rice, E.D. Bukowski, and D.M. Ginsberg*

Department of Physics and Materials Reserach Laboratory, University of Illinois at Urbana Champaign, 1110 W. Street, Urbana, IL 61801, USA
*Present address: AT&T Bell Laboratories, Murray Hill, NJ 07974, USA

Abstract. The room temperature Raman spectrum from single crystals of YBCO and BiSrCCO shows a surprisingly strong continuum that starts at zero energy and extends to beyond 800 cm^{-1}. In YBCO this continuum shows strong anti-resonance-type interactions with two Raman-active phonons and a weaker interaction with a third. In BiSrCCO such interactions are not so well-established. As the temperature is reduced to well below T_c, in both materials the continuum persists, essentially to zero, and it shows a broad peak in the 350-550 cm^{-1} range. This peak may be interpreted as a measure of the gap, 2Δ. In YBCO the strong anti-resonances persist; one at 115 cm^{-1} providing additional proof that the continuum is present at an energy well below the expected value of 2Δ.

1. Introduction

It has been known for some time that electronic Raman scattering from a superconductor provides information about the energy gap 2Δ. The original theoretical predictions assumed that the superconducting coherence length ξ_o was greater that the optical penetration depth δ, but the first experimental observations were on systems such as 2H-$NbSe_2$[1] and Nb_3Sn[2] or V_3Si[3] for which $\xi_o \ll \delta$. In that case BCS-based theory says that for intraband scattering when the coherence length is less than the optical penetration depth there should be a sharp, asymmetric peak in the Raman spectrum at 2Δ [4,5]. If the Fermi surface consists of intersecting sheets, interband Raman scattering should also be observed. The normal state should show a linear rise for its interband Raman response. In the superconducting state at low temperatures, the rise is completely supressed for energy shifts below 2Δ. There is an overshoot at 2Δ, followed by a gradual approach to the normal state response at higher energy shifts. Such a picture using a combination of intraband and interband scattering gives a good quantitative fit to the Raman data from Nb_3Sn.[2] A plausible argument for interband scattering can be based on calculations of band structure.

Springer Series in Solid-State Sciences, Vol. 89 Strong Correlation and Superconductivity
Editors: H. Fukuyama · S. Maekawa · A.P. Malozemoff

2. Results from YBCO

Relatively strong electronic Raman continua have been observed at room temperature in $YBa_2Cu_3O_{7-\delta}$ (YBCO)[6], and in $Bi_2Sr_2CaCu_2O_{8+\delta}$ (BSCCO)[7]. The results from Cooper et al. [6] are shown in Figs. 1 and 2. The notation (_ _) refers to the directions of the incident and scattered photon polarization directions, respectively. The light is scattered back from a surface whose normal is in the c direction, and x and y are along the a and b directions, respectively. Unless otherwise stated, the samples were multiply twinned, so that, e.g., (xx) really denotes an average of (xx) and (yy) spectra. One sees phonon features in the (xx) spectra at 115, 150, 330, 420, and 505 cm^{-1}, assignable to the five expected modes of A_g symmetry in the orthorhombic YBCO structure.

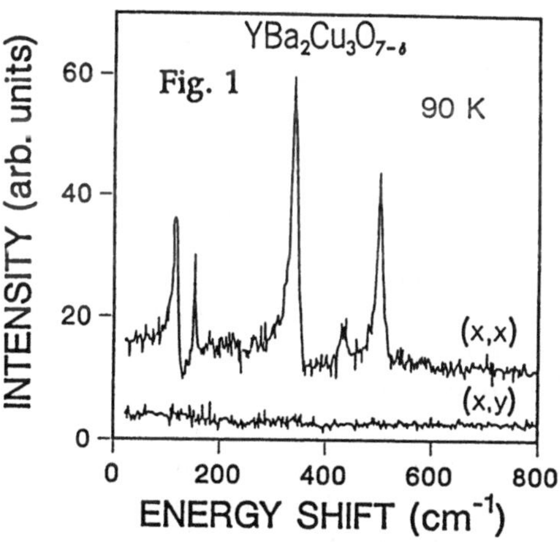

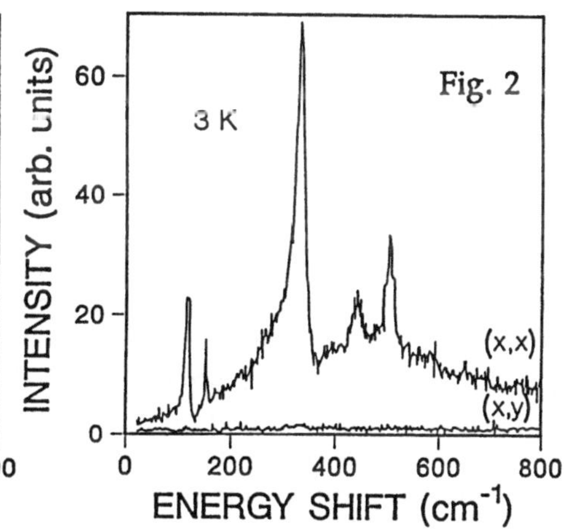

Fig. 1. Polarized [$(xx)=A_g$] and depolarized [$(xy)=B_{1g}$] spectra for single crystal $YBa_2Cu_3O_{7-\delta}$ at 90 K. After Cooper et al. [6].

Fig. 2. Polarized [$(xx)=A_g$] and depolarized [$(xy)=B_{1g}$] spectra for single crystal $YBa_2Cu_3O_{7-\delta}$ at 3K. After Cooper et al. [6].

2.1. Continuum for Temperatures above T_c

Along with the phonon features there is a strong continuum which is quite reproducible from sample to sample. It has also been seen in a high quality film grown by the laser ablation technique on a $LaGaO_3$ substrate; see Fig. 3.[8] Results from an untwinned single crystal of YBCO show that the strength of the continuum is greater when both photon polarizations are along the y (or b) direction than along the x (or a) direction.[9]

The continuum starts at zero energy (shift) and continues beyond 800 cm^{-1} energy. It has the wrong temperature dependence and is too structureless to be a two-phonon continuum. We conclude that the continuum must be electronic in origin.

The fluctuation-dissipation theorem for Raman scattering says that (in units where Planck's constant and Boltzmann's constant are unity) the Raman spectral density is given by

$$S(\omega) = R''(\omega)\,(1 - e^{-\omega/T})^{-1}\,, \qquad (1)$$

where $R''(\omega)$, the imaginary part of an appropriate response function for Raman scattering, obeys

$$R''(-\omega) = -R''(\omega)\,. \qquad (2)$$

For intraband electronic Raman scattering from a clean metal, Fermi liquid theory says that R" should be proportional to energy ω from ω=0, peaking near qv_f, where v_f is the Fermi velocity and where q is the wave-vector transferred. The latter is "smeared" due to the finite penetration depth $\delta \approx 1000$Å, and this leads to an estimate of "q"$v_f \approx 20$ cm^{-1}.

Within a conventional Fermi liquid picture the only other electronic source for the continuum would be interband scattering. This would require the Fermi surface to be multi-sheeted. If the sheets intersect in a line, $q \approx 0$ inter-sheet scattering would be described by an R" that would be proportional to ω at small ω. If the sheets touch in a line, R" would depend on a higher power (cubic) of ω.

In the case of cubic or hexagonal symmetry, it is relatively easy to imagine a Fermi surface with intersecting sheets. For the orthorhombic structure of YBCO, that is not possible. Monien and Zawadowski have recently discussed a model electronic structure for YBCO where one branch of the Fermi surface, derived from the Cu-O planes is essentially tangent to another branch, derived from the Cu-O chains and the bridging oxygen atoms. [10] If this occurs at the zone boundary in the b*-direction, inversion symmetry will prevent the bands from hybridizing.

2.2. Continuum at $T \ll T_c$

As the temperature is lowered below T_c, the continuum changes shape, giving a broad peak in the 300-500 cm^{-1} region (Figs. 2 and 3 and lower two panels in Fig. 4). We associate this peak with some average value of the gap, 2Δ. Results from the untwinned sample are shown in the top two panels of Fig. 4. [9,11] The lower two panels show how the continuum looks using Raman selection rules appropriate for the Cu-O planes, namely tetgragonal.[12] The (y'y') geometry gives tetragonal A_{1g} symmetry, and

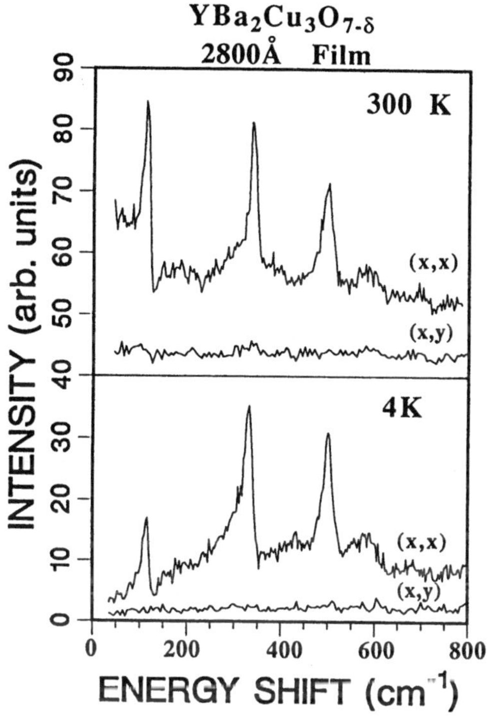

Fig. 3. Polarized Raman spectra at 300 K and 4 K from a superconducting film of $YBa_2Cu_3O_{7-\delta}$ deposited *in situ* by laser ablation on a $LaGaO_3$ substrate.

the (y'x') geometry gives tetragonal B_{1g} symmetry. The 330 cm^{-1} phonon has almost pure tetragonal B_{1g} symmetry, and the 115 cm^{-1} phonon has pure tetragonal A_{1g} symmetry. Note the symmetry-dependence of the continuum. For $YBa_2Cu_3O_{7-\delta}$ the apparent value of 2Δ is higher ($\approx$550 cm^{-1}) in B_{1g} tetragonal symmetry than in A_{1g}, symmetry ($\approx$350 cm^{-1}). We intrepret this as evidence for gap anisotropy.

Surprisingly, the low frequency portion of the Raman response at low temperature shows a linear rise. Within conventional BCS-type theories with a narrow distribution of gaps, the low temperature rise should not be present. Even a nodal structure (lines or points of zeros) of the gap such as discussed for the cases of anisotropic pairing would reasonably be expected to give a frequency dependence stronger than linear. This suggests that some form of gaplessness is present in $YBa_2Cu_3O_{7-\delta}$ and, as we shall see, in $Bi_2Sr_2CaCu_2O_{8+\delta}$. For the former case, Monien and Zawadowski suggest that if the gap function $\Delta(k)$ has the appropriate p-wave or d-wave symmetry, it will vanish at the zone boundary, essentially where the interband Raman transitions are taking place. [10] This picture could explain the persistence of low frequency scattering at low temperatures.

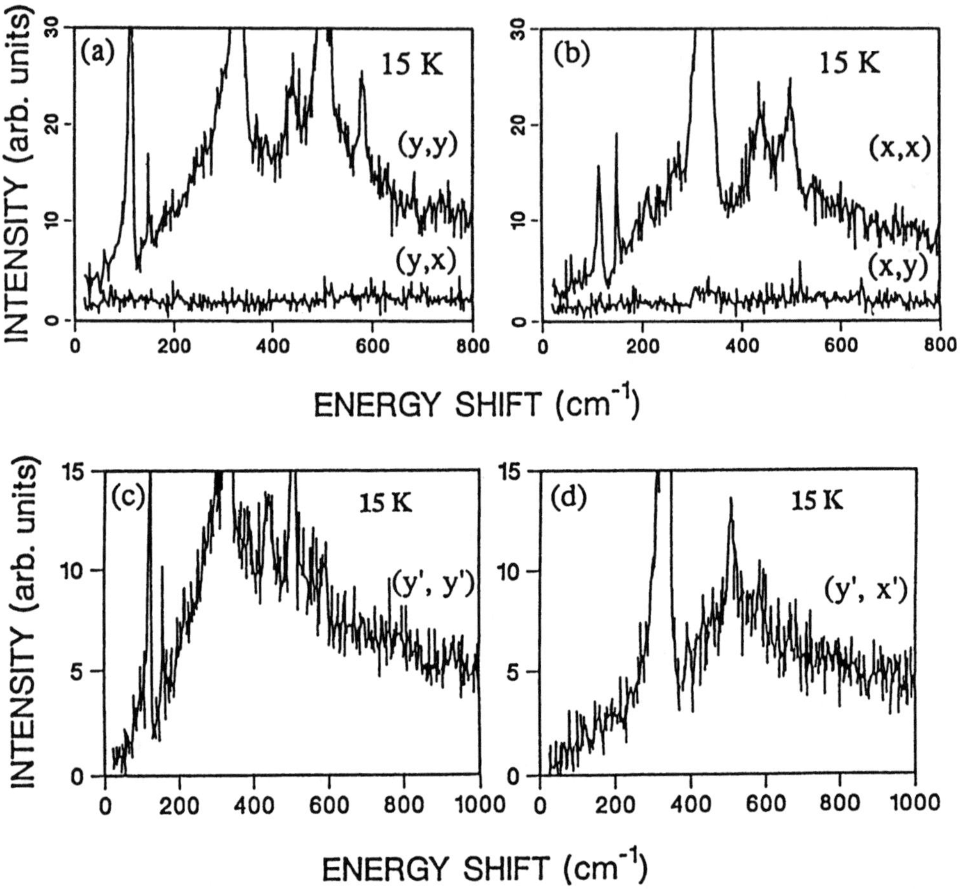

Fig. 4. Low temperature spectra in single-crystal $YBa_2Cu_3O_{7-\delta}$, including (a) polarized (y,y) (=A_g--along Cu-O chain) and depolarized (y,x) spectra at 15K; (b) polarized (x,x) (=A_g--perpendicular to the Cu-O chain) and depolarized (x,y) spectra at 15K; (c) polarized (y',y') (=A_{1g}-tetragonal) spectrum at 15 K; (d) depolarized (y',x') (=B_{1g}-tetragonal) spectrum at 15 K. From Ref. [11].

2.3. Antiresonance with Two Phonons

In $YBa_2Cu_3O_{7-\delta}$ the phonons at 330 cm^{-1} and 112 cm^{-1} show temperature-dependent changes in position and line-shape. [12] This is shown in expanded detail in Fig. 5. The solid lines represent fits to a Fano line shape, wherein R" is given by a Lorentzian multiplied by an anti-resonance factor. The explanation for such a Fano model is the following: The phonon is coupled to the continuum, thereby acquiring a width and suffering a shift in frequency. Light can scatter from the continuum through two channels.

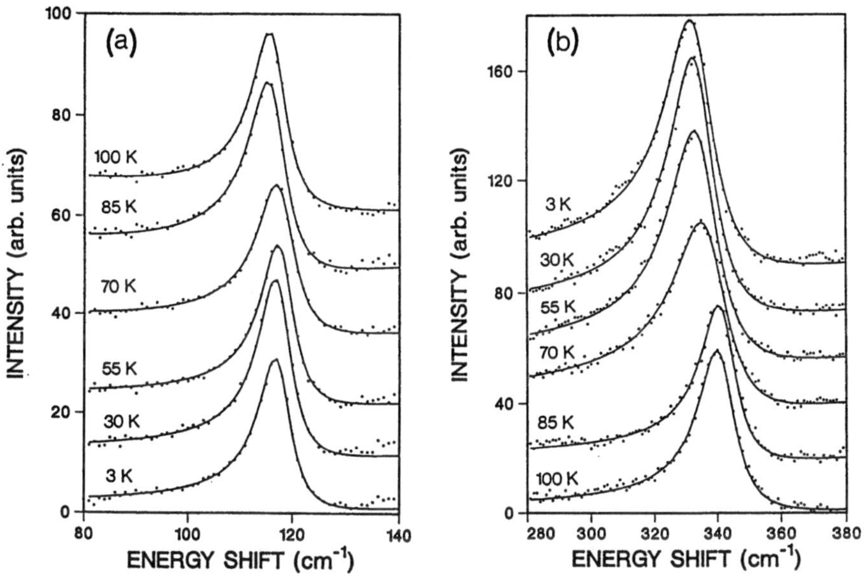

Fig. 5. Temperature dependence of the (a) A_{1g}-tetragonal phonon at 115 cm^{-1} and the (b) B_{1g}-tetragonal phonon at 330 cm^{-1} below T_c. These phonons have been fit to a Fano line shape (solid lines). From Ref. [12].

One channel is direct, and the other is indirect, via the phonon. At the antiresonance frequency the amplitudes for the two channels cancel. For the fits shown in Fig. 5 the antiresonance frequency is greater than the renormalized phonon frequency.

The 115 cm^{-1} phonon, which we have seen has pure A_{1g}-tetragonal symmetry, couples with the A_{1g} part of the continuum. Recalling the lower left panel of Fig. 3, we see that the major strength of the continuum is well above 115 cm^{-1}; thus the renormalization by the coupling to the continuum will soften this mode. As the temperature is reduced through T_c to nearly zero, both the broadening and the softening are seen to reduce. The former effect is a sign that the continuum at 115 cm^{-1} is weakening, and the latter effect is a sign that the effective frequency of the continuum is shifting upwards. Regarding the 330 cm^{-1} phonon, Fig. (5b) shows that it both softens and broadens as the temperature is lowered. Broadening implies that the continuum near 330 cm^{-1} is becoming stronger; softening may imply that the effective frequency of the continuum at some average value of $2\Delta(T)$ is shifting upwards. We see indirectly via these phonon effects and directly from temperature-dependent spectra [Cooper, et al., Ref. 6] that the continuum is redistributing when the temperature is lowered through T_c. We attribute this change to the formation of the superconducting gap, namely an increase in R" in the frequency range of the most probable value of 2Δ, and a reduction in R" at lower frequencies. A theoretical discussion of the effect of superconductivity on the phonon self energy has been given by Zeyher and Zwicknagl.[13]

3. Universality

One naturally wants to know how "universal" is this continuum, its redistribution due the formation of a gap, and its gapless, low-frequency behavior. To our knowledge, the continuum has not been observed in $La_{2-x}Sr_xCuO_4$.

3.1. BSCCO

The continuum has also been seen in $Bi_2Sr_2CaCu_2O_{8+\delta}$ (BSCCO). [7] Results from Slakey et al.,[7] are shown in Fig. 6. The sample was freshly cleaved just before mounting in the cryostat for the measurements. It is multiply twinned and otherwise orientationally disordered, but the c-axis is normal

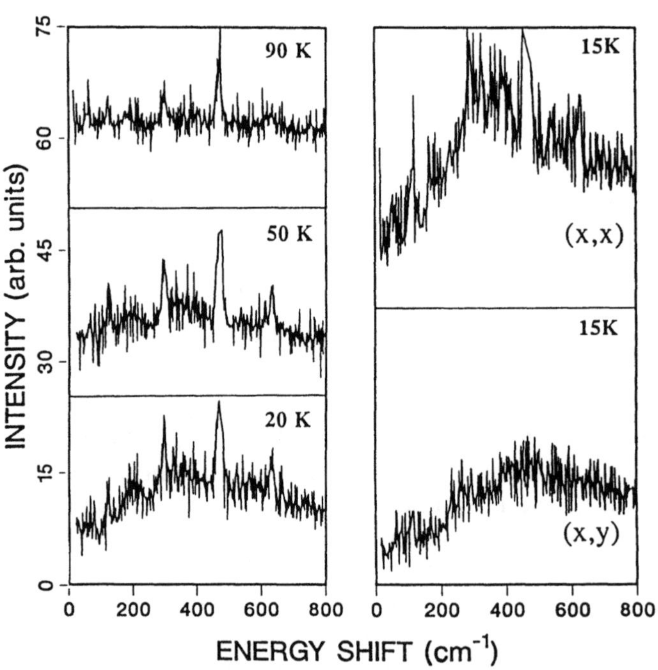

Fig. 6. Low temperature spectra of $Bi_2Sr_2CaCu_2O_{8+\Delta}$, showing the redistribution of the continuum. The more intense phonons have been truncated to make the features of the continuum more apparent. From Slakey et al. Ref. [7].

to the surface. The left three panels show the temperature dependence of the polarized (xx) spectrum, and the right two panels show the polarized and depolarized (xy) spectra well below T_c (≈85K). The latter should be compared with the lower two panels from YBCO in Fig. 3. The two materials give semiquantitatively similar spectra:--broad "2Δ" peaks, the higher one having B_{1g} symmetry, with scattering all the way to (nearly) zero energy.

3.2. Other Observations

A flat continuum for $T>T_c$ has recently been reported for the spectral density in $Tl_2Ca_2Ba_2Cu_3O_{10}$ (2223) by Krantz et al.[14] At low temperatures they observe the familiar linear rise to a rather weak, broad peak near 450 cm^{-1}.

In recent work that is still in progress we have observed a weak continuum in a single crystal of $YBa_2Cu_3O_{6.5}$ having $T_c = 60K$. At room temperature the continuum is flat; at helium temperature it shows a linear rise, but no noticeable maximum.

3.3. Possible Explanations

The result with freshly-cleaved BSCCO means that one can rule out the role of an as-grown near-surface "dead" layer as an explanation for the continuum and its changes with temperature. A similar conclusion may be reached from the results of Fig. 3 showing that an epitaxial thin film shows the same behavior seen with single crystals. The outer surface of the laser-ablated films is quite different from that of the single crystals. The possibility that the continuum results from inhomogeneities or impurities can be almost eliminated. The reason is as follows: The antiresonance of the 115 cm^{-1} phonon is almost complete. This indicates that the excitations that make up the continuum must couple to the phonon displacements in practically every unit cell of the crystal.

A source for the continuum has been suggested by Thomsen and Cardona, [15] namely a process analogous to intervalley scattering in doped indirect-gap semiconductors. In fact, such scattering was observed some time ago in heavily N-doped SiC. [16] In that system a continuum was observed extending to about 1000 cm^{-1} that showed strong Fano antiresonances with Raman-active phonons. The integrated cross-section per electron can be large, of order of the Thomson cross-section multiplied by an appropriate square of the fluctuating part of the reciprocal effective mass tensor. The characteristic energy of the continuum is of the order of the intervalley matrix element of the impurity potential. [17] To apply this idea to the continuum seen in YBCO and BSCCO, one must assume that different regions of the Fermi surface have greatly different reciprocal effective masses, which is probably true and that a rather large number of scattering centers are present. An argument against this picture is the universal presence of the continuum in a wide variety of YBCO samples, including films, having sufficiently high T_c. One would expect that sample-to-sample variations in the nature and concentration of impurities would produce a widely-varying continuum. One would also have to reconcile the strong elastic "intervalley" scattering with the fact that in the normal state of these materials the inelastic transport scattering rate is very strong, of order T, in appropriate units,[18] whereas elastic scattering is much weaker.[19]

We suggest that inelastic scattering from fluctuations in the highly correlated electronic liquid is responsible for the normal-state continuum. The temperature dependence of the continuum for T above T_c has not been studied in great detail, but preliminary results, such as those shown in Figs. 1 and 3 suggest that near ω=0 the spectral density obeys

$$S(\omega) \approx S_0 + S_1\,\omega\,, \tag{3a}$$

and for large ω (~1000 cm^{-1})

$$S(\omega) \approx S_0\,, \tag{3b}$$

where S_0 is slowly varying on a scale of 100's of cm^{-1}, and where $S_1T \ll S_0$ (i.e., the linear term is weak). From Eqs. (1) and (3a,b) we then have

$$R''(\omega) \approx S_0\,\omega/T\,, \text{ for } \omega \ll T, \tag{4}$$

$$R''(\omega) \approx S_0\,, \qquad \text{for } \omega \gg T. \tag{5}$$

Inelastic scattering from fluctuations will be relaxational in nature. If the scattering rate is Γ, then

$$R''(\omega) \propto \omega\Gamma/(\omega^2 + \Gamma^2)\,. \tag{6}$$

The behavior shown in Eqs. (4) and (5) will be obtained if Γ is a function of ω and T of the form

$$\Gamma(\omega,T) = \sqrt{a\omega^2 + bT^2}\,, \tag{7}$$

where a and b are constants of order unity. Equation (7) is consistent with the resistivity measurements and their interpretation which give a scattering rate Γ approximately equal to the greater of T or ω.[18] A similar conclusion has been drawn from fits of the far-infrared conductivity to a Drude model having a relaxation rate depending on T and ω.[20]

This work was supported by the NSF under grants DMR 87-15103, DMR 86-12860, and DMR 87-14555.

---

[1] R. Sooryakumar and M. V. Klein, Phys. Rev. Letters **45**, 660 (1980); R. Sooryakumar and M. V. Klein, Phys. Rev. B **23**, 3213, (1981).
[2] S. B. Dierker, M. V. Klein, G. W. Webb, and Z. Fisk, Phys. Rev. Letters **50**, 853 (1983).
[3] R. Hackl, R. Kaiser, and S. Schicktanz, J. Phys. C **16**, 1729 (1983).
[4] M. V. Klein and S. B. Dierker, Phys. Rev. B**29**, 4976 (1984).
[5] A. A. Abrikosov and L. A. Falkovsky, Physica C **156**, 1 (1988).
[6] A. V. Bazhenov, et al., Pis'ma Zh. Eksp. Teor. Fiz. **46**, 35 (1987) [JETP Lett. **46**, 32 (1987)]; K. B. Lyons et al., Phys. Rev. B **36**, 5592 (1987); S. L. Cooper et al., Phys. Rev. B **37**, 5920 (1988); R. Häckl et al., Phys. Rev. B **38**, 7133 (1988).
[7] A. Yamanaka et al., Jpn. J. Appl. Phys. **27**, L1902 (1988); F. Slakey, M. V. Klein, E. D. Bukowski, and D. M. Ginsberg, submitted to Phys. Rev. B.
[8] Recent unpublished work of F. Slakey, M. V. Klein, T. Venkatesan, A. Inam, X. D. Wu, and L. Nazor. Film from Rutgers Univ., and Bellcore.
[9] F. Slakey, S. L. Cooper, M. V. Klein, J. P. Rice and D. M. Ginsberg, Phys. Rev. B **39**, 2781 (1989).
[10] H. Monien and A. Zawadowski, unpublished work.
[11] S. L. Cooper, F. Slakey, M. V. Klein, J. P. Rice, E. D. Bukowski, and D. M. Ginsberg, J. Opt. Soc. Am. B**6**, 436 (1989).
[12] S. L. Cooper, F. Slakey, M. V. Klein, J. P. Rice, E. D. Bukowski, and D. M. Ginsberg, Phys. Rev. B **38**, 11934 (1988).
[13] R. Zeyher and G. Zwicknagl, Solid State Commun. **66**, 617 (1988).
[14] M. C. Krantz, H. J. Rosen, J. Y. T. Wei, and D. E. Morris, Phys. Rev. B, to be published.
[15] C. Thomsen and M. Cardona, in Physical Properties of High Temperature Superconductors I, edited by D. M. Ginsberg (World Scientific Press, Singapore, 1989) Ch. 8.
[16] P. J. Colwell and M. V. Klein, Phys. Rev. B**6**, 498 (1972).
[17] K. Jain, S. Lai, and M. V. Klein, Phys. Rev. B **13**, 5448 (1976).
[18] P. A. Lee and N. Read, Phys. Rev. Letters **58**, 2691 (1987).
[19] Resistivity measurements are reviewed by P. B. Allen, Z. Fisk, and A. Migliori in Ref. 15, Ch. 5.
[20] R. T. Collins, Z. Schlesinger, F. Holtzberg, P. Chaudari, and C. Feild, Phys. Rev. B **39**, 6571 (1989).

Magnetic Correlations in High T_c Oxides in the Superconducting State

G. Shirane

Brookhaven National Laboratory, Department of Physics, Upton, NY 11973, USA

Abstract. A review is given of recent neutron scattering measurements on $La_{2-x}Sr_xCuO_4$ and $YBa_2Cu_3O_{6+x}$ single crystals. Well defined, modulated magnetic excitations are observed in a high quality single crystal of $La_{1.85}Sr_{0.15}CuO_4$. This crystal exhibits a sharp superconducting transition at $33K$ with a high *Meissner* effect. For magnetic excitation energies of $9meV$ and lower, the intensity drops substantially below $150K$. Similar magnetic scattering of $YBa_2Cu_3O_{6+x}$ in the superconducting region is now being sought.

1. Introduction

The magnetism in the layered high T_c superconductors, $La_{2-x}Sr_xCuO_4$ and $YBa_2Cu_3O_{6+x}$, has been extensively studied over the past two years. Among many experimental techniques, inelastic neutron scattering plays a unique role for detailed studies of magnetic excitations[1]. We have carried out a series of neutron scattering measurements in collaboration with scientists from many research institutions. At present, the basic unanswered question concerns the nature of magnetic correlations in the superconducting samples. This is a brief status report of on-going neutron scattering experiments in high T_c superconductors.

The magnetic phase diagrams of both high T_c systems have been well established. In the case of the $La_{2-x}Sr_xCuO_4$ system (see Fig. 1), the magnetic correlations have been studied by *Birgeneau et al*[2] by two axis, energy integrated techniques. The correlation length follows roughly the average distance between oxygen holes as a function of the doping concentration x.

The inelastic scattering technique was successfully employed[3,4], only very recently, for the measurement of the scattering function $S(Q,w)$ in the superconducting samples. This is due, to a large extent, to the successful growth of larger single crystals. In particular, *Tanaka and Kojima*[5] produced a largely superconducting crystal of $La_{1.85}Sr_{0.15}CuO_4$ with $T_c = 33K$. In the case of $YBa_2Cu_3O_{6+x}$, progress is also being

Springer Series in Solid-State Sciences, Vol. 89 Strong Correlation and Superconductivity
Editors: H. Fukuyama · S. Maekawa · A.P. Malozemoff

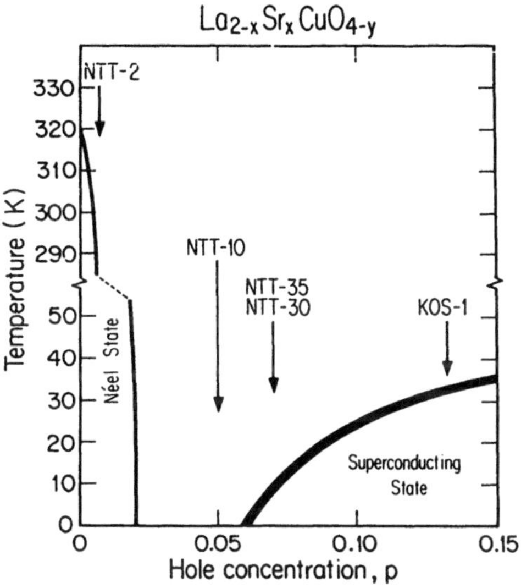

Figure 1. Phase diagram of $La_{2-x}Sr_xCuO_4$ after *Thurston et al*[3]. Hole concentration is approximately equal to Sr content x.

made in characterizing the magnetic correlations in the orthorbombic, superconducting phase.

2. $La_{2-x}SrCuO_4$

Very recently, *Thurston et al*[3] reported the temperature dependence of inelastic neutron cross sections for NTT-10 ($x = 0.06$) and NTT-35 ($x = 0.11$). The latter shows $T_c \cong 10K$ and an 80% *Meissner* effect. Their phases are plotted in Fig. 1 as a function of hole concentration p, under the assumption that the properties depend solely on hole content. The energy loss spectra ($+6meV$) in Fig. 2 show temperature independent cross sections, the characteristic shared by almost all scans for NTT-10 and NTT-35. The energy gain spectra (-$6meV$) were measured to test the internal consistency of the overall data. Note the incommensurate feature previously reported by two axis experiments[6,7] is clearly visible in Fig. 2.

An exceptionally good single crystal of $La_{1.85}Sr_{0.15}CuO_4$ became available recently[5]. This crystal is labelled KOS-1 and Fig. 3 shows resistivity data parallel and perpendicular to the CuO_2 planes. ($T_c = 33K$). Magnetic measurements on the entire neutron sample (6mm diameter, 17mm long) imply a flux exclusion of 80 ± 20%. Effects of the supercon-

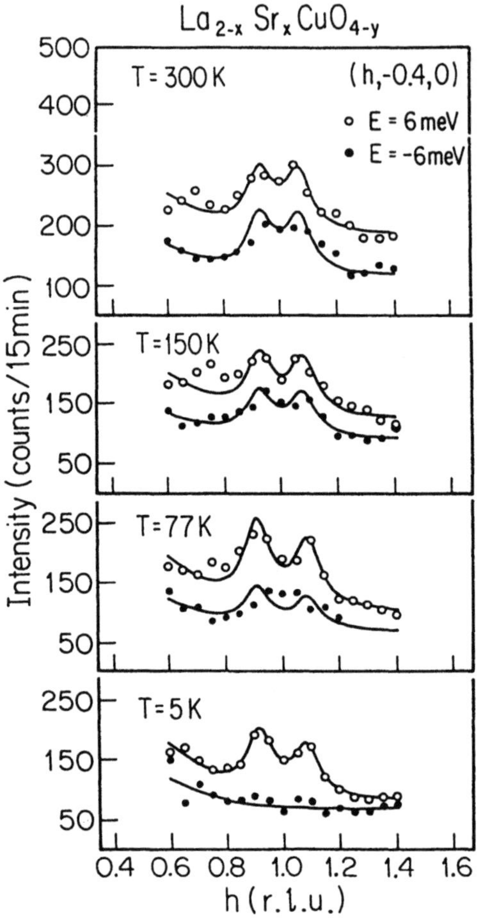

Figure 2. Inelastic neutron profiles for $x = 0.11$ sample, *NTT*-35. Neutron energy loss data are shown by open circles and energy gain data by solid circles[3].

ductivity may be seen directly in neutron depolarization measurements as shown in the insert of Fig. 3. Fig. 4 gives a direct comparison between the parent $x = 0$ crystal (*NTT*-2) and the superconducting $x = 0.15$ crystal (*KOS*-1). These profiles reveal a few important characteristics. First, integrated magnetic intensities at a constant energy transfer of $-6meV$ are nearly identical, though a considerable broadening is noted for $x = 0.15$. Second, the modulated nature of the excitations, as seen in *NTT*-35, is clearly demonstrated. The double peaks are missing in $x = 0.03$ (*NTT*-10).

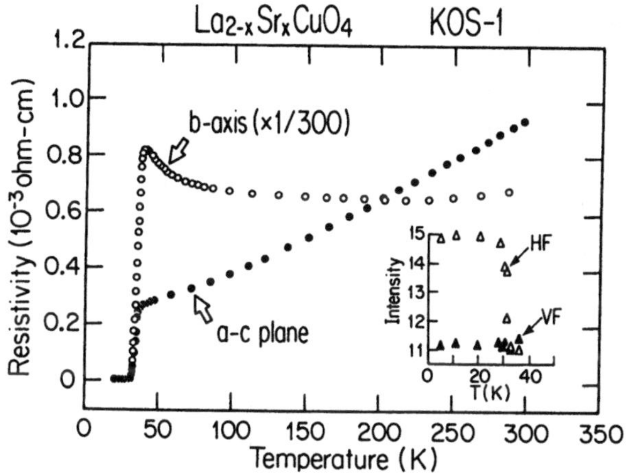

Figure 3. Registivity of KOS-1 crystal grown by *Kojima and Tanaka*[5]. Insert shows spin flip neutron intensity[4] with the applied vertical (VF) and horizontal field (HF).

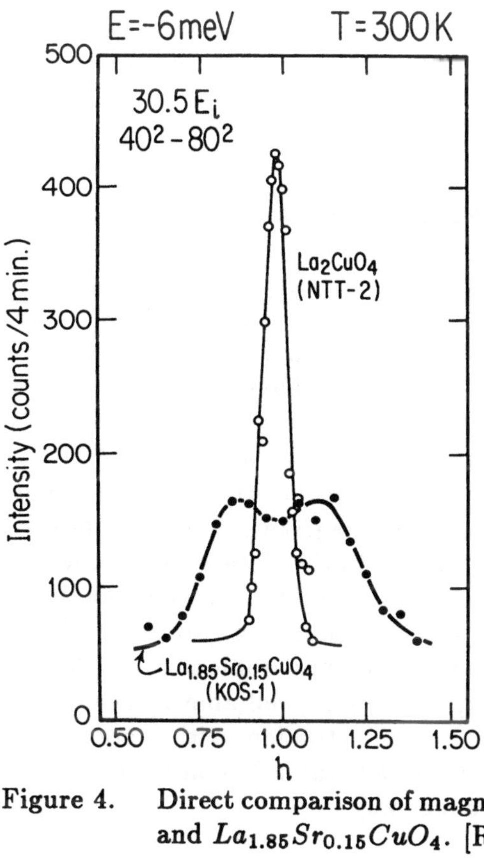

Figure 4. Direct comparison of magnetic excitations of La_2CuO_4 and $La_{1.85}Sr_{0.15}CuO_4$. [Ref. 4].

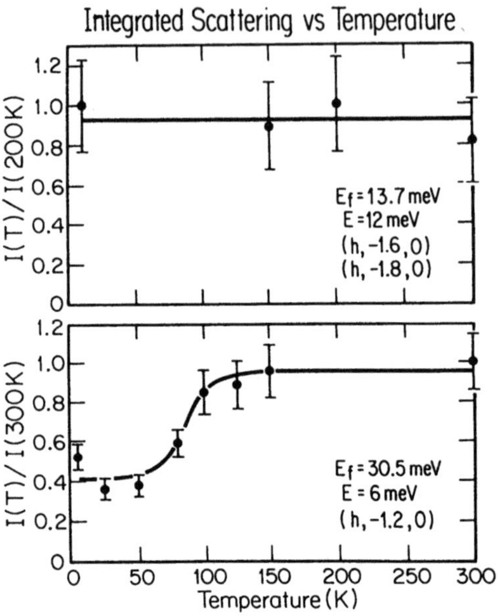

Figure 5. Temperature dependence of normalized integrated intensities at energies of $6meV$ and $12meV$ for $La_{1.85}Sr_{0.15}CuO_4$ (Ref. 4). Solid Lines are guides to the eye.

Fig. 5 shows the most significant results obtained so far in this series of experiments. It depicts the integrated intensities as functions of temperature for scans at 6 and $12meV$. The intensity at $12meV$ is independent of temperature, as was the case for $6meV$ data in NTT-35 (Fig. 2). However, in KOS-1, there is a dramatic diminition in the intensity between $\approx 150K$ and $50K$ at $6meV$. Similar behavior was observed at $3meV$ and $9meV$. For $T_c = 10K$ (NTT-35) and $T_c = 33K$ (KOS-1) the BCS gaps, $2\Delta = 3.5K_BT_C$, are $\approx 3meV$ and $10meV$ respectively. The energy $6meV$ is thus above the superconducting gap in NTT-35 and below it in KOS-1. It is therefore strongly suggestive that the difference in the temperature dependencies of the excitation intensities in the two samples is connected with the energy scales for the superconductivity. The data of KOS-1 are suggestive of the opening of a gap in the Cu^{2+}-hole spin excitation spectrum with an energy comparable to the BCS gap value. We note that the apparent gap develops well above T_c and that there is some residual intensity for the spin fluctuations with energy below the BCS gap value in the superconducting state. Further experiments on the KOS series of crystals are under way.

3. $YBa_2Cu_3O_{6+x}$

The magnetic phase diagram of $YBa_2Cu_3O_{6+x}$ is now well established[8] and shown by solid lines in Fig. 6. Magnetic excitations below $x = 0.4$ were studied by *Sato et al*[9] and *Rossat-Mignod et al*[10]; they show many essential similarities to those observed in $La_{2-x}Sr_xCuO_4$. The major difference comes from the bi-layer structure of CuO_2 in the 123 system. At present, however, the magnetism in the superconducting region ($x \geq 0.4$) is in a somewhat confused state. *Brüchet et al*[11] reported negligible magnetic scattering below $25meV$ for $YBa_2Cu_3O_7$. At the same time, light scattering studies [12] indicate that magnetic correlations are still present in the orthorhombic phase. *Mezei et al*[13] also reported a weak, low-energy magnetic scattering in a powder sample with $x = 0.59$.

Recently, *Tranquada et al*[8] were engaged in extensive studies of the magnetic excitations of this system: the concentrations studied are marked with bars in Fig. 6. Fig. 7 shows the bilayer modulations of the magnetic cross-sections along the tetragonal c direction for the $x = 0.3$ sample. The magnetic intensities are nearly temperature independent above $T_N = 260K$. And very recently, we have seen reasonably strong spin-wave-like scattering in a larger orthorhombic crystal with $x \sim 0.45$ and $T_c \simeq 15K$.

However, all other superconducting crystals shown in Fig. 6 did not reveal any magnetic scattering. Most of the measurements have been

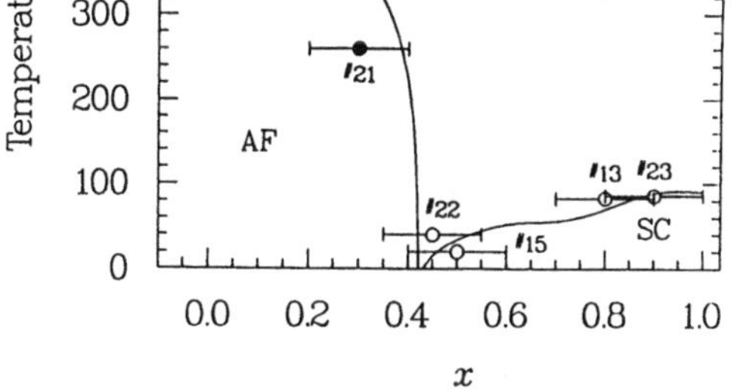

Figure 6. $YBa_2Cu_3O_{6+x}$ shown by smooth curves. The single crystal compositions studied by *Tranquada et al*[8] are shown by sample numbers with horizontal bars.

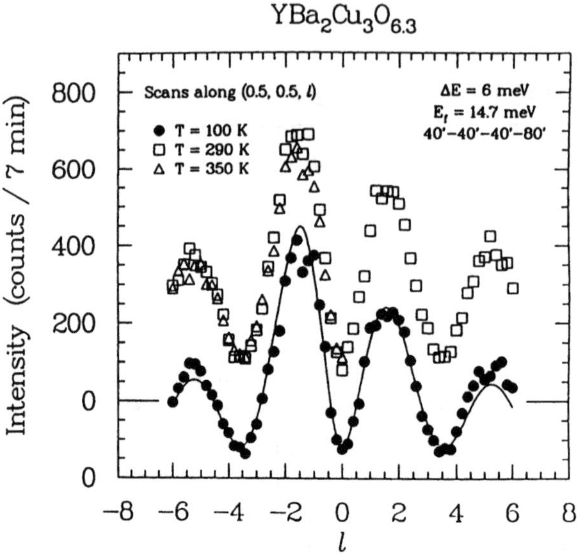

Figure 7. Intensity modulations due to bilayer magnetic correlations in $YBa_2Cu_3O_{6+3}$ with $T_N = 260K$. After *Tranquada et al*[8].

made at 3 and $6meV$ energy transfers over a temperature range of 5 to $300K$. What are the implications of these negative results? One possibility is that there are no magnetic moments present in the superconducting, orthorhombic phase of $YBa_2Cu_3O_{6+x}$. However, there is mounting experimental evidence, particularly in NMR studies[14], against such a conclusion. We would like to speculate that we have somehow missed the proper window for observing inelastic neutron scattering in $(Q,\ w,\ T)$ space. Suppose a similar drastic change in intensity takes place here, as in the Sr doped case, in some temperature range above T_c; then we could have easily missed the window. The search for a magnetic cross-section in this system was made before we discovered the characteristic temperature dependence in the $Sr_{0.15}$ sample. Further work is required to properly characterize the magnetic fluctuations in orthorhombic $YBa_2Cu_3O_{6+x}$ and any relationship they may have to superconductivity.

These studies have been carried out in collaboration with many scientists including, R. J. Birgeneau, Y. Endoh, P. Gehring, Y. Hidaka, H. Kojima, M. Sato, T. R. Thurston, T. M. Tranquada, and K. Yamada. The major part of the work was supported by the U.S.-Japan Cooperative Neutron Scattering Program. Research at Brookhaven National Laboratory was supported by the Division of Materials Sciences, U.S. Department of Energy, under Contract No. DE-AC02-76CH00016.

3. References

[1] G. Shirane, Y. Endoh, R. J. Birgeneau, M. A. Kastner, Y. Hidaka, M. Oda, M. Suzuki, and T. Murakami, Phys. Rev. Lett. 59, 1613 (1987); Y. Endoh, K. Yamada, R. J. Birgeneau, D. R. Gabbe, H. P. Jenssen, M. A. Kastner, C. J. Peters, P. J. Picone, T. R. Thurston, J. M. Tranquada, G. Shirane, Y. Hidaka, M. Oda, Y. Enomoto, M. Suzuke, and T. Murakami, Phys. Rev. B 37, 7443 (1988).

[2] R. J. Birgeneau, D. R. Gabbe, H. P. Jenssen, M. A. Kastner, P. J. Picone, T. R. Thurston, G. Shirane, Y. Endoh, M. Sato, K. Yamada, Y. Hidaka, M. Oda, Y. Enomoto, M. Suzuki, and T. Murakami, Phys. Rev. B 38, 6614 (1988).

[3] T. R. Thurston, R. J. Birgeneau, M. A. Kastner, N. W. Preyer, G. Shirane, Y. Fujii, K. Yamada, Y. Endoh, K. Kakurai, M. Matsuda, Y. Hidaka, and T. Murakami, Phys. Rev. B (submitted).

[4] G. Shirane, R. J. Birgeneau, Y. Endoh, P. Gehring, M. A. Kastner, K. Kitazawa, H. Kojima, I. Tanaka, T. R. Thurston, and K. Yamada, Phys. Rev. Letts. (submitted).

[5] I. Tananka and H. Kojima, Nature 337, 21 (1989).

[6] R. J. Birgeneau, Y. Endoh, Y. Hidaka, K. Kakurai, M. A. Kastner, T. Murakami, G. Shirane, T. R. Thurston, and K. Yamada, Phys. Rev. B 39, 2868 (1989).

[7] H. Yoshizawa, S. Mitsuda, K. Kitazawa, and K. Katsumata, J. Phys. Soc. Japan 57, 3686 (1988).

[8] J. M. Tranquada, G. Shirane, B. Keimer, S. Shamota, and M. Sato, Phys. Rev. B (submitted).

[9] M. Sato, S. Shamoto, J. M. Tranquada, G. Shirane, B. Keimer, Phys. Rev. Lett. 61, 1317 (1988).

[10] J. Rossat-Mignod, P. Burlet, M. J. Jurgens, L. P. Regnault, J. Y. Henry, C. Ayache, L. Forro, C. Vettier, H. Noel, M. Potel, P. Gougeon, and J. C. Levet, J. Phys. (Paris) Collog. 49, C8-2119 (1988); M. J. Jurgens P. Burlet, C. Vettier, L. P. Regnault, J. Y. Henry, J. Rossat-Mignod, H. Noel, M. Potel, P. Gougeon, and J. C. Levet, Physica B 156 & 157, 846 (1989).

[11] T. Brückel, H. Capellmann, W. Just, O. Schärpf, S. Kemmler-Sack, R. Kiemel, and W. Schaefer, Europhys. Lett. 4, 1189 (1987).

[12] K. B. Lyon, P. A. Fleury, L. F. Schneemeyer, and J. V. Waszczak, Phys. Rev. Lett. 60, 732 (1988).

[13] F. Mezei, B. Faragó, C. Pappas, Gy. Hutiray, L. Rosta, and L. Mihály, Physica C 153-155, l669 (1988).

[14] W. W. Warren, Jr., R. E. Walstedt, G. F. Brennert, R. C. Cava, R. Tycko, R. F. Bell, G. Dabbagh, Phys. Rev. Lett. 62, 1193 (1989).

NMR Studies of $YBa_2Cu_3O_{7-\delta}$ (T_c=90 K)

D.J. Durand, S.E. Barrett, C.H. Pennington, C.P. Slichter, E.D. Bukowski, T.A. Friedmann, J.P. Rice, and D.M. Ginsberg

Department of Physics and Materials Research Laboratory,
University of Illinois at Urbana-Champaign,
1110 W. Green St., Urbana, IL61801, USA

Abstract. The authors report Cu NMR studies of $YBa_2Cu_3O_{7-\delta}$ (T_c = 90K) in the normal state (using oriented single crystals) and superconducting state (using oriented powders). In the normal state they measure the electric field gradient tensor, the magnetic shift tensor, the spin-lattice relaxation tensor, and the time dependence of the transverse decay (spin-echo envelope) and draw conclusions about the charge and spin states of the Cu atoms. In the superconducting state they measure the shift tensor, using the ^{89}Y resonance to measure the internal field in the sample. The data are compared with various detailed models of spin pairing.

1. Introduction

We report ^{63}Cu and ^{65}Cu Nuclear Magnetic Resonance (NMR) frequency and relaxation time measurements made in the normal and superconducting states of $YBa_2Cu_3O_{7-\delta}$ (T_c = 90K) and the use of ^{89}Y NMR to measure the field inside the sample in the superconducting state. Using the results of these experiments we have concluded: 1) The 22 MHz Nuclear Quadrupole Resonance (NQR) line corresponds to the chain [Cu(1)] site and the 31.5 MHz NQR line corresponds to the plane [Cu(2)] site [1]. 2) The Cu(2) [Cu(1)] atom is essentially a Cu^{2+} ion with a single hole in the $d_{x^2-y^2}$ [$d_{y^2-z^2}$] state [2]. 3) In a fluctuation picture of the electron spin magnetic moment, the Cu(2) atom is at or close to the local moment regime in the normal state [2]. 4) The Cu(2) local moments are exchange coupled, at least in the normal state [3]. 5) The Cu(1) shift and spin-lattice relaxation rate are determined by an interaction with the conduction band (or Fermi liquid) probably made up of holes on the chain and bridge oxygens. 6) The Cu(2) shift and spin-lattice relaxation rate are each caused by an interaction with the copper moment and oxygen holes. 7) In the superconducting state, the behavior of the chain copper is described by the conventional, weak coupling BCS theory (assuming an isotropic gap ($2\Delta(0) = 3.52\ k_BT_c$), but the behavior of the Cu(2) indicates that the pairing state may be either s-wave or d-wave spin singlet with $2\Delta(0)$ larger than $3.52\ k_BT_c$.

All of our samples are single phase $YBaCu_3O_{7-\delta}$ with δ near zero (T_c = 90K). The superconducting state shift measurements have been made on a

Springer Series in Solid-State Sciences, Vol. 89 **Strong Correlation and Superconductivity**
Editors: H. Fukuyama · S. Maekawa · A.P. Malozemoff

uniaxially aligned powder sample described in Ref. 5. All other measurements have been made on samples made up of one or more single crystals. These samples are described in Ref. 1 and 2. We have put together a fully aligned single crystal sample for the relaxation measurements in the superconducting state. It is basically the same as the sample of Ref. 1 except in it the a and b axes have been aligned using polarized light.

2. The Normal State

The identification of the resonance lines and the determination of the charge state are done by examining the electric field gradient (EFG) and magnetic shift tensors of the two families of NMR lines. Nuclei at each site have three transitions for each of the three orientations of the external magnetic field, H_o. These nine frequencies are determined in relation to H_o by the two independent parameters of the electric field gradient tensor and the three independent parameters of the magnetic shift tensor. We adjust these five parameters to fit the nine measured frequencies. The resulting tensors are given in Table I where a, b, and c refer to the crystal axes.

Table I. The shift and electric field gradient tensors

	Cu (2)	Cu (1)
Shift		
K_{aa} (%)	0.59 ± 0.04	1.32 ± 0.07
K_{bb}	0.57 ± 0.04	0.56 ± 0.07
K_{cc}	1.267 ± 0.001	0.59 ± 0.04
Electric field gradient		
ν_{aa} (MHz)	+ 15.95 ± 0.05	-19.01 ± 0.07
ν_{bb}	+ 15.54 ± 0.05	+ 19.18 ± 0.07
ν_{cc}	-31.49 ± 0.05	- 0.17 ± 0.03

We use the symmetries of the tensors in table I to associate each with its crystallographic site. Thus, the identification of the site we have labelled Cu(2) arises because both the shift and quadrupole tensors are axially symmetric about the crystal c-axis, in keeping with the fact that the Cu(2) surroundings have approximately 4-fold symmetry about that axis. We then calculate the expected NQR frequencies using

$$\nu_Q^2 = \nu_{zz}^2 + \frac{(\nu_{xx} - \nu_{yy})^2}{3} \tag{1}$$

to find that the Cu (1) [Cu(2)] site should have an NQR frequency of 22 MHz [31.5 MHz].

We have also calculated the EFG tensors at the two sites assuming that the field gradients have two sources: 1) the copper hole in the Cu(2) [Cu(1)] d_{x2-y2} [d_{y2-z2}] state, and 2) the other ions in the crystal.We assume an axial contribution, ν_{axial}, of unknown magnitude to estimate the copper hole contribution. We calculate the contribution from the other ions by treating them as point charges whose contribution must be multiplied by the (unknown) Sternheimer

factor γ_S. We then adjust ν_{axial} and γ_S to fit the two values of ν_{cc}. This calculation gives reasonable values of ν_{axial} and γ_S only if there is a hole in the Cu(1) [Cu(2)] y^2-z^2 [x^2-y^2]d-state.

Figure 1 shows spin-lattice relaxation data for Cu(2) with the static field $H_0 \| c$ following a disturbance from thermal equilibrium. Plotted is the deviation of the signal from its thermal equilibrium value ($S(\infty) - S(t)$) versus recovery time t. For the 1/2, - 1/2 transition (Fig. 1a), the deviation should follow a curve

$$S(\infty) - S(t) = A\,[0.9 \exp(-4W_1t) + 0.1 \exp(-2W_1t/3)] \quad (2)$$

Where A is the deviation at t = 0. The theoretical curve is Eq (2) in which only A and W_1 have been adjusted to fit the data. Fig. 1b shows the recovery for the 1/2, 3/2 and -1/2, -3/2 transitions. The theoretical curve is drawn using W_1 obtained from Fig. 1a. The excellent fit to the data shows that the relaxation mechanism is magnetic <u>and</u> the correlation time is short compared to the Larmor period.

The values of W_1 found in this way are shown in Table II.

Table II. The W_1 tensors

	Cu(2)	Cu(1)
W_{1a} (msec^{-1})	4.7 ± 0.3	1.1 ± 0.1
W_{1b}	4.7 ± 0.3	0.9 ± 0.1
W_{1c}	1.05 ± 0.1	0.8 ± 0.1

The anisotropy of $W_{1\alpha}$ (α = a,b,c) for the Cu(2) site is what one expects from the fluctuations in magnetic field at the nucleus due to the hole in the x^2 - y^2 state if the moment fluctuates with an isotropic correlation time. This is a strong argument that the moment on the Cu(2) atom is local and of fixed magnitude. Further support comes from a detailed analysis of the relaxation time and shift data in normal and superconducting states.

We have made precise measurements of the temperature dependence of W_1 for the Cu(1) sites. We find that from 100K to 200K W_1 is proportional to T, a result reminiscent of a conduction band. We return to this point below.

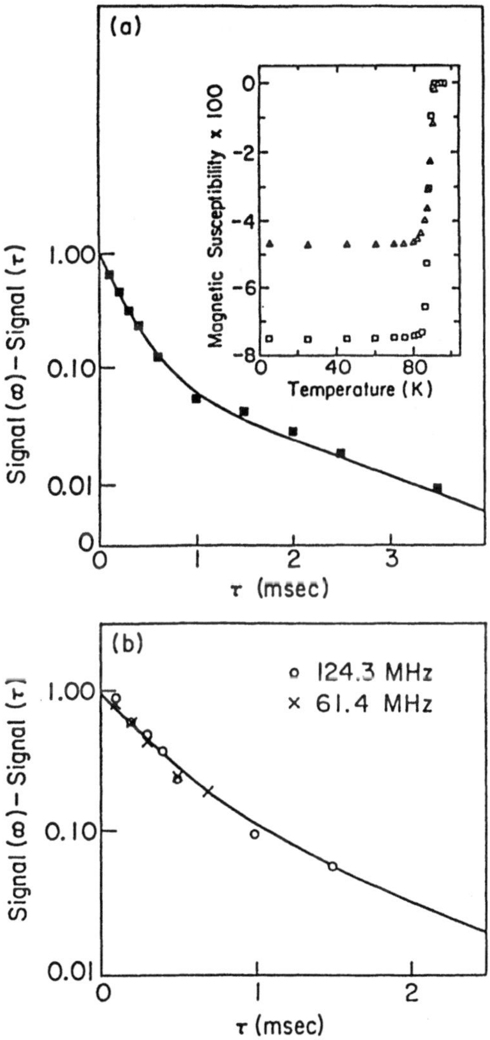

Fig. 1. Inset: magnetic susceptibility of the crystal vs temperature showing a sharp superconducting transition at 90 K and a complete Meissner effect, $H = 16$ Oe $\perp$ c axis (Δ, field cooled; ❒, zero-field cooled). Main figure: The spin-lattice relaxation of the ^{63}Cu(2) nuclei for the transitions (a) $1/2 \to (-1/2)$, (b) $3/2 \to 1/2$ and $(-3/2) \to (-1/2)$. The points are the data plotted as the $\log_{10}$ [signal (∞) - signal (τ)], where τ is the time after an inversion pulse. The solid line in (a) is a theoretical fit to the data assuming a magnetic relaxation mechanism independent of frequency (see text) where the fundamental rate W_1 has been adjusted to yield the best agreement with the data ($W_1 = 1.05$ msec^{-1}). The solid line in (b) is the theoretical prediction for these transitions with the same W_1 as deduced in (a).

We have studied the transverse relaxation of the Cu nuclei by plotting the envelope of the size of the spin echo signal. We find that the envelope of the Cu(2) nuclei with $H_0 \parallel c$ is described by

$$\exp(-t/T_2) \quad \exp[-t^2/(2\tau^2)] \tag{3}$$

The origin of the Gaussian (t^2) component of Eq. 3 is a nuclear spin-spin coupling but the size of τ is an order of magnitude too short to arise from direct nuclear-nuclear magnetic dipole coupling.. To isolate this coupling, eliminate the effects of T_2, and measure τ, we measured the ^{63}Cu-^{65}Cu double resonance described by the pulse sequence shown in Fig. 2. The fit in the main figure of Fig. 2 is for τ = 110 msec. We show in Ref. 3 that this implies an exchange coupling between neighboring Cu(2) moments of $J_{eff} \approx$ 1100 cm^{-1}. Such coupling is known to exist in the antiferromagnetic phases of the 123 materials, but we believe [3] is the first demonstration of its existence in the 90K material.

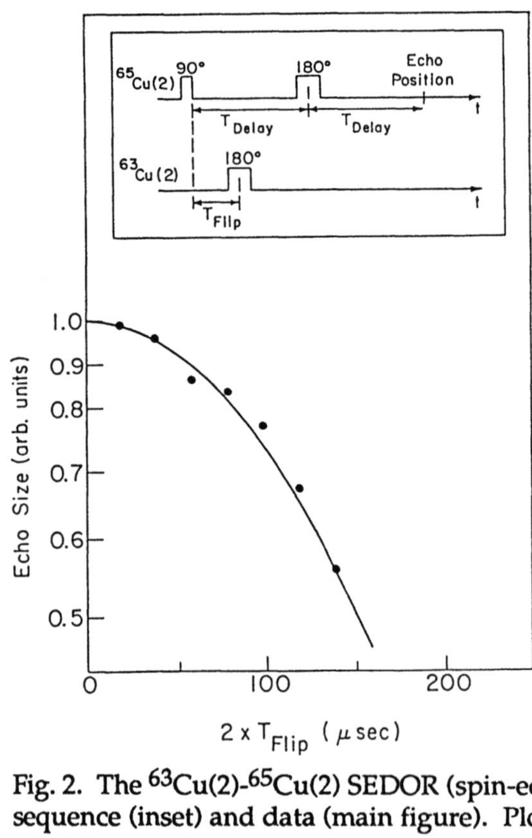

Fig. 2. The ^{63}Cu(2)-^{65}Cu(2) SEDOR (spin-echo double-resonance) pulse sequence (inset) and data (main figure). Plotted is the echo vs $2T_{flip}$. The solid line is a Gaussian with τ= 130 μsec.

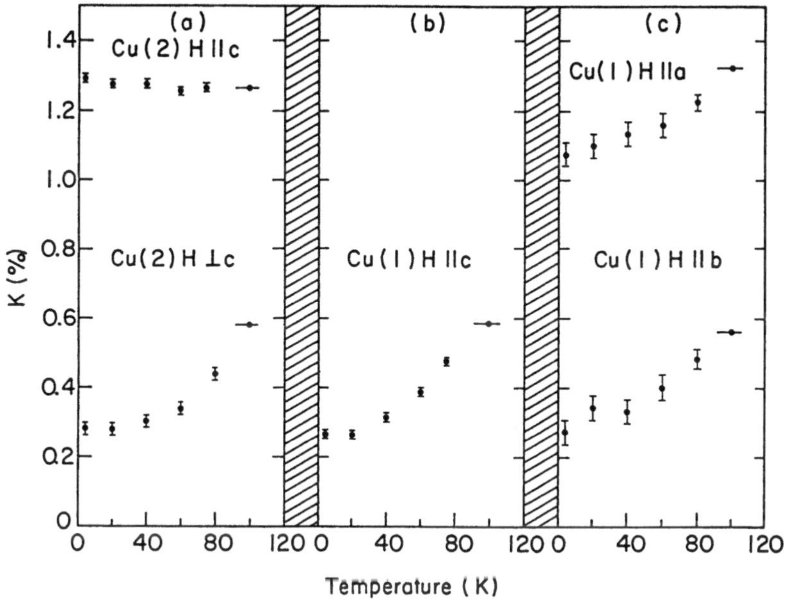

Fig. 3. The total magnetic shift as a function of temperature for (a) Cu(2) with (upper) $H_0 \| c$ and with (lower) $H0 \perp c$, (b) Cu(1) with $H_0 \| c$, (c) Cu(1) with (upper) $H_0 \| a$ and with (lower) $H_0 \| b$.

3. The Superconducting State

We have measured the shifts of the Cu(1) and Cu(2) nuclei in the superconducting state using an oriented powder. Our procedure for determining the magnetic shift, K(T) as a function of temperature in the superconducting state is as follows [5]. At each temperature and field orientation the resonance frequencies are calculated by exact diagonalization of the Hamiltonian. To do so we need to know K, the electric field gradient, and the magnitude of the static magnetic field inside the sample. Knowing the field inside the sample in the superconducting state is complicated by the fact that demagnetizing supercurrents alter the internal field. We have used ^{89}Y NMR to measure the internal field (the ^{89}Y frequency shift is known to be small on the scale of the ^{63}Cu shifts).

The details of how we find K(T) are in Ref. 5. Fig. 3 shows the results. At any temperature, K(T) is the sum of an orbital (chemical shift) term, $K^L(T)$, and a spin (Knight shift) term, $K^S(T)$. In our analysis of these data we assume that the zero temperature spin, $K^S(O)$, shift is zero, or

$$K^L(100K) = K(O), \text{ and}$$
$$K^S(100K) = K(100K) - K(O) \quad (4)$$

In reference [5] we give arguments which rule out other possibilities as far as we can tell.

Table III gives our values of K^L and K^S assuming Eq. 3 is valid

Table III. K^L and K^S

	Cu(2)	Cu(1)
K^L_{aa}(%)	0.28 ± 0.02	1.08 ± 0.04
K^L_{bb}	0.28 ± 0.02	0.27 ± 0.04
K^L_{cc}	1.29 ± 0.01	0.27 ± 0.01
K^S_{aa} (%)	0.30 ± 0.02	0.24 ± 0.04
K^S_{bb}	0.30 ± 0.02	0.29 ± 0.04
K^S_{cc}	-0.02 ± 0.01	0.32 ± 0.01

The first thing to notice about the values given in Table III is the anisotropy of $K^L_{\alpha\alpha}$ (α = a,b,c) at both sites. We show in Ref. 2 that a hole in the $x^2 - y^2$ [$y^2 - z^2$] state will lead to $K^L_{aa} = K^L_{bb}$; $K^L_{cc} \geq 4K^L_{aa}$ [$K^L_{bb} = K^L_{cc}$; $K^L_{aa} \geq 4K^L_{bb}$]. The equality sign holds for the case that the excited hole states xy and xz [yz and yx] are degenerate. We note next that the $K^L_{\alpha\alpha}$ values for the two Cu sites are very similar if one replaces the c-axis of the Cu(2) with the a-axis of the Cu(1). This result is just what one expects for a Cu^{2+} with a single hole in the x^2-y^2 [y^2-z^2] state for Cu(2) [Cu(1)], lending further support to this picture of the ion electronic structures. From Table III we also note that $K^S\alpha\alpha$ of the Cu(1) is within experimental error almost isotropic, as is $W_{1\alpha}$. This result suggests that the shift and spin-lattice relaxation come from the same source. The fact that the spin-lattice relaxation rate is proportional to T, as with a conduction band, suggests trying these quantities in the famous Korringa relationship [7] for relaxation of nuclei by s-state conduction electron spins:

$$T_1T(K^s)^2 = \frac{h}{8\pi^2 k_B} \frac{\gamma_e^2}{\gamma_n^2} B \qquad (5)$$

where B is a parameter introduced by Pines to include electron-electron coupling. Using $1/T_1 = (2W_1/3)$ we find B = 0.4 fits our data. This value corresponds to conduction electrons which interact only weakly. For the Cu(1), the s-state electrons must represent the presence of Cu 4s wave functions in the conduction band since a different expression holds true for core polarization effects from inner s-shells. A possible source of this coupling is transferred hyperfine coupling to an O hole band. A likely candidate is the nearly one dimensional band related to chain and bridge oxygens. An alternative phrasing is that the Cu(1) nuclei are relaxed by quasi-particles of a Fermi liquid. This description is somewhat more general (though correspondingly at this stage less specific). In any event, the value B = 0.4 corresponds to weakly interacting fermions.

The shift data $K_c^s(T)$ for the Cu(1) spin shown in Fig. 4. Also plotted is a theoretical fit to the Yosida function for a conventional BCS weak coupling spin-singlet s-wave superconductor. The data are seen to fit this curve within experimental error.

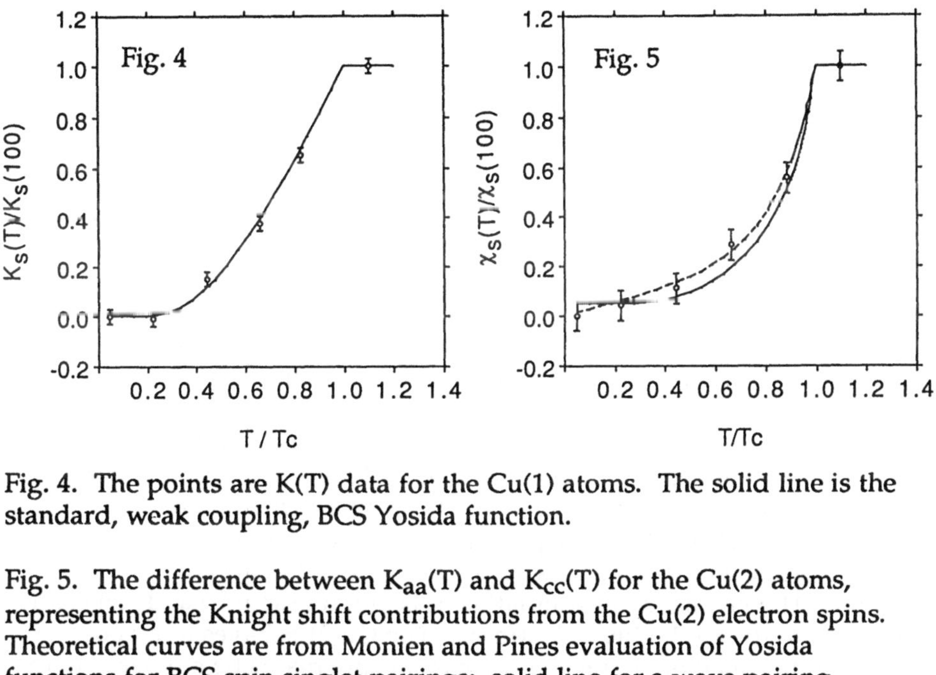

Fig. 4. The points are K(T) data for the Cu(1) atoms. The solid line is the standard, weak coupling, BCS Yosida function.

Fig. 5. The difference between $K_{aa}(T)$ and $K_{cc}(T)$ for the Cu(2) atoms, representing the Knight shift contributions from the Cu(2) electron spins. Theoretical curves are from Monien and Pines evaluation of Yosida functions for BCS spin singlet pairings: solid line for s-wave pairing, dashed line for d-wave pairing.

The Cu(2) $K^S(T)$ arises from Cu(2) spin and from the spin of oxygen holes [5]. It may be decomposed into two components, one isotropic, the other axial about c. We expect the oxygen hole contribution to be isotropic. Hence, to isolate the Cu(2) spin contribution, we need to eliminate the isotropic component. We can isolate the axial component by subtracting $K^S{}_{cc}(T)$ from $K^S{}_{aa}(T)$. We find that the axial component cannot be fit by the standard Yosida function. We show in Fig. 5 that our data can be fit by either of two theoretical curves of Monien and Pines [8]. These curves are deduced as part of their analysis of shift and T_1 data in the normal and superconducting states.

The solid line is for spin-singlet with s-wave pairing. In order to analyze T_1 data, Monien and Pines considered both an isotropic and an anisotropic energy gap. For the isotropic gap they can fit $2\Delta(0)$, the energy gap at T = 0, by 4.94 k_BT_c. For the anisotropic gap they have $2\Delta(0)_{max}$ = 5.60 k_BT_c, and $2\Delta(0)_{min}$ = 4.48 k_BT_c.

The dashed curve corresponds to singlet spin-pairing with a single d-wave, or a sum of a d and higher orbital angular momentum (the multiple wave) energy gap. For the single d-wave, $2\Delta(0)_{max}$ = 10.24 k_BT_c, whereas for the multiple wave $2\Delta(0)_{max}$ = 5.28 k_BT_c.

These gap values are reminiscent of those reported by J. Tsai in his tunneling experiments. He reported two gaps, one near 3.5 k_BT_c, the other near 6.0 k_BT_c.

4. Summary

We now summarize the picture of the electronic character of both Cu(1) and Cu(2) sites.

The Cu(1) nuclei have normal state spin-lattice relaxation (Table II) and hyperfine shifts (Table III) which are essentially isotropic. In the temperature region just above Tc, the spin-lattice relaxation rate is proportional to temperature. In the superconducting state, $K^S(T)$ obeys the conventional, weak coupling BCS theory. We take this to mean that the Cu(1) nuclear resonance properties are determined by the interaction of the nuclei with holes on the chain and bridge oxygens and that those holes form a weakly interacting conduction band. More generally, we can say that the Cu(1) nuclei are relaxed by electrons of a Fermi liquid.

The Cu(2) nuclear resonance properties are a result of interactions both with the local Cu moment and with the holes on the plane and bridge oxygens. The normal state spin-lattice relaxation (Table II) and hyperfine shifts (Table III) are highly anisotropic. The Cu(2) behaves to a good approximation as though it had a permanent spin magnetic moment. In the superconducting state, the Cu(2) Knight shift looks much like the case of a strong coupling BCS spin-singlet state with either s or d wave pairing. This fact suggests that the Cu(2) moments are intimately involved in the superconducting state, hence with the charge carriers. This situation might be an expression of the inexactness of the local moment picture in the normal state, or be a manifestation of an essential role being played by the Cu(2) spins in the transition into the superconducting state. In any event, the Cu(2) spins particpiate intimately in the superconducting state. They are not passive bystanders.

It is a pleasure to thank H. Monien and D. Pines for their theoretical help. D.J.D. gratefully acknowledges receipt of an IBM predoctoral fellowship. This research is supported through the University of Illinois Materials Research Laboratory by the Department of Energy Division of

Materials Science under Contract No. DE-AC02-76ER01198 and the National Science Foundation Division of Materials Research under Contract No. DMR 86-12860 and DMR 87-14555. This paper is based on a talk presented by CPS at the IBM Japan Symposium.

5. References

1. C.H. Pennington et al., Phys. Rev. B, **37**, 7944 (1988).
2. C.H. Pennington et al., Phys. Rev. B, **39**, 2902 (1989).
3. C.H. Pennington et al., Phys. Rev. B, **39**, 274 (1989).
4. C.H. Pennington et al., Adv. Mag. Res. (to appear).
5. S.E. Barrett et al., Phys. Rev. B (to be published).
6. Bleaney et al., Proc. R. Soc. London, Ser. A **228**, 166 (1955).
7. C.P. Slichter, *Principles of Magnetic Resonance*, 2nd edition, 2nd printing (Springer, Berlin, 1980).
8. H. Monien and D. Pines, Phys. Rev. B (to be published).

NMR and NQR in Highly Correlated Metallic and Superconducting Cu Oxides

H. Yasuoka, T. Imai, and T. Shimizu

Institute for Solid State Physics, University of Tokyo, Roppongi, Tokyo 106, Japan

Abstract. In order to understand the exotic Cu spin dynamics in the normal (metallic) state of the high-T_c and related metallic Cu oxides, the measurements of the NQR frequency spectra and the nuclear spin-lattice relaxation have been utilized. A sizable charge differentiation in the Cu-O plane sites is found even in the metallic phases of $YBa_2Cu_3O_y$ and $La_{2-x}Sr_xO_4$. From the temperature dependence of the relaxation rate $1/T_1$, it is found that highly enhanced relaxation process exists in the high-T_c oxides. The data are also exhibiting of the opening of a gap in the low-lying excitations with an energy comparable to the superconducting gap.

1. Introduction

It is well recognized by now that the understanding of the normal state metallic properties of high-T_c Cu oxides is the most essential part of the problem to make up a theoretical model for the mechanism of their superconductivity. The major question here is how we can describe the magnetism associated with the highly correlated Cu 3d electrons in the metallic state. The magnetic susceptibility has quite characteristic signatures, in most of the cases it has essentially no temperature dependence, in contrast to other metallic oxides like vanadium oxides where the amplitude of longitudinal spin fluctuations are saturated and the Curie-Weiss like temperature dependence has been observed at finite temperatures. Therefore, any experimental and theoretical arguments must be conformed to this most fundamental magnetic property. It should be also recalled that most of the superconducting Cu oxides are located at just near the insulating phase and the metallic state is realized by doping a small amount of holes or electrons under alloying. This alloying produces chemically several Cu valence states in the metallic phase and transfer of charge between sites of different valence should be of lower energy to invoke the metallic nature. Then the copper-oxygen interatomic Coulomb interaction would appear to be an important driving force for the transition to the superconducting state.

The NMR and NQR techniques offer an attractive way to investigate the microscopic magnetic properties since the response of inequivalent sites can, in principle, be resolved in the nuclear-resoncance spectra and the dynamical properties of those can be studied by the nuclear magnetic relaxation at respective sites. Particularly, it is quite important for understanding the low-laying elementary excitations to make a systematic measurement of the nuclear spin-lattice relaxation time T_1. Among variety of our NMR/NQR investigations, we restrict ourselves to review and to discuss following two subjects in this paper. The first is the charge differentiation in the Cu-O plane found in the Cu NQR spectra for $YBa_2Cu_3O_y$ and $La_{2-x}Sr_xO_4$. The second is the unprecedented

Springer Series in Solid-State Sciences, Vol. 89 **Strong Correlation and Superconductivity**
Editors: H. Fukuyama · S. Maekawa · A.P. Malozemoff

behavior of the temperature dependence of T_1 in a series of high-T_c Cu oxides. The results are compared with those for non-superconducting but metallic oxides, $La_4BaCu_5O_{13}$ and $La_{1.7}Sr_{0.3}CuO_4$.

2. Charge differentiation in Cu-O plane

Full investigations of Cu NQR spectra in $YBa_2Cu_3O_y$ ($6.0 \le y \le 6.98$, YBCO) have been reported [1,2]. Here, we show only our data for the oxygen concentration dependence of ^{63}Cu NQR frequencies in Fig. 1. The characteristic features of this results are summarized as follows: (1) All the NQR associated with the Cu(1) sites, $Cu(1)_2$, $Cu(1)_3$ and $CU(1)_4$ (subscripts denote the number of oxygen coordination), have a sharp spectrum width (FWHM being typically 0.3 MHz) and essentially no y dependence. (2) Only the NQR frequency associated with the $Cu(1)_2$ sites has a discontinuous change between tetragonal ($y<0.3$) and orthorombic ($y>0.3$) phases. Reducing y from 7.0, an NQR peak due to "magnetic" Cu(2) sites was replaced by a "non-magnetic" $Cu(1)_2$ signal at almost the same frequency. (3) A clear site differentiation for the NQR signals has been observed in the Cu(2) sites with reducing y from 7.0. This is a quite surprising result in the sense that most of the experimental and theoretical arguments are based on a homogeneous electronic state in the Cu-O planes.

The same kind of site differentiation has also been observed in the metallic phase of $La_{2-x}Sr_xCuO_4$ ($0.1<x<0.3$, LCO) system [3]. The typical examples of ^{63}Cu NQR spectra are shown in Fig. 2, where the plotted data points were obtained by a numerical deconvolution of the full NQR spectra.

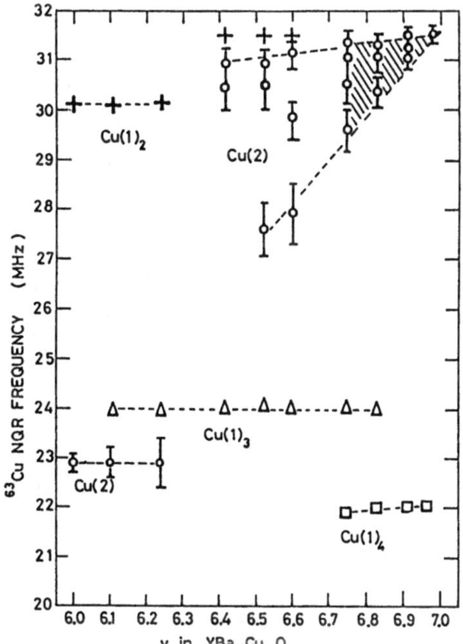

Fig. 1 Oxygen concentration dependence of the ^{63}Cu NQR frequencies for the inequivalent Cu sites in $YBa_2Cu_3O_y$.

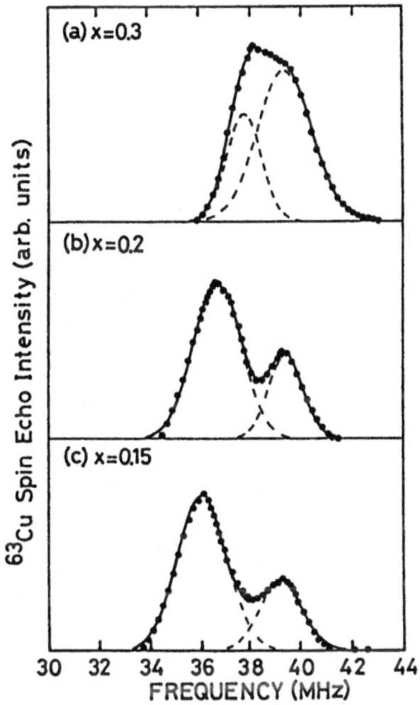

Fig. 2 ^{63}Cu NQR spectra for charge-differentiated Cu sites in $La_{2-x}Sr_xCuO_4$.

It is evident from these data that there exist two inequivalent Cu sites and the relative population depends on x, higher frequency sites being more pronounced in higher Sr doped materials. The existence of two Cu sites has also been proved by the NMR measurement for a c-axis oriented powder sample. The observed slower relaxation rates, both $1/T_1$, and $1/T_2$, for the higher frequency sites indicate that these sites are also differentiated magnetically.

Although the conclusive microscopic origin for the site differentiation is not clear at present, the following speculative arguments may be drawn. When oxygen atoms along the b-axis in the Cu-O chain of YBCO system, O(1), are depleted with decreasing y from 7.0, there exist two and three oxygen coordinated Cu(1) sites, $Cu(1)_2$ and $Cu(1)_3$. Those sites may modify the charged state of oxygen atoms in the Ba-O layer, O(4), hence the copper atoms in the Cu-O plane, Cu(2). Therefore, the site differentiation in the Cu(2) sites of YBCO system depends on what kind of oxygen coordination is realized at the Cu(1) sites. For the LCO system, the site differentiation is introduced by the substitution of Sr for La. This substitution may create a change in the charged state of oxygen atoms located nearby the Sr atoms, hence modify the electronic state of corresponding Cu atoms. Were these arguments correct, the experimental evidences suggest the primary location of the doped holes; namely, oxygen sites in Ba-O layers for YBCO and oxygen sites near by Sr atoms in La-O layers for LCO. The former has also been suggested from the fact that the $Cu(1)_4$ sites have a large value of the asymmetry factor of the electric field gradient ($\eta \sim 1$) with negligible component along the c-axis [1].

3. Temperature dependence of nuclear spin-lattice relaxation

A great number of data for the temperature dependence of T_1 in high-T_c Cu oxides has been published already [4-12]. The following qualitative characteristics for the planar Cu sites are generally accepted by now. (1) The relaxation rate, $1/T_1$, in the normal state is highly enhanced and its temperature dependence is not accord with the Korringa Process, which is due to the electron hole pair excitations at the Fermi surface. (2) There exist no "isotropic BCS" characteristics in $1/T_1(T)$ just below T_c. Irrespective of the system, $1/T_1(T)$ has been found to decline much more steeply, without any peak, just below T_c than the BCS prediction. (3) The low temperature behavior ($T \lesssim 20K$) is still sample dependent where the relaxation process is limited by a small amount of magnetic impurities.

We have been pointing out that, before arguing the properties of the superconductivity from T_1 measurements, we must understand the normal state behaviors. Here, we have made a systematic investigation on the temperature and material dependences of T_1 at the normal state, including non-superconducting but metallic Cu oxides.

In Fig. 3, $1/T_1(T)$ associated with the planar Cu sites is presented for the normal states. To make this plot, we have selected materials which exhibit "homogeneous" relaxation process, except for $La_{1.7}Sr_{0.3}O_4$ case where the longest relaxation component has been extracted from non exponential recovery of the nuclear magnetization. It has also been confirmed at several temperatures that the dominant relaxation process is to be "magnetic" from the measurements of isotope effect of T_1 (the ratio of $1/T_1$ for the two Cu isotopes being nearly equal to the square of the ratio of their γ_n values).

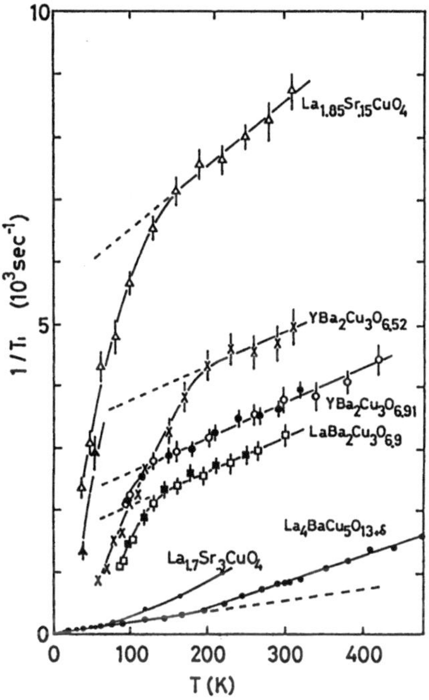

Fig. 3 Temperature dependence of ^{63}Cu nuclear spin-lattice relaxation rate $1/T_1$ in the normal state of high-T_c oxides for the Cu-O plane sites. For comparison, data for metallic Cu oxides are also shown. The dashed lines are the best fits of the high-temperature data to $1/T_1 = a + bT$ (see text).

In general, the quantity $1/T_1$ may be defined in terms of autocorrelaton function of the fluctuating local field, δH, acting on the nucleus in concern [13],

$$(1/T_1) = (1/2)(\gamma_n)^2 \int_{-\infty}^{+\infty} <\{\delta H^+(t), \delta H^-(0)\}> \exp[i\omega_0 t]\, dt, \qquad (1)$$

where γ_n and ω_0 are the nuclear gyromagnetic ratio and the frequency for resonance, respectively. If we assume the isotropic hyperfine field operator, $H = AS$, and make use of the fluctuation and dissipation theorem, $1/T_1$ can be reduced to,

$$(1/T_1) = (1/2)(\gamma_n A)^2 \sum_q S_\perp(q, \omega_0) \qquad (2)$$

$$= 2(\gamma_n A)^2 kT \sum_q \mathrm{Im}[\chi_\perp(q, \omega_0)]/\omega_0, \qquad (3)$$

where A is the hyperfine coupling constant, and $S_\perp(q, \omega_0)$ and $\chi_\perp(q, \omega_0)$ are the transverse components of the spin correlation function and the dynamical susceptibility, respectively. Then the problem is reduced to how we can describe the $\chi(q, \omega)$ in our system. Unfortunately, however, we can not relay on any kind of existing theory to describe $\chi(q, \omega)$ at present. Therefore, only phenomenological interpretations for the experimental results are given here.

The following characteristic features are immediately extracted from Fig. 3. (1) For the non-superconducting materials, the Korringa process is really seen at low temperatures, the rate of which can roughly be accounted for the estimated Cu site density of states.

(2) With increasing temperature, an excess relaxation process starts to develop at around 100 K for all the cases. This is particularly remarkable for the superconducting materials. Since the uniform static susceptibility, $\chi(0,0)$, has essentially no enhancement in these systems and does not vary so much with material to material, the excess relaxation process is considered to be due to a peak in $\chi(q,\omega)$ at the finite value of q with negligible enhancement near q=0. This case arises when there are antiferromagnetic correlations, that seem appropriate for high-T_c oxide system because of the antiferromagnetic phase being present for the materials with zero or small carrier densities. (3) In the temperature well above T_c, $1/T_1(T)$ for the superconducting materials has an asymptotic functional form on temperature, i.e. $1/T_1 = aT + b$. Here, we associate the first and the second terms with the Korringa, $(1/T_1)_K$, and the process characteristic to the Cu spin dynamics, $(1/T_1)_d$, at finite temperatures, respectively.

There may be several ways to interpret the above characteristic features, ranging from localized to itinerant picture for the Cu 3d holes. We start to discuss the results from a localized regime, since some of the resonance experiments have indicated the existence of exchange coupled localized moments on the Cu sites in the normal state [14,15]. In this case, the temperature dependence of the d-spin contribution, $(1/T_1)_d$, to the relaxation process may be extracted by subtracting the Korringa term from the observed value. This is shown in Fig. 4. Here, we have assumed the Korringa term is due to the contact interaction with the conduction holes and not depend on temperature.

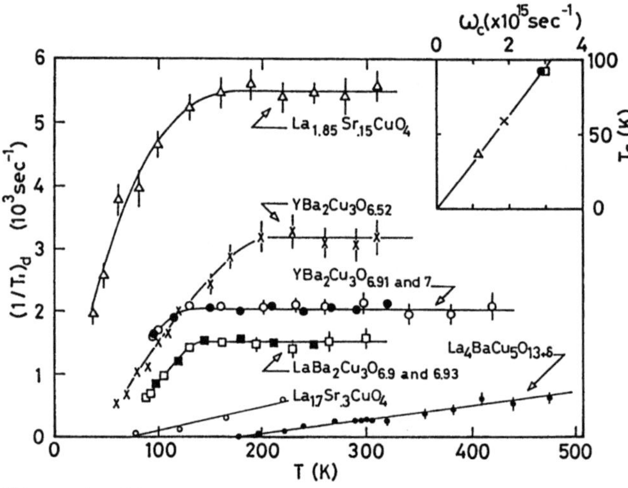

Fig. 4 Temperature dependence of the Cu d-spin contribution $(1/T_1)_d$ to the relaxation rate. Inset shows a relation between the T_c and the characteristic correlation frequency of the d-spin fluctuations.

It is clearly seen in Fig. 4 that the temperature independent $(1/T_1)_d$ tends to diminish below about 150 K which is well above T_c. The high temperature value of $(1/T_1)_d$ may be analyzed by the same analogy of purely local moment limit with a characteristic correlation frequency, ω_c, for the local spin fluctuations. If we assume that the local field spectra have a Gaussian distribution centered about zero frequency, hence $S(q,\omega)$ in eq. (2) has a Gaussian decay with ω, we have [16].

$$(1/T_1)_d = (\pi/2)^{1/2} (A)^2 S(S+1)/(3\omega_c). \tag{4}$$

Using reasonable estimates of A from hyperfine fields observed in the antiferromagnetic phases [17, 18], ω_c is estimated from the high temperature values of $(1/T_1)_d$. In the inset of Fig. 3, ω_c thus obtained is plotted against T_c. It is seen interestingly that T_c is proportional to the correlation frequency of the d-spin fluctuations within this model. It is also noted that the estimated ω_c is by an order of magnitude larger than the exchange frequency $\omega_{ex}=1.3\times10^{14}$ (rad/sec) estimated from generally accepted value of J~1000 K. This means that the frequency spectrum of $S(q,\omega)$ extend much higher frequency than the ω_{ex}, hence high energy excitations may be playing an important role on the Cu spin dynamics.

An alternative interpretation can be given from the basis of itinerant picture for the 3d holes. In this model, $\chi(q,\omega)$ in eq. (3) may be treated by the self-consistent renormalization theory of spin density fluctuations [19]. For nearly itinerant antiferromagnetic cases with two dimenensional character of the Fermi surface, as is the present materials, this theory predicts a temperature independent relaxation process if χ_Q (Q being the antiferromagnetic wave vector) obeys the Curie-Weiss like temperature dependence. Although the quantitative account is rather difficult to make since it depends on the detailed nature of the Fermi surface and degree of the enhancement of $\chi(Q,\omega_o)$, the high temperature behavior may be explained in this regime.

The remaining question having arisen even within the two limited discussions above is how we can interpret the drastic decrease of $1/T_1$ below T~150 K. We believe this is one of the most important findings in the present relaxation studies. Irrespective of the model, the experimental data are exhibiting of the opening of a gap in the low laying excitation spectrum with an energy comparable to the value of BCS gap. It should be stressed, however, that this gap is not directly due to the superconducting gap, since a sharp decrease of $1/T_1$ associated with the superconducting gap opening does exists at T_c for all the superconducting materials (see Fig. 5). The similar gap opening around 150 K has recently been observed in the Cu^{2+}-hole spin excitation spectrum in $La_{1.85}Sr_{0.15}CuO_4$ by the neutron scattering experiments [20].

Finally, we should comment on the possible interpretation based on the proposed quasi-particle excitations (e.g. Spinons in the RVB picture) in these system [21, 22]. If these particles obey Fermi statics, the value of $1/T_1T$ is a good measure of the density of states at the Fermi level, $N(E_f)$, in such excitations. The temperature dependence of $1/T_1T$ is shown in Fig. 5. It is seen for the cases of $YBa_2Cu_3O_{6.5}$ and $LaBa_2Cu_3O_{6.9}$ that $N(E_f)$ increases with decreasing temperature having a plateau around 150 K and decreases with farther decreasing temperature. The latter is again suggestive of the opening of a gap in the density of states.

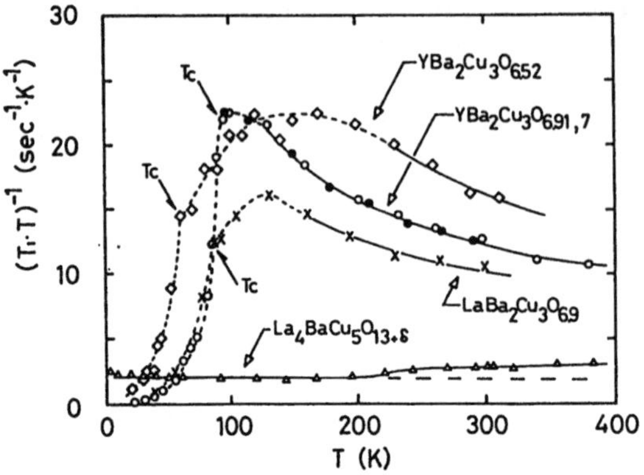

Fig. 5 Temperature dependence of $1/T_1T$ in $YBa_2Cu_3O_7$, $LaBa_2Cu_3O_{6.9}$ and $YBa_2Cu_3O_{6.5}$. For comparison, the data in the non-superconducting $La_4BaCu_5O_{13}$ are also shown.

4. Summary

In the analysis of the NQR spectra, a sizable Cu charge differentiation has been observed in Cu-O planes for $YBa_2Cu_3O_y$ and $La_{2-x}Sr_xO_4$. The temperature dependence of copper spin dynamics has been investigated by T_1 measurements for high-T_c and related oxides. Highly enhanced relaxation process with a characteristic temperature dependence was found in the normal state of high-T_c oxides. This has been discussed by the manifestation of either antiferromagnetic Cu spin fluctuations or quasi-particle excitations. The experimental data are also exhibiting of the opening of a gap in the low-laying excitation spectrum with an energy comparable to the value of superconducting gap.

Acknowledgments. We thank Professors T. Moriya and H. Fukuyama for valuable theoretical discussions, and Professors Y. Ueda, K. Kosuge and Dr. K. Yoshimura for the material synthesis. This work is partially supported by a grant in aid for scientific research on priority areas from the Ministry of Education, Science and Culture of Japan.

References

[1] H. Yasuoka: In Springer Series in Materials Science, Vol. 11, Mechanisms of High Temperature Superconductivity, edited by H. Kamomura and A. Oshima (Springer-Verlag, 1989) p. 156
[2] A. J. Vega, W. E. Farneth, E. M. McCarron and R. K. Bodia: Phys. Rev. B39, 2322 (1989)
[3] K. Yoshimura, T. Imai, T. Shimizu, Y. Ueda, K. Kosuge and H. Yasuoka: submitted to J. Phys. Soc. Jpn. (1989)
[4] M. Mali, D. Brinkmann, L. Pauli, J. Roos, H. Zimmermann and H. Hulliger: Physic Letters A124, 112 (1987)
[5] W. W. Warren, Jr., R. E. Walstedt, R. F. Bell, G. F. Brennert, R. J. Cava, G. P. Espinosa, J. P. Remeika and E. A. Rietmann: Phys. Rev. Lett. 59, 1860 (1987)
[6] Y. Kitaoka, S. Hiramatsu, T. Kondo and K. Asayama: J. Phys. Soc. Jpn. 57, 30 (1988)

[7] T. Imai, T. Shimizu, T. Tsuda, H. Yasuoka, T. Takabatake, Y. Nakazawa and M. Ishikawa: J. Phys. Soc. Jpn. 57, 1771 (1988)
[8] T. Imai, T. Shimizu, H. Yasuoka, Y. Ueda and K. Kosuge: J. Phys. Soc. Jpn. 57, 2280 (1988)
[9] R. E. Walstedt, W. W. Warren. Jr., R. F. Bell, G. F. Brennert, G. P. Espinosa, R. J. Cava, L. F. Schneemeyer and J. V. Waszczak: Phys. Rev. B38 9299 (1988)
[10] W. W. Warren, Jr., R. E. Walstedt, G. F. Brennert, R. J. Cava, R. Tycko, R. F. Bell and G. Dabbagh: Phys. Rev. Lett. 62 1193 (1989)
[11] K. Ishida, Y. Kitaoka and K. Asayama: J. Phys. Soc. Jpn. 58, 36 (1989)
[12] T. Imai, T. Shimizu, H. Yasuoka, Y. Ueda, K. Yoshimura and K. Kosuge: submitted to Phys. Rev. B (1989)
[13] T. Moriya: J. Phys. Soc. Jan. 18, 516 (1963)
[14] M. Horvatic, P. Segransan, C. Berthier, P. Butaud, J. Y Henry, M. Couach and J. P. Chaminade: Phys. Rev. B39, 7332 (1989)
[15] C. P. Pennington, D. J. Durand, C. P. Slichter, J. P. Rice, E. D. Bukowski and D. M. Ginsberg: Phys. Rev. B39, 274 (1989)
[16] T. Moriya: Prog. Theor. Phys. (Kyoto), 16, 641 (1956)
[17] H. Yasuoka, T. Shimizu, Y. Ueda and K. Kosuge: J. Phys. Soc. Jpn. 57, 2659 (1988)
[18] T. Tsuda, T. Shimizu H. Yasuoka, K. Kishio and K. Kitazawa: J. Phys. Soc. Jpn. 57, 2908 (1988)
[19] T. Moriya: Springer Series in Solid-State Sciences 56, Spin Fluctuations in Itinerant Electron Magnetism, (Springer-Verlag, 1985)
[20] G. Shirane, R. J. Birgeneau, Y. Endoh, P. Gehring, M. A. Kastner, K. kitazawa, H. Kojima, I. Tanaka, T. R. Thurston and K. Yamada: preprint
[21] P. W. Anderson: Science 235, 1196 (1987)
[22] P. W. Anderson, G. Baskaran, Z. Zou and T. Hsu: Phys. Rev. Lett. 58, 2790 (1987)

NMR Study of High-T_c Superconductors

Y. Kitaoka[1], *K. Ishida*[1], *F. Fujiwara*[1], *T. Kondo*[1], *K. Asayama*[1], *M. Horvatic*[2], *Y. Berthier*[2], *P. Butaud*[2], *P. Segransan*[2], *C. Berthier*[2], *H. Katayama-Yoshida*[3], *Y. Okabe*[3], *and T. Takahashi*[3]

[1]Department of Material Physics, Faculty of Engineering Science, Osaka University, Toyonaka, Osaka 560, Japan
[2]Laboratoire de Spectrometrie Physique, Université Joseph-Fourier Grenoble I, BP 87, F-38402 St. Martin d'Hères, Cedex, France
[3]Department of Physics, Tohoku University, Sendai 980, Japan

Abstract. Cu NQR and oxygen NMR studies of high T_c materials are reported. The behavior of the Cu nuclear spin-lattice relaxation in CuO_2 planer site is demonstrated to be unconventional above and below T_c for $(La_{0.925}Sr_{0.075})_2CuO_4$ (T_c=38 K), $YBa_2Cu_3O_7$ (T_c=92K) and Bi-Pb-Sr-Ca-Cu-O (T_c=109 K) compared with BCS superconductor. From the hole content dependence of T_1 in $(La_{1-x}Sr_x)_2CuO_4$, the normal state of high T_c materials is found to be dominated by the antiferromagnetic spin fluctuations. The Knight shift of ^{17}O has been measured for the O(1) site in CuO_2 plane and O(2) site in the LaO layer in $(La_{0.925}Sr_{0.075})_2CuO_4$. In normal state, the shift for the O(1) site decreases with decreasing temperature in proportion to the bulk susceptibility, while the shift for O(2) site is temperature independent. Below T_c, both shifts decrease with a moderate rate compared to the shift for O(2,3) site in $YBa_2Cu_3O_7$ which shows a faster reduction than the BCS prediction. The coupling between the localized spins at Cu(2) sites and the mobile oxygen p-holes is discussed in the light of these results.

1. Introduction

Since the discovery of high-T_c superconductors (1), many experimental and theoretical studies have been carried out in order to elucidate the mechanism for the high T_c superconductivity. As for the overall electronic structure, the photoemission studies have revealed that the doped holes are predominantly oxygen 2p-like rather than Cu-3d like due to the strong on-site Coulomb interaction among Cu-3d electrons. There is a general consensus that the presence of CuO_2 plane with a bridging or apex oxygen is responsible for the superconductivity. The important problem is hence to explore separately the respective roles of Cu and oxygen sites in the normal and superconducting state from the microscopic point of view. The nuclear magnetic (NMR) and quadrupole (NQR) resonance are powerful tools which provide an information on the electronic state at each atomic site to clarify the magnetic and superconducting properties of the high-T_c oxides.

As shown by many groups (2,3,4,5), the nuclear spin-lattice relaxation behavior of Cu in CuO_2 plane in $YBa_2Cu_3O_7$ tells us that both the normal and superconducting states are unconventional compared to the BCS

Springer Series in Solid-State Sciences, Vol. 89 Strong Correlation and Superconductivity
Editors: H. Fukuyama · S. Maekawa · A.P. Malozemoff

superconductor. Recent NMR works have also established that the Cu spins are fluctuating with an antiferromagnetic exchange coupling (6,7). Furthermore, the recent development in the Cu NQR observed in $(La_{1-x}Sr_x)_2CuO_4$ (8) and Bi-Pb-Sr-Ca-Cu-O (9) has yielded an extensive understanding of the normal state property in high-T_c materials. In particular, from the hole content dependence of T_1 in La system, the disappearance of the superconductivity is found to be associated with the depression of the antiferromagnetic spin fluctuations in the normal state (10). These results of Cu NQR have drawn attention to the role of the doped p-holes in the normal and superconducting states and raised the interest in the NMR study of ^{17}O in these materials. In this paper, we focus on the results of Knight shift measurement in $(La_{0.925}Sr_{0.075})_2CuO_4$.

2. Cu nuclear relaxation behavior

Figure 1 shows the temperature dependence of the nuclear spin lattice relaxation rate $1/T_1$ of ^{63}Cu for $(La_{0.925}Sr_{0.075})_2CuO_4$ (La compound)(8), $YBa_2Cu_3O_7$ (Y compound)(4) and Bi-Pb-Sr-Ca-Cu-O (Bi compound)(9) with T_c= 38 K, 92 K and 109 K, respectively. As for the Y compound, several groups have reported almost the same result (2,3,5), while the T_1 data of La and Bi compound have recently been presented for the first time by our group.

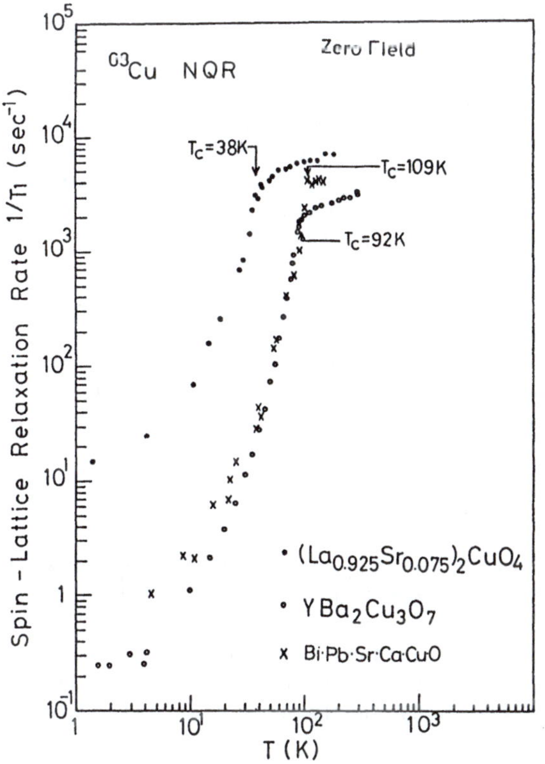

Fig. 1. Temperature dependence of $1/T_1$ of ^{63}Cu of High-T_c superconductors.

As seen in the figure, the relaxation behaviors are similar to each other, showing that a common mechanism for T_1 works in all compounds. $1/T_1$ above T_c shows a weak temperature dependence. It is evident that the Korringa law (T_1T=const.) characteristic for a usual metal is not observed for high T_c materials.

Below T_c, $1/T_1$ of all compounds decreases markedly without the enhancement just below T_c which is a unique feature for the BCS superconductor. At low temperature $1/T_1$ is saturated, implying that the intrinsic magnetic relaxation through the hyperfine interaction is masked. As for the saturation of $1/T_1$ at lower temperature, the quadrupole relaxation process was confirmed to come up in our samples of La and Y compound (4,8). As reported already (4), we can, however, discriminate the intrinsic magnetic relaxation from the observed T_1. Then $1/T_1$ for Y compound was reported to approach a T^3 behavior below 40 K (4) in contrast with the exponential type expected in a BCS case. The behavior of T_1 below T_c for high T_c materials reminds one of the behavior observed for the heavy fermion superconductors (HFS) which obey a T^3 law at low temperature without the enhancement just below T_c (11). In this case, the unusual relaxation behavior is understood in terms of an anisotropic energy gap model with the gap zeros of lines on the Fermi surface due to the strong on-site Coulomb interaction among the renormalized heavy electrons (11). For the high T_c materials, it is also likely that the order parameter cannot possess the amplitude at the same site due to the strong on siteCoulomb interaction among Cu d-electrons and hence a s-wave model is not favorable but a d-wave model may be applicable like in the case of heavy Fermion superconductors (4). In the analysis of T_1 below T_c based on the anisotropic energy gap model, the Korringa law of T_1T=const. is implicitly involved in the normal state. In fact, the Korringa behavior has been observed just above T_c for HFS (11), although the temperature range where the Korringa law is valid, is narrow. On the other hand, this is not necessarily the case for high-T_c materials reported here, though $1/T_1$ for La compound seems to approach the linear temperature dependence below 60 K (8). Accordingly, first of all, we should discuss about the origin of the relaxation process in the normal state which is not of the type expected for a non-magnetic metal. The successful observation of Cu NQR for La compound enables us to investigate the hole content dependence of T_1 of Cu affected by spin fluctuations (8).

3. Hole content dependence of T_1 of Cu in $(La_{1-x}Sr_x)_2CuO_4$ compound

Torrance et al. have shown that T_c in $(La_{1-x}Sr_x)_2CuO_4$ decreases with increasing Sr content i.e. hole content and then the normal metallic state is stabilized for $x > 0.15$ (12).

The Cu NQR spectrum includes the important information. Kumagai et al. proposed a possibility of a coexistence of the superconductivity and the magnetism (13). However, our NMR study on the oriented powder proved that there is no static magnetic order down to 1.4 K (10). The details can be found in the ref. 10. Here we focus on the result of $1/T_1$ of ^{63}Cu measured in the temperature range of 1.4 K- 250 K. $1/T_1$ for x=0.075 is indicated by closed circle in Fig. 2. The recovery curve of the nuclear magnetization after saturation is single-exponential above 30 K, which enables us to evaluate uniquely T_1, while non-single-exponential behavior appears slightly below 30 K. The longest component of T_1 below 30 K is plotted in Fig. 2 . With further increasing x, the multi-exponential behavior due to the distribution of T_1 becomes remarkable. In order to see the temperature dependence of each component of the distributed T_1, the log m(t) is plotted against tT for x=0.15 in Fig. 3 where T is the temperature. Here $m(t)=(M(\infty)-M(t))/M(\infty)$, and M(t) is the nuclear magnetization at time t after the saturating pulse. The plotted points are distributed on a line within the experimental scattering, showing that the relation of T_1T=const. is valid for each component in a wide temperature range of 1.3 K - 90 K, while roughly above 110 K (10).

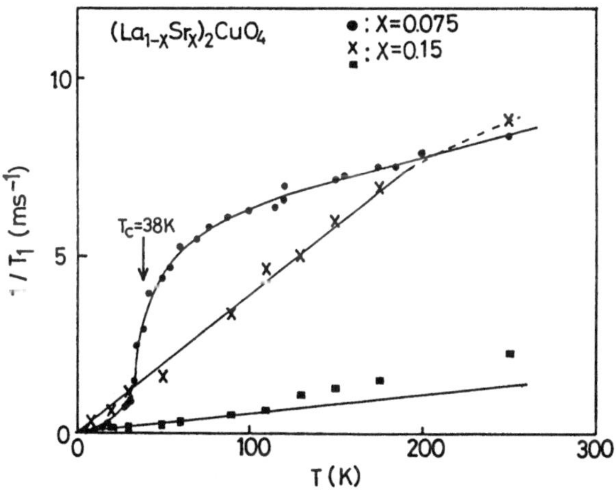

Fig. 2. Temperature dependence of $1/T_1$ of ^{63}Cu in $(La_{1-x}Sr_x)_2CuO_4$. x=0.075 (T_c=38 K):● x=0.15 (T_c=0);T_{1s} (shorter component):x T_{1L} (longer component) : ■

In Fig.2, we show the temperature dependence of T_{1L} (cross marks) and T_{1s} (rectangle marks), which are tentatively extracted from a two exponential fitting to the recovering data. Although the longer component T_{1L} seems to deviate gradually from the relation of T_1T=const above 110 K (14), it may not be safely said whether or not this deviation is intrinsic because the quality of the recovery curve becomes worse with increasing temperature. The important feature is that the T_{1s} for x=0.15 approaches the value for x=0.075 at high temperature. Furthermore, it has been confirmed that the T_{1s} for x=0.125 (T_c=10 K) exhibits nearly the same values as those for x=0.075 above 38 K followed by a 1/T dependence below 38 K (10). This means that even when the Sr contents increased, the contributions similar to that of T_1 observed for x=0.075 above T_c are still present in some part of sample, causing the distribution of T_1. We suppose

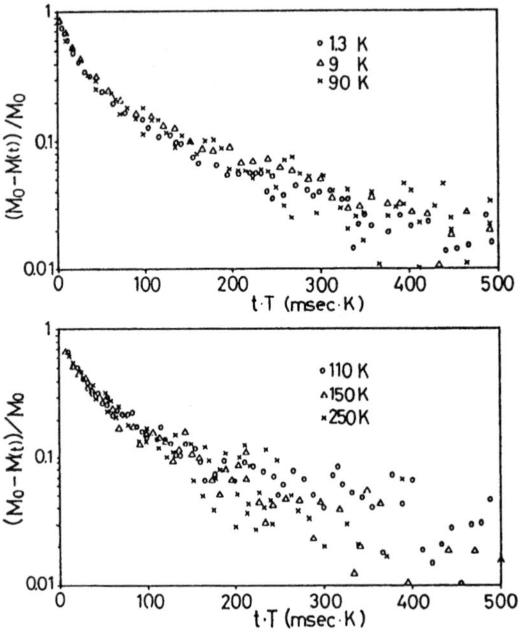

Fig. 3.
Recovering behavior of the nuclear magnetization. $(M(\infty)-M(t))/M(\infty)$ is plotted against tT, where M(t) is the nuclear magnetization at time t after the saturating pulse and T is the temperature.

that due to the local distribution of Sr and La and/or oxygen vacancies, the hole concentration is not fully homogeneous in the sample and in some region not large enough to suppress the A.F. spin fluctuations. In contrast, the longer component T_{1L} is longer with increasing x from x=0.125 to 0.15, following the relation of T_1T=const. In this sense, the longer component seems to reflect the intrinsic variation of T_1 with increasing the hole concentration. We can explore from the present experiment an important correlation between the antiferromagnetic (A.F.) fluctuations of Cu spin and T_c. Namely, as T_c decreases, the temperature dependence of $1/T_1$ changes from a weak dependence enhanced by A.F. spin fluctuations to a linear dependence after $1/T_1$ was suppressed. The important feature is that $1/T_1$ for high-T_c materials is enhanced by A.F spin correlation down to T_c, while that for non-superconducting materials decreases linearly with temperature, less affected by A.F. spin fluctuation.

Furthermore, from the experiment of the transverse relaxation time (spin echo decay time), T_2 (10,13), we have other evidence for the presence of the A.F. spin fluctuations for high-T_c materials. At 4.2 K the decay curve is of simple exponential. T_2 decreases from 52 μsec for x=0.15 to 15 μsec for x=0.075. For $x < 0.05$, T_2 is quite short with a smaller value than 10 μsec which makes the observation of Cu NQR difficult. From this strong dependence of T_2 on Sr content, it is suggested that the Cu nuclear spins couple strongly with each other through the Cu electronic spin fluctuations, which are depressed by the increase of Sr content (10,13). It is expected that the magnetic coherence length of A.F. spin fluctuation is developing with decreasing Sr content, which yields a strong coupling among Cu nuclear spins and decreases T_2.

Pennington et al. demonstrated that the transverse relaxation behavior of Cu(2) in $YBa_2Cu_3O_7$ can be explained by a strong indirect coupling among Cu(2) nuclear spins, which is caused by the fluctuations of d spin at Cu(2) site exchange-coupled to one another (6). The shorter T_2 in La compound than that in Y compound is attributed to the stronger indirect interaction among nuclear spins in La compound than that in Y compound.

Hence combining the results of T_1 and T_2 for La compound, we can conclude that the short range A.F. spin correlation is present even in the superconducting materials and the increase of Sr content reduces the magnetic coherence length of A.F. spin fluctuations, yielding the decrease of T_c at the same time. We consider that the high-T_c superconductivity is realized with the background of A.F. spin correlations (10). For La compound, when the A.F. spin correlation becomes less effective as Sr content increases, T_c decreases and furthermore vanishes for x=0.15. The relaxation behavior of Cu provides an evidence that there exists a strong correlation between T_c and A.F. spin fluctuations of Cu(2).

4. ^{17}O NMR

Since the doped holes are mainly introduced to oxygen site, NMR of oxygen is expected to give an important information. Measurements have been carried out on ^{17}O isotope exchanged in place of ^{16}O isotropically. Up to now, there are several reports on ^{17}O in Y compound (15,16,17,18,19) We have measured the temperature dependence of T_1 and reported the different behavior from that of T_1 of Cu(2). The striking result was an appearance of an enhancement of $1/T_1$ just below T_c. In our first report (15), this behavior was considered to be associated with the oxygen sites in CuO_2. Takigawa et al.(18) and Horvatic et al.(19), showed that the various oxygen sites i.e. O(2,3) in CuO_2 plane, O(1) in CuO chain and O(4) in BaO layer possess different electric field gradients and magnetic hyperfine tensors. Takigawa et al reported that T_1 of O(2,3) and O(1) sites are markedly shorter than T_1 of O(4) in BaO layer. Comparing our results with theirs, our T_1 should be associated with O(4) site. Recently, Wzietek et al. also observed a distinct enhancement of $1/T_1$ in a powder sample in which ^{17}O gas-exchange was made at a low temperature of 300 C (17). Here ^{17}O is substituted not only into the O(1) but also in O(4) site as shown by Horvatic et al (19). Considering the magnitude of their T_1, the enhancement observed by Wzietek et al. should also be associated with O(4) site, in contrary to their conclusions (17). On the other hand, Takigawa et al. have reported that T_1 of all oxygen sites including the O(4) site show no enhancement just below T_c (20). Accordingly, the results of T_1 are not consistent with each other at present. Further systematic experiment on the external field and sample dependence of T_1 are required and now currently in progress.

As far as the magnetic hyperfine shift (MHS) tensors in Y compound are concerned, only the main conclusions are reported here. Details can be found in Ref.18 and 19. The striking features we want to stress are as following.

(1) The anisotropy of the MHS tensor indicates that the spin densities for both O(2,3) (in the plane) and O(4) sites (bridging oxygen) are along the Cu-O-Cu bonding axis suggesting a p_σ symmetry for the doping holes; (2) For both sites, the shift decreases below T_c (Fig. 4) (19). For O(2,3) site, this decrease is much faster than expected from the BCS prediction (19).

Here, we present the result of ^{17}O NMR in $(La_{0.925}Sr_{0.075})_2CuO_4$ with T_c=38 K, obtained on an oriented powder sample ($H_0 \parallel$ c-axis) (21). In Fig.5 are shown the typical line shapes of the central lines (±1/2) transition

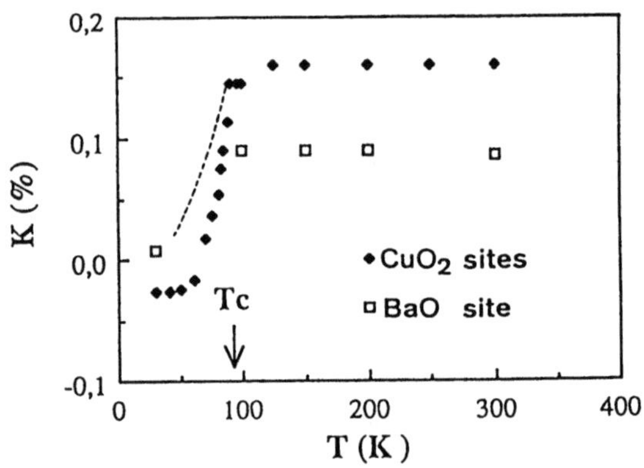

Fig. 4. Temperature dependence of ^{17}O Knight shift (c-axis component) on the CuO_2 (◆) and the bridging oxygen sites (□) in $YBa_2Cu_3O_7$. A dashed line represents the Yosida-BCS prediction.

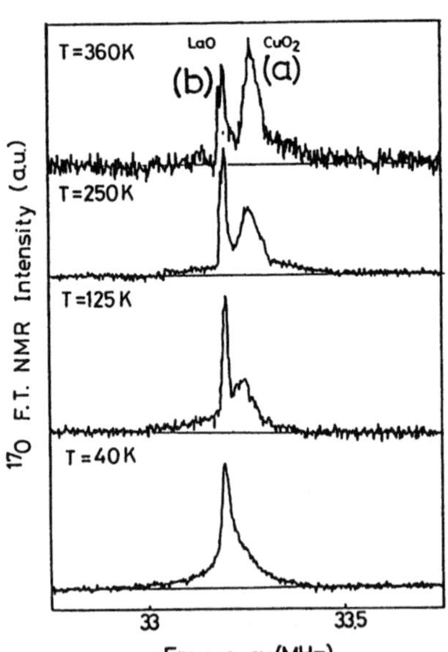

Fig. 5. Spectra of the (±1/2) transitions of ^{17}O in oriented powders of $(La_{0.925}Sr_{0.075})_2CuO_4$.

of ^{17}O observed in the temperature range of 40 K-360 K. The spectra have been obtained by Fourier transform of the spin echo. As seen in the figure, two lines have been detected for ($\pm 1/2$) transition. The lines (a) and (b) with larger and smaller positive shift are respectively assigned to the oxygen site, O(1) in the CuO_2 plane and O(2) in the LaO plane. Details can be found in Ref. 19 and 21. From the satellites associated with the line (a), we obtain the value of e^2qQ/h at 300 K as 3.9 MHz, assuming tentatively the axis of the field gradient to be parallel to a or b and η =0.26 as in Y compound. As shown in Fig. 6, $K_c(O(1))$ decreases linearly from 0.15 % at 300 K to 0.09 % at 40 K, while $K_c(O(2))$ keeps a constant value of 0.052 %. Here K_c is the shift parallel to the c axis. Below T_c, the shifts corresponding to both sites O(1) and O(2) do not vanish suggesting an important orbital contribution. This situation contrasts with the result in $YBa_2Cu_3O_7$ (Fig. 4) (19), $YBa_2Cu_3O_{6.65}$ (22) and $Bi_2Sr_2CaCu_2O_8$ (21) compounds, where the orbital contributions to the MHS tensor are negligibly small. This temperature dependence of $K_c(O(1))$ and $K_c(O(2))$ contrasts with that of the MHS of copper $K_c(Cu)$ which was roughly temperature independent having the same order of magnitude as for the Cu(2) site in $YBa_2Cu_3O_7$ (19) and $YBa_2Cu_3O_{6.65}$ (22). Although there is only one copper site in this compound, the random distribution of Sr atoms introduces a serious broadening of the line corresponding to the Cu ($\pm 1/2$) transition. This limits the precision of our $K_c(Cu)$ data. Nevertheless, it is evident that $K_c(Cu)$ for the CuO_2 plane is dominated by the orbital contribution reflecting the localized Cu^{2+} configuration.

Let us now discuss the local contributions to the susceptibility from the behavior of the MHS at each atomic site, which should bring valuable information on the electronic structure of the high T_c superconducting oxides. After subtraction of a small extrinsic Curie term (C/T), the magnetic susceptibility χ_{obs} measured on powder and oriented powder sample from the same batch as our NMR sample was found in good agreement with the data previously reported. The susceptibility is decreasing with decreasing temperature. Looking at our results for O(1) and O(2) in Fig. 6, one can infer that this temperature dependence of χ is closely related to the spin density at the O(1) in the CuO_2 plane. In Fig. 7, $K_c(O(1))$ is plotted against χ_{obs} with the temperature as an implicit parameter. From the coefficient of the linear dependence, the hyperfine coupling constant A_{hf} is estimated to be 137 kOe/μ_B (21).

Generally, the Knight shift consists of spin and orbital parts, $K=K_{spin} + K_{orb}$. The magnetic susceptibility can be decomposed: $\chi = \chi_{spin} + \chi_{orb} + \chi_{dia}$, the last term being the core diamagnetic contribution. For the present case, χ_{spin} is divided into two contributions: $\chi_{spin} = \chi_s(1) + \chi_s(2)$, where $\chi_s(1)$ arises from the nearly localized d-hole spins and the mobile doped hole spins at CuO_2 plane site and $\chi_s(2)$ is due to the doped hole spins at LaO layer. It is expected that $\chi_s(1)$ is dominated by the Cu-d spin susceptibility as shown by ^{17}O NMR study in Y compound (18,19). χ_{orb} is also decomposed: $\chi_{orb} = \chi_{orb}(Cu) + 2\chi_{orb}(O(1)) + 2\chi_{orb}(O(2))$. Taking account of the result in Y compound, we assume that both O(1) and O(2) sites possess an axial symmetry around the axis of Cu-O-Cu bonding, i.e. the a or b axis for O(1) and the c-axis for O(2). Thus, the K_{spin} can be divided into the isotropic and axial component:

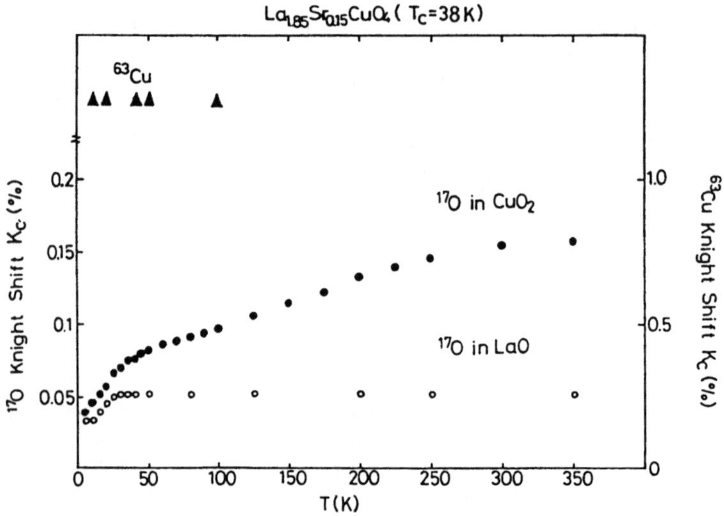

Fig. 6. Temperature dependence of the ^{17}O Knight shift on the CuO_2 (●) and the LaO layers (o) and of the ^{63}Cu (▲) in $(La_{0.925}Sr_{0.075})_2CuO_4$.

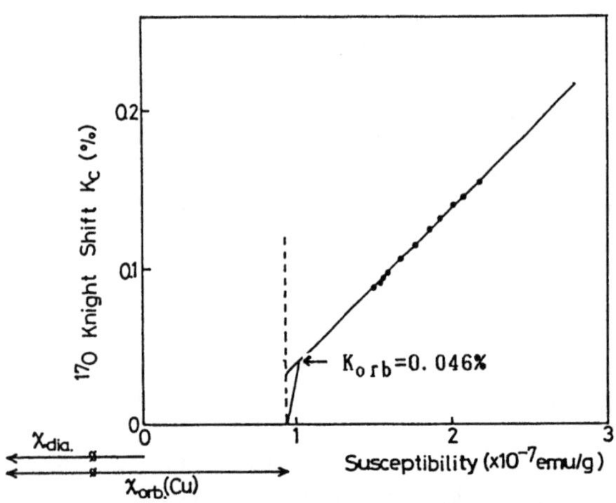

Fig. 7. The susceptibility vs. Knight shift plot in $(La_{0.925}Sr_{0.075})_2CuO_4$. From the slope of the soild line, A_{hf} is determined to be 137 kOe/μ_B.

$$K = K_{iso} + K_{ax}(3\cos^2\theta - 1)/2, \qquad (1)$$

where θ denotes the angle between the magnetic field H_0 and the direction of the Cu-O-Cu bonding. Since the shift $K_c(O(1))$ is temperature dependent, whereas the shift $K_c(O(2))$ is not as seen in Fig. 7, we suppose that $\chi_s(1)$

is temperature dependent, while $\chi_s(2)$ is not. Thus the isotropic and axial shift, K_{iso} and K_{ax} are described as

$$K_{iso}=(A^{2s}/N\mu_B)(2f_s\chi_s(1)) \quad (2)$$

$$K_{ax}=(A^{2p}/N\mu_B)(2f_p\chi_s(1)) \quad (3)$$

for O(1) site and

$$K_{iso}=(A^{2s}/N\mu_B)(2f_s'\chi_s(1)) \quad (4)$$

$$K_{ax}=(A^{2p}/N\mu_B)(2f_p'\chi_s(1)+\chi_s(2)) \quad (5)$$

for O(2) site where f_s and f_p are the fractions of the unpaired spin in the 2s and 2p orbitals at oxygen site, resulting from the overlap with the Cu 3d orbital. f_p' and f_s' are those at LaO layer. Here we suppose the p_σ symmetry for the doped holes as in the case of Y compound. A^{2s} and A^{2p} $=4/5\,\mu_B\langle r^{-3}\rangle$ are reported to be 6000 kG/μ_B and 190 kG/μ_B, respectively (23). From the temperature independent behavior of $K_c(O(2))$, f_p' and f_s' in LaO layer are to be smaller than those in CuO_2 plane.

Now, we can analyze $K_c(O(1))$ vs. X_{obs} plot shown in Fig. 7. From eq. (1), (2) and (3), the following relation is obtained for the O(1) site in the CuO_2 plane with $\theta=90$:

$$K_c(O(1))-(2A^{2s}f_s - A^{2p}f_p - A_{dip}/2)/(N\mu_B)X_s(1) + K_{c,orb}(1) \quad (6)$$

where $A_{dip}=2.3$ kG/μ_B is the dipole field from the Cu-d spin. The temperature dependence of $K_c(O(1))$ is coming from the Cu-d spin component involved in $\chi_s(1)$. From the slope of Fig. 7, we have a relation of $A_{hf}=137$ KG/$\mu_B=(2A^{2s}f_s - A^{2p}f_p - A_{dip}/2)$. From the present experiment, we cannot estimate f_s and f_p without deducing the K_{iso} and K_{ax}. However, if we assume the values of f_s and f_p being 1 % and 10 % extracted from the experiment in $YBa_2Cu_3O_7$, A_{hf} is calculated to be about 100 kG/μ_B from eq. (6) which is close to the experimental value of 137 kG/μ_B. The important point is that the large transferred hyperfine coupling with the positive value results from the Fermi-contact hyperfine interaction through the overlap between the Cu-d(x^2-y^2) and oxygen 2s wave function. By taking a tentative values of χ_{dia} and χ_{orb}(Cu) being -5.84×10^{-5} emu/mole and 9.6×10^{-5} emu/mole as evaluated in $YBa_2Cu_3O_7$ together with the relation of $K_{orb}(O(1))=A_{orb}\chi_{orb}(O(1))/N\mu_B$ with $A_{orb}=606$ kG/μ_B, $K_{c,orb}(I)$ is estimated to be 0.046 %. This is in good agreement with the residual shift at 10 K below T_c being 0.04 %. This agreement implies that the above estimation is not so bad.

Next we discuss the result of $K_c(O(2))$. From eq. (1), (4) and (5), we obtain a relation with $\theta=0$:

$$K_c(O(2))=(2A^{2s}f_s'+2A^{2p}f_p'+A_{dip})/(N\mu_B)\chi_s(1)+A^{2p}\chi_s(2)/(N\mu_B)+K_{c,orb}(2) \quad (7)$$

where A_{dip} = 2.1 kG/μ_B is the dipole field. From the fact that $K_c(O(2))$ does not depend on the temperature above T_c, the first term of eq. (5), which is the transferred hyperfine interaction from the CuO_2 plane site, is considered to be negligible. Furthermore, from the fact that the shift decreases below T_c, the second term of eq. (5), the spin term, is considered to be important. If we evaluate the $K_{c,orb}(2)$ to be 0.03 % at 10 K as in the case of $K_{c,orb}(1)$ mentioned above, $\chi_s(2)$ is then estimated to be 1.2×10^{-5} emu/mole, which is by one order of magnitude smaller than χ_{obs}. This demonstrates that a larger spin susceptibility in La compound originates also from the copper spins as well as in the case of Y compound. The temperature dependence of the susceptibility is associated with the temperature dependence of $\chi_s(1)$ in the CuO_2 plane. The behavior of $\chi_s(1)$, being temperature dependent, differs from that observed in $YBa_2Cu_3O_7$, but is similar to that observed in $YBa_2Cu_3O_{6.65}$ (22). The decrease of $\chi_s(1)$ with decreasing temperature may be relevant to the development of a short range spin correlation.

5. Conclusion

First, we have presented the systematic measurement of the relaxation time of Cu in the CuO_2 plane in high T_c materials. From both behaviors of T_1 and T_2, the normal state of the high T_c materials has been found to be strongly affected by the fluctuations of the Cu spins which are exchange-coupled to one another. In La compound, it has been clarified that the strong enhancement of $1/T_1$ for superconducting materials is significantly suppressed by the increase of hole content, which accompanies the reduction of T_c. This result suggests a possible evidence that the superconductivity needs the background of the antiferromagnetic spin fluctuations.

Below T_c, the dramatic collapse of the largely enhanced $1/T_1$ at CuO_2 site, which is owing to the pairing of superconducting carriers, has been commonly observed for various types of compounds. The most significant feature of this result is that the behavior is in strong contrast to BCS prediction and the 3d-holes should actually be itinerant, though the Cu(2) may in some respects be local-moment-like.

The oxygen NMR results of La compound tell us that most of the spin density is located within the CuO_2 plane and to some extent on the oxygen in the LaO. This result is consistent with that in the Y compound. Both shifts in the CuO_2 and LaO layers have been found to decrease below T_c. From the analysis of the Knight shift and the susceptibility, it was demonstrated that the spin part of the shift in the plane is enhanced by the hybridization between the Cu 3d and the oxygen 2p and 2s orbitals, reflecting sensitively the temperature dependence of the macroscopic susceptibility.

This work is partially supported by Special Project Research on High Temperature Oxides Superconductors of Grant-In-Aid for Scientific Research from the Ministry of Education, Science and Culture.

References

(1) J.G. Bednorz and K.A. Muller: Z. Phys. B64 (1986) 189.

(2) W.W. Warren, Jr., R.E. Walstedt, G.F. Brennert, G.P. Espinosa and J.P. Remeika, Phys. Rev. Lett. 59 (1987) 1860.

(3) M. Mali, D. Brinkmann, L. Pauli, J. Roos, H. Zimmermann and J. Hulliger, Phys. Lett. A124 (1987) 112.

(4) Y. Kitaoka, S. Hiramatsu T. Kondo and K. Asayama, J. Phys. Soc. Jpn. 57 (1988) 30. Y. Kitaoka, S. Hiramatsu, Y. Kohori, K. Ishida, T. Kondo, H. Shibai, K. Asayama, H. Takagi, S. Uchida, H. Iwabuchi, and S. Tanaka, Physica C153-155 (1988) 83.

(5) T. Imai, T. Shimizu, T. Tsuda, H. Yasuoka, T. Takabatake, Y. Nakazawa and M. Ishikawa, J. Phys. Soc. Jpn. 57 (1988) 1771.

(6) C.H. Pennington, D.J. Durand, C.P. Slichter, J.P. Rice, E.D. Bukowski, and D.M. Ginsberg, Phys. Rev. B39 (1989) 274; ibid, (1989) 2902.

(7) M. Horvatic, P. Segransan, C. Berthier, Y. Berthier, P. Butaud, J.Y. Henry, M. Couach, J.P. Chaminade, Phys. Rev. B39 (April 1989).

(8) K. Ishida, Y. Kitaoka and K. Asayama, J. Phys. Soc. Jpn. 58 (1989) 36.

(9) K. Fujiwara, Y. Kitaoka, K. Asayama, H. Sasakura, S. Minamigawa, K. Nakahigashi, S. Nakanishi, M. Kogachi, Nobuo Fukuoka and A. Yanase, J. Phys. Soc. Jpn. 58 (1989) 380.

(10) K. Ishida, T. Kondo, Y. Kitaoka and K. Asayama, J. Phys. Soc. Jpn. to be published.

(11) K. Asayama, Y. Kitaoka and Y. Kohori, J. Magn. Magn. & Mat. 76 & 77 (1988) 449.

(12) J.B. Torrance, Y. Tokura, A.I. Nazzal, A. Bezinge, T.C. Huang and S.S.P. Parkin Phys. Rev. Lett. 61 (1988) 1127.

(13) K. Kumagai and Y. Nakamura, Physica C157 (1989) 307.

(14) H. Yasuoka, et al., in this Proceeding.

(15) K. Ishida, Y. Kitaoka, K. Asayama, H. Katayama-Yoshida, Y. Okabe and T. Takahashi, J. Phys. Soc. Jpn. 57 (1988) 2897.

(16) C. Coretsopoulos, H.C. Lee, E. Ramli, L. Reven, T.B. Rauchuss and E. Oldfield,: Phys. Rev. B39 (1989) 781.

(17) P. Wzietek, D. Kongeter, P. Auban, D. Jerome, J.P. Coutures, B. Dubuis, Ph. Odier, Europhys. Lett. 8 (1989) 363.

(18) M. Takigawa, P.C. Hammel, R.H. Heffner, Z. Fisk, K.C. Ott and J.D. Thompson, preprint.

(19) M. Horvatic, Y. Berthier, P. Butaud, Y. Kitaoka, P. Segransan, C. Berthier, H. Katayama-Yoshida, Y. Okabe, and T. Taskahashi, Physica C, to be published

(20) M. Takigawa, private communication.

(21) Y. Kitaoka, et al. in preparation.

(22) P. Butaud, et al.: in preparation.

(23) J.S.M. Harvey, Proc. Roy. Soc. A 285 (1965) 581.

Tunneling Spectroscopy and Energy Gap of High T_c Oxide Superconductors

J.S. Tsai and I. Takeuchi

Microelectronics Research Laboratories, NEC Corporation,
34 Miyukigaoka, Tsukuba 305, Japan

Abstract Problems associated with the tunneling experiments of the high-T_c oxidesuperconductors are discussed in this paper. Local variations in T_c, multiple gap like structures, gap anisotropy, and the broadened and leaky gap structures are considered. New results of the experiments that reported the gap anisotropies are reviewed here. They all give the anisotropic factor of about 1.5~2. Several probable tunneling models that give the broadened gap structures are considered. The gap value of $2\Delta(0) \simeq 4\sim6\ k_BT_c$ is inferred from these models in Nd-Ce-Cu-O, Y-Ba-Cu-O, Bi-Sr-Ca-Cu-O and Tl-Ba-Ca-Cu-O compounds.

Since its first discovery by Giaever[1], quasiparticle tunneling spectroscopy of superconductors has become one of the major tools to probe the superconducting states of materials. Tunneling characteristics of S/N junctions at low temperatures reveal the quasiparticle density of states of the superconductors[2], and they can be used to directly measure the superconducting energy gap as well as the distribution of quasiparticle density of states outside the gap. Also, observed phonon structures on the conductance characteristics[3] have played a key role in establishing the BCS theory. According to the BCS theory, the state of the quasiparticles inside superconductors is characterized by the energy gap ($1.76k_BT_c$ at T=0) that exists at the Fermi surface. The ratio of this energy gap Δ to k_BT_c is a parameter of singular importance, at least within the realm of the conventional theory, in that it represents the strength of the undelaying mechanism of the superconductivity, and therefore proper measurement and evaluation of this parameter is crucial. The energy gap can also be studied using other methods such as NMR, specific heat measurements, infrared absorption spectroscopy, and Raman scattering[4]. But the tunneling spectroscopy probably yields the most reliable and accurate measurements.

For this reason, since the first discovery of the high T_c oxide superconductors, there have been enumerous reports of the tunneling studies on these materials. Some preliminary review articles on these experiments by a number of authors such as Gray, Hawley, Moog[5], and Barone[6] have appeared. These tunnel junctions include point-contact types[7], thin film types[9], break junction types[10], and STM types[8]. In the beginning, there seemed to be a significant scattering in the reported values of the normalized energy gap $\gamma=2\Delta/k_BT_c$ (with γ ranging from 3 to 21!), and there were much confusion over how to interpret the data obtained by these tunneling spectroscopy measurements of the high T_c superconductors. The main problems here can be attributed to following factors:

1) Appearance of multiple conductance peaks. This effect is often observed when there exist dc tunneling currents modulated[11] by Coulomb force accumulated at grain boundaries within the sample. This can also be caused by internal grain boundary junctions in series[12]. Even when there are no multiple peaks on the conductance characteristics, it is difficult to tell whether the observed peak is really the Δ from the intended N/S junction or 2Δ, or even $n\Delta$ from the internal grain boundary junctions. But this particular problem has become less and less dominant and noticeable as the general quality of the samples has increased (especially with the use of epitaxial and single crystal films).

2) Anomaly on the T_c values, which are used to normalize the gap. The conventional metallic superconductors have relatively well-defined values of T_c. In

Springer Series in Solid-State Sciences, Vol. 89 **Strong Correlation and Superconductivity**
Editors: H. Fukuyama · S. Maekawa · A.P. Malozemoff

comparison, the oxide superconductors possess and show transition temperatures in some-what large range of temperatures. The point at which the superconducting fluctuation begins is much higher than the R=0 temperature. La-Sr-Cu-O and Y-Ba-Cu-O systems can have a continuous variation of the carrier concentration, depending on which the T_c can also vary continuously. The very surface with which the tunneling is conducted on might have local T_c variations which are quite different from the values inferred from resistive and magnetic susceptibility measurements. Tunnel junctions utilize surfaces which tend to suffer from excessive oxygen deficiency and segregations of off-stoichiometric domains at grain boundaries. In this respect, these surfaces have T_c's which are almost always different from the resistively determined T_c's. As a result, the estimation of γ using $T_c(R=0)$ would not yield a very accurate value. One tunneling experiment on cleaved surfaces of Y-Ba-Cu-O thin films in cryogenic environment[13] revealed that the local T_c's of the tunneling surfaces were consistently lower than that of the films themselves by about 10K. Some reports have suggested that the surfaces of the bulk and films of the Y-Ba-Cu-O compound consist of 60K phase materials rather than 90K phase[9,14,15]. Therefore, in the tunneling spectroscopy of materials such as Y-Ba-Cu-O, Bi system compounds, and the 2212 phase of the Tl system compounds, it is essential to obtain $T_c(\Delta=0)$ in order to calculate γ.

3) Anisotropy of the energy gap. It is well known and established that owing to their layered structures, the oxide superconductors exhibit marked anisotropic characteristics in such parameters as the carrier density, coherence length, and the magnetic penetration depth. Some recent results point to the possibility that the anisotropy exists even in the superconductive energy gap. In a system with an anisotropic gap, it is expected that the tunneling study would reveal an "averaged" value of the gap due to the scattering in the tunneling directions of the electrons. In the past, anisotropic energy gap have been observed in such systems as Pb[16], NS[17], and organic superconductors[18]. In some sense, it almost seems natural that such anisotropy should exist in these oxide superconductors, considering their three-dimensional crystal structure. Figure 1 is a result from tunneling experiment with epitaxial films of Y-Ba-Cu-O which showed the probable existence of the energy gap anisotropy for the first time[19,13]. In this study, using (001) and (110) oriented epitaxial films, directions of the tunneling of the quasiparticles were selectively controlled by preparing the surfaces with the c-axis of the crystal either parallel or perpendicular to the substrate. Data from eight tunneling attempts with four films of different T_c's are presented in the graph, where Δ_0 (peak values of the conductance characteristics) and T_c are plotted against each other. Because of the difficulty in determining the $T_c(\Delta=0)$, the error is relatively large here, but the points clearly suggest the existence of two distinct gaps, one for tunneling in the direction of c-axis and another for tunneling in ab-plane. Subsequent measurements by other

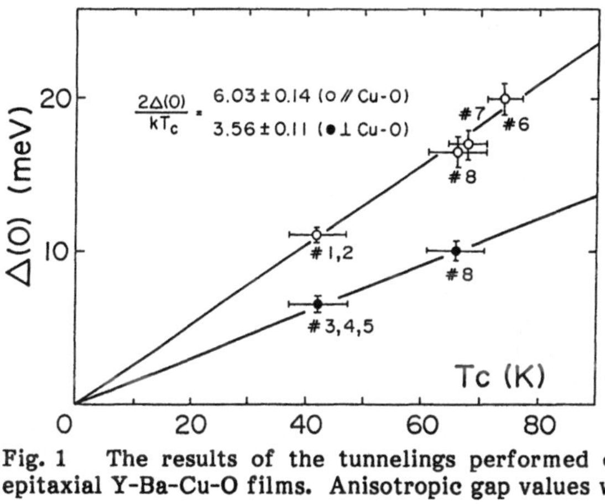

Fig. 1 The results of the tunnelings performed on the 4.2K cleaved surfaces of epitaxial Y-Ba-Cu-O films. Anisotropic gap values were obtained. (Ref.19)

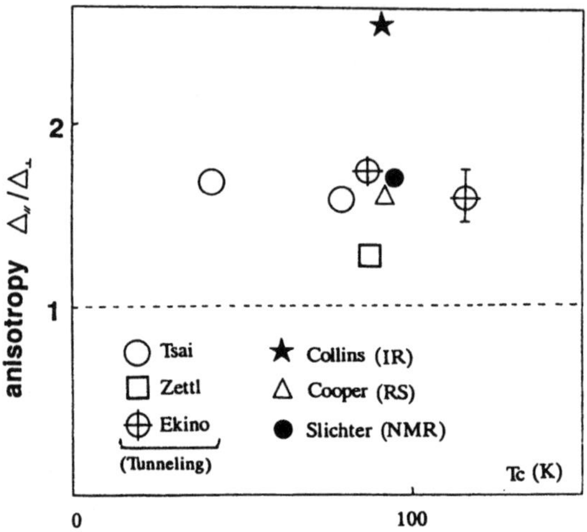

Fig. 2 The gap anisotropies observed in tunnelings, infrared reflection, Raman scattering and NMR.

groups using various techniques (tunneling[20,21], far-infrared reflectance[22], Raman scattering[23], and NMR[24]) have also found the anisotropy in energy gap. Figure 2 summarizes these data in terms of the anisotropic factor $\Delta_{\parallel}/\Delta_{\perp}$. In regard to the result from the Raman scattering, the axis of the anisotropic factor does not exactly (necessarily) correspond to $\Delta_{\parallel}/\Delta_{\perp}$, and the NMR result is the ratio of the Cu-O plane to the Cu-O chain in Y-Ba-Cu-O. Except for the value from the far-infrared result which is slightly larger than the rest, these six measurements from four different methods, remarkably enough, gave reasonably similar values. It has been predicted that for systems with extremely short coherence lengths, the anisotropic energy gap at the surface of the tunnel junctions might be influenced by the anisotoropic coherence length due to the proximity effect at low temperatures[38]. But methods other than tunneling which probe the bulk properties well inside of the surface layer have revealed about the same degree of anisotropy in gap. While it is possible that the difference in gap between the Cu-O layers and the Cu-O chains observed by NMR is caused by the short coherence length, at present it is not clear how this effect is reflected on other methods such as tunneling. If one assumes that the space between the planes and the chains with localized energy gap consists of insulators, it should not make any difference on the tunneling measurement where the cleaved surface lies, and its result should show the anisotropy of the gap. Even though, we can only speculate as to what is really taking place, experiments show reproducibly what appear to be the anisotropic energy gap. Some authors have reported that tunneling (STM) on single crystals revealed no anisotropy in gap[25,26,27], but this is generally blamed on the poor quality of the probed surfaces[26,27] or blockade of tunneling current by non-superconducting top layer of the sample[25].

4) Accurate determination of the gap energy. Majority of the tunnel conductance results reported until now show energetically broadened gap like structures. According to the s-wave BCS theory, at low temperatures no quasiparticle density of states exists inside the gap, and they possess dispersed density of states at the energy gap. In such a case, SIN type tunneling spectroscopy usually gives a very sharp peak on the conductance curve corresponding to the gap energy, and the task of gap determination remains relatively simple. But it is difficult to judge whether the above-mentioned conductance curves which clearly show peak structures with broadening that persist down to much lower energy regions really reflect the true density of states of the materials. Then how would one go about estimating the gap value from such broadened structures? If the broadening is due to the thermal energy, the generally accepted method consists of taking the intersection point of the

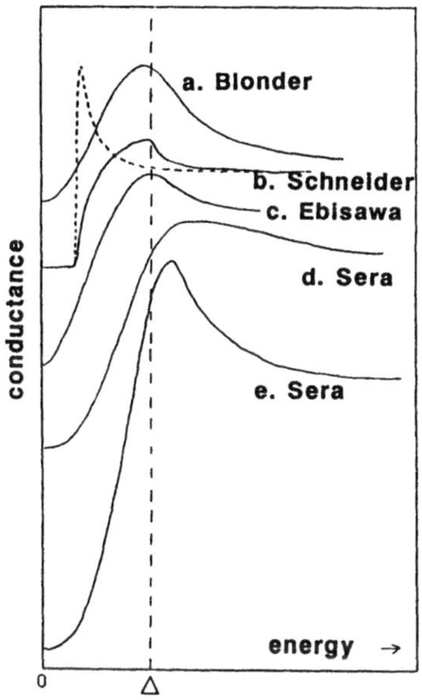

Fig. 3 σ-V characteristics of Various models which provide the broadened gap structures. Models a, d and e assume the BCS density of states, while model b and c assume the anisotropic energy gap.

conductance curves from when the the junction is NIN like and when the junction is NIS like[28]. However, the observed broadening seems to stem from some other mechanism(s)[28]. In Figure 3 are conductance-voltage (σ-V) characteristics based on several models depicting the broadened gap like structures. For the simple purpose of crude comparison, the details of each model and parameter are neglected here, but the curves capture the main features of each model. a is a characteristic observed in typical NS point contact junctions[29]; b gives the anisotropic tunneling characteristic on anisotropic gap (the dotted line is for direction parallel to the c-axis); c shows the characteristic of directionally-averaged tunneling on anisotropic gap; d considers the non-equilibrium effect due to excessive injection of quasiparticles into superconductors[32,33]; e is from the model that non-uniform distribution of the gap exists at the junction surface[33]. Models b, c and perhaps e predict a zero conductivity at V=0 and T=0, while models a and d give a finite conductivity at V=0 even at low temperature. In reality, many junctions indicate the presence of rather large ohmic type leakage currents. Some detailed studies have shown that for characteristics with relatively wide peaks, model d with lifetime broadening energy Γ best fits the experimental data[33,34]. In comparing the similar models, there is a report which pointed out that for characteristics with relatively sharp peaks, model e which takes into account the distribution of the gap is more suited than model d[35]. Models b and c assume the anisotropic gap, but the detailed characteristics depend on the types of anisotropy. The difference in the sharpness of the peak structures in the σ-V characteristics of epitaxial samples (sharper) and polycrystal samples (broader) also suggest the influence of the anisotropy[36]. The questions to ask are: how significant or substantial are the broadened gap structures? Are they really signifying the density of states? In some of the La-Sr-Cu-O intergranular SIS junctions, sharp gap structures were observed[37]. On the other hand, a vacuum STM tunneling on a Bi-Sr-Ca-Cu-O sample was reported to be accompanied by broadened peaks[27]. Models a, d and e assume a BCS density of states, while models b and c assume an anisotropic energy gap. At present, the results of the tunneling experiments do not give completely reproducible σ-V spectroscopic characteristics to determine the most probable model or the exact gap value γ. So, perhaps it is safe and reasonable to assign relatively large errors to the estimations of the gap energy at this point. Setting the peak value as the upper limit as indicated by models a, b and c, and setting the lower bound as defined by the model d, tunneling data for various oxide superconductors could be reduced to Fig. 4.

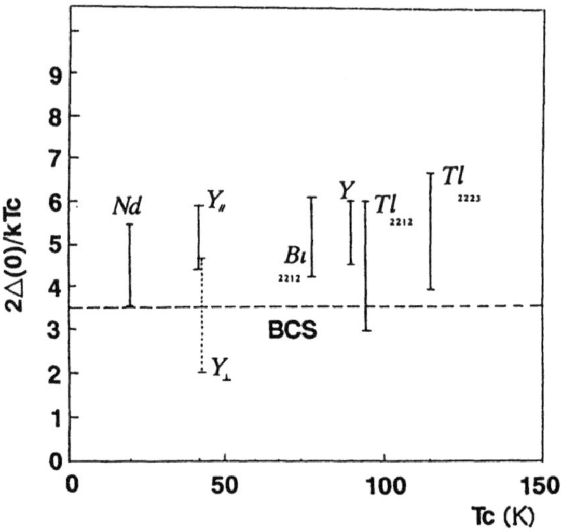

Fig. 4 Energy gap γ of Nd-Ce-Cu-O, Y-Ba-Cu-O, Bi-Sr-Ca-Cu-O and Tl-Ba-Ca-Cu-O. Relatively large errors are assigned to accord the various proposed models.

To summarize the above discussions, several conclusions can be stated. We still do not have a completely reliable tunneling method which gives the reproducible tunneling spectroscopy. So the quasiparticle density of states of these material is not quite established yet. Nevertheless, the size of the gap structures are quite reproducible for many of the oxide materials. The exact evaluation of the gap is model dependent, but it could be estimated that $2\Delta(0)\simeq 4\sim 6k_BT_c$ for Y-Ba-Cu-O, Bi-Sr-Ca-Cu-O, Tl-Ba-Ca-Cu-O and Nd-Ce-Cu-O compounds. Such values are consistent with those found by others[39]. The anisotropy in the gap of Y-Ba-Cu-O and Bi-Sr-Ca-Cu-O was found to be about 1.5~2. This anisotropy was determined by several methods beside the tunneling spectroscopy. The linearity between the gap $\Delta(0)$ and T_c was established in a wide critical temperature range.

Acknowledgments We are in debts to J. Fujita, T. Yoshitake, T. Miura, Y. Shimakawa, T. Manako, Y. Kubo, H. Igarashi for the supply of the various samples as well as helpful discussions. We would also like to thank Y. Wada and K. Utsumi for the support of this work.

References

[1] I. Giaever, Phys. Rev. Lett. 5, 147 and 464 (1960)
[2] M. Tinkham, "Introduction to superconductivity," p.46, McGrow Hill, New York (1975)
[3] W.L. McMillan, J.M. Rowell, Phys. Rev. Lett, 14, 108 (1965)
[4] G. Rickayzen, "Superconductivity" edited by R.D. Parks, P.1 Marcel Dekker, Inc., New York (1969)
[5] K.E. Gray, M.E. Hawley, E.R. Moog, "Novel Superconductivity" edited by S.A. Wolf and V.Z. Kresin, P.611, Plenum Press, New York (1987)
[6] A. Barone, Physica C, 153-155, 1712 (1988)
[7] T. Ekino, J. Akimitsu, Jpn. J. Appl. Phys. 26, L452 (1987)
[8] J.R. Kirtley, C.C. Tsuei, S.I. Park, C.C. Chi, J. Rozen, M.W. Shafer, Phys. Rev. B35, 7216 (1987)
[9] I. Iguchi, H. Watanabe, Y. Kasai, T. Mochiku, A. Sugishita, E. Yamaka, Jpn. J. Appl. Phys., 26, L645 (1987)
[10] J. Moreland, A.F. Clark, H.C. Ku, R.N. Shelton, Cryogenics 27, 227 (1987)
[11] J.B. Barner, S.T. Ruggiero, Phys. Rev. Lett. 59, 807 (1987)
[12] Ref.[9] and private communication, I. Iguchi

[13] J.S. Tsai, I. Takeuchi, J. Fujita, S. Miura, T. Terashima, Y. Bando, K. Iijima, K. Yamamoto, Physica C, 157, 537 (1989)
[14] M. Lee, M. Naito, A. Kapitulnik, M.R. Beasley, Solid State Comm. 70, 449 (1989)
[15] M. Lee, D.B. Mitzi, A. Kapitulnik, M.R. Beasley, Phys. Rev. B39, 801 (1989)
[16] B.L. Blackford, R.H. March, Phys. Rev. 186, 397 (1969)
[17] G. Binnig, H.E. Hoenig, Z. Physik, B32, 23 (1978)
[18] M.E. Hawley, K.E. Gray, B.D. Terris, H.H. Wang, K.D. Carlson, J.M. Williams, Phys. Rev. Lett. 57, 629 (1986)
[19] J.S. Tsai, I. Takeuchi, J. Fujita, T. Yoshitake, S. Miura, S. Tanaka, T. Terashima, Y. Bando, K. Iijima, K. Yamamoto, Physica C, 153-155, 1385 (1988)
[20] A. Zettl, A. Behrooz, G. Briceno, W.N. Creager, M.F. Crommie, S. Hoen, P. Pinsukanjana, "Mechanisms of High Temperature Superconductivity," edited by K. Kamimura, A. Oshiyama, Springer-Verlag, 249 (1989)
[21] T. Ekino, J. Akimitsu, preprint
[22] R.T. Collins, Z. Schlesinger, F. Holtzberg, C. Feild, preprint
[23] S.L. Cooper, F. Slakey, M.V. Klein, J.P. Rice, E.D. Bukowski, D.M. Ginsberg, Phys. Rev. B38, 11934 (1988)
[24] C.P. Slichter, to be published in Proceeding of "IBM Japan International Symposium on Strong Correlation and Superconductivity," Mt. Fuji, Japan (1989)
[25] J.R. Kirtley, R.T. Collins. Z. Schlesinger, W.J. Gallager, R.L. Sandstrom, T.R. Dinger, D.A. Chance, Phys. Rev. B35, 8845 (1987)
[26] J. Moreland, A.F. Clark, Physica C, 153-155, 1383 (1988)
[27] T. Hasegawa, H. Suzuki, S. Yaegashi, H. Takagi, K. Kishio, S. Uchida,K. Kitazawa, K. Fucki, Jpn. J. Appl. Phys. 28, L179 (1989)
[28] I.Giaever, Phys. Rev. Letts, 5, 147 (1960)
[29] G.E. Blonder, M. Tinkham, Phys. Rev. B27, 112 (1983)
[30] T. Schneider, H. De Raedt, M. Frick, preprint
[31] H. Ebisawa, Y. Isawa, S. Maekawa, Jpn. J. Appl. Phys. 26, L992 (1987)
[32] R.C. Dynes, J.P. Garno, G.B. Hertel, T.P. Orland, Phys. Rev. Lett, 53, 2437 (1984)
[33] M. Sera, S. Shamoto, M. Sato, Solid State Comm. 65, 997 (1988)
[34] Ref.[33] or K.W. Ng, S. Pan, A.L. de Lozanne, A.J. Panson, J. Talvacchio, Jpn. J. Appl. Phys. Supplement 26-3, 993 (1987)
[35] J.R. Kirtley, C.C. Tsuei, S.I. Park, C.C. Chi, J. Rozen, M.W. Shafer, W.J. Gallager, R.L. Sandstrom, T.R. Dinger D.A. Chance, Jpn. J. Appl. Phys. Supplement 26-3, 997 (1987)
[36] I. Takeuchi, J.S. Tsai, Y. Shimakawa, T. Manako, Y. Kubo, Physica C, 158, 83 (1989)
[37] P.J. van Bentum, L.E.C. van de Leemput, L.W.M. Schreurs, P.A.A. Teunissen, H. van Kempen, Phys. Rev. B36, 843 (1987)
[38] G. Deutscher, K.A. Müller, Phys. Rev. Lett. 59 (1987)
[39] See review of recent experiments in Ref.[15] and Ref.[8]

Optical Absorption in $La_{2-x}Sr_xCuO_4$ Epitaxial Thin Films

M. Suzuki

NTT Opto-Electronics Laboratories, 162 Tokai, Ibaraki 319-11, Japan

An experimental study on optical absorption is described for $La_{2-x}Sr_xCuO_4$ ($x = 0 \sim 0.36$) c-axis oriented epitaxial films. In addition to the fundamental absorption corresponding to the energy gap of 2 eV, an anomalous absorption band is seen in the optical transmission spectra. This absorption, the intensity of which increases with Sr doping, implies Sr doping induces electronic states within the gap of La_2CuO_4. With further Sr doping, the density of states forms a band about 1 eV wide and centered at about 0.5 eV above the top of the original valence band. This emerging density of states within the gap is totally at variance with the conventional rigid band model, and seems to be inherent to the high T_c superconductors.

1. Introduction

Since the discovery of high T_c superconductivity in 1986 by Bednorz and Müller [1], the $La_{2-x}Sr_xCuO_4$ system has been the subject of extensive experimental and theoretical investigations for the following reasons. The $La_{2-x}Sr_xCuO_4$ system changes its properties from being an insulator to a superconducting metal, and then to a nonsuperconducting metal by successive Sr doping. This systematic change in electronic and magnetic properties together with its simple cubic crystal structure makes this material the prototype for understanding high T_c superconductivity in copper oxide superconductors.

Experimentally, insight into the electronic structure is obtained by means of, *e.g.*, electron energy loss spectroscopy (EELS), x-ray photoemission spectroscopy (XPS) or optical spectroscopy. Recent experimental studies using the former two methods [2,3] revealed that the electronic states at the Fermi level come from O 2p states, providing evidence that conducting carriers are holes at oxygen sites in nature. Optical spectroscopy, on the other hand, also provides important information about the band structure near the Fermi level. This paper describes optical absorption of $La_{2-x}Sr_xCuO_4$, with an aim to clarify the change of electronic band structure, when this system changes its property from a semiconductor to metallic superconductor by Sr doping.

2. Experimental

Samples used are $La_{2-x}Sr_xCuO_4$ epitaxial thin films grown on $SrTiO_3$ (100) face substrates with the c-axis normal to the substrate by rf magnetron sputtering with a single target [4]. Epitaxial growth of these films was confirmed by the x-ray diffraction and reflection electron diffraction analysis. The x-ray diffraction patterns showed that other phases are, even if they exist, less than a part per 10^3. The ion channeling analysis using 7.5 MeV α particles showed that the minimum back-scattering yield from La ions is 10 %, indicating that the films are of nearly single crystalline quality [5]. The Rutherford

Springer Series in Solid-State Sciences, Vol. 89 Strong Correlation and Superconductivity
Editors: H. Fukuyama · S. Maekawa · A.P. Malozemoff

backscattering analysis showed the tendency that for $x < 0.2$ Sr concentration of the films nearly coincides with the nominal one, while for $x > 0.2$ it becomes smaller. In what follows an x value indicates nominal composition, although it is subjected to an error to a small extent for $x > 0.2$. These films were annealed in air or in flowing oxygen in order to minimize remaining oxygen vacancies. Film thickness measured with a stylus profilometer ranged from 3000 to 4000 Å and the surface roughness is below 100 Å. Optical transmission and reflection spectra were measured with a monochromater in the wavelength range from 0.4 to 4.5 μm. The reflectivity was determined with an evaporated silver thin film.

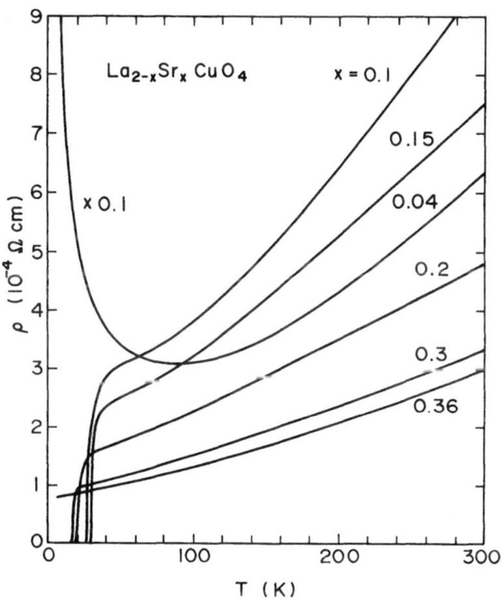

Fig. 1 Temperature dependence of resistivity along the CuO planes for the $La_{2-x}Sr_xCuO_4$ (001) epitaxial thin films with x from 0.04 to 0.36.

Figure 1 shows the temperature dependence of resistivity $\rho(T)$ as a function of x. For $0.1 \leq x \leq 0.3$, superconductivity is observed, with the highest T_c of 29 K at $x = 0.15$, as observed in previous works [6,7]. At $x = 0.36$, ρ becomes lowest, but no sign of superconductivity was observed down to 4.2 K. For $x = 0.1$, the temperature dependence is metallic over the temperature range measured, showing absence of upturn, which is often observed in sintered ceramic specimens with the same x or below 0.1. This indicates that the amount of remaining oxygen vacancies is sufficiently reduced. For $x = 0.04$, the temperature dependence is metallic at temperatures above 90 K, while it is semiconductive below 90 K. The further examination of this semiconducting temperature dependece showed that the relation $\rho(T) = \rho_0 \exp(T_0/T^{1/3})$ holds in the lower temperature range, indicating that the resistivity is governed by the two-dimensional (2D) variable hopping conduction in the CuO planes [4].

3. Optical Properties

3.1 Reflectance

Room temperature reflectance spectra for $La_{2-x}Sr_xCuO_4$ epitaxial thin films are shown in Fig. 2 as a function of x. The results are basically the same as previously reported by Tajima *et al.* [8] except for a little enhanced structure in the shorter than 1.5 μm wavelenghth range. This difference probably comes from the form of thin films and from the total alignment of the CuO planes in the electrical field direction. The structure seen in the shorter than 1.4 μm range reflects the electronic structure inherent in the $La_{2-x}Sr_xCuO_4$ system. This structure corresponds to the interband transition, which appears much clearer in the transmission spectra. It should be noted that this structure was also recognized in the spectra observed in the sintered polycrystalline specimens [9-11].

The solid lines in Fig. 2 are the Drude fits to the experimental results using the formula,

$$\epsilon(\omega) = \epsilon_\infty\left(1 - \frac{\omega_p^2}{\omega^2 + i\gamma\omega}\right). \tag{1}$$

As clearly seen , the fits are reasonably good only in the wavelength region greater than 1.5μm. The value of the plasma frequency ω_p of 0.87 eV obtained by this fitting is basically the same as obtained by Orenstein *et al.* [9] and by Tajima *et al.* [8], and shows no significantly systematic increase with the increase in carrier density. The optical scatter-

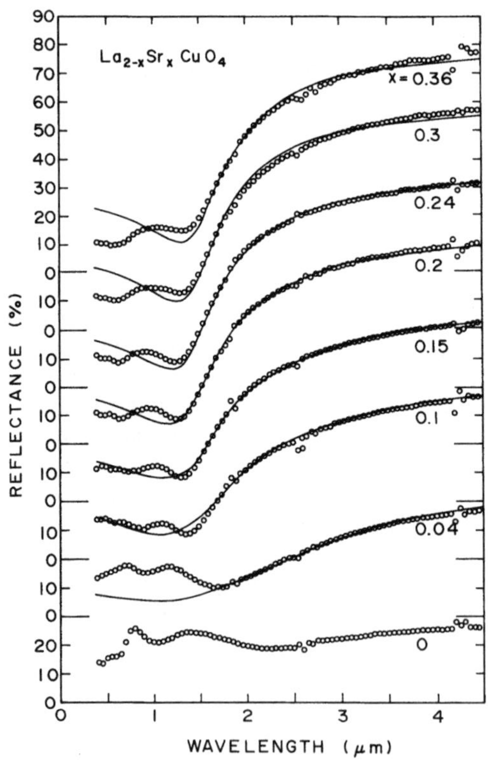

Fig. 2 Reflectance spectra for the $La_{2-x}Sr_xCuO_4$ (001) epitaxial thin films with x from 0 to 0.36. Solid lines are the Drude fits. The fitting parameters (ω_p[eV], γ[eV], ϵ_∞) are (0.77, 0.77, 3.2), (0.85, 0.58, 5.1), (0.87, 0.48, 5.1), (0.87, 0.38, 5.6), (0.86, 0.34, 6.1), (0.84, 0.32, 8), (0.82, 0.32, 8.5) for x from 0.04 to 0.36 in order of increasing x.

ing time ($\tau = \gamma^{-1}$) increases systematically with x up to 1.3×10^{-14} s at $x = 0.36$, which is consistent with the change in the electrical properties from an insulator to a metal.

3.2 Optical absorption

Figure 3 shows the optical transmittance $T(\lambda)$ in the λ range from 0.4 to 4.5 μm for the $La_{2-x}Sr_xCuO_4$ epitaxial films. In contrast to the modest change in reflectance spectra with x, the transmittance spectrum shows a dramatic change by Sr doping, as clearly seen in Fig. 3. The most prominent change is the development of an absorption peak centered at 0.85 μm (1.46 eV). For $x = 0$, where the system is a semiconductor, the spectrum shows only an absorption corresponding to an energy gap of about 2 eV. With the systematic doping with Sr, a decrease in transmittance is seen in the infrared region ($\lambda > 2\,\mu$m) and in a λ region centered at $0.85\,\mu$m. The former absorption corresponds to the intraband absorption of free carriers generated by Sr doping, while the latter indicates an interband electronic transition, which is not immediately evident within the electronic structure of La_2CuO_4. The strength of the latter absorption exhibits a systematic increase with the Sr doping. This behavior is quite unusual when taking into accout a transmission spectrum of an ordinary oxide superconductor such as $BaPb_{1-x}Bi_xO_3$, where the transmission spectrum shows only a fundamental absorption edge [12].

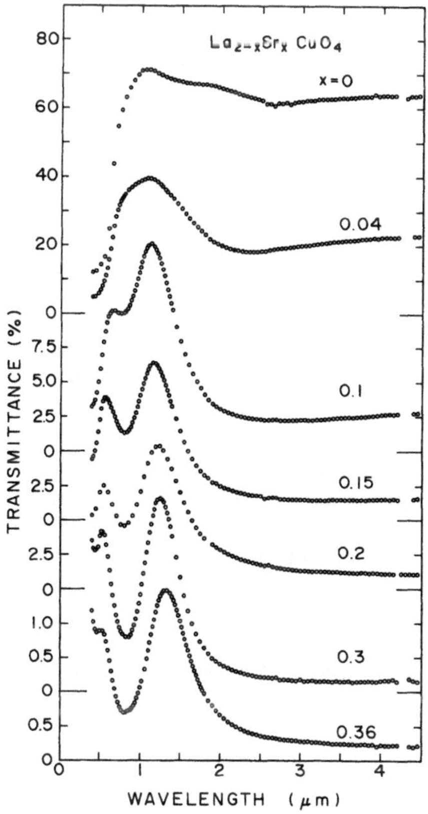

Fig. 3 Transmittance spectra for the $La_{2-x}Sr_xCuO_4$ (001) epitaxial thin films with x from 0 to 0.36. Scales are shifted by appropriate values.

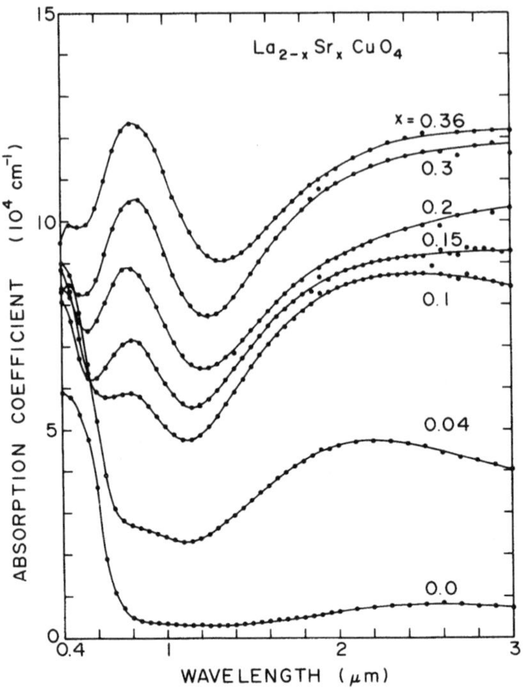

Fig. 4 Optical absorption coefficient α for the $La_{2-x}Sr_xCuO_4$ (001) epitaxial films with x from 0 to 0.36.

Figure 4 shows absorption coefficient $\alpha(\omega)$ for the same x values, calculated from reflectivity $R(\hbar\omega)$ and $T(\hbar\omega)$ values in Fig. 2 and Fig. 3 using the approximate formula, $T = (1 - R)e^{-\alpha d}$. Approximately, α is proportional to the joint density of states $N(E)N(E - \hbar\omega)$, and reflects the density of states near the Fermi level E_F. If it is assumed that there is no particular structure in the upper conduction band (the final states of the electronic transition), then α roughlty reflects the DOS below E_F, or the valence band structure. Bearing this in mind, it is evident that Sr doping induces a development of DOS within the energy gap of semiconducting La_2CuO_4. Furthermore, the DOS increases in proportion to x, forming a band about 1 eV wide for the x range where the system is superconductive. This behavior is totally at variance with the rigid band model, where the Fermi level is expected to lie within the original valence band and shift to lower energies in accordance with the doping level.

Since absorption coefficient is proportional to the frequency-dependent conductivity $\sigma(\omega)$, the present results are compared with other experimental results of $\sigma(\omega)$, obtained from reflectivity data based on the Kramers-Kronig analysis. A small peak corresponding to the absorption peak at 0.85 eV is recognized in the $\sigma(\omega)$ obtained by Herr *et al.* [10] or by Etemad *et al.* [11]. In the electron energy loss spectroscopy, Nücker *et al.* [3] observed a peak just below E_F, which is again compared with the spectrum in Fig. 4. Also Ginder *et al.* [13] observed photoinduced electronic states 0.5 – 0.6 eV above the valence band top. The energy levels of these states are again compared with the optical absorption results and both the experiments seem to be closely related to each other [13].

4. Discussion

The systematic change in $\alpha(\omega)$ in Fig. 4 provides useful information about the band structure in the $La_{2-x}Sr_xCuO_4$ system. For $x = 0$, the system is a charge-transfer insulator with an energy gap between the upper Hubbard band and the O 2p band. The optical absorption results imply that doping with Sr generates electronic states within the band gap of La_2CuO_4. These states are considered to be O2p states coming from original valence band, as Nücker *et al* [3] indicated. When the doping level is low, these states are highly localized, which is well reflected by the transport experiment results, where the conduction is dominated by the 2D variable hopping conduction at low temperatures for $x < 0.1$. When the doping level becomes high and the system becomes superconductive, these electronic states form a band, the center of which lies about 0.5 eV above the original valence band top. Contrary to the conventional rigid band model, the Fermi level lies in this doping-induced band. As the doping level increases, the Fermi level shifts to lower energies and, at the same time, the density of states increases nearly proportioanlly to the Sr concentration.

If we view the optical reflectivity spectra in Fig. 2 from the standpoint of the electronic structure mentioned above, we can deduce further useful information about the palsma frequency. In previous works [4,8], ω_p shows an extraordinarily small change when the doping level is varied. This eventually led to a breakdown of the simple Drude model in high T_c superconductors, since in that model ω_p^2 is proportional to the carrier density n. For example, Etemad *et al.* [11] ascribed this behavior to an 0.5 eV excitation responsible for superconductivity. However, if we take an interband excitation of the kind shown in Fig. 4 into account, it is immediately evident that the simple Drude model without any excitation is insufficient to explain the characteristics in Fig. 2. A more appropriate model is the one which includes this interband excitation as an additional oscillator, the strength of which increases with x. Then we take a dielectric function having another oscillator at 0.85 μm, after the method by Tajima *et al* [14] or by Orenstein *et al.* [9], as expressed in the following.

$$\epsilon(\omega) = \epsilon_\infty\left[1 + \omega_p^2\left(\frac{f_G}{\omega_G^2 - \omega^2 - i\omega\gamma_G} - \frac{1 - f_G}{\omega^2 + i\omega\gamma_p}\right)\right], \tag{2}$$

$$\omega_p^2 = 4\pi ne^2/\epsilon_\infty m^*, \tag{3}$$

where ω_G is the excitation energy of about 1.4 eV associated with the emerged density of states within the energy gap, γ_G is roughly the width of the band formed within the gap formed by hole doping, and f_G is the fraction of the electronic oscillator strength. Figure 5 shows the fits to the data using Eq. (2), where the circles are again the experimental data. For $x > 0.2$ the fits are reasonably good, while for $x < 0.2$ the fit starts to deviate as x decreases, probably due to the existence of the band gap excitation of 2 eV. Although adding another oscillator might improve fits to the data, espicially for smaller x, this simple model itself is quite useful to reveal the systematic change in the electronic properties which otherwise remain almost the same irrespecitve of doping levels.

Figures 6 (a) and (b) show the composition dependence of ω_p, ω_G, $\gamma_p(= 1/\tau)$, and γ_G. Contrary to the previous data, ω_p increases with x, and the apparent insensitivity of ω_p to x, as reported earlier [4,8], can be ascribed to the screened plasma frequency, which shifts to lower energies in the presence of an additional oscillator, as indicated by Orenstein *et al.*[9]. The inset in Fig. 6(a) plots ω_p^2 vs x, showing roughly linear dependence, which is quite reasonable in view of the relation of Eq.(3). The scatter in these data can be ascribed to the deviation of the Sr composition from the nominal one for

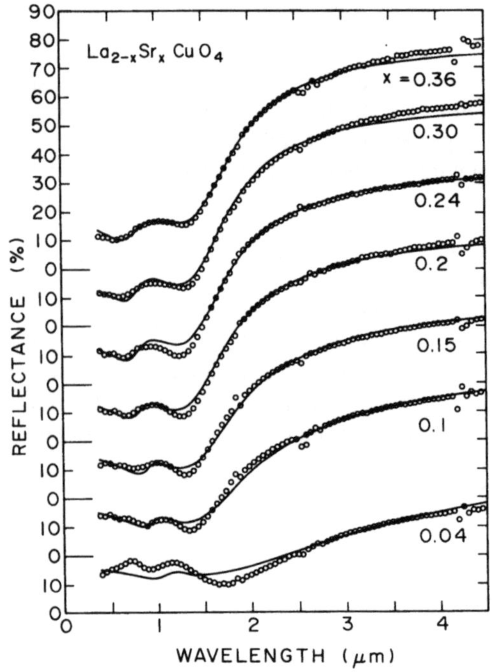

Fig. 5 Reflectance spectra for the $La_{2-x}Sr_xCuO_4$ (001) epitaxial thin films with x from 0 to 0.36. Circles are the same data as in Fig. 2. Solid lines are the fits using Eq. (2). The fitting parameters are shown in Fig. 6, except for (ϵ_∞, f_G), which are (5.3, 0.09), (5.3, 0.11), (5.1, 0.15), (4.9, 0.24), (5.0, 0.32), (5.2, 0.36), (5.5, 0.40) for x from 0.04 to 0.36 in order of increasing x.

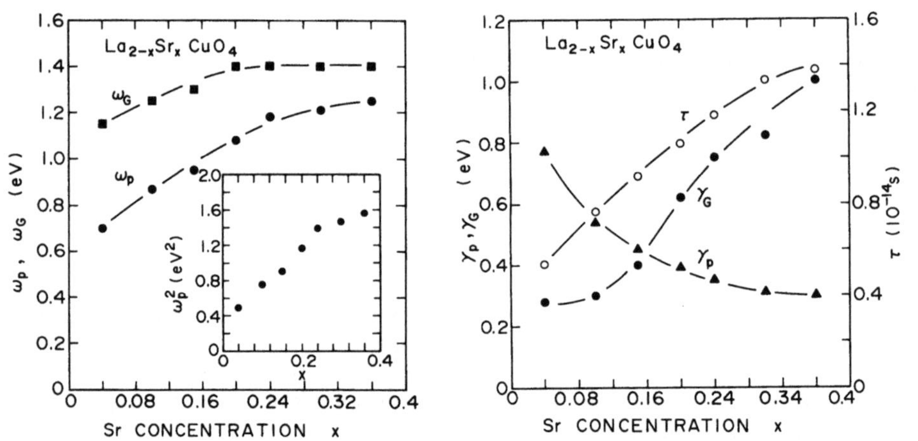

Fig. 6 (a) (left) Sr concentration dependence of ω_p and ω_G derived by the fitting using Eqs.(2) and (3). The inset shows ω_p^2 vs Sr concentration. (b) (right) Sr concentration dependence of $\tau = 1/gamma_p$ and γ_G derived by the fitting in Fig. 5.

larger x's, and to the poor fitting for smaller x's. When we notice that γ_G roughly reflects the width of the density of states at E_F, the x dependence of γ_G in Fig. 6(b) shows that the band forms a sharp peak at E_F at small x and then its width increases significantly as x increases. This also implies that for small x the electronic states generated by hole doping is almost localized, which agrees with the transport properties described in the former section.

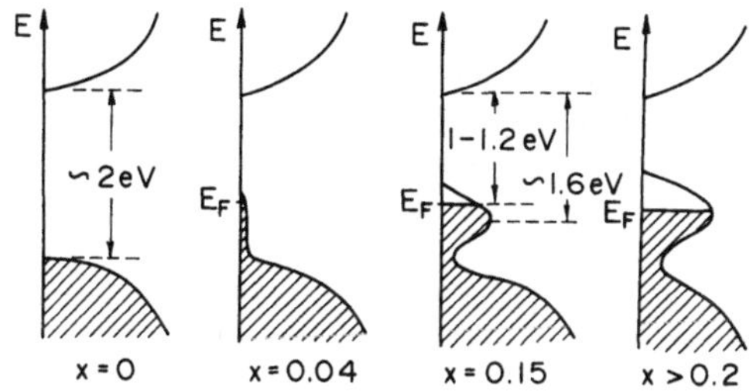

Fig. 7 Schematic illustration of the change in the band diagram in the Reflectance spectra for the $La_{2-x}Sr_xCu$ O_4 system inferred from the optical absorption experiments.

Figure 7 shows a schematic band diagram for the $La_{2-x}Sr_xCuO_4$ system when it changes from an insulator to a superconductor, and further to a nonsuperconducting metal. When Sr is doped into this system, electronic states emerge at about 0.5 eV above the valence band top. At a small doping level, these electronic states are almost localized. Further doping with Sr increases the density of states at E_F, forming a band about 1 eV wide centered at 0.5 eV above the original valence band top. The Fermi level lies within this band, giving rise to the hole conduction. With the further doping, it is likely that the Fermi level almost reaches the midband, as reflected by the x dependece of the Hall effect measurements [4].

5. Conclusion

From the optical absorption measurements on $La_{2-x}Sr_xCuO_4$ epitaxial thin films, it is found that Sr doping generates electronic states within the band gap at 0.5 eV above the valence band top. The Fermi level is located withn these levels. These electronic states develops with the further Sr doping, forming a band about 1 eV wide and centered at 0.5 eV with the Fermi level almost unchanged. Analysis of the reflectivity data using a dielectric function having the Drude term and an oscillator associated with the electronic transition involving these electronic states revealed that with increasing x the density of states and its width increases systematically. The plasma frequency also increases with x, showing reasonable agreement with the increase in carrier density by Sr doping.

Acknowledgements

The author would like to acknowledge useful discussions with Profs. S. Uchida and S. Tanaka at the University of Tokyo, and valuable comments by Prof. H. Fukuyama at the University of Tokyo, Profs. M. Tachiki and H. Katayama-Yoshida at Tohoku University. He is also obliged to his colleagues at NTT Laboratories for discussions and encouragement.

References

[1] J. G. Bednorz and K. A. Müller, Z. Phys. B **64**, 189 (1986).
[2] A. Fujimori, E. Takayama-Muromachi, Y. Uchida, and B. Okai, Phys. Rev. B **35**, 8814 (1987); A. Fujimori, E. Takayama-Muromachi, Y. Uchida, Solid State Commun. **63**, 857 (1987).
[3] N. Nücker, J. Fink, J. C. Fuggle, P. J. Duhram, and W. M. Temmerman, Phys. Rev. B **37**, 5158 (1988).

[4] M. Suzuki, Phys. Rev. B **39**, 2312 (1989).
[5] H. Yonezawa, Y. Hidaka, M. Suzuki, and T. Shigematsu, in Proceedings of the Materials Research Society Spring Meeting, San Diego, California, 1989 (to be published).
[6] J. M. Tarascon, L. H. Greene, W. R. McKinnon, G. W. Hull, and T. H. Geballe, Nature (London) **235**, 1373 (1987).
[7] S. Uchida, Physica B **148**, 185 (1987).
[8] S. Tajima, T. Nakahashi, S. Uchida, and S. Tanaka, Physica C **156**, 90 (1988); S. Tajima, S. Uchida, S. Tanaka, S. Kanbe, K. Kitazawa, and K. Fueki, Jpn. J. Appl. Phys. **26**, L432 (1987).
[9] J. Orenstein, G. A. Thomas, D. H. Rapkine, C. G. Bethea, B. F. Levine, R. J. Cava, E. A. Rietman, and D. W. Johnson, Jr., Phys. Rev. B **36**, 729 (1987).
[10] S. L. Herr, K. Kamarás, C. D. Porter, M. G. Doss, D. B. Tanner, D. A. Bonn, J. E. Greedan, C. V. Stager, and T. Timusk, Phys. Rev. B **36**, 733 (1987).
[11] S. Etemad, D. E. Aspnes, P. Barboux, G. W. Hull, M. K. Kelly, J. M. Tarascon, R. Thompson, S. L. Herr, K. Kamarás, C. D. Porter, and D. B. Tanner, Proc. Materials Research Society, Boston, 1987.
[12] M. Suzuki, Y. Enomoto, T. Murakami, and T. Inamura, J. Appl. Phys. **53**, 1622 (1982).
[13] J. M. Ginder, M. G. Roe, Y. Song, R. P. McCall, J. R. Gaines, and E. Ehrenfreunt, Phys. Rev. B **37**, 7506 (1988), The author is much obliged to H. Katayama-Yoshida for his pointing out the important relationship between this reference and the present experiment.
[14] S. Tajima, S. Uchida, A. Masaki, H. Takagi, K. Kitazawa, S. Tanaka, and A. Katsui, Phys. Rev. B **32**, 6302 (1985).

Infrared Studies of Oxide Superconductors in the Normal and Superconducting States

R.T. Collins[1], *Z. Schlesinger*[1], *F. Holtzberg*[1], *C. Feild*[1], *G. Koren*[1], *A. Gupta*[1], *D.G. Hinks*[2], *A.W. Mitchell*[2], *Y. Zheng*[2], *and B. Dabrowski*[2]

[1]IBM T.J. Watson Research Center, Yorktown Heights, NY 10598, USA
[2]Materials Science Division, Argonne National Laboratory, Argonne, IL 60439, USA

Abstract. We present a summary of the infrared properties of the cubic oxide superconductor $Ba_{1-x}K_xBiO_3$ and the layered oxide superconductor $Y_1Ba_2Cu_3O_7$ in the normal and superconducting states. The normal state conductivity of $Ba_{0.6}K_{0.4}BiO_3$ exhibits a broad peak near 5000cm^{-1} which is similar to a peak previously observed in $BaPb_{1-x}Bi_xO_3$. The presence of this feature in both systems may provide an important clue to the mechanism of superconductivity in the BiO_3 superconductors. The normal state conductivity of $Y_1Ba_2Cu_3O_7$ is characterized by a sharp temperature dependent peak near the origin and a broad mid-infrared background. This background has been interpreted both as evidence for low lying charge excitations and in terms of strong coupling between carriers and excitations of magnetic origin. In the superconducting state of $Ba_{0.6}K_{0.4}BiO_3$ we obtain an energy gap of conventional magnitude. In contrast we find an extremely unconventional energy scale of 7-8kT_c for the a-b plane of $Y_1Ba_2Cu_3O_7$ and also find evidence for substantial gap anisotropy between the a-b plane and c-axis.

1. Introduction

The first reports of the infrared properties of high T_c oxide superconductors closely followed the discovery of these unusual compounds.[1] To a great extent, the IR work was stimulated by the success of infrared measurements in uncovering the properties of conventional superconductors. Here, infrared provided one of the first direct measurements of the energy gap, and was shown to yield information about the coupling spectrum of the phonons ($\alpha^2F(\omega)$) which mediate the superconductivity. The initial studies of the high T_c superconductors were complicated by materials problems and by the large anisotropies present in the Cu-O systems which made it difficult to determine intrinsic a.c. properties. The availability of high quality single phase samples, oriented films, and single crystals has helped to resolve many of the early controversies and is currently leading to a reliable characterization of the dynamics of these materials.

Here we summarize our infrared studies of $Ba_{1-x}K_xBiO_3$ and $Y_1Ba_2Cu_3O_7$ which are representative of the cubic and layered perovskite superconductors, respectively. First we discuss the infrared reflectivity and conductivity in the normal state where a key question remains determining the origin of the mid-infrared conductivity and its relationship to the strength of the coupling and the existence of low lying electronic excitations. We then discuss the changes in the conductivity which occur at the superconducting transition. For cubic $Ba_{1-x}K_xBiO_3$ we observe an energy gap of conventional size, consistent with moderate coupling ($\lambda \lesssim 1$). Similar measurements on $Y_1Ba_2Cu_3O_7$ suggest the presence of substantial gap anisotropy and provide evidence for an extremely large energy scale (7-8KT_c) associated with the Cu-O planes.

2. Normal State Conductivity

We begin with the normal state properties of $Ba_{1-x}K_xBiO_3$ which is a variant of the previously studied $BaPb_{1-x}Bi_xO_3$ alloy system ($T_c \lesssim 11$K) and in which transition temperatures in excess of 30K have recently been demonstrated.[2] Figure 1 presents the reflectivity from 75-15000cm^{-1} for a ceramic $Ba_{0.6}K_{0.4}BiO_3$ sample[3] with $T_c \simeq$ 26K along with the real part of the conductivity $\sigma_1(\omega)$ and the real part of the dielectric constant $\varepsilon_1(\omega)$ obtained from a Kramers-Kronig transform of the reflectivity. In the inset, the reflectivity extending to 40000cm^{-1} shows a minimum in the red-yellow part of the spectrum which is responsible for the bluish color of the sample. The conductivity spectrum shows a broad peak at $\simeq$ 5000cm^{-1} which is reminiscent of spectra previously obtained for $BaPb_{0.8}Bi_{0.2}O_3$ by Tajima et al.[4] in which a broad maximum at $\simeq$5000cm^{-1} was also observed. Based on a study of the evolution of the 5000cm^{-1} peak throughout the $BaPb_{1-x}Bi_xO_3$ alloy system, Tajima et al. associated this feature with the persistence of local charge-density-wave order into the metallic region of $BaPb_{1-x}Bi_xO_3$. Still others have taken this feature as evidence for low lying charge excitations. The observation that the maximum is also present in $Ba_{1-x}K_xBiO_3$ is potentially significant both for understanding the normal state metal insulator transition and the superconducting mechanism in the BiO_3 superconductors. We note that the fraction of the oscillator strength below 15000cm^{-1} consumed by the mid-infrared mode is about the same for both $BaPb_{0.8}Bi_{0.2}O_3$ and $Ba_{0.6}K_{0.4}BiO_3$ even though the later is closer to half filling which supports the view that doping on the Ba site pushes the metal-insulator transition closer to half filling than doping on the Bi site.

The nearly cubic structures of the Bi-O superconductors makes the interpretation of measurements on ceramic samples relatively straightforward in comparison to the highly anisotropic Cu-O systems. Many of the early studies of $La_{2-x}Sr_xCuO_4$ and $Y_1Ba_2Cu_3O_7$ were based on ceramic samples and the conductivities reported in these studies were often dominated by peaks near 0.5eV that seemed to resemble the broad mode seen in $Ba_{1-x}K_xBiO_3$ and $BaPb_{1-x}Bi_xO_3$. Although subsequent work on oriented films and crystals did not show this feature,[5,6] the question of how to interpret the mid-infrared conductivity of the Cu-O superconductors remains a central one.

In Fig. 2 we show the reflectivity and conductivity at 250K and 100K for the a-b plane of a twinned crystal of $Y_1Ba_2Cu_3O_7$ in which $T_c \simeq 92$K and the transition width was less than 0.5K. The conductivity exhibits a sharp temperature dependent peak at the origin and then remains finite throughout the entire mid-infrared. The peak at the origin is closely associated with the d.c. transport properties of the system. Within experimental error, the conductivity we obtain from an extrapolation of our data to the origin is consistent with the d.c. conductivities reported for the a-b plane of $Y_1Ba_2Cu_3O_7$ and with the linear temperature dependence of the d.c. resistivity which is observed for $T > T_c$. From the standpoint of understanding the mechanism

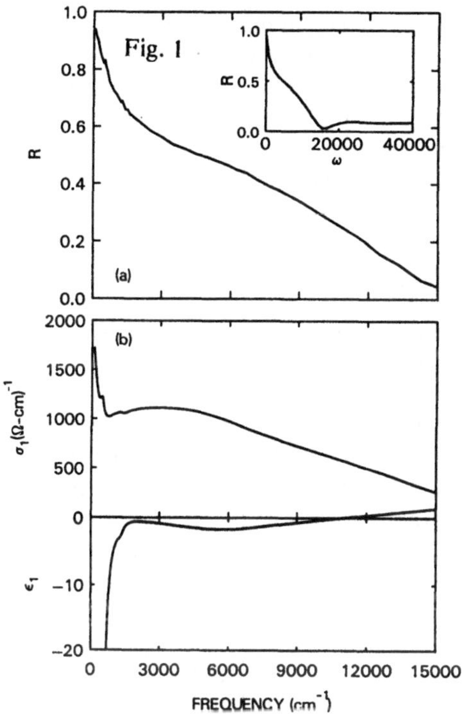

Fig. 1 a) The normal state reflectivity and b) the conductivity and dielectric function for $Ba_{0.6}K_{0.4}BiO_3$. The insert shows the reflectivity up to 40000cm^{-1}.

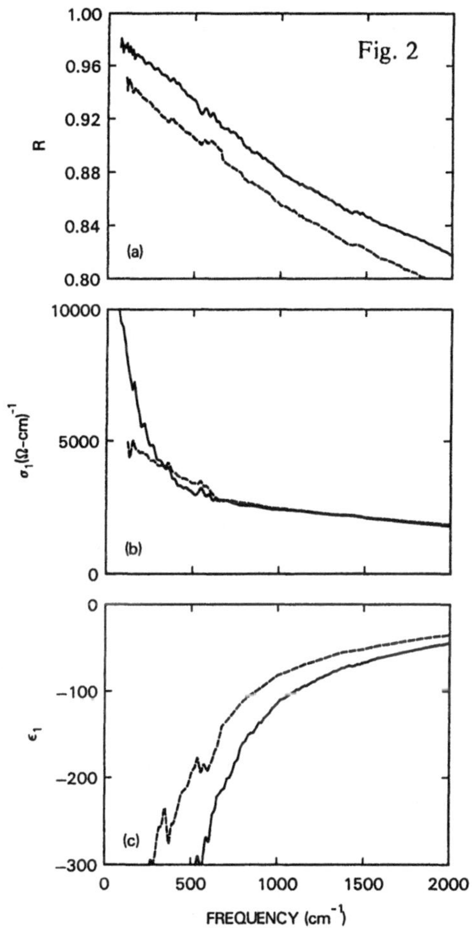

Fig. 2 a) The a-b plane normal state reflectivity at 100K (solid) and 250K (dashed) for a $Y_1Ba_2Cu_3O_7$ crystal. b) The conductivites at 100K and 250K. c) The dielectric function at 100K and 250K.

of superconductivity in the copper-oxides, it is highly desirable to determine the intrinsic properties of the Cu-O planes, since, presumably, these are the essential ingredients in the superconductivity. The conductivity of $Y_1Ba_2Cu_3O_7$ shown in Fig. 2 is therefore not the ideal starting point since holes present on both the chains and planes could lead to two distinctly different contributions to the conductivity. Reflectivity measurements in the near infrared already indicate that the a and b axis are inequivalent.[7] Recent measurements on $Bi_2Sr_2Ca_1Cu_2O_{8-y}$[8] which does not possess chains give a conductivity which strongly resembles the $Y_1Ba_2Cu_3O_7$ conductivity suggesting that the conductivity of $Y_1Ba_2Cu_3O_7$ is at least qualitatively representative of the planes.

Two different approaches have primarily been used to model the $Y_1Ba_2Cu_3O_7$ conductivity in Fig. 2. In the first, the conductivity is assumed to be a superposition of a low frequency free carrier peak and one or more bound mid-infrared excitations at higher frequencies (with roughly 1/4 of the area below 12000cm^{-1} in the low frequency peak).[9,10,11] In the second the presence of a dynamic interaction is responsible for the form of the conductivity.[10,11] In this latter approach the conductivity divides naturally into a coherent peak at frequencies less than some characteristic excitation energy and an incoherent background at higher frequencies. The ratio of the total area to the area in the coherent peak is an approximate measure of the mass enhancement and therefore the coupling which in this case would be $\simeq$3 at 100K. While both the temperature dependence of the reflectivity and of the width of the low frequency conductivity peak suggest the presence of dynamic scattering, the temperature dependences one calculates from frequency dependent scattering approaches assuming $\lambda \simeq 3$ are larger than observed in our measurements. The small isotope shift in $Y_1Ba_2Cu_3O_7$ suggests that electron-phonon coupling is not unusually large. If a coupling of 3 were present in this system, the scattering would presumably be of magnetic origin and it is not clear if the frequency dependent scattering models, which are generally applied to phonons, would give the proper temperature dependence (we note that the characteristic cutoff energy of 700cm^{-1} one would obtain in the frequency dependent scattering model is close to J as would be expected for magnetic scattering). By dividing the conductivity between a mid-infrared mode and a free carrier band with $\lambda \simeq 0.5$-1.0 it is possible to obtain a reasonable fit to both the temperature and frequency dependence. Although such a fit is suggestive, it is possible to obtain equivalent fits with several different choices for the mid-infrared contribution. In addition, some of the uncertainties in the data, such as the effect of chains and the differences between oxygen deficient[11] and fully oxygenated samples,[10] make a separation of the conductivity into free and bound contributions very difficult. Future measurements on other oxide superconductors will hopefully lead to a definitive picture of the normal state conductivity in the planes.

A complete understanding of the electrodynamics of the Cu-O compounds also requires a study of the response of the system when the electric field of the incident radiation is perpendicular to the planes. Fig. 3 gives the reflectivity and conductivity at 100K and 45K from 50-700cm^{-1} with the incident electric field polarized along the c-axis of $Y_1Ba_2Cu_3O_7$.[12] The measurements were made on a stack of $\simeq$12 of the plate-like crystals which were potted in epoxy and polished to obtain sufficient area for the measurements. Here we concentrate on the normal state properties (100K) and in the next section will discuss the changes which occur as the sample passes through the superconducting transition. The reflectivity along the c-axis is considerably lower than with the field in the a-b plane and infrared active phonons are quite visible in the spectrum. The low reflectivity and presence of phonons are both a result of the low free carrier conductivity in this polarization which leads to poor screening of the infrared active phonons. The bottom panel of Fig. 3 shows that the conductivity extrapolates to a d.c. value of $\simeq$400(Ωcm)$^{-1}$ which is approximately 40 time smaller than the a-b plane value at 100K. This conductivity anisotropy is in the range of the lowest values obtained from d.c. resistivity studies.[13] We also find that the width of the free carrier conductivity along this direction appears to be substantially larger than for the a-b plane. This suggests a larger scattering rate and hence a dirtier conductor, although the meaning of the width in this orientation depends on the conduction mechanism which is still the subject of debate.

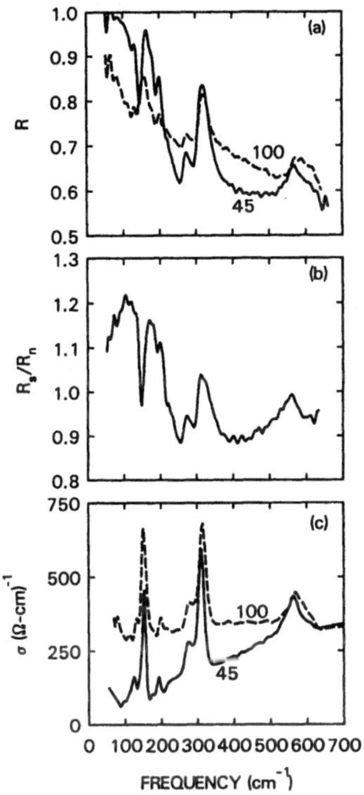

Fig. 3

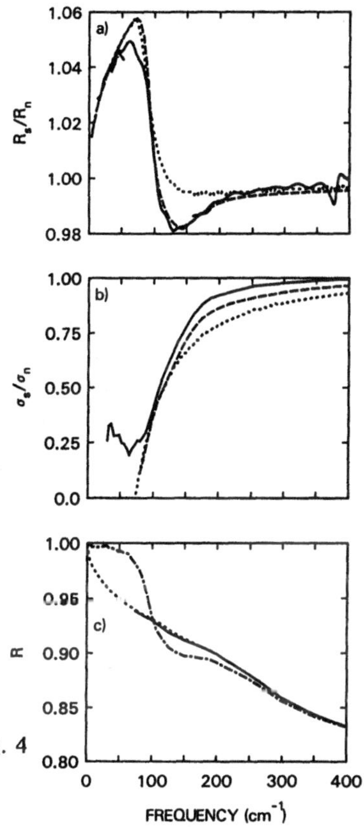

Fig. 4

Fig. 3 a)The reflectivity of $Y_1Ba_2Cu_3O_7$ (T_c=92K) in the superconducting (T=45K) and normal (T=100K) states for the incident electric field along the c-axis direction. b) The c-axis reflectivity in the superconducting state divided by that in the normal state. c) The real part of the conductivity in the superconducting (T=45K) and normal (T=100K) states.

Fig. 4 a) The reflectivity of $Ba_{0.6}K_{0.4}BiO_3$ with $T_c \simeq 29$K in the superconducting state divided by that in the normal state (solid line). b). The conductivity in the superconducting state divide by that in the normal state (σ_{1s}/σ_{1n}) (solid line). The dotted and dashed lines are the results of Mattis-Bardeen and Eliashberg calculations, respectively, with $2\Delta \simeq 70\mathrm{cm}^{-1}$. c) The reflectivity in the normal state (solid line) and its Hagen-Rubens extrapolation to d.c. (dotted line) are shown. The reflectivity in the superconducting state (dot-dashed line) is obtained from the product of the normal state reflectivity and R_s/R_n from the top of the figure.

3. Superconducting Conductivity

We now discuss the changes in the infrared response of the oxide superconductors which occur at the superconducting transition. Figure 4a presents the ratio of the reflectivity deep in the superconducting state ($\lesssim$10K) to the reflectivity in the normal state ($\simeq$35K) for a ceramic $Ba_{0.6}K_{0.4}BiO_3$ sample with $T_c \simeq 29K$.[3] The reflectivity below $100cm^{-1}$ is enhanced in the superconducting state relative to the normal state. Temperature dependent measurements near T_c demonstrate that this enhancement becomes visible as the sample passes through T_c. We also find that a magnetic field in excess of 10T is sufficient to quench the enhancement confirming that this feature is associated with the superconducting state. In Fig. 4c we show the absolute reflectivity in the normal state and in the superconducting state where the latter was obtained by multiplying the normal state reflectivity by R_s/R_n. We apply a Kramers-Kronig transform to the reflectivities to obtain $\sigma_1(\omega)$ and plot the ratio of the superconducting to normal conductivity (σ_{1s}/σ_{1n}) in Fig. 4b. It is clear from σ_{1s}/σ_{1n} that the reflectivity enhancement below T_c is caused by a gap-like depression of the conductivity at low frequencies. We may obtain an estimate for the size of the energy gap either from the frequency at which R_s/R_n is a maximum or from an extrapolation of σ_{1s}/σ_{1n} from frequencies above the gap to the horizontal axis. Fits to the Mattis-Bardeen form (dotted) and using a more sophisticated calculation based on the Eliashberg equations (dashed) also give estimates of the gap.[3] In all of these cases we find $2\Delta \simeq 3\text{-}4kT_c$ suggesting weak to moderate coupling ($\lambda \lesssim 1$) in this system. We also find a gap of conventional size ($\simeq 3.5kT_c$) in $BaPb_{1-x}Bi_xO_3$[14] in agreement with previous tunneling studies of this system.[15]

Ideally the conductivity should be identically zero (unity reflectivity) at frequencies less than 2Δ. Although the gap estimates given above seem reasonable, we find a conductivity which is finite below this frequency. While this observation could be taken as evidence for unconventional pairing, it seems more likely that imperfect sample surfaces and uncertainty in the absolute reflectivity measurements are responsible for any residual absorption below the gap in these materials. More will be said below about the difficulties encountered in using infrared reflectivity to determine whether a true energy gap exits in a superconductor.

When $Y_1Ba_2Cu_3O_7$ is cooled below its superconducting transition temperature the low frequency reflectivity also exhibits an enhancement.[6] This is shown for the a-b plane in Fig. 5a where absolute reflectivities in the superconducting state (45K) and in the normal state (100K and 150K) are plotted. Figure 5b gives $\sigma_1(\omega)$ and Fig. 5c presents $\varepsilon_1(\omega)$. The conductivity in the superconducting state is depressed relative to the normal state at low frequencies showing a gap-like onset of absorption at $500cm^{-1}$ ($8kT_c$). $\varepsilon_1(\omega)$ shows an increase at $\simeq 800cm^{-1}$ in the superconducting state which is responsible for the excess absorption in the superconducting state relative to the normal state at this frequency. We also point out that $\varepsilon_1(\omega)$ is quite negative throughout the frequency range of Fig. 5 which rules out the possibility that a plasma edge may responsible for the drop in reflectivity at $\simeq 500cm^{-1}$ in the the superconducting state.

Recently there has been some controversy over whether this absorption edge is related to the occurrence of superconductivity or merely a continuation of normal state temperature dependences. The primary evidence for the latter view comes from a study of oxygen deficient $Y_1Ba_2Cu_3O_{7-x}$ where a feature in the conductivity at approximately $500cm^{-1}$ which resembles that shown in Fig. 5b is observed at temperatures well above T_c.[16] To clarify this point we have measured the temperature dependence of the reflectivity at temperatures near T_c. Figure 6 shows ratios of the reflectivity at a sequence of temperatures to the reflectivity at 150K. The enhancement of the reflectivity below $500cm^{-1}$ is easily observed at temperatures below T_c

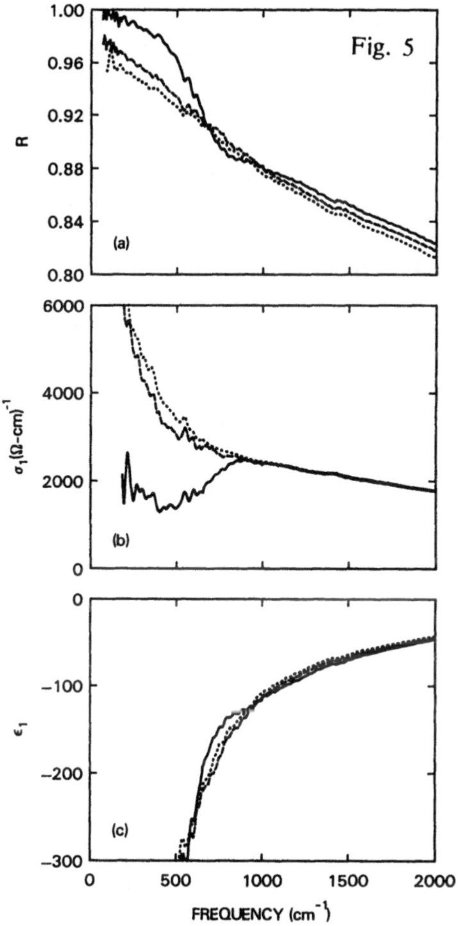

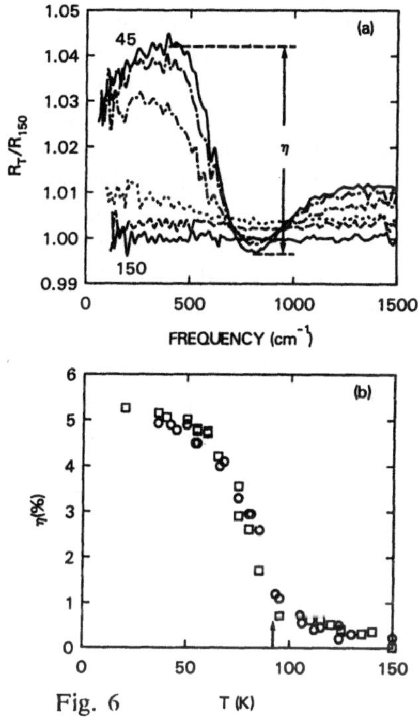

Fig. 5 a) Reflectivity of the a-b plane of $Y_1Ba_2Cu_3O_7$ ($T_c = 92K$) in the superconducting state at 45K (solid) and in the normal state at 100K (dashed) and 150K (dotted). b) The conductivity at 45K, 100K and 150K. c) The dielectric constant at each of these three temperatures. The superconducting conductivity is depressed at low frequencies and shows a gap-like edge at $\simeq$500cm^{-1}.

Fig. 6 a) a-b plane reflectivity ratios for $Y_1Ba_2Cu_3O_7$ in the vicinity of T_c. R_T/R_{150} is the reflectivity at a temperature T divided by that at T = 150K, where T = 150, 125, 105, 80, 65, and 45K. b)The drop in the reflectivity ratio between 400 and 800cm^{-1}(η) is plotted as a function of temperature for the sample used in part (a) (dots) and a second sample (squares) illustrating the relationship between the $\sim 8kT_c$ reflectivity enhancement and T_c (indicated by the arrow).

and is quite consistent with our original report of this feature.[6] In the bottom half of the figure we have plotted η (defined as the drop in the reflectivity ratio between 400 and 800cm^{-1}) as a function of temperature. η is directly related the onset of absorption, and its temperature dependence clearly demonstrates the relationship between our 500cm^{-1} absorption edge and the transition to superconductivity in $Y_1Ba_2Cu_3O_7$. Similar conclusions follow from the temperature dependence of the a.c. conductivity near T_c. We note that several other techniques including NQR,[17] Raman scattering,[18] and most recently photoemission[19] now provide evidence for an energy scale near $8kT_c$ in the superconducting state of the Cu-O superconductors. Many tunneling studies also report gap-like features with energies in excess of the BCS value although there is considerable scatter in the reported values (see for example Ref. 20). It is not known at this point if the feature observed in the oxygen deficient samples is indicative of the persistence of a psuedo-gap or self energy anomaly into the normal state. This question remains an interesting area for future studies of $Y_1Ba_2Cu_3O_{7-x}$ and the other layered Cu-O superconductors.

We can use the conductivities in Fig. 5b to estimate the penetration depth in the superconducting state. The conductivity sum rule tells us that area which is missing from the superconducting conductivity relative to the normal state conductivity should show up in the delta function at the origin. Once we determine this area, A, we can estimate the penetration depth in the superconducting state which is given by $\lambda = c/\sqrt{8A}$. We find a value of $\simeq$1700Å for the a-b plane which is reasonably close to the value of 1400Å reported in d.c. magnetization measurements on single crystals.[21]

We now return to Fig. 3 which shows the reflectivities (Fig.3a), reflectivity ratio (Fig. 3b) and conductivities (Fig. 3c) in the superconducting (45K) and normal states (100K) when the electric field is polarized along the c-axis. The rapid drop in the reflectivity ratio at about 200cm^{-1} suggests the possibility of a c-axis gap in this vicinity. Changes in the shapes of the phonon lines in the reflectivity and small shifts in the phonon energies are consistent with this interpretation as are the changes which occur in the conductivity at the superconducting transition.[12] While the presence of phonons in the spectra makes it difficult to pin down the exact energy at which the superconducting conductivity begins to rise back to the normal state value, it is quite clear that the energy scale associated with the c-axis is much lower than the 7-$8kT_c$ value we find for the a-b plane suggesting substantial gap anisotropy. From the phonon behavior, reflectivity ratios, and calculated fits to the conductivity[12] we estimate a superconducting frequency scale of 170cm^{-1} or $\simeq 3kT_c$. From the missing area in the superconducting conductivity in Fig. 4c we estimate a c-axis penetration depth of $\simeq 1\mu$m which is more than five times larger than the a-b plane value given above. We point out that the changes in reflectivity at the superconducting transition which we see for the c-axis are virtually identical to those observed in ceramic samples.[5] This is not surprising since the large changes in reflectivity associated with the c-axis (about 20 per cent) will dominate the a-b plane contribution (about 4 per cent) in an unoriented sample as we have previously proposed.[22]

While the absorption edge we observe in the a-b plane conductivity and its temperature dependence indicate an energy scale of $\simeq 8kT_c$ in the superconducting state, the conductivity is not zero below this value. Because of the reproducibility of this behavior in our samples and in other infrared studies,[9,23] it is quite possible that a significant fraction of this absorption is intrinsic to $Y_1Ba_2Cu_3O_7$. The most interesting explanation for this residual absorption is that it arises from anisotropy within the a-b plane energy gap (such as might be expected for non-s-wave pairing), although our polarization dependent studies of twinned and untwinnded $Y_1Ba_2Cu_3O_7$ crystals have failed to show any anisotropy of R_s/R_n within the a-b plane. The absorption may also arise from contributions to $\sigma_1(\omega)$ which do not have a 500cm^{-1} gap such as the conductivity of chains. An alternative explanation involves the anisotropy we observe between the a-b plane and c-axis. Even with the field polarized within the Cu-O planes, impurity scattering will tend to mix, to some degree, contributions from all directions in k-space into the infrared conductivity measurement. While the magnitude of this effect is not known, one should only expect the conductivity to be identically zero below the smallest gap in the system.

In light of the above discussion a question which naturally arises is whether the conductivity actually becomes zero (unity reflectivity) at some finite frequency below $8kT_c$. Thomas et al. have suggested the possibility of an energy gap of conventional magnitude ($3.5KT_c$) based on their observation of unity reflectivity in the superconducting state of oxygen deficient $Y_1Ba_2Cu_3O_{7-x}$ crystals.[11] Schützmann et al.[23] report unity reflectivity up to 130cm^{-1} in their study of an oriented $Y_1Ba_2Cu_3O_7$ film with T_c of 91K They also see the 500cm^{-1} absorption edge discussed above and propose the existence two a-b plane gaps to explain their spectra. Our present results neither refute nor support the presence of a low frequency region with unity reflectivity, since we obtain reflectivity in excess of 99%, which we cannot distinguish from 100% within our experimental error, below 150cm^{-1}. Setting the reflectivity equal to unity below this frequency does give a gap-like region in which the conductivity is zero, but, reducing the reflectivity below unity by an amount consistent with our experimental error tends to remove any evidence of this feature. We also find that as the reflectivity approaches unity and frequency approaches zero, the noise in the conductivity (obtained from the Kramers-Kronig transform of the reflectivity) increases dramatically. While the criterion of unity reflectivity to identify the energy gap is in principle a sound one, we find that it is difficult to implement in a meaningful way. Direct absorption measurements would probably be a better way to determine if a region of zero conductivity actually exits.

As a summary of our observations in the superconducting state of the oxide superconductors we refer to Fig. 7 where we present reflectivity ratios (R_s/R_n) and conductivity ratios (σ_{1s}/σ_{1n}) for $Ba_{1-x}K_xBiO_3$, $BaPb_{1-x}Bi_xO_3$, and the a-b plane and c-axis of $Y_1Ba_2Cu_3O_7$ plotted in units of kT_c. The conventional size of the energy gap in the cubic BiO_3 superconductors is clearly visible in both R_s/R_n and σ_{1s}/σ_{1n}. Although the changes in the a-b plane reflectivity and conductivity of $Y_1Ba_2Cu_3O_7$ at the superconducting transition are qualitatively similar to those of $Ba_{1-x}K_xBiO_3$ and $BaPb_{1-x}Bi_xO_3$, the unusually large energy scale associated with the direction is quite obvious. The much smaller energy scale of the c-axis response is also easily seen. The layered cuprates differ from the cubic bismuthates in terms of their highly anisotropic nature and their proximity to antiferromagnetism. Either or both of these differences may be relevant to understanding the anisotropic superconducting properties of $Y_1Ba_2Cu_3O_7$ and the large a-b plane energy scale. Our data indicates this large value of $2\Delta_{a-b} \simeq 7 - 8kT_c$ is not associated with conventional strong coupling,

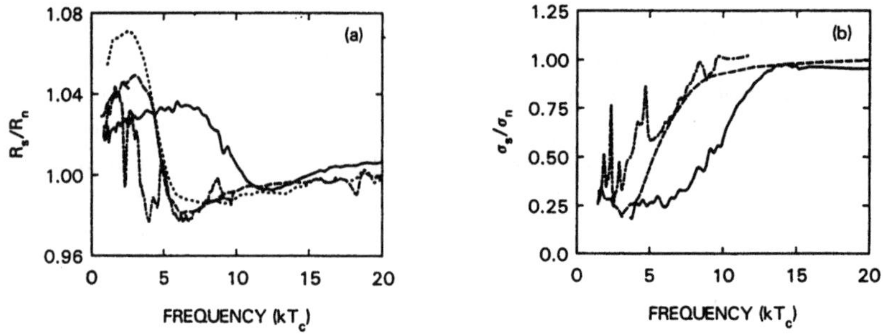

Fig. 7 A comparison of the (a) reflectivity ratios (R_s/R_n) and (b) conductivity ratios (σ_{1s}/σ_{1n}) for $BaPb_{0.8}Bi_{0.2}O_3$ (dotted), $Ba_{0.6}K_{0.4}BiO_3$ (dashed), the a-b plane (solid) and c-axis (dot-dash) of $Y_1Ba_2Cu_3O_7$. The c-axis reflectivity ratio for $Y_1Ba_2Cu_3O_7$ has been multiplied by 0.2.

since a modification of the density of states above 2Δ would be clearly visible in both the reflectivity and conductivity ratios in this case.

Acknowledgements

We thank N. E. Bickers, D. J. Scalapino, B. A. Scott and G. V. Chandrashekhar for important discussions and J. A. Calise for technical assistance.

References

1. J. G. Bednorz and K.A. Müller, Z. Phys. B **64** 189 (1986); J.G. Bednorz, M. Takashige and K.A. Müller, Europhys. Lett. **3**, 379 (1987).
2. R. J. Cava, B. Batlogg, J.J. Krajewski, R. C. Farrow, L. W. Rupp, A. E. White, K. T. Short, W. F. Peck, T. Y. Kometani, Nature **332**, 814 (1988); D. G. Hinks, D. R. Richards, B. Dabrowski, D. T. Marx and A. W. Mitchell, Nature **335**, 419 (1988).
3. Z. Schlesinger, R. T. Collins, J. A. Calise, D. J. Hinks, A. W. Mitchell, Y. Zheng, B. Dabrowski, N. E. Bickers, and D. J. Scalapino, submitted Phys. Rev. B.
4. S. Tajima, S. Ushida, A. Masaki, H. Tagaki, K. Kitazawa, S. Tanaka, and S. Katsui, Phys. Rev. **B32**, 6302(1985); S. Tajima, S. Ushida, A. Masaki, H. Tagaki, K. Kitazawa, S. Tanaka, and S. Sugai, Phys. Rev. **B35**, 696(1987).
5. R. T. Collins, Z. Schlesinger, R. H. Koch, R. B. Laibowitz, T. S. Plaskett, P. Freitas, W. J. Gallagher, R. L. Sandstrom and T. R. Dinger, Phys. Rev. Lett.**59**, 704 (1987).
6. Z. Schlesinger, R. T. Collins, D. L. Kaiser, and F. Holtzberg, Phys. Rev. Lett. **59**, 1958 (1987); Z. Schlesinger, R. T. Collins, D. L. Kaiser, F. Holtzberg, G. V. Chandrashekhar, M. W. Shafer, and T. M. Plaskett, Physica C **153**, 1734 (1988).
7. M. P. Petrov, A. I. Grachev, M. V. Krasin'kova, A. A. Nechitailov, V. V. Prokofiev, V. V. Poborchy, S. I. Shagin, and N. F. Kartenko, Solid State Commun. **67**,1197 (1988).

8. M. Reedyk, D. A. Bonn, J. D. Garrett, J. E. Greedan, C. V. Stager, T. Timusk, K. Kamaras, and D. B. Tanner, Phys. Rev. **B38**, 11981 (1988).
9. T. Timusk, S. L. Herr, K. Kamaras, C. D. Porter, D. B. Tanner, D. A. Bonn, J. D. Garrett, C. V. Stager, J. E. Greedan, and M. Reedyk, Phys. Rev. **B38**, 6683 (1988).
10. R. T. Collins, Z. Schlesinger, F. Holtzberg, P. Chaudari, and C. Field, Phys. Rev.**B39**, 6571 (1989).
11. G. A. Thomas, J. Orenstein, D. H. Rapkine, M. Capizzi, A. J. Millis, R. N. Bhatt, L. F. Schneemeyer and J. V. Waszczak, Phys. Rev. Lett. **61**, 1313 (1988).
12. R. T. Collins, Z. Schlesinger, F. Holtzberg, and C. Feild, submitted Phys. Rev.
13. Yasuhiro Iye, Journal of Modern Physics B.
14. Z. Schlesinger, R. T. Collins, B. A. Scott, and J. A. Calise, Phys. Rev. **B38**, (1988).
15. B. Battlogg, Physica **B 126**, 275 (1984).
16. G. A. Thomas, J. Orenstein, S. L. Cooper, D. H. Rapkine, M. Capazzi, T. Timusk, A. J. Millis, L. F. Schneemeyer and J. V. Waszczak, Preprint.
17. H. Seidel, F. Hentsch, M. Mehring, J. G. Bednorz, and K. A. Müller, Europhys. Lett. **5**, 647 (1988).
18. S. L. Cooper, F. Slakey, M. V. Klein, J. P. Rice, E. D. Bukowski, and D. M. Ginsberg, Phys. Rev. **B38**, 11934 (1988).
19. J.-M. Imer, F. Patthey, B. Dardel, W. -D. Schneider, Y. Baer, Y. Petroff, and A. Zettl, Phys. Rev. Lett. **62**, 336 (1989).
20. M. Lee, A. Kapitulnik, and M. R. Beasley, Proceedings of the NEC symposium on Mechanisms of High Temperature Superconductivity, Oct. 24-27, 1988, Tokyo, Japan.
21. L. Krusin-Elbaum, R. L. Greene, F. Holtzberg, A. P. Malozemoff, and Y. Yeshrun, Phys. Rev. Lett. **62**, 217 (1989).
22. Z. Schlesinger, R. T. Collins, M. W. Shafer and E. M. Engler, Phys. Rev.**B36**, 5275 (1987).
23. J. Schützmann, W. Ose, J. Keller, K. F. Renk, B. Roas, L. Schultz, G. Saemann-Ischenko, accepted for publication in Europhys. Lett.

Nature of the Electronic States near the Fermi Level in High-T_c Superconductors: High-Energy Spectroscopic Evidence

A. Fujimori

Department of Physics, University of Tokyo,
Hongo, Bunkyo-ku, Tokyo 113, Japan

Abstract. Based on the results of recent photoemission and other spectroscopic studies on pure and doped Cu and Ni oxides, the origin of the electronic states near the Fermi level (E_F) in high-T_c superconductors is discussed. It is pointed out that the location and symmetry of localized holes or electrons at low doping levels may not be necessarily the same as those of itinerant carriers responsible for superconductivity. The singlet state $^1A_1$ of the CuO_n local cluster or the impurity Anderson model is likely to be the origin of itinerant carriers in the doped Cu oxides.

1. Introduction

High-energy spectroscopic studies on high-T_c Cu-oxide superconductors have revealed that (i) electron correlation is important for Cu d electrons, (i) doped holes enter oxygen p orbitals rather than Cu d orbitals, (iii) hybridization between the oxygen p and Cu d orbitals is unusually strong in the ground state [1,2]. In particular, one can extract quantitative informnation on the magnitudes of the Coulomb correlation energy of the Cu d electrons U, p-d charge-transfer energy Δ, and p-d hybridization T [$= -\sqrt{3}$ $(pd\sigma)$], from analyses of photoemission spectra using the cluster [2,3] or impurity Anderson models [4,5]: One obtains typically $U \sim 5$–7 eV, $\Delta \sim 0$–3 eV, and $T \sim 2.5$ eV. These parameters define the electronic structure on a large energy scale as noted above. However, the low-energy properties, i.e., thermodynamic and transport properties of the normal and superconducting states of the doped materials are also dependent on more subtle electronic structure near the Fermi level (E_F). Namely, they are dependent on the symmetry of the doped p holes, hence on the other parameters such as the energies of various oxygen p orbitals and hopping integrals bewteen these [6-8], which are difficult to estimate. This complexity arises from the complicated crystal structures with inequivalent oxygen (and Cu) sites and **also** from structural defects associated with the doping process.

The nature of the "metallic" state of the doped matrials has also been controversial: There is no general consensus even on whether the doped carriers form a Fermi liquid or not. Even though there is increasing experimental evidence that the metallic state is a Fermi liquid, the description of the Fermi liquid has not yet been established. In particular, it is important to know whether the dispersive states within ~ 1 eV of E_F in $Bi_2Sr_2CaCu_2O_8$ [9] can be considered as states analogous to the renormalized f bands in mixed-valence and heavy-fermion compounds.

In this article, we discuss the location and symmetry of the doped carriers and the nature of the Fermi liquid state based on the photoemission and other high-energy spectroscopic results on the high-T_c and related compounds and on cluster-model calculations. In order to clarify the microscopic nature of the doped carriers, we start from the low carrier concentration limit where the extra holes or electrons are supposed to be localized around structural defects such as substitutional metal ions or

Springer Series in Solid-State Sciences, Vol. 89 Strong Correlation and Superconductivity
Editors: H. Fukuyama · S. Maekawa · A.P. Malozemoff

excess oxygens. It will then be pointed out that these localized carriers may not necessarily be the same as the itinerant carriers which are responsible for the metallic conductivity. Considering the localized carriers provides a natural starting point in the initial stage of doping. As for hole doping, we will show that the Fermi liquid state is most likely to be formed from the "local singlet" in which an oxygen pσ hole in the CuO_2 plane is antiferromagnetically coupled to the Cu spin, although an "impurity band" which is formed from overlapping defect orbitals may not be completely excluded.

2. Energy Levels of CuO_n Clusters

It has been recognized from the analyses of photoemission spectra that small clusters such as CuO_4, CuO_5, and CuO_6 provide an appropriate starting point in describing the ground state and single-particle excited states of divalent Cu oxides if electron correlation at the Cu site is exactly treated [2-6]. In order to study the character of doped holes, therefore, it is relevant to consider the lowest energy states of the cluster with one extra hole, i.e. two holoes with respect to the closed-shell d^{10} configuration. Figure 1 shows low-lying energy levels of the CuO_5 cluster with one extra hole $[(CuO_5)^{7-}]$ as a function of the energy difference $\Delta\varepsilon_p$ between the CuO_2-plane and apex oxygen p orbitals, $\Delta\varepsilon_p = \varepsilon_p(apx)-\varepsilon_p(pl)$. Here, three p orbitals at each oxygen site are assumed to have the same energy. That is, anisotropic potentials are neglected for each oxygen site. This may not necessarily be the case for the Cu oxides: Anisotropy will become particularly important near defects created in the doping process.

If the spin of the x^2-y^2-symmetry Cu 3d hole is ↑, an extra hole goes into either a x^2-y^2↓- or a $3z^2-r^2$↑-symmetry orbital of primary oxygen p character depending on the parameters to form a $^1A_1$ or a $^3B_1$ state, respectively (see Fig. 1). Here, the x^2-y^2 orbital consists of $p\sigma_{x,y}$ orbitals in the CuO_2 plane, whereas the $3z^2-r^2$ orbital consists of the $p\sigma_z$ orbital of the apex oxygen as well as of the CuO_2-plane $p\sigma_{x,y}$ orbitals. Figure 1 also shows that a ^3E-symmetry state is localted only 0.2 eV above the $^3B_1$ state. This means that the ^3E state which involves $p\pi_{x,y}$ orbitals of the apex oxygen could be lowered below the $^3B_1$ state and could become the ground state of the hole-doped cluster as a result of the anisotropic potential at the apex oxygen site.

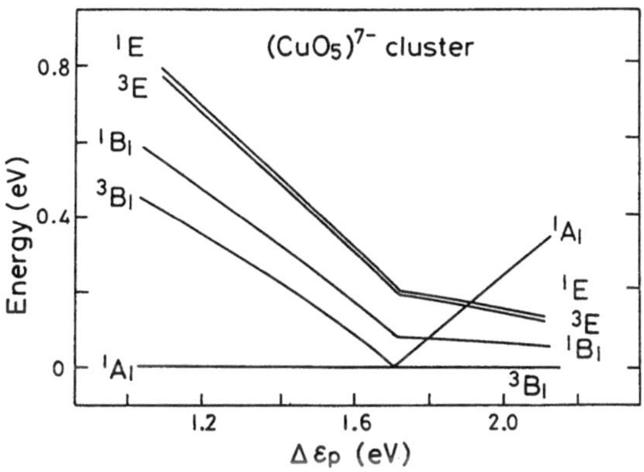

Fig.1 Low-lying energy levels of the CuO_5 local cluster with an extra hole calculated with U = 6.5 eV, Δ = 1 eV, and T = 2.34 eV [6]. The axial Cu-O distance relative to the in-plane one has been taken to be 1.27.

3. Hole Doping in NiO and La_2NiO_4

Before discussing the character of doped holes in the superconductors, it is illustrative to consider a prototypical case of Li-doped NiO, $Li_xNi_{1-x}O$, where holes are doped by Li^+ substitution for Ni^{2+}. This system has recently been studied by oxygen K-edge absorption by Kuiper et al.[10], who have demonstrated that the doped holes go into localized states that are split off from the top of the O 2p valence band into the p-d charge-transfer gap. The strong electrostatic potential of the substitutional Li^+ ion raises those oxygen p orbitals whose lobes are pointed toward the Li^+ ion as shown in Fig.2. Kuiper et al. have pointed out that the spins of these holes are coupled to neighboring Ni^{2+} spins ($^3A_{1g}$) to form low-spin states through strong antiferromagnetic interaction. This interaction arises from the pdσ interaction between the oxygen p holes and the Ni 3d (e_g) holes. Also, we can see from Fig. 2 that other neighboring Ni ions are coupled **ferromagnetically** with the localized state through weaker pdπ interaction.

The lowest ionization state of pure NiO, on the other hand, is found to be a high-spin state, $^4T_1$, in which the extra hole is located in oxygen pπ orbitals hybridizing with the Ni d(t_{2g}) orbitals through pdπ interaction. This has been concluded from the cluster-model [11] and impurity Anderson-model [12] analyses of the valence-band photoemission spectra. In the periodic lattice, the local $^4T_1$ state propagates between "NiO_6 clusters" with a well-defined wave vector. The ^2E state, in which a pσ hole is antiferromagnetically coupled to the Ni^{2+} spins as in the impurity state of $Li_xNi_{1-x}O$, has somewhat higher energy in pure NiO [13].

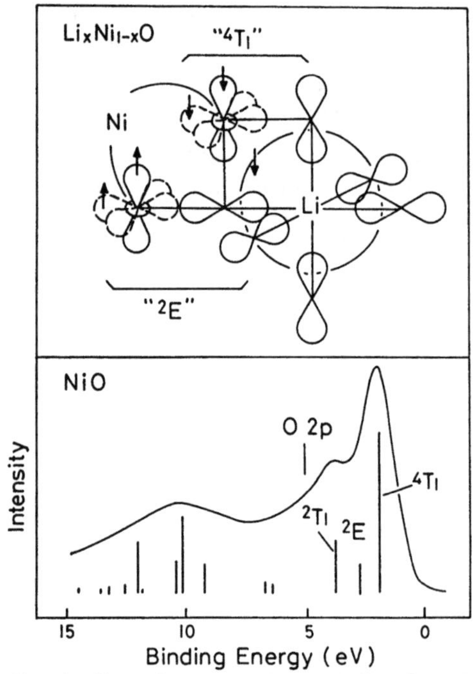

Fig.2 Top: Oxygen p-derived localized orbital around the Li^+ ion substituting Ni^{2+} in NiO. The Ni^{2+} ion has two e_g holes with $x^2-y^2\uparrow(\downarrow)$ and $3z^2-r^2\uparrow(\downarrow)$ symmetry. Bottom: Valence-band photoemission spectrum of NiO ($h\nu$ = 1486.6 eV). The vertical bars are the result of a cluster-model calculation [11].

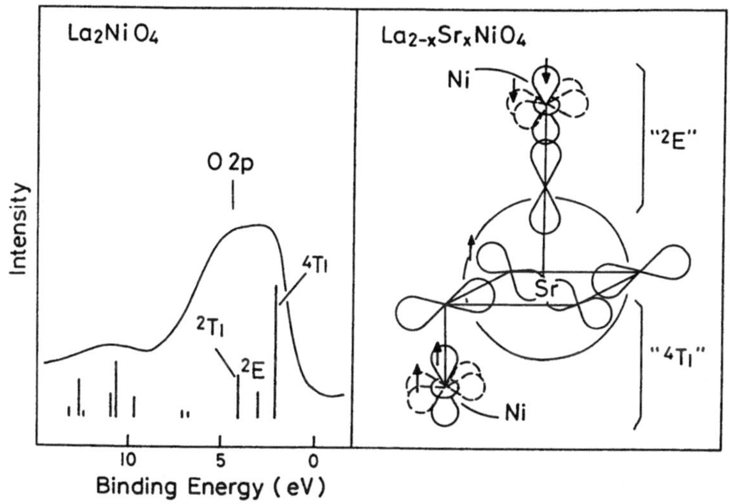

Fig.3 Right: Oxygen p-derived localized orbital around the Sr^{2+} ion substituting La^{3+} in La_2NiO_4. Left: Valence-band photoemission spectrum of La_2NiO_4 ($h\nu$ = 1253.6 eV) taken at liquid nitrogen temperature. Shallow core levels have been subtracted. The vertical bars are the result of a cluster-model calculation.

One can proceed in a similar way for La_2NiO_4. When a La^{3+} ion is replaced by Sr^{2+}, the electrostatic potential due to the Sr^{2+} ion will raise p orbitals on surrounding oxygen atoms. Thus we expect that a localized state formed from these orbitals is split off from the O 2p band into the charge-transfer gap of La_2NiO_4 and accomodates the extra hole. As illustrated in Fig.3, the localized hole is largely located within the LaO plane, i.e. on the apex oxygens of the NiO_6 clusters. This localized hole can be coupled to the Ni^{2+} ions through pdσ or pdπ interaction as shown in Fig.3. For the pdπ interaction, the doped hole forms a high-spin state (or a "local $^4T_1$"-like state) with the Ni^{2+} ion whereas the pdσ interaction leads to a low-spin state (or a "local ^2E"-like state). (Here, we retain the notation of cubic symmetry for the distorted NiO_6 octahedra because the distortion is much smaller than that for La_2CuO_4.) By ignoring the distortion of the NiO_6 cluster, the first ionization state of La_2NiO_4 would be a high-spin local $^4T_1$ state but not a low-spin ^2E state.

4. Hole Doping in La_2CuO_4

In the case of $La_{2-x}Sr_xCuO_4$ also, essentially the same oxygen p orbitals as those of $La_{2-x}Sr_xNiO_4$ will be split off from the O 2p band into the charge-transfer gap and to accomodate doped holes (Fig.4). As for the Cu^{2+} ion, only a $d_{x^2-y^2}\uparrow$ hole is present, making pdσ interaction with the localized hole impossible hence antiferromagnetic coupling between the hole and the Cu^{2+} spins. Thus the p hole-d^9 coupling is certainly ferromagnetic for the localized p hole (corresponding to a "local ^3E" or "$^3B_1$" state) although the coupling strength is predicted to be small as ca be seen form the $^3B_1$-$^1B_1$ and ^3E-^1E splittings shown in Fig.1 [6].

On the other hand, the first ionization state of the $(CuO_6)^{10-}$ cluster is most likely to be $^1A_1$ for $\Delta\varepsilon_p \sim 1$ eV which is reasonable for the undoped materials: The energy rise of the apex oxygen orbitals due to the repulsive potential due to the substitutional Sr^{2+} ions is necessary to stabilize the

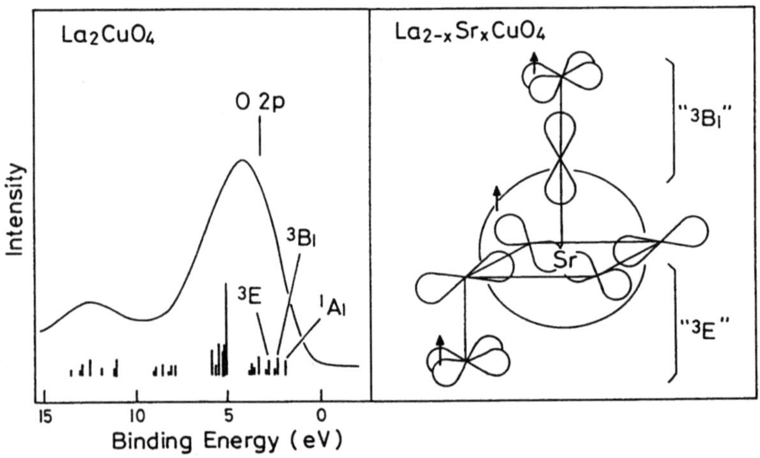

Fig.4 Right: Oxygen p-derived localized orbital around the Sr^{2+} ion substituting La^{3+} in La_2CuO_4. Left: Valence-band photoemission spectrum of La_2CuO_4 ($h\nu$ = 1253.6 eV) taken at liquid nitrogen temperature. The vertical bars are the result of a cluster-model calculation [30].

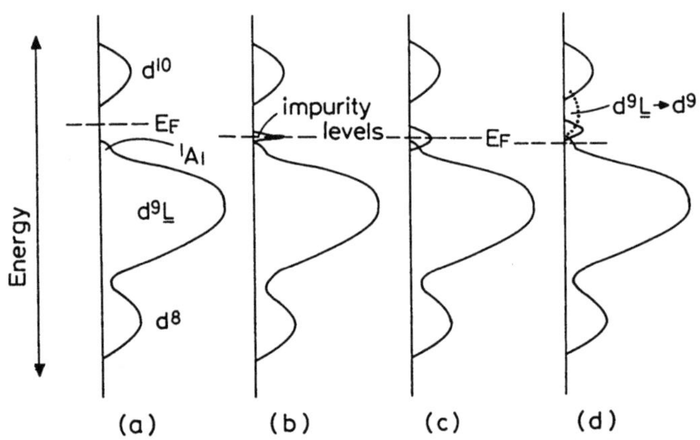

Fig.5 Schematic densities of states for the pure and doped Cu oxides. (a) Undoped; insulator. (b) Lightly doped; semiconductor showing impurity conduction. The impurity levels are actually split into the unoccupied and occupied bands due to electron correlation. (c) Highly doped; metal with E_F within the impurity band. (d) The same as (c), but with E_F within the $^1A_1$ band. An additional feature ($d^9\underline{L} \rightarrow d^{10}$) appears above E_F and fills the band gap.

local triplets, $^3B_1$ or ^3E [6]. Thus we may conclude that the symmetry of the localized doped holes in the low Sr-concentration regime ($x < 0.02$) is not the same as that of the first ionization state of the undoped materials.

Figure 5 schematically illustrates possible changes in the electronic structure of $La_{2-x}Sr_xCuO_4$ as a function of Sr concentration. The undoped La_2CuO_4 [Fig.5(a)] has a charge-transfer gap between $d^9\underline{L}$ ($^1A_1$) and d^{10}. At low doping levels, the localized states are formed within the gap [Fig.5

(b)]. Above a critical concentration x 0.02, the doped holes become itinerant leading to a metallic behavior. The itinerant states at E_F arise either from increased overlap between the localized defect orbitals at high Sr concentration ["impurity band", Fig.5(c)] or from the $^1A_1$-hole band of the periodic lattice [Fig.5(d)]. The latter situation occurs if E_F is lowered from the impurity band into the A_1 band in a rigid-band sense. Or the random potentials of the Sr impurities will be screened by metallic conduction holes, which may effectively corresponds to the picture of Fig.5(d).

5. Hole Doping in $YBa_2CuO_3O_y$ and $Bi_2Sr_2CaCu_2O_8$

Hole doping in the $YBa_2CuO_3O_y$ system is realized by adding oxygen atoms into the CuO chains [Cu(1)-O(1)]. Addition of oxygens above y = 6.5 leads to doping the system with extra oxygen p holes. The hole doping process in this system has been discussed based on a tight-binding model of the CuO chains by Zaanen et al [14]. According to these calculations, doped holes are largely localized on the O(1) atoms whereas part (∿20 %) of the doped holes go into the CuO_2 planes and become responsible for the metallic conductivity. Such a hole distribution is supported by self-consistent band-structure calculations on $YBa_2CuO_3O_6$ and $YBa_2CuO_3O_7$ by Park et al.[15], who have shown that states induced above E_F in going from y = 6 to 7 have character of O(1) and BaO-plane oxygen [O(4)] but have only small weight at the CuO_2-plane oxygens.

An oxygen hole around Cu(1) can be viewed as a localized $^1A_1$ configuration consisting of a $d_{y^2-z^2}\downarrow$ hole of the Cu(1) atom and a $d_{y^2-z^2}\uparrow$-symmetry hole made of O(1) p_y and O(4) p_z atomic orbitals (see Fig.6). The projection of this localized hole onto the $Cu(2)O_5$ pyramidal cluster leads to a $^3B_1$ state which involves an apex oxygen $p\sigma_{z1}$ (= O(4) p_z) orbital. Therefore, if itinerant holes correspond to the $^1A_1$ state of the $Cu(2)O_5$ cluster as suggested by Zaanen et al.[15] and many others [16], one has to distinguish between the symmetry of the localized and itinerant holes as in the case of $La_{2-x}Sr_xCuO_4$.

As for $Bi_2Sr_2CaCu_2O_8$, there has been no clear evidence that substitutional impurities or excess oxygens play an essential role in the hole dopin process, while Y substitution for Ca decreases the hole concentration [17]. Thus, we first consider a periodic crystal with holes populating the itinerant $^1A_1$ band and assume that the Y substitution simply depopulates the hole band as shown in Fig.7. (We are aware that such a rigid-band description has only limited meaning since the "$^1A_1$ band" is essentially of many-body nature.) This has been confirmed by a comparative photoemission study of $Bi_2Sr_2CaCu_2O_8$ and $Bi_2Sr_2Ca_{1-x}Y_xCu_2O_8$ (Fig.8) [18]. The p-hole orbitals have been found to be directed within the ab plane by O 1s core-level electron-energy-loss spectroscopy [19], consistent with the $^1A_1$ state. Indirect support for the assigment that the states near E_F is $^1A_1$ is provided by the photoemission study of an insulator $Sr_2CuO_2Cl_2$, which consists of SrCl and CuO_2 planes thereby ruling out holes on the apex oxygens [18]: As in Fig.8, the photoemission spectrum clearly shows a weak emission feature above the top of the O 2p band as in the case of $Bi_2Sr_2Ca_{1-x}Y_xCu_2O_8$, indicating that this emission is indeed associated with the CuO_2 planes hence with $^1A_1$.

6. Electron Doping in Nd_2CuO_4

The orbitals which doped electrons go into in the n-type superconductor $Nd_{2-x}Ce_xCuO_4$ have also been controversial since its discovery. Initially it was believed that the electrons enter the unoccupied $d_{x^2-y^2}\downarrow$ orbital of Cu^{2+}: A Cu K-edge x-ray absorption study [20] showed an additional peak

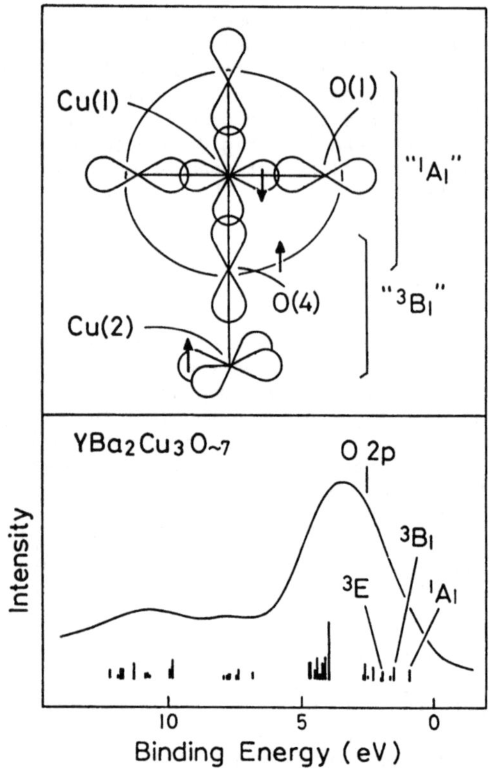

Fig.6 Top: Oxygen p-derived localized orbital in the Cu-O chain of $YBa_2Cu_3O_{6.5+\delta}$. Bottom: Valence-band photoemission spectrum of $YBa_2Cu_3O_y$ [6].

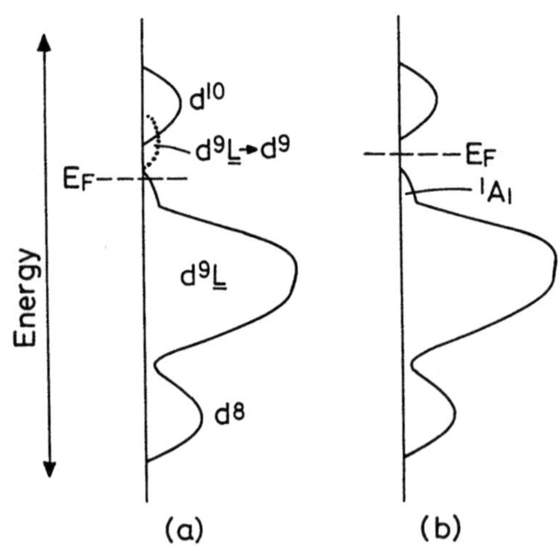

Fig.7 Schematic densities of states for (a) $Bi_2Sr_2CaCu_2O_8$ and (b) $Bi_2Sr_2Ca_{1-x}Y_xCu_2O_8$.

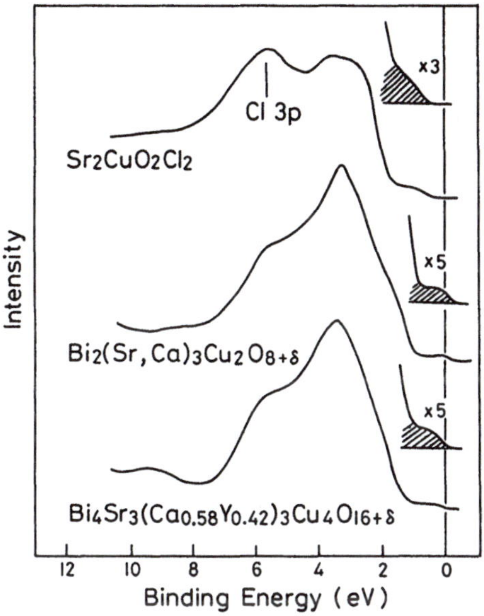

Fig.8 Valence-band photoemission spectra of $Bi_2Sr_2CaCu_2O_8$, $Bi_2Sr_2Ca_{1-x}Y_x Cu_2O_8$ ($h\nu$ = 21.2 eV), and $Sr_2CuO_2Cl_2$ ($h\nu$ =40.8 eV) [18]. Shaded areas represent $^1A_1$ final states.

similar to that of Cu_2O growing in intensity with Ce concentration. Later studies, however, pointed out that such a feature cannot readily be associated with Cu^+ [21], and so far no sizeable amount of Cu^+ (d^{10}) has been confirmed yet.

We have studied valence-band and core-level photoemission of $Nd_{2-x}Ce_xCuO_4$ and found no clear evidence for the presence of Cu^+ (d^{10}) configuration in the ground state [22]: As shown in Fig.9, the valence-band spectra are dominated by $d^9 \rightarrow d^8$, $d^9\underline{L}$ emission as expected for the Cu^{2+} compounds and the position of E_F relative to the top of the O 2p band is higher than $La_{2-x}Sr_xCuO_4$, consistent with the n-type conductivity. However, $d^{10} \rightarrow d^9$ emission which would be expected to appear from E_F to below it for a Cu^{2+}-Cu10+ mixed-valence system is not observed. Further, the Cu $2p_{3/2}$ core level does not show changes that would be expected for the population of d^{10} configuration.

Thus we are lead to speculate that the doped electons go into the bottom of the wide conduction band which may become localized due to the impurity potential of the Ce^{4+} ions substituting the Nd^{3+} sites in the semiconducting region. According to band-structure calculations [23], the bottom of the conduction band is located at the Γ point of the Brillouin zone and should therefore has s-symmetry with respect to the Cu site. Such states would have much (almost two orders of magnitude) lower photoionization cross sections than those of O 2p or Cu 3d [24], consistent with the experimental results which show very low emission intensity at E_F even in the doped samples. However, we note again that the character of itinarant electrons in the metallic region may not necessarily be the same as that of the localized states in the semiconducting region. In this context, it is probabale that the Cu $3d^{10}$ states or local $^1A_1$ states form a coherent electron band near E_F and become itinerant states responsible for the superconductivity.

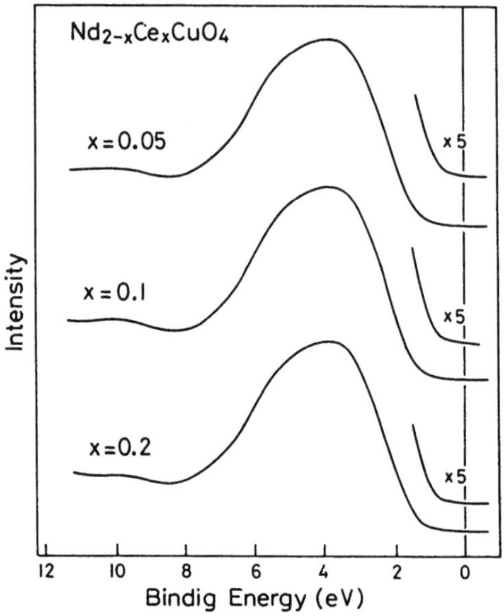

Fig.9 Valence-band photoemission spectra of $Nd_{2-x}Ce_xCuO_4$ ($h\nu$ =40.8 eV) taken at liquid nitrogen temperature [22].

7. Fermi Liquid or Not ?

Recent photoemission results performed at low temperatures on single crystals show sharp Fermi edges (within the resolution of photoemission spectrometers). Angle-resolved photoemission studies on $Bi_2Sr_2CaCu_2O_8$ have revealed dispersive states within $\sim$ 1 eV of E_F which cross E_F at certain points in the Brillouin zone [9]. Furthermore, high resolution studies have demonstrated the opening of a BCS gap below T_C [25]. These results seem to point to Fermi liquids rather than exotic metals like RVB states.

Even though Fermi liquids are formed in the high-T_C superconductors, there remains a question of whether these Fermi liquids can be treated within the same framework as in the case of f-electron systems. In the conventional Fermi-liquid approach as has been employed for 4f and 5f heavy Fermion systems, one assumes renormalized "one-electron" f bands whose widths are narrowed due to electron correlation [26]. In this case, states in the vicinity of E_F has f character rather than that of the wide conduction band states. The renormalized f band can be viewed as a coherent superposition of the single-site Kondo peaks. Thus one may be tempted to treat the dispersive states as renormalized Cu d bands in analogy with the f-electron systems. This is not the case, however, since these states remain essentially unchanged in the semiconducting $Bi_2Sr_2Ca_{1-x}Y_xCu_2O_8$ [18] (although their dispersions have not been measured). That is, the real Kondo peak is a result of infrared divergence of the Fermi sea and should not be observed in the insulator.

The $^1A_1$ state is analogous to the lower binding energy component of the well-known double f peaks observed in the photoemission spectra of Ce pnictides in that it is a bonding state between the d(f)-hole and ligand-hole states [27,28]. This antibonding state crosses E_F when the system is chemically doped with holes in the high-T_C superconductors or when the host band gap is collapsed in the rare-earth pnictides. Thus, in the case of the rare-earth pnictides, renormalization of quasiparticles occurs in two steps. First, the bare f hole on a high energy scale (U or ε_f) is renormalized by the f-ligand hybridization resulting in the f^{n-1}-$f^n\underline{L}$ bonding hole state. In the second step, the hole states cross E_F giving a single-site Kondo peak, which develops into a coherent f band in the periodic lattice. There is a general consensus that the resulting f-band states form a Fermi liquid. The bare d hole in high-T_c superconductors on the high energy scale (U or ε_d) is also screened through d-ligand hybridization to form the $^1A_1$ hole. In the next step, coherent band states are formed out of the $1A_1$ before the formation of a single-site "Kondo" peak, because the (oxygen p-mediated d-d) intersite interaction is much larger than in the case of the f-electron systems. No second renormalization has been postulated, the quasiparticles have oxygen p character rather than Cu d, and the metallic carriers are not so heavy. If the f-ligand hybridization is increased in the rare-earth pnictide pnictide, the system becomes less magnetic or more Fermi-liquid like and finally distinction between the f^n-$f^{n-1}\underline{L}$ antibonding peak and the real Kondo peak disappears. This might be analogous to the case of the Cu oxides where the d-ligand hybridization is so strong that no "Kondo peak" has been considered. On the other hand, there has been many arguments against the Fermi liquid approaches, mainly on the basis of the unique statistics of the $^1A_1$ holes [29]. Finally, a remarkable feature of the doped Cu oxides which has not been noticed in most of theoretical studies may be that the carrier number is small and the Coulomb interaction may be long ranged. Further studies are necessary to solve these problems.

Acknowlegement.

I would like to thank H. Eisaki, H. Takagi, S. Uchida, Y. Tokura, S. Hosoya, H. Katayama-Yoshida, M. Sato, S. Takekawa, and E. Takayama-Muromachi for collaboration.

REFERENCES

1. For recent review, see J. C. Fuggle, J. Fink, and N. Nucker, Int. J. Mod. Phys. B (in press); G. Wendin, Phys. Scr. (in press).
2. A. Fujimori, E. Takayama-Muromachi, Y. Uchida, and B. Okai, Phys. Rev. B **35**, 8814 (1987); Z.-X. Shen, J. W. Allen, J.-J. Yeh, J.-S. Kang, W. Ellis, W. Spicer, I. Lindau, M. B. Maple, Y. D. Dalichaouch, and M. S. Torikachvili, ibid **36**, 8414 (1987).
3. H. Eskes L. H. Tjeng, and G. A. Sawatzky, Phys. Rev. B (in press).
4. H. Eskes and G. A. Sawatzky, Phys. Rev. Lett. **61**, 1415 (1988).
5. K. Okada and A. Kotani, to be published.
6. A. Fujimori, Phys. Rev. B **39**, 793 (1989).
7. E. B. Stechel and D. R. Jennison, Phys. Rev. B **38**, 8873 (1989).
8. A. K. McMahan, R. M. Martin, and S. Satpathy, Phys. Rev. B **38**, 6650 (1988).
9. T. Takahashi, H. Matsuyama, H. Katayama-Yoshida, Y. Okabe, S. Hosoya, K. Seki, H. Fujimoto, M. Sato, and H. Inokuti, Nature **334**, 691 (1988).
10. P. Kuiper, G. Kruizinga, J. Geisen, G .A. Sawatzky, and H. Verweij, Phys. Rev. Lett. **62**, 221 (1989).
11. A. Fujimori and F. Minami, Phys. Rev. B **30**, 957 (1984).
12. J. Zaanen, G. A. Sawatzky, and J. W. Allen, J. Magn. Magn. Mater. **54–57**, 607 (1986).
13. The ^2E state is indeed the lowest ionization state of NiS [A. Fujimori, K. Terakura, M. Taniguchi, S. Ogawa, S. Suga, M. Matoba, and S. Anzai, Phys. Rev. B **37**, 3109 (1988)], which is traced back to the strong sd interaction between the Ni 3d and S 3s orbitals.

14. J. Zaanen, A. T. Paxton, O. Jepsen, and O. K. Andersen, Phys. Rev. Lett. **60**, 2685 (1988).
15. K. T. Park, K. Terakura, T. Oguchi, A. Yanase, and M. Ikeda, J. Phys. Soc. Jpn. **57**, 3445 (1988).
16. Y. Tokura, J. B. Torrance, T. C. Huang, and A. I. Nazzal, Phys. Rev. B **38**, 7156 (1988).
17. Y. Ando, K. Fukuda, S. Kondoh, M. Sera, M. Onoda, and M. Sato, Solid State Commun. **67**, 815 (1988).
18. A. Fujimori, Y. Tokura, H. Eisaki, H. Takagi, S. Uchida, and M. Sato, Phys. Rev. B (submitted).
19. N. Nucker, H. Romberg, X. X. Xi, J. Fink, B. Gegenheimer, and Z. X. Zhao, Phys. Rev. B **39**, 6619 (1989).
20. J. M. Tranquada, S. M. Heald, and A. R. Moodenbaugh, Nature **337**, 720 (1989).
21. N. Kosugi, private communication.
22. A. Fujimori, Y. Tokura, H. Eisaki, H. Takagi, S. Uchida, and E. Takayama-Muromachi, Phys. Rev. B (submitted).
23. S. Massida, N. Hamada, J. Yu, and A. J. Freeman, preprint; K. Takegahara and T. Kasuya, Solid State Commun. (in press).
24. J.-J. Yeh and I. Lindau, At. Data Nucl. Data Tables **32**, 1 (1985).
25. J.-M. Imer, F. Pathy, B. Darel, W.-D. Schneider, Y. Baer, Y. Petroff, and A. Zettl, Phys. Rev. Lett. **62**, 3366 (1989).
26. P. A. Lee, T. M. Rice, J. W. Serene, L. J. Sham, and J. W. Wilkins, Comments Cond. Mat. Phys. **12**, 99 (1986).
27. A. Fujimori, Phys. Rev. B **27**, 3992 (1983).
28. O. Sakai and T. Kasuya, Solid State Commun. **52**, 997 (1984).
29. F. C. Zhang and T. M. Rice, B **37**, 3759 (1988); H. Fukuyama, H. Matsukawa, and Y. Hasegawa, J. Phys. Soc. Jpn. **58**, 364 (1989).
30. A. Fujimori, S. Takekawa, E. Takayama-Muromachi, Y. Uchida, A.Ono, T. Takahashi, Y. Okabe, and H. Katayama-Yoshida, Phys. Rev. B **39**, 2255 (1989).

Photoemission, Inverse Photoemission and X-Ray Absorption Spectroscopies of $Bi_2Sr_2CaCu_2O_8$

T. Takahashi

Department of Physics, Tohoku University, Sendai 980, Japan

A systematic high-energy-spectroscopic study (angle-resolved photoemission, inverse photoemission and x-ray absorption) combined with scanning-tunneling spectrocopy has been performed on $Bi_2Sr_2CaCu_2O_8$ to elucidate the nature and origin of the electronic states at the Fermi level. The experimental results indicate that the Fermi-liquid states do exist in the high-T_c superconductor and they have a dominant O $2p_{x-y}$ character with their major weight on CuO_2 planes. The Fermi-liquid states would be a kind of "impurity states (bands)" formed in the charge-transfer gap through the strong hybridization between doped O $2p_{x-y}$ hole orbitals and surrounding empty Cu $3d_{x^2-y^2}$ orbitals.

1. Introduction

High-energy spectroscopies such as photoemission have revealed a basic framework of the electronic structure of high-T_c superconductors. It is well established that the cuprate high-T_c superconductor is a highly-correlated electron system with the on-site Coulomb repulsion energies of Cu $3d$ and O $2p$ electrons of about 7 and 5 eV, respectively [1]. The next step to approach the high-T_c mechanism is to elucidate the nature and origin of the electronic states in the vicinity of the Fermi level which is directly related to the occurrence of high-T_c superconductivity.

In this paper, we report a comprehensive result of our systematic high-energy-spectroscopic study on the electronic structure of Bi_2Sr_2Ca-Cu_2O_8. Using photoemission, inverse photoemission, and x-ray absorption spectroscopies as well as scanning-tunneling spectroscopy, we reached the conclusion that the Fermi-liquid states do exist in the high-T_c superconductor and they are a kind of "impurity bands" formed in the charge-transfer gap through the strong hybridization between doped O $2p_{x-y}$ hole orbitals and surrounding empty Cu $3d_{x^2-y^2}$ orbitals.

2. Evidence for Fermi-Liquid States

Figure 1 shows angle-resolved photoemission spectra of $Bi_2Sr_2CaCu_2O_8$ in the vicinity of the Fermi level measured for a high-symmetry direction (ΓX) in the Brillouin zone. The experimental details have been described elswhere [2,3].

We find in Fig. 1 that there are two bands A and B in the vicinity of the Fermi level. Band A is located at about 0.2 eV at θ=0° which corresponds to the center of the Brillouin zone (Γ point), and it disappears into the unoccupied states (above the Fermi level) at θ=7.5°. Band B, located at 0.5 eV at θ=0°, gradually approaches the Fermi level as the polar angle (θ) and crosses the Fermi level at θ=15-17.5°. This is a direct observation of dispersive bands which intersect the Fermi level, i.e., a direct evidence for the existence of Fermi surface and Fermi-liquid states in the high-T_c superconductor.

Figure 2 shows the "band structure" of $Bi_2Sr_2CaCu_2O_8$ determined from the present angle-resolved photoemission measurement. A representative band structure calculation [4] is included in the figure for comparison. We find at first an overall disagreement between the experiment and the calculation. Experimental bands C, D, and E are almost dispersionless in sharp contrast with highly dispersive features of calculated bands. Further band F exhibits an entirely opposite dispersive feature against the calculated Cu3*d*-O2*p* bonding band although they are in the same energy region.

In the vicinity of the Fermi level, there are two dispersive bands A and B in the experiment as described above (Fig. 1). While band A has no correspondence to the calculation, band B may be related to the calculated Cu3*d*-O2*p* anti-bonding band which crosses the Fermi level. This may suggest that the one-electron approximation is qualitatively correct at least in the vicinity of the Fermi level. However, the dominant O 2*p* character of the electronic states at the Fermi level as revealed by resonant photoemission and x-ray absorption spectroscopies (described below) is not consistent with the heavily hybridized Cu3*d*-O2*p* nature predicted from the band-structure calculation.

Figure 3 shows an inverse photoemission spectrum of $Bi_2Sr_2CaCu_2$-O_8 [5] compared with those of $YBa_2Cu_3O_7$ [6] and $La_{1.8}Sr_{0.2}CuO_4$ [7]. The spectrum of Bi-system has a clear Fermi edge (band A*) indicative of the existence of Fermi-liquid states, supporting the result of the angle-resolved photoemission described above. Band B* is ascribed to the empty Cu 3*d* state with a small admixture of the O 2*p* state (so-called $3d^{10}$ state). A remarkable enhancement of band B in Y-system is due to the difference in the photon energy detected in the experiment; 9.5 eV

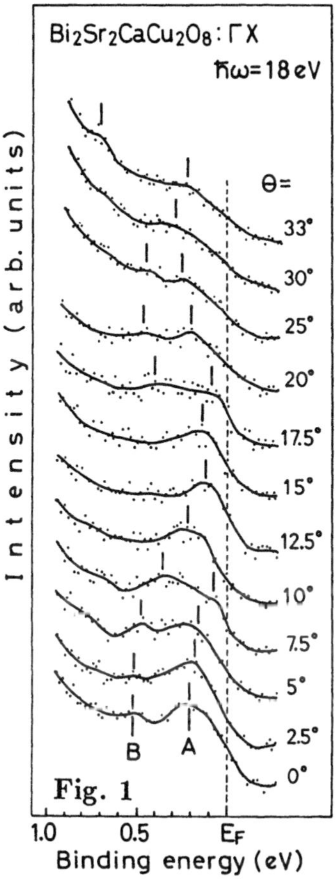

Fig. 1

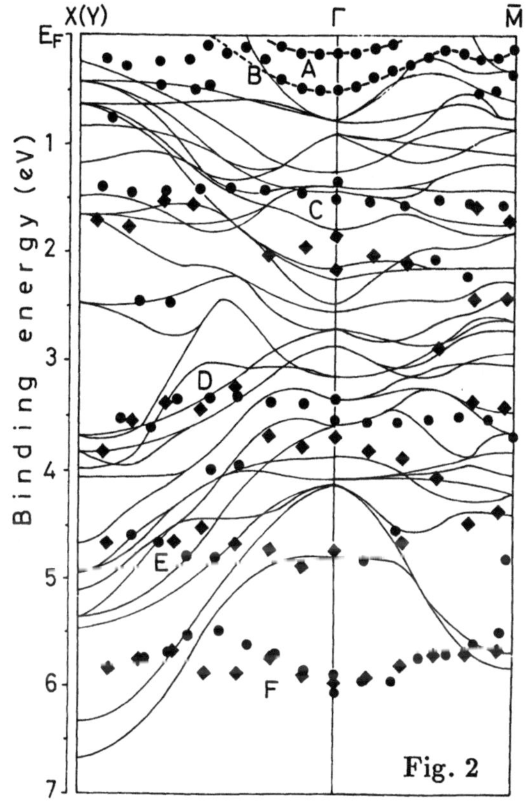

Fig. 2

Fig. 1 Angle-resolved photoemission spectra of $Bi_2Sr_2CaCu_2O_8$ in the vicinity of the Fermi level measured with photon energy of 18 eV for a high-symmetry direction (ΓX) in the Brillouin zone. Polar angle referred to the surface normal is indicated on each spectrum.

Fig. 2 "Band structure" of $Bi_2Sr_2CaCu_2O_8$ determined from the angle-resolved photoemission. Solid circles and rectangles represent the measurements with $\hbar\omega$=18 and 40 eV, respectively. A band-structure calculation [4] is shown by thin solid lines for comparison.

for Bi- and La-systems and 29 eV for Y-system. The photo-ionization crosssection of a Cu 3*d* orbital considerably decreases in the low photon-energy region such as 10 eV. The absence or remarkable reduction of the Fermi-edge-intensity for Y- and La-systems may be due to desorption of oxygen atoms from the sample surface into vacuum which gives rise to an insulating behavior of the surface [8,9].

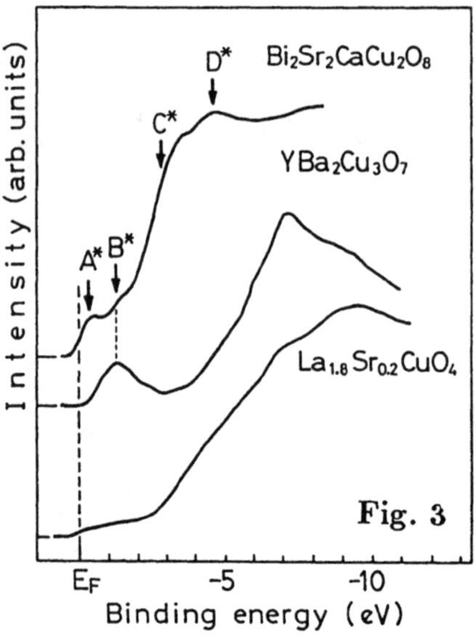

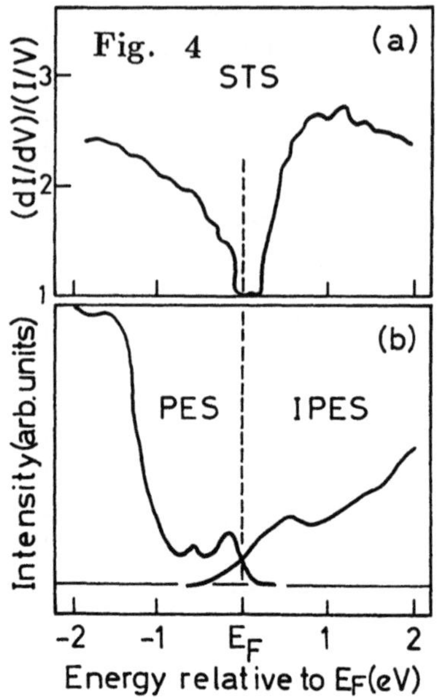

Fig. 3 Inverse photoemission spectra of $Bi_2Sr_2CaCu_2O_8$ [5], YBa_2Cu_3-O_7 [6] and $La_{1.8}Sr_{0.2}CuO_4$ [7].

Fig. 4 Scanning-tunneling-spectroscopy (STS) spectrum compared with photoemission (PES) and inverse photoemission (IPES) spectra.

Thus, the existence of Fermi-liquid states in $Bi_2Sr_2CaCu_2O_8$ is established by the angle-resolved photoemission and the inverse photoemission. The next problem to be solved is where in the crystal the Fermi-liquid states exist. Band structure calculations have predicted that BiO layers provide a substantial density of electronic states at the Fermi level. Some photoemission studies [10,11] proposed that the Fermi-edge observed by photoemission is ascribed to the Bi6p-O2p bonding state, supporting the band calculation. If it is the case, CuO_2 layers have few density of electronic states and resultingly the occurrence of high-T_c superconductivity does not necessarily need Fermi-liquid states. This point is very crucial in modeling the high-T_c mechanism. In order to clarify this point, we have performed scanning-tunneling spectroscopy (STS) on $Bi_2Sr_2CaCu_2O_8$ [12]. Figure 4 shows a STS spectrum of a cleaved surface of $Bi_2Sr_2CaCu_2O_8$ compared with photoemission [2] and inverse photoemission [3] spectra. The scanning-tunneling-microscope image of the cleaved surface shows a clear incommensurate superstructure characteristic of BiO layer [13], indicating that the cleaved surface is a BiO layer. As shown in Fig. 4, the STS spectrum has little intensity at the

Fermi level while both the photoemission and inverse photoemission spectra have a substantial intensity there. When we take into account of the fact that STS probes a top-most atomic layer, namely BiO layer in this case, while photoemission and inverse photoemission probe a few atomic layers involving BiO and CuO_2 layers, we conclude that BiO layers are almost insulating and the Fermi-liquid states observed by photoemission and inverse photemission spread on CuO_2 layers.

3. Nature of Fermi-Liquid States

We studied the nature of the Fermi-liquid states using resonant photoemission and x-ray absorption spectroscopies.

Figure 5 shows photoemission spectra of $Bi_2Sr_2CaCu_2O_8$ in the vicinity of the Fermi level measured at various photon-enegries. While the spectral intensity shows almost no change even when the photon energy passes the Cu $3p$ core threshold (about 74 eV), it is remarkably enhanced at the O $2s$ core threshold (about 18 eV). The observed resonance strongly suggests that the electronic states in the vicinity of the Fermi level have a dominant O $2p$ nature with far less Cu $3d$ character.

Figure 6 shows oxygen K absorption spectra of $Bi_2Sr_2Ca_{1-x}Y_xCu_2O_8$ (x=0.0 to 1.0) measured with the total-photoemission yield method. In $Bi_2Sr_2Ca_{1-x}Y_xCu_2O_8$, we can control the concentration of holes and resultingly the superconducting transition temperature by changing the ratio (x) between divalent Ca and trivalent Y atoms [14]. $Bi_2Sr_2CaCu_2O_8$ has the T_c of about 85 K as well known and the T_c gradually decreases upon addition of Y, then finally the sample does not show superconductivity around x=0.6-0.7. As shown in Fig. 6, the oxygen K absorption edge, which corresponds to the absorption from the O $1s$ level to the empty O $2p$ states at the Fermi level, has a prominent peak-structure at about 528 eV for x=0.0. The intenisity of peak is gradually reduced as the addition of Y and completely disappears at x=1.0. This is a clear evidence for that doped holes are transferred to the O $2p$ states at the Fermi level, supporting the result of the resonant photoemission (Fig. 5). We should remark in Fig. 6 that the change in the oxygen K absorption spectrum upon addition of Y is not a rigid shift of the spectrum but is a reduction of the intensity of the absorption edge. This point is very important to study the origin of the Fermi-liquid states and will be discussed later.

Figure 7 shows oxygen K absorption spectra of single crystal $Bi_2Sr_2CaCu_2O_8$ measured by changing the direction of the electric vector (E) of incident photons relative to the crystal surface as shown in the figure. On decreasing the incident angle ($\theta_{inc.}$), the intensity of the absorption

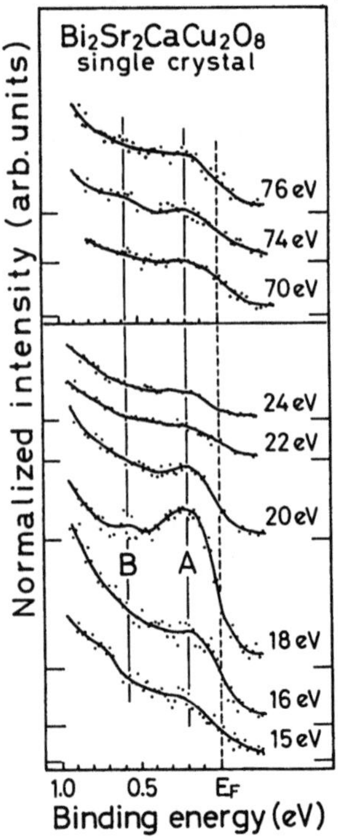

Fig. 5

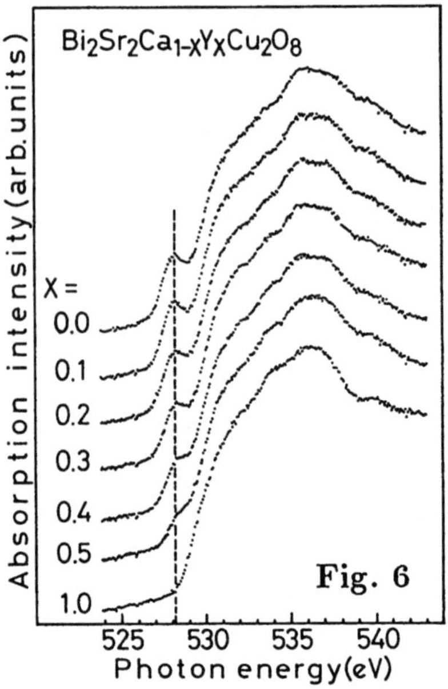

Fig. 6

Fig. 5 Photoemission spectra of $Bi_2Sr_2CaCu_2O_8$ in the vicinity of the Fermi level measured at various photon energies near the Cu 3p and O 2s core thresholds.

Fig. 6 Oxygen K absorption spectra of $Bi_2Sr_2Ca_{1-x}Y_xCu_2O_8$ (x=0.0 to 1.0) measured with the total-photoemission-yield method.

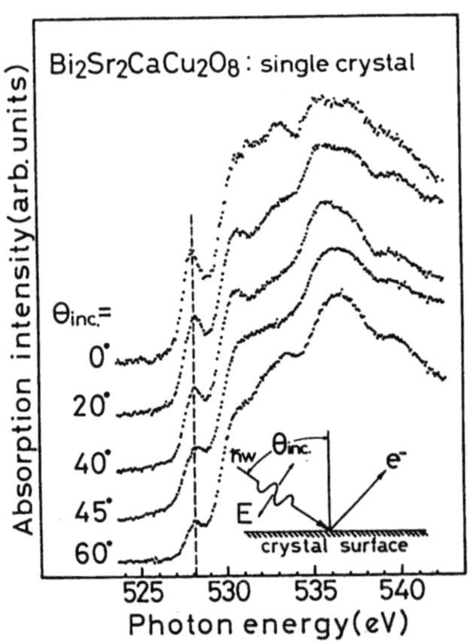

Fig. 7 Oxygen K absorption spectra of single crystal $Bi_2Sr_2CaCu_2O_8$ measured at various incident angles ($\theta_{inc.}$) with linearly polarized incident light.

edge at 528 eV is gradually reduced. This indicates that the O $2p$ orbital at the Fermi level is mostly oriented along the *a*-*b* plane, i.e., CuO_2 plane.

4. Origin of Fermi-Liquid States

So far, it has been revealed that Fermi-liquid states do exist in the high-T_c superconductor and have a dominant O $2p_{x-y}$ nature with its major weight on CuO_2 plane. The next step to approach the high-T_c mechanism is to elucidate the origin of the Fermi-liquid states.

Figure 8 shows valence-band photoemission spectra of $Bi_2Sr_2Ca_{1-x}Y_xCu_2O_8$ (x=0.0 and 0.5). As described above, replacement of Ca with Y atoms reduces the concentration of holes. As shown in Fig. 8, the photoemission spectral intensity in the vicinity of the Fermi level decreases upon the addition of Y. We should remark that the change of valence-band structure due to a decrease of hole concentration is not a simple rigid shift of the Fermi level, but is a decrease of the density of states in the vicinity of the Fermi level. This experimental result is quite consistent with the result of the oxygen K absorpotion experiment in Fig. 6. Thus, it is established that the change in the hole concentration does not move the Fermi level but creates or annihilates electronic states in the vicinity of the Fermi level. This picture entirely contradicts to the band-filling model proposed from the band-structure calculation. The observed behavior of Fermi-edge states quite resembles to that of Kondo states in heavy-electron system. However, the possibility of a Kondo-like state as an origin of the Fermi liquid states is ruled out by the observed dominant O $2p$ nature, since a Kondo-like state should have a strong

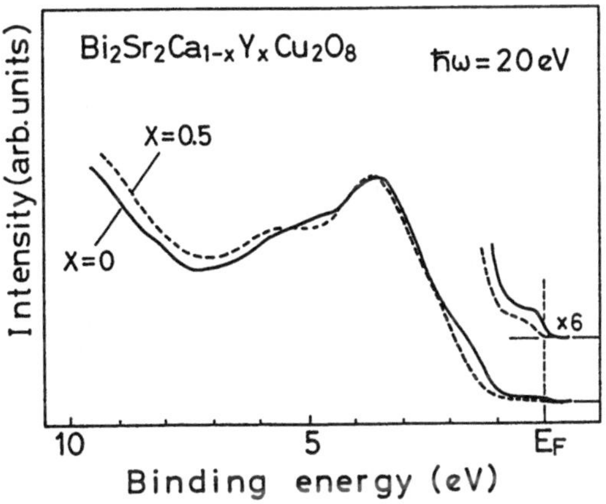

Fig. 8 Valence-band photoemission spectra of $Bi_2Sr_2Ca_{1-x}Y_xCu_2O_8$ (x=0.0 and 0.5).

Cu $3d$ character. We conclude that the Fermi-liquid states consist of a kind of "impurity states (or bands)" formed in the charge-transfer gap through the strong hybridization between doped O $2p_{x-y}$ hole orbitals and surrounding empty Cu $3d_{x^2-y^2}$ orbitals.

5. Conclusion

The angle-resolved photoemission spectroscopy on $Bi_2Sr_2CaCu_2O_8$ has revealed that there are two bands having a small energy dispersion of 0.2-0.5 eV in the vicinity of the Fermi level and they intersect the Fermi level midway between the center and the boundary of the Brillouin zone. This is a ditect evidence for the existence of Fermi-liquid states in the high-T_c superconductor. This result is supported by the inverse photoemission study which found a clear Fermi edge characteristic of normal metal. The scanning-tunneling spectroscopy has elucidated that the Fermi-liquid states spread on the CuO_2 planes but not on the BiO planes.

The nature and origin of the Fermi-liquid states have been studied by photoemission and x-ray absorption spectroscopies on $Bi_2Sr_2Ca_{1-x}Y_x$-Cu_2O_8. The experimental results indicate that the Fermi-liquid states are a kind of "impurity states (bands)" with a dominant O $2p_{x-y}$ nature and they are created through the strong hybridization of doped O $2p_{x-y}$ orbitals and empty Cu $3d_{x^2-y^2}$ orbitals.

The electronic structure of the high-T_c superconductor derived from the present systematic study is depicted in Fig. 9. In undoped compounds such as La_2CuO_4, $YBa_2Cu_3O_{6.5}$ and $Bi_2Sr_2CaCu_2O_8$, a charge-transfer gap (Δ) opens because of the large Coulomb repulsion energy between Cu $3d$ electrons (U_{dd}) compared with the relatively small charge-transfer energy (Δ) between oxygen and copper atoms. Although the valence band is heavily hybridized states of Cu $3d$ and O $2p$ orbitals, the upper part close to the Fermi level has a larger O $2p$ weight and the lower has a dominant Cu $3d$ nature. The unoccupied states close to the Fermi level consists mainly of Cu $3d$ orbital with a small admixture of O $2p$ states. When holes are introduced into this system by substituting the trivalent atom (La or Y) with the divalent one (Sr or Ca) or introducing excess oxygen atoms, the doped holes are accommodated by the O $2p$ orbital because of the large on-site Coulomb repulsion energy on copper atom (U_{dd}). The doping of holes does not necessarily move the Fermi level, but creats new electronic states at the Fermi level. This new electronic state has a dominant O $2p_{x-y}$ nature and would be a kind of "impurity state" produced through the strong hybridization between the doped O $2p_{x-y}$ hole orbital and surrounding empty Cu $3d_{x^2-y^2}$ orbitals. As long as the doping concentration is small, the newly introduced states

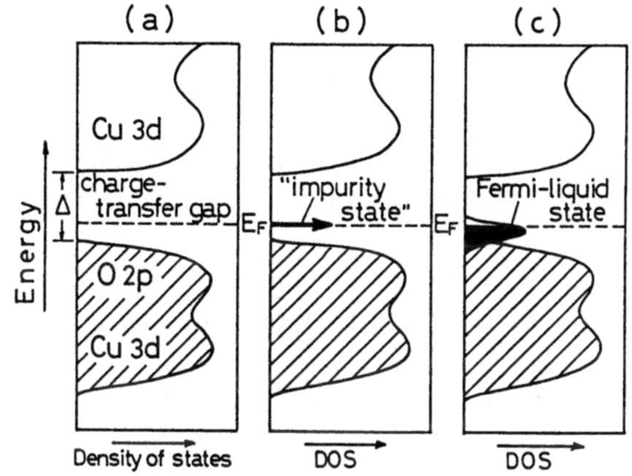

Fig. 9 Schematic diagram of the electronic structure of high-T_c superconductor. (a) charge-transfer semiconductor; (b) "impurity states" are created at the Fermi level by hole-doping; (c) Fermi-liquid states are formed by overlapping of the "impurity states".

would behave literally as impurity states like in semiconductors. The variable range-hopping conduction observed for the small hole-doping-concentration-region is due to carrier-hopping between these "impurity states". When the hole concentration becomes large, the overlapping between the "impurity states" transform them into bands with a small energy dispersion of about 0.5 eV in the vicinity of the Fermi level, i.e., Fermi-liquid states. The superconductivity would be driven by Cooper-pairing of holes with a dominant O $2p_{x-y}$ nature in the Fermi-liquid states spreading on the CuO_2 plane.

Acknowledgment

The author thanks H. Katayama-Yoshida, H. Matsuyama, Y. Okabe, S. Hosoya, H. Ohta, K. Murata, H. Fujimoto, N. Kosugi, S. Suzuki, A. Yagishita, M. Tanaka, S. Yamazaki, and M. Fujinami for their collaboration in the experiments. He is also grateful to Profs. M. Sato, H. Inokuchi, Y. Kuramoto, M. Tachiki, and A. Kotani for their stimulating discussions.

References

1. J.C. Fuggle, J. Fink, and N. Nücker, Int. J. Mod. Phys. B **1**, 1185 (1988).
2. T. Takahashi, H. Matsuyama, H. Katayama-Yoshida, Y. Okabe, S. Hosoya, K. Seki, H. Fujimoto, M. Sato, and H. Inokuchi, Nature (London) **334**, 691 (1988).
3. T. Takahashi, H. Matsuyama, H. Katayama-Yoshida, Y. Okabe, S. Hosoya, K. Seki, H. Fujimoto, M. Sato, and H. Inokuchi, Phys. Rev. B **39**, 6636 (1989).
4. S. Massidda, J. Yu, and A.J. Freeman, Physica C **52**, 251 (1988).
5. H. Ohta, T. Takahashi, K. Murata, H. Matsuyama, S. Suzuki, Y. Okabe, and H. Katayama-Yoshida, Phys. Rev. B **39**, 7354 (1989).
6. A.J. Viescas, J.M. Tranquada, A.R. Moodenbaugh, and P.D. Johnson, Phys. Rev. B **37**, 3738 (1988).
7. B. Reihl, T. Riester, J.G. Bednorz, and K.A. Müller, Phys. Rev. B **35**, 8804 (1987).
8. T. Takahashi, F. Maeda, S. Hosoya, and M. Sato, Jpn. J. Appl. Phys. **26**, L349 (1987).
9. R.S. List, A.J. Arko, Z. Fisk, S.-W. Cheong, S.D. Conradson, J.D. Thompson, C.B. Pierce, D.E. Peterson, R.J. Bartlet, N.D. Shinn, J.E. Shirber, B.W. Veal, A.P. Paulikas, and J.C. Campuzano, Phys. Rev. B **38**, 11961 (1988).
10. Z.-X. Shen, P.D.P. Lindberg, P. Soukiassian, C.B. Eom, I. Lindau, W.E. Spicer, and T.H. Geballe, Phys. Rev. B **39**, 823 (1988).
11. F.U. Hillebrecht, J. Fraxedas, L. Ley, H.J. Trodahl, J. Zaanen, W. Braun, M. Mast, H. Petersen, M. Schaible, L.C. Bourne, P. Pinsukanjan, and A. Zettl, Phys. Rev. B **39**, 236 (1989).
12. M. Tanaka, T. Takahashi, H. Katayama-Yoshida, S. Yamazaki, M. Fujinami, Y. Okabe, W. Mizutani, K. Kajimura, and M. Ono, to be published.
13. Y. Matsui, H. Maeda, Y. Tanaka, and S. Horiuchi, Jpn. J. Appl. Phys. **27**, L361 (1988).
14. T. Tamegai, A. Watanabe, K. Koga, I. Oguchi, and Y. Iye, Jpn. J. Appl. Phys. **27**, L1074 (1988).

A New Family of High-T_c Superconductors: $(Ln_{1-x}Ce_x)_2(Ba_{1-y}Ln_y)_2Cu_3O_{10-z}$ (Ln: Nd, Sm, Eu and Gd)

J. Akimitsu and H. Sawa

Department of Physics, Aoyama-Gakuin University,
6-16-1, Chitosedai, Setagaya-ku, Tokyo 157, Japan

We found superconductivity with the highest transition temperature of about 48K in the Ln-Ce-Ba-Cu-O system, where Ln is Nd, Sm, Eu and Gd. We carried out structural analysis on this system by the combined use of high-resolution electron microscopy and Rietveld analysis of the X-ray powder diffraction technique. This oxide was found to be a new family of solid solutions $(Ln_{1-x}Ce_x)_2(Ba_{1-y}Ln_y)_2Cu_3O_{10-z}$. The crystal structure is body centered tetragonal with the space group I4/mmm. Its structure consists of tetragonal $Ba_2YCu_3O_{7-y}$- and Nd_2CuO_4-type units stacked alternatively along the c-axis.

1. INTRODUCTION

Despite the tremendous efforts aiming at understanding of the high-T_C superconductivity, the mechanism lying behind is not yet clear. It is worth pointing out that the progress in the elucidation of the mechanism of high-T_C superconductivity has always been coupled with the discovery of new materials.

Recently, our group found a new class of superconductors in the Nd-Ce-Sr-Cu-O system[1]. We found that the crystal structure of this system consists of alternating slabs of the K_2NiF_4 and Nd_2CuO_4 structures, and that all the copper atoms are crystallographically identical, surrounded by oxygens in square-pyramidal coordination[2]. Ce^{4+} atoms are a crucial role in this compounds.

Here, we report another new Ce^{4+} contained family of superconducting copper oxides $(Ln_{1-x}Ce_x)_2(Ba_{1-x}Ce_x)_2Cu_3O_{10-z}$ with the highest onset transition temperature of 48K[3]. The crystal structure of this system consists of alternating slabs of the tetragonal $Ba_2YCu_3O_y$ and Nd_2CuO_4 structures. This structure of our new material is similar to that of tetragonal $Ba_2YCu_3O_y$ in Cu-O frame. It is important, therefore, to elucidate the similarity and difference between $Ba_2YCu_3O_{7-y}$ and this new compound.

2. SAMPLE PREPARATION AND SUPERCONDUCTING PROPERTIES

The samples were prepared by the solid state reaction of rare-earth metal oxides (Nd_2O_3, Sm_2O_3, Eu_2O_3, Gd_2O_3 and CeO_2), $BaCO_3$ and CuO in appropriate ratio. They are mixed, pressed into pellets and heated in O_2 atmosphere at 1020°C for 15h. The reacted mixtures were reground, pressed into pellets and reheated under the same conditions. Finally, they are annealed at 650°C for 24h under high-pressure(80Kg/cm^2) O_2 atmosphere. Single-phase samples were obtained at a nominal composition of Ln:Ba:Ce:Cu=6:4:2:9 for Ln=Eu atom. However, the solubility ratio is different for each rare earth oxides, nominal compositions are slightly different for the other rare earth compounds.

A standard four-probe technique was employed for the resistivity measurements. The temperature was determined with a Au+0.07at%Fe-chromel thermocouple attached to the sample. Figure 1 shows the temperature dependences of resistivities for $Ln_6Ba_4Ce_2Cu_9O_y$ with Ln:Nd, Sm, Eu. Although the samples are single-phased in according to their powder X-ray diffraction patterns, the semiconducting behaviors are observed at low temperature. This seems to be due to the oxygen vacancies, since such a tendency tends to disappear after the long-term annealing under the high-pressure oxygen as seen in the case of Eu-compound. The maximum onset superconducting transition temperature at present is about 48K. The superconducting characteristics of this materials strongly depend on the annealing conditions. Figure 2 shows the temperature dependences of the A.C. magnetic susceptibilities in $Nd_6Ba_4Ce_2Cu_9O_y$. The curve(2) for the sample annealed under O_2 atmosphere improved superconducting behavior with regard to

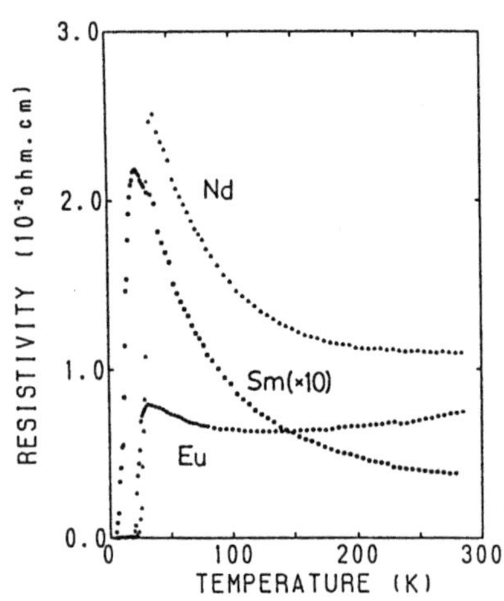

Fig. 1. Temperature dependences of the resistivities in $Ln_6Ba_4Ce_2Cu_9O_y$ (Ln : Nd, Sm, Eu).

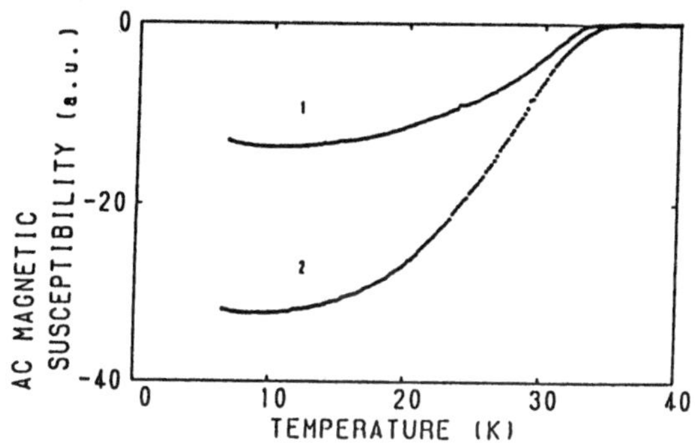

Fig. 2. Temperature dependences of the A.C. magnetic susceptibilities in $Nd_6Ba_4Ce_2Cu_9O_y$. Curves (1) and (2) correspond to the non-annealed and the annealed samples, respectively.

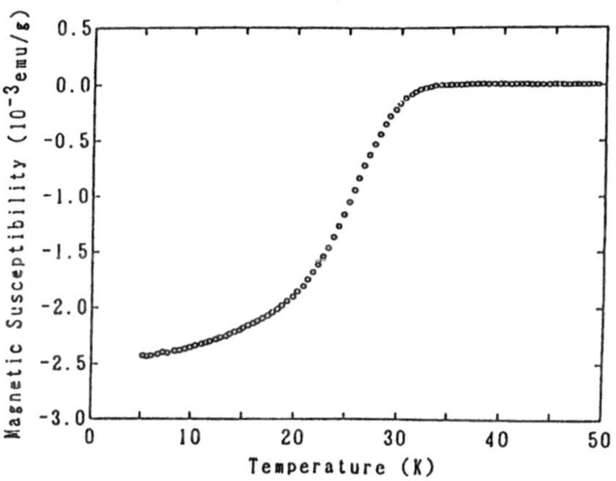

Fig. 3. Meissner signals in $Eu_6Ba_4Ce_2Cu_9O_y$.

both the Meissner volume fraction and transition temperature compared to the curve(1) for a non-annealed sintered sample.

Figure 3 shows a typical example of the Meissner diamagnetism in $Eu_6Ba_4Ce_2Cu_9O_y$ as measured with the SQUID magnetometer in an applied field of 1.14 Oe under field-cooled condition. The Meissner volume fraction is about 22% at 4.2K, confirming the bulk nature of superconductivity.

Recently, Tamegai et al.[4] measured the Hall effect on the Eu and Nd compounds(Figure 4). Hall coefficients for these compounds are positive, indicating that carriers are holes. As shown in Figure 4, Hall coefficients are slightly temperature dependent, with a broad maximum at around 120K. Average hole number per copper ion in the Cu-O planes at just above T_C is estimated to be 0.09. This value is much smaller the value 0.23 in $Ba_2YCu_3O_{7-y}$. The difference in the carriers concentrations may be the reason why T_C in this system is as low as about 40K.

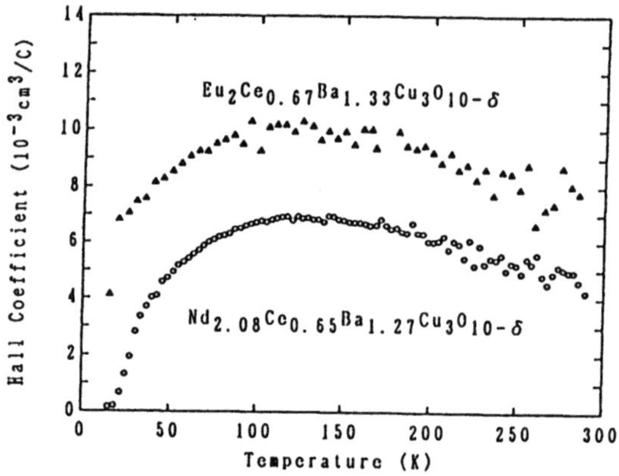

Fig. 4. Temperature dependences of Hall coefficients for $Eu_6Ba_4Ce_2Cu_9O_y$ and $Nd_6Ba_4Ce_2Cu_9O_y$.

3. CRYSTAL STRUCTURE DETERMINATION

To determine the crystal structure of this system, we performed the Rietveld analysis of X-ray powder diffraction as well as high-resolution electron imaging technique.

Electron diffraction patterns and electron micrographs were taken with a JEM-2000EX-type electron microscope operated at 200kV[3]. Electron diffraction patterns of the $Eu_6Ba_4Ce_2Cu_9O_y$ indicate that the unit cell is body-centered tetragonal ($a \simeq 3.8Å$ and $c \simeq 28Å$); hkl reflections with odd values of h+k+l are systematically absent. Tsuda et al.[5] determined the space group to be I4/mmm (No. 139) by using the convergent-beam electron diffraction.

The high-resolution image given in Figure 5 was taken with the incident beam parallel to [100]. Two types of arrangements were observed in the image. One is the large two dark dots marked with R. This may be R_2O_2 layers (Nd_2CuO_4-type slabs). Between two R_2O_2 layers, zig-zag arrangement of three perovskite layers was observed. They are considered to be perovskite layers consisting of almost the same heavy ions (marked as R') and Cu ions. Local structure must be similar to that of the tetragonal $Ba_2YCu_3O_{7-y}$. Therefore, the layer sequence in the structure of our new compounds is shown in Figure 6.

Based on the above lattice image and referring to cationic order in $Ba_{1.8}Nd_{1.2}Cu_3O_{7-y}$[6] and $(Nd,Ce,Sr)_2CuO_4$[2], the structural model was constructed. The R site (8-coordinated site in Nd_2CuO_4-type) is occupied by Eu^{3+} and Ce^{4+} ions, because both ions have nearly the same radius (ionic radii in

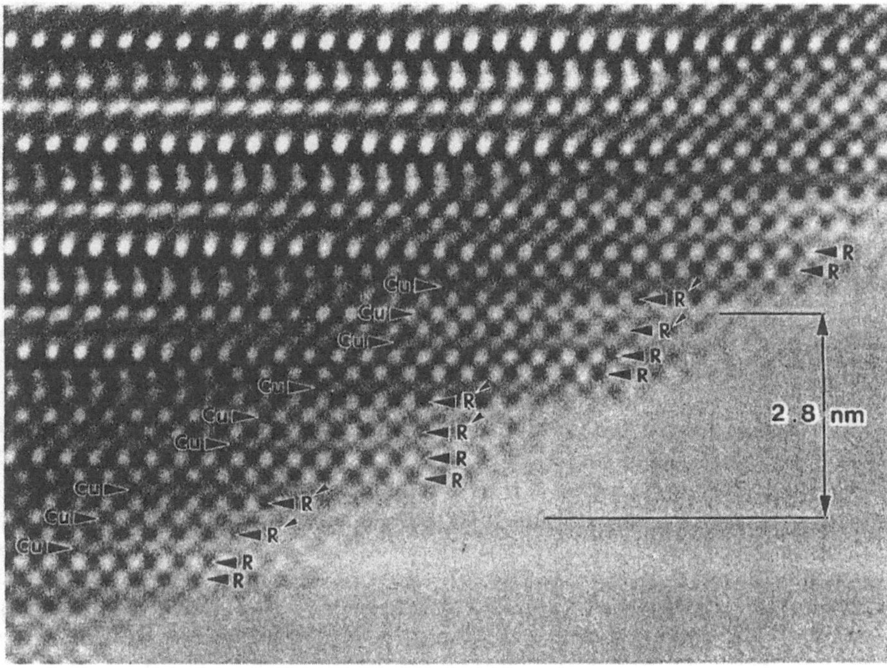

Fig. 5. A high resolution electron microscope image of $Eu_6Ba_4Ce_2Cu_9O_y$ taken with an incident electron beam along the a axis. Dots indicated by R and R' correspond to the (Eu,Ce) and (Ba,Eu) sites, respectively.

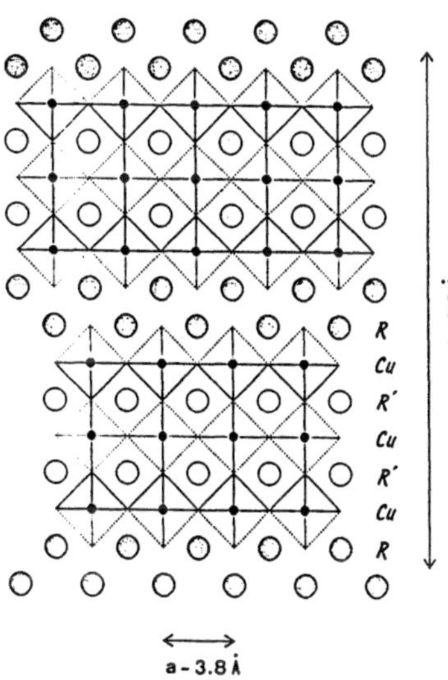

Fig. 6. Schematic structural model of $R_2R'_2Cu_3O_y$. Circles indicated by R and R' correspond to the same construction in Fig.5.

the 8-coordinated site are $^{VIII}Eu^{3+}$:1.066Å and $^{VIII}Ce^{4+}$:0.97Å[7]). On the other hand, Ba^{2+} ionic radii are too large in the coordination ($^{VIII}Ba^{2+}$:1.42Å). The coordination number in the R' site is difficult to fix because of the oxygen vacancy. Analogous to $Ba_2YCu_3O_{7-y}$, this site is supposed to be 10-coordinated. We assume that the relatively large Ba^{2+} and Eu^{2+} ions prefer the R' site ($^{X}Ba^{2+}$:1.52Å and $^{X}Eu^{2+}$:1.35Å). Therefore, we assume the structural model to be $(Eu_{1-x}Ce_x)_2(Ba_{1-y}Eu_y)_2Cu_3O_{10-z}$ in a chemical formula for the Rietveld analysis.

Powder X-ray diffraction was performed on a Rigaku diffractometer RAD-rC with a rotating anode source and a graphite monochromator. Intensity data were collected with Cu K_α radiation at 0.02° intervals over the 2θ range from 20° to 110°. The diffraction pattern indicates that the sample is single phased. The crystal structure was refined by Rietveld analysis with the RIETAN program[8]. The preferred orientation was corrected using a gaussian preferred-orientation function on the assumption that the cleavage plane is (001). Preliminary calculations reveal that the thermal parameters of each oxygen atom could not be refined independently because of the strong correlation between the occupation factors and thermal parameters of oxygens. Therefore, the thermal parameters of all the oxygen sites were approximated to be the same for the refinements.

Final structure parameters are listed in Table I. The refined lattice parameters are a = 3.85044(6)Å and c = 28.4598(5)Å. The R factors are R_{wp}=6.6%, R_p=5.2%, R_I=2.9%, R_F=2.8% and R_e=4.2%. Figure 7 shows the final profile fit and difference patterns. Good agreement between the observed and

Table I.
Structure parameters in $(Eu_{0.67}Ce_{0.33})_2(Ba_{0.67}Eu_{0.33})_2Cu_3O_{10-z}$ (Z=2). Estimated standard deviations in parentheses refer to the last digit printed. B is the isotropic thermal parameter. The occupation factors, g, of Eu, Ce and Ba are fixed at 0.67, 0.33 and 0 for site R, and at 0.33, 0 and 0.67 for site R'.

Atom	Site	x	y	z	g	B/Å^2
R	4e	0	0	0.2942(3)	1	0.6(2)
R'	4e	0	0	0.4245(3)	1	1.1(2)
Cu(1)	2a	0	0	0	0.85(5)	2.2(12)
Cu(2)	4e	0	0	0.1423(8)	1	1.1(3)
O(1)	4c	0	1/2	0	0.27(15)	2.7(8)
O(2)	4e	0	0	0.066(3)	1	2.7(8)
O(3)	8g	0	1/2	0.354(2)	1	2.7(8)
O(4)	4d	0	1/2	1/4	1	2.7(8)

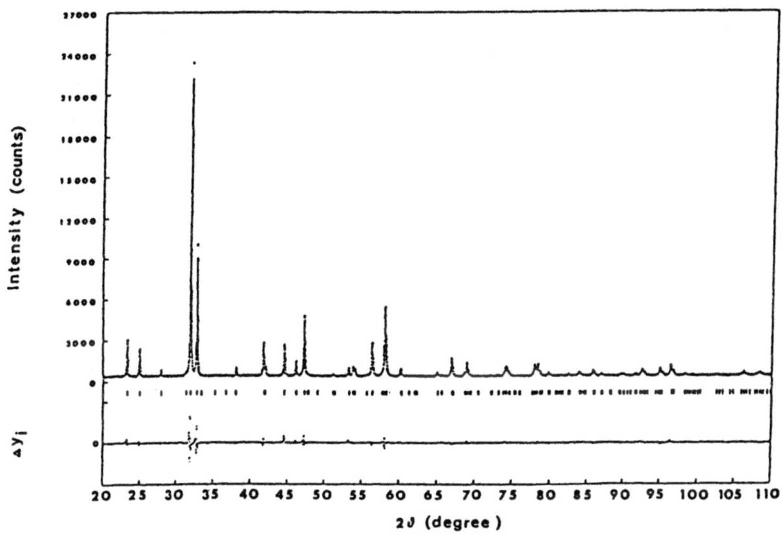

Fig. 7. Rietveld refinement of an X-ray diffraction pattern of $(Eu_{0.67}Ce_{0.33})_2(Ba_{0.67}Eu_{0.33})_2Cu_3O_{10-z}$. The solid lines are the calculated intensities, the crosses overlying them are the observed intensities, and Δy_i is the difference between the observed and calculated intensities. The short vertical lines mark the positions of the 236 peaks.

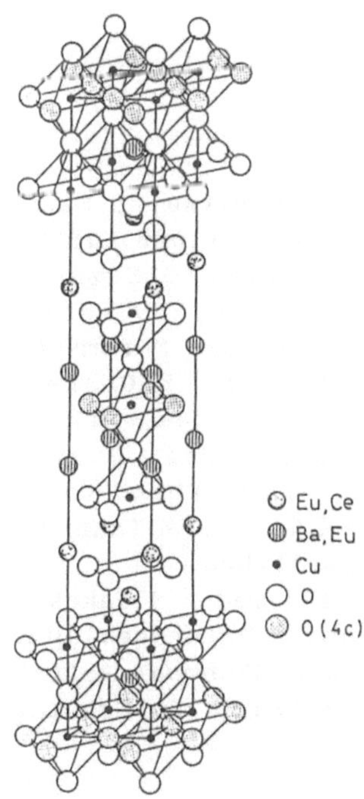

Fig. 8. Crystal structure of the $(Eu_{0.67}Ce_{0.33})_2(Ba_{0.67}Eu_{0.33})_2Cu_3O_{10-z}$. O(4c) site is oxygen-deficient (see Table I).

calculated patterns strongly supports the proposed structural model. Although the X-ray data were sufficient to determine the arrangement of the metal ions, to obtain reliable structure parameters for oxygen a precise structure refinement on neutron powder diffraction data will be necessary.

The crystal structure of the Eu-Ce-Ba-Cu-O system is shown in Figure 8. The unit consists of two parts: the upper and lower slabs correspond to the Nd_2CuO_4-type structure and the center part exhibits Cu-O pyramids and a deficient perovskite structure. In the Cu-O frame, the structure is very similar to tetragonal $Ba_2YCu_3O_{7-y}$.

We wish to thank Prof. K. Kitazawa and Dr. K. Kishio (The University of Tokyo) for the SQUID measurements. We also wish to thank Dr. Y. Matsui, Dr. S. Horiuchi and Dr. F. Izumi (National Institute for Research in Inorganic Materials) for their invaluable help and useful discussions. Prof. D. J. Kim is deeply acknowledged for his critical reading of this manuscript.

This work was supported by a Grant-in-Aid for Scientific Research on Priority Areas "Mechanism of Superconductivity".

REFERENCES

1. J.Akimitsu, S.Suzuki, M.Watanabe and H.Sawa: Jpn. J. Appl. Phys. **27**(1988)L1859.
2. H.Sawa, S.Suzuki, M.Watanabe, J.Akimitsu, H.Matsubara, H.Watabe, S.Uchida, K.Kokusho, H.Asano, F.Izumi and E.Takayama-Muromachi: Nature **337**(1989)347.
3. H.Sawa, K.Obara, J.Akimitsu, Y.Matsui and S.Horiuchi: J. Phys. Soc. Jpn. **58**(1989).
4. T.Tamegai, Y.Iye, M.Ogata, K.Obara and J.Akimitsu:Jpn. J. Appl. Phys. to be published.
5. K.Tsuda, M.Tanaka and J.Akimitsu:Jpn. J. Appl. Phys. to be published.
6. F.Izumi, S.Takekawa, Y.Matsui, N.Iyi, H.Asano, T.Ishigaki and N.Watanabe: Jpn. J. Appl. Phys. **26**(1987)L1616.
7. R.D.Shannon: Acta Crystallogr. **A32**(1976)751.
8. F.Izumi: J. Crystallogr. Soc. Jpn. **27**(1985)23.

Magnetic Interactions in High-T_c Superconductors and Related Oxides

H.R. Ott[1], *P. Birrer*[2], *F.N. Gygax*[2], *B. Hitti*[2], *E. Lippelt*[2], *A. Schenck*[2], *M. Weber*[2], *S. Barth*[1], *F. Hulliger*[1], *P. Allenspach*[3], *S.-W. Cheong*[4], *P. Fischer*[3], *Z. Fisk*[4], *A. Furrer*[3], *and B. Rupp*[5]

[1]Laboratorium für Festkörperphysik, ETH-Hönggerberg, CH-8093 Zürich, Switzerland
[2]Institut für Mittelenergiephysik, ETH Zürich, CH-5232 Villigen PSI, Switzerland
[3]Labor für Neutronenstreuung, ETH Zürich, CH-5232 Villigen PSI, Switzerland
[4]Los Alamos National Laboratory, Los Alamos, NM 87545, USA
[5]Paul-Scherrer Institut, CH-5234 Villigen PSI, Switzerland

Zero-field μ^+SR measurements in a ceramic sample of $Ba_2HoCu_3O_7$ reveal 3D antiferromagnetic order below 5 K and some incoherent magnetic correlations up to 50 K in a decreasing volume fraction of the sample. A change of the magnetic-moment arrangement around 100 mK indicates competition between isotropic and anisotropic exchange interactions. The interpretation of the data also allows for a fairly consistent determination of the μ^+-site in $Ba_2HoCu_3O_7$. Since Ho^{3+} adopts a singlet ground state, the onset of magnetic order in the superconducting state is not à priori expected.
Neutron- and magnetization studies on a parent compound of n-type high-T_c superconductors, Pr_2CuO_4, reveal antiferromagnetic order among the Cu moments below 190 K in a single crystal and below 250 K in a powder sample. Some of the magnetic features of Pr_2CuO_4 are similar to those reported for La_2CuO_4. The most obvious difference between the two materials is the temperature dependence of the Cu sublattice magnetization below T_N. The large value of T_N of Pr_2CuO_4 is surprising in view of the tendency towards frustration of the Cu-atom sublattice in the T structure. Inelastic neutron-scattering data indicate a singlet ground state for the Pr^{3+} ions in this substance.

1. Introduction

Some of the oxide superconductors with elevated critical temperatures are obtained by suitably doping materials that are insulators and order magnetically. With increasing doping, magnetic order is continuously

Springer Series in Solid-State Sciences, Vol. 89 Strong Correlation and Superconductivity
Editors: H. Fukuyama · S. Maekawa · A.P. Malozemoff

suppressed and above a critical substitution level, the substances turn into metals which are high-temperature superconductors. This must come about by characteristic changes of the electronic subsystem and its interactions. Experimental investigations probing these interactions or their influence on ground-state properties are therefore much desired in view of modelling the anomalous features of the normal state and to single out possible causes for transitions to a superconducting state at relatively high temperatures. Below we describe two microscopic studies aiming at these goals.

2. Magnetic Correlations in Superconducting $Ba_2HoCu_3O_{6.9}$

Substitution of Y in $Ba_2YCu_3O_7$ by most rare-earth elements does not affect the superconducting properties of these compounds in the sense that still critical temperatures of the order of 90 K are observed [1]. Some of these rare-earth compounds order antiferromagnetically at low temperatures and superconductivity coexists with magnetic order, a previously observed phenomenon. Among these compounds, $Ba_2HoCu_3O_7$ is interesting because inelastic neutron-scattering data [2] reveal a singlet ground state of the crystal-field split Ho^{3+} Hund's rule ground-state multiplet. From previous work in rare-earth intermetallics it is known that over-critical exchange interactions may drive such a singlet unstable by admixture of excited states via a non-diagonal matrix element [3].

Preliminary neutron-diffraction experiments on $Ba_2HoCu_3O_7$ indicate some kind of magnetic ordering at very low temperatures [4] and the present μ^+SR study was aimed to reveal more information on this phenomenon. Previous μ^+SR experiments [5-7] indicated an effective-spin freezing below about 3 K but no evidence for static magnetic order was detected. An induced-moment is, of course, not à priori expected in a superconducting background matrix.

The zero-field μSR measurements were made at the Paul Scherrer Institute's low-energy πM3 beam line. Use of the low-temperature facility (LTF) enabled to do measurements between 0.039 and 2.8 K and with a conventional spectrometer the temperature range between 2.4 and 50 was covered. The polycrystalline sample showed a magnetically-probed superconducting transition at 90 K and consisted, according to a neutron-diffraction test, of essentially single-phase material.

Figure 1 shows the time dependence of the μ^+ polarization at various temperatures. A strongly damped oscillatory component is the most prominent feature at low-temperatures and is still faintly dis-

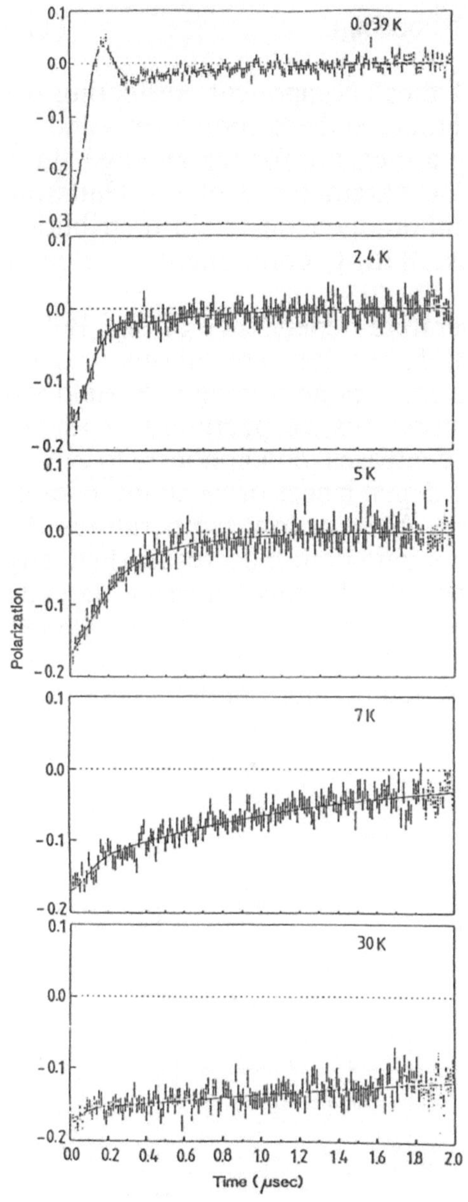

Fig. 1 - Time dependence of the μ^+ polarization $P_\mu(t)$ at different temperatures in $Ba_2HoCu_3O_7$

cernible at 2.4 K. At all temperatures, a two-component approximation is needed to fit the experimental data. The solid lines in Fig. 1 are fits using the functions

$$P_\mu(t) = A \quad e^{-\lambda t} + A \quad e^{-(1/2)\sigma^2 t^2} \cos\omega_\mu t. \tag{1}$$

A and A are the amplitudes of these components reflecting the respective fraction of muons contributing to these components and ω_μ is the μ^+ Larmor frequency giving a measure for the average static magnetic field at the μ^+ sites. The parameter σ of the Gaussian relaxation function indicates an inhomogeneous field distribution across the μ^+ sites. The non-oscillating component is clearly governed by exponential damping.

In Fig. 2 we show the temperature dependence of ω_μ. Below 75 mK, it amounts to about 2.5 MHz but decreases abruptly to 1.5 MHz at 0.15 K. These frequencies correspond to magnetic-field values of 185 and 111 G at the muon sites, respectively. With increasing temperature the frequency decreases monotonously to zero at about 5 K. In Fig. 3 the temperature dependences of the amplitudes of the two components are displayed. At very low temperatures the oscillatory component is clearly dominating but it decreases distinctly above 1.5 K. Although ω_μ vanishes at 5 K, this Gaussian damped component persists at least up to temperatures of 50 K. In this higher-temperature range most of the muons relax exponentially and this component decreases rapidly in the temperature range where, with decreasing temperature, ω_μ adopts non-zero values.

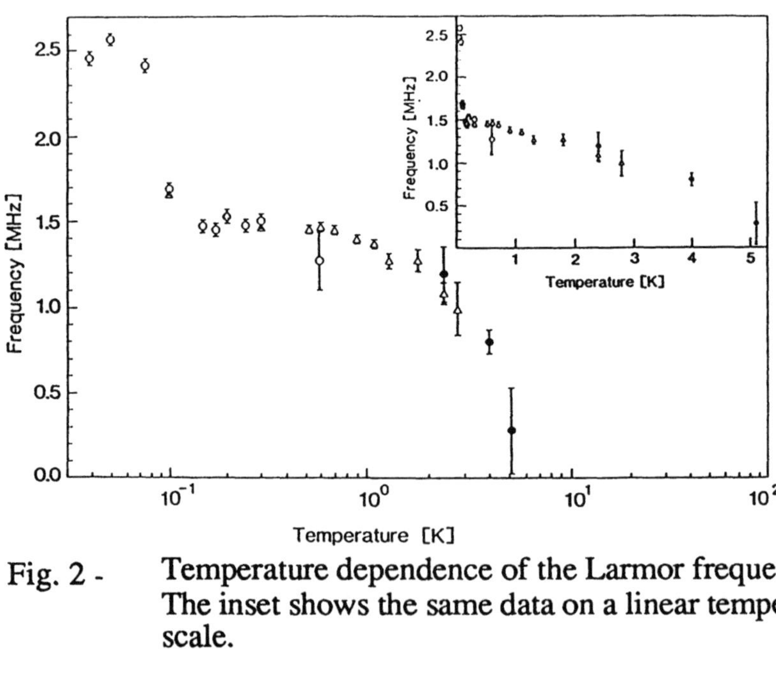

Fig. 2 - Temperature dependence of the Larmor frequency $\omega_\mu/2\pi$. The inset shows the same data on a linear temperature scale.

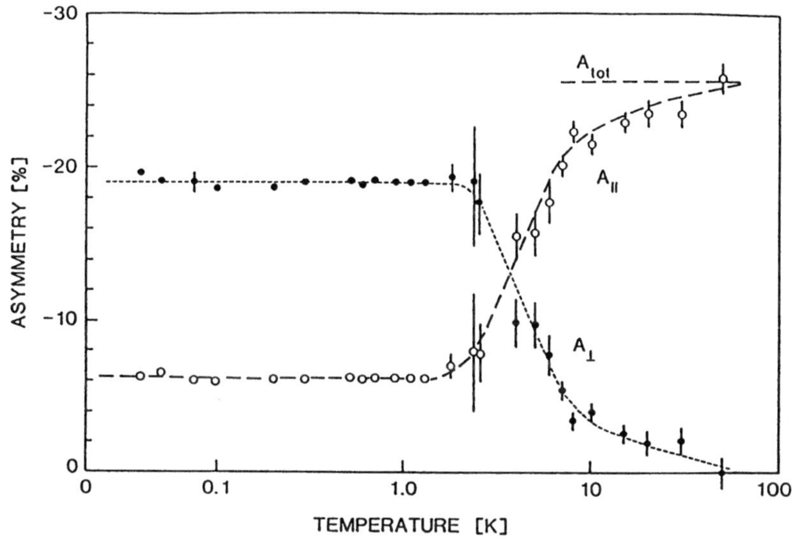

Fig. 3 - Temperature dependence of the fitted amplitudes A and A . The lines are to guide the eye.

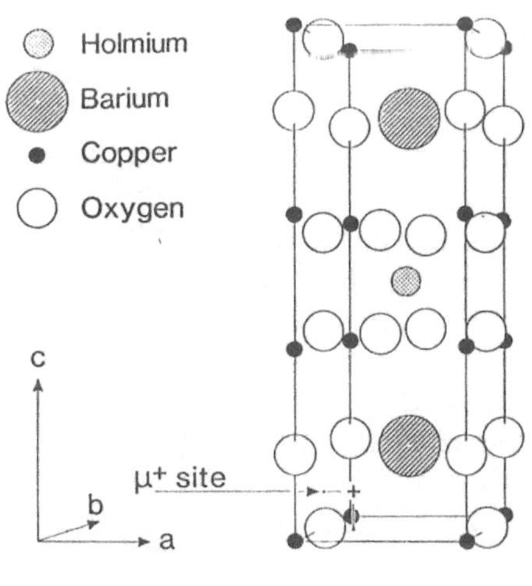

Fig. 4 - Proposed μ^+ site within the unit cell of $Ba_2HoCu_3O_7$.

Before we briefly discuss these data we comment on the most likely μ^+ site in the $Ba_2YCu_3O_7$ lattice. This site, which is indicated as a cross in Fig. 4, provides agreement between calculated and experimentally determined values for σ and ω_μ in $Ba_2YCu_3O_7$ and $Ba_2GdCu_3O_7$, respectively [8,9], where these quantities are well defined. Provided that this site is also adopted in $Ba_2HoCu_3O_7$ our ω_μ values can now be transformed into effective moments of the Ho^{3+} ions. Assuming the antiferromagnetic structure suggested in ref. 4, we obtain an effective Ho moment of 2.56 μ_B for T < 0.1 K. In view of the μ^+ site being situated halfway between Ho layers, the observation of well defined oscillations is, in our opinion, good evidence for three dimensional magnetic order. The drastic change of ω_μ around 0.1 K is interpreted as a rotation of the Ho moments from being parallel to the c axis below 0.1 K into the [a,b] plane above. This moment arrangement and a doubling of the unit cell in all three crystallographic directions results in a Ho moment of 2.65 μ_B, essentially unchanged with respect to the low-temperature value. It is this consistency which makes this interpretation to be the most likely one although the experimentally determined crystal-field symmetry would favour the easy axis in the basal plane. Therefore a reorientation of the moments at very low temperatures is not à priori expected but may be a result of competing isotropic and anisotropic exchange interactions.

The temperature dependences of the other parameters (for λ and for σ see ref. [10]) indicate that the volume fraction with static magnetic order decreases rapidly with increasing temperature before $\omega_\mu = 0$. However, the width of the field distribution as determined by σ does not change significantly from the mK range up to 15 K. The exponential relaxation component is due to internal-field fluctuations perpendicular to the quasistatic field direction. At the lowest temperatures this contribution probably arises from uncorrelated precessions of the Ho moments around the respective preferred orientations mentioned above.

What is particularly intriguing is that the onset of magnetic correlations is not, as far as μSR can tell, accompanied by a cooperative phase transition. The moment-freezing in the paramagnetic phase sets in at temperatures below 50 K and smoothly develops into a quasi-statically-ordered phase which is complete throughout the sample at approximately 2 K. Since this onset of magnetism occurs in a superconducting matrix, it remains to be analyzed which kind of exchange interactions might be responsible for these observations. In particular it needs to be explained how the large distance between the Ho-planes and the atomic arrangements between them allow for a three dimensional ordering among these moments. A more complete discussion of the low-temperature magnetism in $Ba_2HoCu_3O_7$ is presented in ref. [10].

3. Magnetic Ordering in Pr_2CuO_4

With the discovery of superconductivity in suitably doped compounds of the form Ln_2CuO_4 (Ln = Pr, Nd, Sm) where the doping may be done with Ce or Th on the Ln sites [11,12] or F on the O sites [13], studies of the physical properties of the respective parent compounds are of some importance. As before in the case of La_2CuO_4 and the well known doped varieties of high-T_c superconductors it is particularly the structural and magnetic properties of the parent compounds that may give hints for reasons of the occurrence of superconductivity in the doped materials. It is known for some time [14] that La_2CuO_4 and Ln_2CuO_4 (Ln = Lanthanides) adopt distinctly different crystal structures and it seems obvious to check whether this difference also manifests itself in the magnetic properties of these compounds. Particularly intriguing is the difference in oxygen arrangements around the Cu atoms, which is of octahedral type in La_2CuO_4.but of quadratic-planar nature in Ln_2CuO_4. Below we describe an investigation on magnetic properties of Pr_2CuO_4.

Our experiments included macroscopic magnetization measurements on single-crystalline Pr_2CuO_4 between 4.2 K and room temperature. External fields up to 100 kOe were applied both parallel and perpendicular to the basal plane. A large single crystal was also investigated by means of neutron diffraction for studying the crystallographic structure and searching for ordered magnetic moments as a function of temperature between 10 and 300 K in zero magnetic field. The same set of neutron experiments was repeated on a polycrystalline powder sample. On this same sample, inelastic neutron scattering was used to probe possible excitations between crystal-field split energy levels of the trivalent Pr-ion 4f electrons.Our structural data confirm the $Nd_2Cu_2O_4$ T -type structure at room temperature and indicate that this structure persists down to 10 K. Additionally recorded low-temperature x-ray diffraction patterns confirm this conclusion.

No indication for a magnetic phase transition is discernible from the macroscopic susceptibilities χ and χ , shown in Fig. 5. Both the magnitude and the anisotropy are governed by the magnetic response of the localized 4f-electron moments (see also below). Close inspection of the low-temperature neutron-diffraction patterns, however, reveals weak additional peaks that are not observed in the spectra obtained at room temperature. As an example we show the additional (1/2, 1/2, 1)-signal on an extended angular scale in Fig. 6. Similar additional peaks were observed at orientations (1/2, 1/2, 2) and (1/2, 1/2, 4) but no analogous signal was obtained at (1/2, 1/2, 0). These observations are compatible with an antiferromagnetically-ordered arrangement of Cu moments where the moment orientation is along (-1, 1, 0). In spite of the low intensities, the calculated and experimental neutron intensities are in fair agreement when assuming the magnetically-ordered structure shown in Fig. 7. The data from both the powder spectrum and the single-crystal measurements are compatible and an ordered Cu moment of 0.45 ±

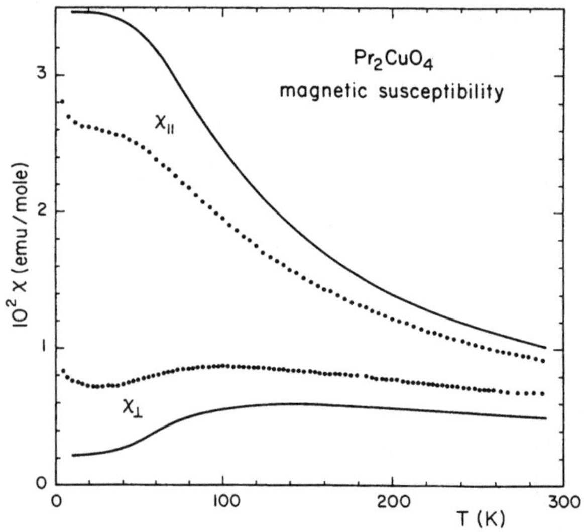

Fig. 5 - Temperature dependence of the magnetic susceptibilities χ and χ of Pr_2CuO_4. The solid lines are calculations of the single-ion susceptibility based on an experimentally determined crystal-field energy level scheme.

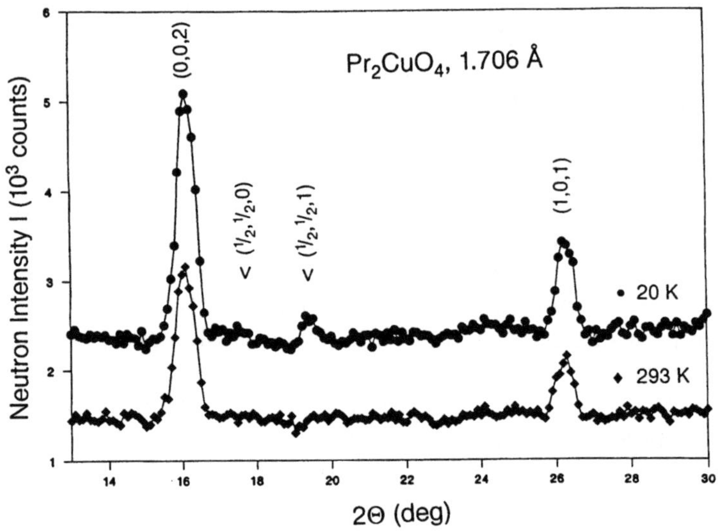

Fig. 6 - Details of the neutron-diffraction pattern obtained from polycrystalline Pr_2CuO_4 at 20 K and room temperature.

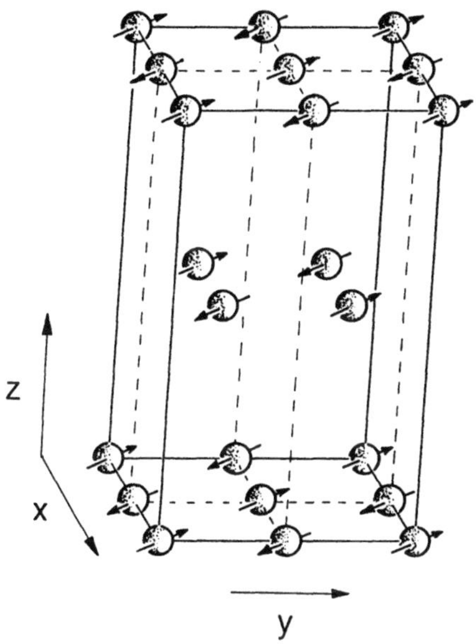

Fig. 7 -
The magnetic unit cell of Pr_2CuO_4 below T_N.

0.12 μ_B is calculated. This is obviously much less than the expected value of approximately 1 μ_B /Cu atom.

In Fig. 8 we show the temperature dependence of the most intense magnetic (1/2, 1/2, 1) peak obtained in the single-crystal measurements in reduced coordinates. All other peaks from single-crystal or powder experiments show similar temperature behaviour. The inset of Fig. 8 demonstrates the signal intensity above and well below the ordering temperature. The plot shown in Fig. 8 but drawn on absolute temperature scale yields a Néel temperature of (190 ± 2) K. This is distinctly different to the observation using the powder sample, where T_N is approximately 250 K. This discrepancy is most likely a result of different oxygen content in these specimens. According to the neutron structure analysis, our powder sample shows no measureable amount of oxygen vacancies but we suspect this to be the case for the single crystal. As for La_2CuO_4, the cooperative ordering of the Cu moments is obviously influenced by the occupancy of the oxygen sites in a very essential manner.

An obvious feature of Fig. 8 is the peculiar temperature dependence of the sublattice magnetization which is distinctly different from an $S = 1/2$ mean-field behaviour as indicated by the dotted line in Fig. 8. A rough estimate of the critical parameter β results in a value of $\beta = 0.34 \pm 0.04$. Although this indicates 3-dimensional character of the ordering process, no more-detailed conclusions can be drawn so far.

In view of the crystallographic differences between La_2CuO_4 and Pr_2CuO_4 it is somewhat surprising that the magnetic properties of

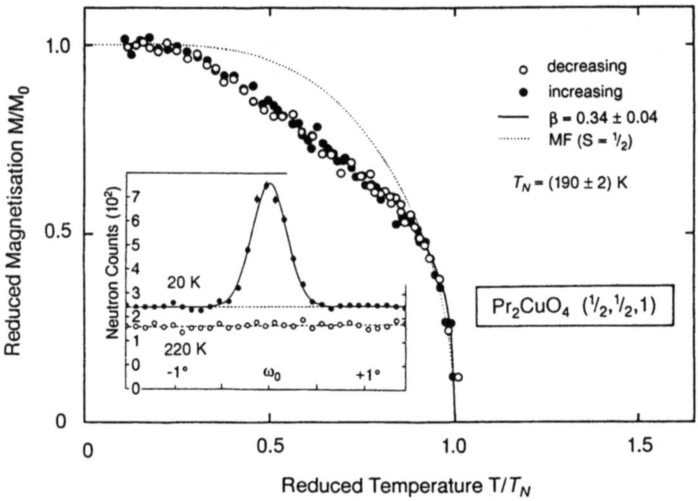

Fig. 8 - Temperature dependence of the Cu-sublattice magnetization of Pr_2CuO_4. The inset demonstrates the peak intensity above and below T_N.

these two compounds are, at least as far as the Cu-atom subsystem is concerned, almost identical. In spite of the tendency of the Cu sublattice to frustration in Pr_2CuO_4, the Néel temperature is found to be of the same order of magnitude as in La_2CuO_4. A major issue in the magnetic behaviour of La_2CuO_4, the claimed dominant 2-dimensional correlations [15], has not yet been investigated for Pr_2CuO_4. The differences in oxygen coordination around the Cu atoms and in the temperature dependences of the sublattice magnetization do not allow for a straightforward conclusion by comparison for the interpretation of the data on Pr_2CuO_4.

The analysis of the inelastic neutron-scattering spectra reveal a singlet ground state of the crystal-field split J = 4 multiplet of the Pr^{3+} ions. It is therefore not surprising that no indication for magnetic ordering of the Pr ions above 4.2 K has been found. In Fig. 5 we show, as solid lines, the calculated single-ion susceptibilities that are obtained from the experimentally derived crystal-field energy-level scheme. It may be seen that the observed anisotropy and the salient temperature-dependent variations are well reproduced.

A more detailed account of the magnetic properties of Pr_2CuO_4 is given in ref. [16].

4. Acknowledgement

Part of this work was financed by the Schweizerischer Nationalfonds zur Förderung der wissenschaftlichen Forschung. Work at Los Alamos was done under the auspices of the US Department of Energy.

References

1. see, e.g., F. Hulliger and H.R. Ott, Z. Phys. B 67, 291 (1987)
2. A. Furrer, P. Brüesch, and P. Unternährer, Phys. Rev. B 38, 4616 (1988)
3. K. Andres, E. Bucher, S. Darack, and J.P. Maita, Phys. Rev. B 6, 2716 (1972)
4. P. Fischer, K. Kakurai, M. Steiner, K.N. Clausen, B. Lebech, F. Hulliger, H.R. Ott, and P. Unternährer, Physica C 152, 145 (1988)
5. N. Nishida, H. Miyatake, S. Okuma, Y. Kuno, Y. Watanabe, R. Yamazaki, S. Hikami, M. Ishikawa, T. Takabatake, Y. Nakazawa, S. F. Kreitzman, J.H. Brewer, and Ch.-Y. Huang, Jap. J. Appl. Phys. 27, L94 (1988)
6. V.N. Duginov, I.E. Graboj, V.G. Grebinnik, I.I. Gurevitch, A.R. Kaul, B.F. Kirilov, E.P. Krasnoperov, A.B. Lazarev, B.A. Nikolsky, V.G. Olshevsky, A.V. Pigorov, V.Yu. Pomjakushin, A.N. Ponomarev, S.N. Shilov, V.A. Suetin, and V.A. Zhukov, JINR Rapid Comm. 4 [30], 63 (1988)
7. Y. Kumo, N. Nishida, H. Miyatake, S. Okuma, Y. Watanabe, T. Yamazaki, M. Ishikawa, T. Takabatake, Y. Nakazawa, J.H. Brewer, S.R. Kreitzman, and T.M. Riseman, Phys. Rev. B 38, 9276 (1988)
8. F.N. Gygax, B. Hitti, E. Lippelt, A. Schenck, S. Barth, and K. Kwasnitza, Phys. Lett. 127A, 447 (1988)
9. A. Golnik, Ch. Niedermayer, E. Recknagel, M. Rossmanith, A. Weidinger, J.I. Budnick, B. Chamberland, M. Filipkowsky, Y. Zhang, D.P. Yang, L.L. Lynds, F.A. Otter, and C. Baines, Phys. Lett. 125A, 71 (1987)
10. P. Birrer, F.N. Gygax, B. Hitti, E. Lippelt, A. Schenck, M. Weber, S. Barth, F. Hulliger, and H.R. Ott, Phys. Rev. B, in print.
11. Y. Tokura, H. Takagi, and S. Uchida, Nature 337, 345 (1989)
12. J.T. Markert, E.A. Early, T. Bjørnholm, S. Ghamaty, B.W. Lee, J.J. Neumeier, R.D. Price, C.L. Seaman, and M.B. Maple, Physica C 158, 178 (1989)

13. A.C.W.P. James, S.M. Zahurak, and D.W. Murphy, Nature 338, 240 (1989)
14. H. Müller-Buschbaum and W. Wollschläger, Z. Anorg. Allg. Chemie 44, 76 (1975)
15. R.J. Birgeneau and G. Shirane, in Physical Properties of High Temperature Superconductors, ed. D.M. Ginsberg (World Scientific Publishing, Singapore 1989), to be published
16. P. Allenspach, S.-W. Cheong, A. Dommann, P. Fischer, Z. Fisk, A. Furrer, H.R. Ott, and B. Rupp, Z. Phys. B, submitted

Physical Properties of $(Ba,K)BiO_3$

M. Sato, S. Kondoh, and M. Sera*

Institute for Molecular Science, Myodaiji, Okazaki 444, Japan
*Permanent address: Central Research Division,
Asahi Glass Co., Ltd., Kanagawa-ku, Yokohama 221, Japan

Abstract. In order to see if the mechanism of the superconductivity of $(Ba,K)BiO_3$ is common to that of Cu-oxide systems, the transport and magnetic properties of $(Ba,K)BiO_3$ were measured and the oxygen isotope effect was also studied. All these results suggest that the superconductivity can be explained by the classical phonon-mediated mechanism, which seems to be unable to explain the high temperature superconductivity of the Cu-oxides.

1. Introduction

In the study on the mechanism of the superconductivity found in Cu-oxide systems, anomalous magnetic properties in their normal states have been emphasized to be closely connected with their very high transition temperatures T_c. However, because it is very difficult to show a direct experimental evidence for the real mechanism, possibility of other kinds of mechanisms cannot easily be excluded and studies from rather broad view points are always required. In this sense, the discovery of the superconductivity with $T_c \simeq 30K$ in the Cu-free oxide $(Ba,K)BiO_3$ with perovskite structure [1, 2] presented a good opportunity to compare its normal state properties with those of the Cu-oxide systems.

In $(Ba,K)BiO_3$, there exists, formally one outer electron per Bi site and the oxide is usually expected to have a half-filled metallic band. However, it is insulating. By the substitution of K for Ba which introduces the reduction of the electron number, the metallic phase appears and the superconductivity is realized. These behaviors are similar to those of the Cu-oxides and therefore it is tempting to imagine that a common mechanism of the superconductivity exists in these two kinds of systems. However, the situation may not be so simple. In case of Cu-oxides, the superconducting phase always has a neighboring antiferromagnetic and nonmetallic phase as a function of the outer electron number in CuO_2 planes. For example, La_2CuO_4 which has one electron per Cu atom site, is insulating and exhibits antiferromagnetic transition at $T_N \simeq 240K$. These behaviors are now understood to be due to the strong electron correlation caused by the large intraatomic Coulomb repulsion energy U at Cu sites. The hole doping to this phase by the Sr or Ba substitution for La induces the superconducting phase. An argument made by Kasuya et al. before the discovery of the high-T_c superconductivity of the Cu-oxide systems [3], that the insulating nature of $BaBiO_3$ can be understood by a large correlation energy of the Bi 6s electrons, suggested the similarity of the electronic natures

Editors: H. Fukuyama · S. Maekawa · A.P. Malozemoff

of both oxides with Bi and Cu. However, as is well known, $BaBiO_3$ has a lattice distortion with a type of the BiO_6 breathing [4, 5], which is expected to strongly couple to the charge density fluctuation. Then, it seems to be natural to consider that $BaBiO_3$ is in a charge density wave (CDW) state, even if further complicated facts should be taken into account for the understanding of its whole behaviors. It has a clear contrast to the case of the Mott-Hubbard type insulators with antiferromagnetic ordering. If the strong electron-phonon interaction expected in the neighborhood of the CDW instability, is essential to its high-T_c, the mechanism of the superconductivity can be considered to be usual phonon-mediated one, which seems, on the other hand, to be unable to explain the T_c values of the Cu-oxide systems.

The present article reviews results of the magnetic and transport measurements and the oxygen isotope effect study on $(Ba,K)BiO_3$ carried out by the authors' group [6-8] and indicates the nonnecessity of introducing any novel type mechanism for the explanation of the superconductivity of $(Ba,K)BiO_3$.

2. Experiments

For the details of the experimental procedures, one can refer our previous papers [6-8]. Here, only a brief description is given. Because of the volatile nature of K atoms, the synthesis of $(Ba,K)BiO_3$ is rather difficult. We adopted the method essentially same as that Cava et al. used: Nonstoichiometric mixtures of Bi_2O_3, BaO or BaO_2 and KO_2 were sealed in evacuated Ag tubes. Then, the tubes were also sealed in the evacuated quartz and heat-treated. In the present case, the most conductive specimens with blue-black color were obtained by annealing the starting mixture of $Ba_{0.7}K_{0.6}BiO_y$ without any additional heat treatment. Hereafter, the formula of the starting mixture is used nominally and sometimes, the actual K-concentration x of $Ba_{1-x}K_xBiO_3$ determined by EPMA is also specified. The X-ray powder diffraction patterns of these specimens indicated that they were single phase with perovskite structure. For the transport measurements, electrodes were attached with silver paint in a dry condition and then sealed by a cover glass with epoxy resin. An ac resistance bridge was used in the resistivity measurements. Hall coefficients R_H were measured under the magnetic field of ±15kOe. The normal state magnetic susceptibility χ was measured by BTI SQUID magnetometer under the field of 20kOe. The thermoelectric power was measured by using an apparatus similar to that developed by Chaikin and Kwak [9].

In the study of the oxygen isotope effect, we did not adopt the gas exchange method, which was used in the previous studies carried out by Batlogg et al. [10] and Hinks et al. [11]. At first, by the parallel oxidation in $^{16}O_2$ and $^{18}O_2$, each set of $(Ba^{16}O_2, Ba^{18}O_2)$, $(K^{16}O_2, K^{18}O_2)$ and $(Bi_2{}^{16}O_3, Bi_2{}^{18}O_3)$ were prepared. All oxides of Bi_2O_3 and BaO_2 were found to be single phase by X-ray powder diffraction study. It was also found that for each kind of these oxides, the qualities of the powders with ^{16}O and ^{18}O were almost equal. The qualities of $K^{16}O_2$ and $K^{18}O_2$ were almost equal, too. However, they contained a small amount of KOH, the origin of which was not clear. Because the amount of the oxygens from KOH is estimated to be not more than a few percent of the total oxygens in the mixtures, we did not take into account the effect of the oxygens in the latter analyses. By mixing the proper amounts of raw oxides with ^{16}O and ^{18}O prepared above, three mixtures of $Ba_{0.7}K_{0.6}BiO_y$ with the average oxygen mass number M of 16, 17 and 18 were prepared and used in this study, which guaranteed us that the accurate value of M could be used.

The X-ray powder patterns of the finally obtained samples indicated that they were single phase (see Fig. 1). The Meissner volume fraction estimated from the diamagnetic moment measured by dc SQUID under the field of 20 Oe in a field cooling condition was about 10%.

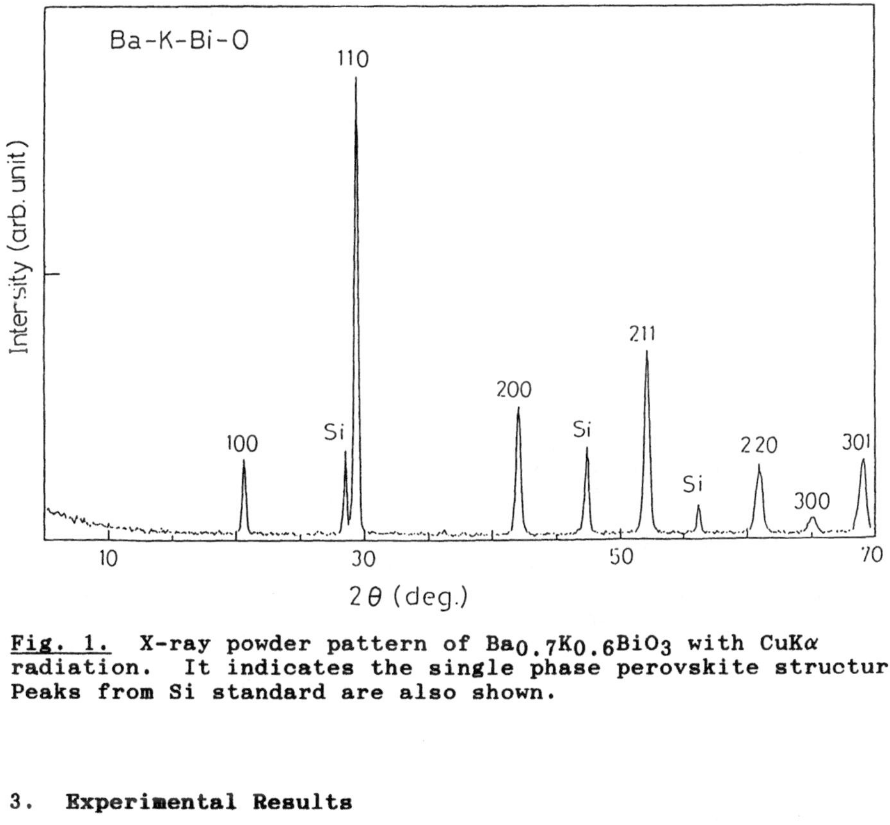

Fig. 1. X-ray powder pattern of $Ba_{0.7}K_{0.6}BiO_3$ with CuKα radiation. It indicates the single phase perovskite structure. Peaks from Si standard are also shown.

3. Experimental Results

The actual K-concentration x was determined for several specimens by EPMA. The lattice constant a of the pseudo cubic cell was found to be linear in x, which were consistent with the data reported previously [5, 11 and 12]. Although we could not determine the clear boundary between the superconducting and nonsuperconducting regions, it is expected to be in the region $0.1<x<0.2$. Even for the superconducting specimens, the resistivity above T_c did not show the metallic temperature dependence, which clearly indicated the significant effect of grain boundaries and/or inhomogeneities. However, the transition width was found to be rather sharp and the zero resistivity was realized at about 30K.

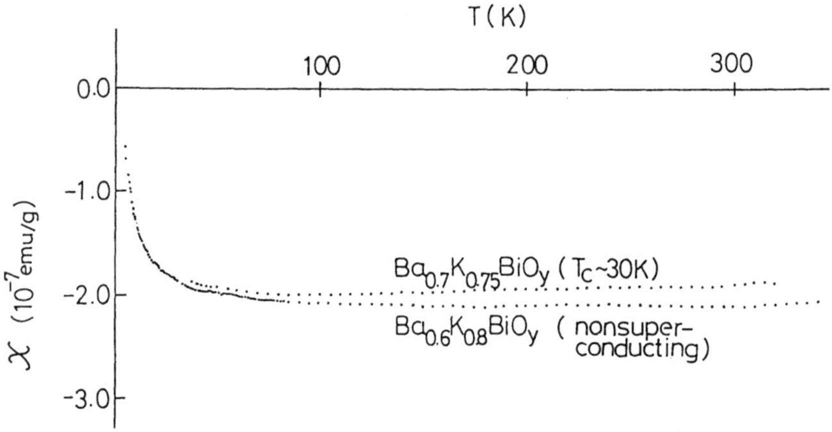

Fig. 2. Temperature dependence of the magnetic susceptibilities of a superconducting specimen and an insulating one are shown.

Fig. 2 shows the temperature dependence of the magnetic susceptibilities of a superconducting specimen $Ba_{0.7}K_{0.75}BiO_y$ ($Ba_{0.77}K_{0.23}BiO_y$ from EPMA) and an insulating one $Ba_{0.6}K_{0.8}BiO_y$ [6]. From these results, the paramagnetic contribution from the mobile electrons are estimated to be of the order of 10^{-8}emu/g, which is much smaller than the paramagnetic contributions of the outer electrons in Cu-oxides. Similar values was reported by Batlogg et al. later [10]. As will be discussed later, this result indicates that the (Ba,K)BiO_3 system is nonmagnetic and their electronic nature should be quite different from that of Cu-oxides, in which the spins at Cu sites seem to have a very important role even in the superconducting phase [13-15].

Fig. 3 shows the T-dependence of the Hall coefficient R_H of $Ba_{0.7}K_{0.6}BiO_y$ ($Ba_{0.8}K_{0.2}BiO_y$ from EPMA). Its negative sign indicates that the charge carriers are not holes but electrons,

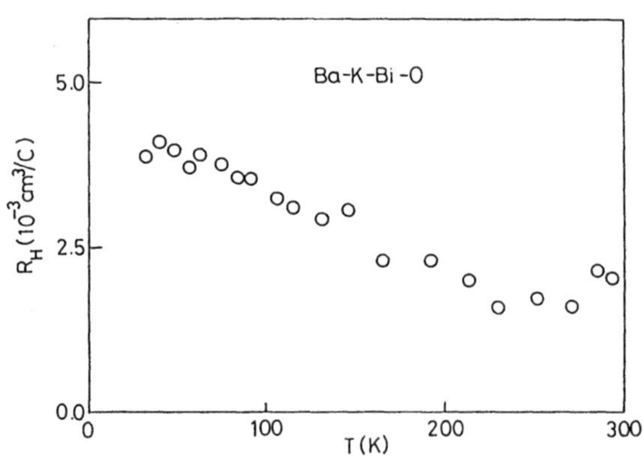

Fig. 3. Temperature dependence of the Hall coefficient R_H of $Ba_{0.7}K_{0.6}BiO_y$.

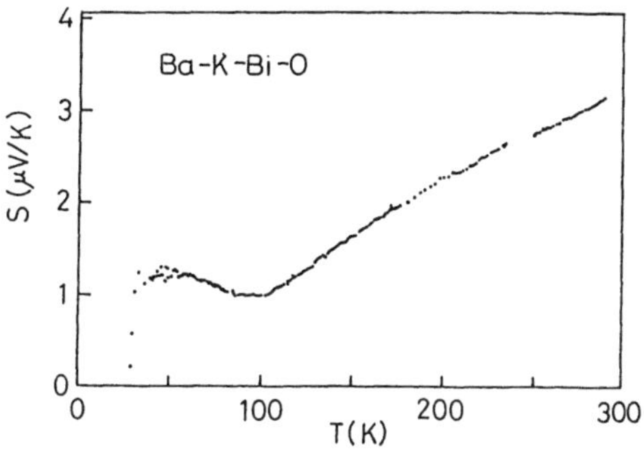

Fig. 4. Temperature dependence of the thermoelectric power S of $Ba_{0.7}$ $K_{0.6}BiO_y$. The hump below 100K is due to the error in subtracting the contribution of the gold electrodes.

which again has a clear contrast to the case of Cu-oxides, where the substitution by atoms with less outer electrons leads to the hole conduction. Fig. 4 shows the T-dependence of the thermoelectric power S of $Ba_{0.7}K_{0.75}BiO_y$ ($Ba_{0.77}K_{0.23}BiO_y$ from EPMA). Although the grain boundary effect is significant in the T-dependence of the resistivity of this specimen [7], S does not seem to have such a serious effect and T-linear behavior can be seen above 100K. The complicated T-dependence at lower temperatures was found to be due to the error in subtracting the contribution of the gold electrodes. The sign of S is positive. We do not know the clear reason why the sign of S disagrees with that of R_H. However, it can be said that the sign of S does not always indicate the sign of the charge carriers, because it is not determined solely by the shape of the Fermi surface.

In the study on the oxygen isotope effect, three specimens of $Ba_{0.7}K_{0.6}BiO_y$ ($Ba_{0.8}K_{0.2}BiO_y$ from EPMA) with the average M of 16, 17 and 18 were simultaneously prepared [8]. The X-ray diffraction study showed that they had equal lattice parameters. The T_c values are determined by the measurement of the Meissner diamagnetism, the results of which are shown in Fig. 5. It is quite clear that T_c shifts to lower temperature as M increases. Fig. 6 shows the ℓnT_c-ℓnM plot and from the slope of the fitted line, the relation $T_c \propto M^{-0.35}$ is obtained or the parameter α is determined to be 0.35 ($\pm$0.05), which should be compared with 0.21 and 0.41 reported by Batlogg et al. [10] and Hinks et al. [11], respectively. Here, we emphasize that we did not use the gas exchange method, which guaranteed us that we could use the accurate value of M. Note that the gas exchange process may bring ambiguous factors for the present kind of inhomogeneous specimens with nonstoichiometric metal atom ratios. It may also induce the change of the K-concentration x because of the volatile nature of K atoms.

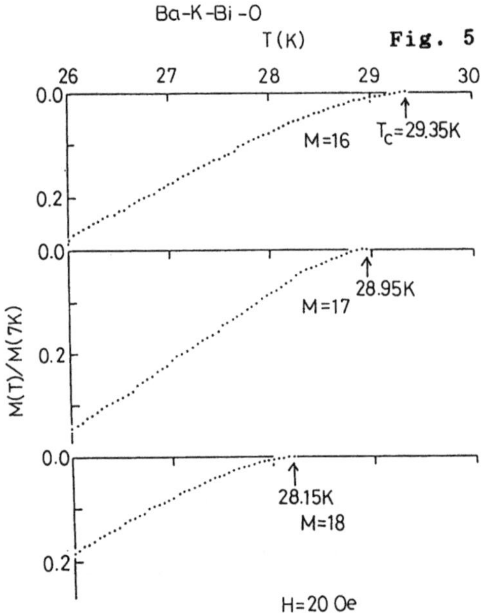

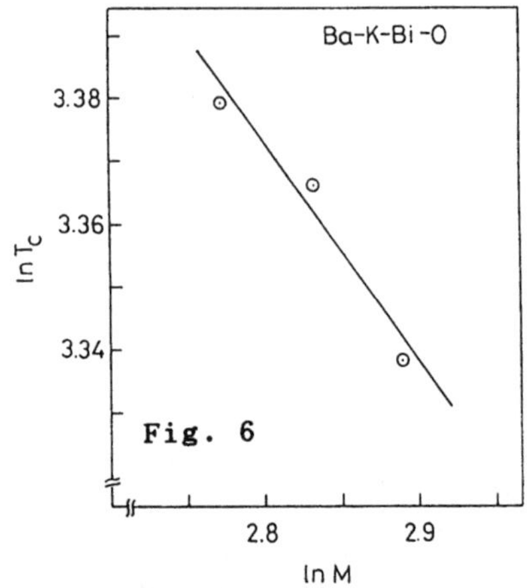

Fig. 5. Meissner diamagnetisms of $Ba_{0.7}K_{0.6}BiO_y$ specimens with the average oxygen mass number of 16, 17 and 18 are plotted as a function of temperature. For each specimen, the data are normalized by the value at 7K.

Fig. 6. Values of ℓnT_c are plotted against ℓnM. From the slope, α is estimated to be 0.35±0.05.

4. Discussion

The data of the magnetic susceptibility and the Hall coefficient shown in the previous section clearly indicate that the electronic nature of $(Ba,K)BiO_3$ is quite different from that of Cu-oxides. The present oxide is found to be nonmagnetic, which is consistent with the finding of the μSR [16]. The negative sign of R_H indicates that the reduction of certain amount of the outer electrons by the K^+ substitution for Ba^{2+} leads to the delocalization of all outer electrons, while in case of Cu-oxides, only the holes doped into the Matt-Hubbard insulators become mobile and the high temperature superconductivity appears in this hole conduction region. Moreover, the nonsuperconducting and superconducting phases seem to correspond to the x regions with and without lattice distortions, respectively. All these results suggest that the high temperature superconductivity in $(Ba,K)BiO_3$ occurs near the metal-CDW insulator boundary, where the electron-phonon interaction is expected to be large. Recent optical data taken by Sato et al. of Tokyo Univ. are also consistent with the present picture.

The large value of $\alpha\simeq0.35$ obtained here also support the present picture. Batlogg et al. [10] pointed out the existence of the nonphonon mechanism based on their small value of $\alpha\simeq0.21$. They estimated the electronic density of states from the experimentally obtained values of H_{c1} and H_{c2} and reached the conclusion that the electron-phonon coupling constant λ should be less than unity. On the other hand, with the Debye temperature $\theta_D \simeq$ (150-200)K, they could not explain the T_c value of 30K for $\lambda \lesssim 1$, which was their main reason why they introduced the nonphonon mechanism. However, if we use the ω_D value of 500-600K, the T_c value can be explained within their constraint mentioned above. Recently, Shirai et al. [17] presented the electron-phonon spectral function $\alpha^2_{gp}F(\omega)$ based on their band calculation. Even if the band theory may not be able to describe the detalied electronic state of the insulating phase of $BaBiO_3$, their result seems to indicate that the value ω_D should be of the order of 500-600K. Thus, it seems to be natural to conclude that the transition temperature of $(Ba,K)BiO_3$ can be essentially explained by the classical phonon-mediated mechanism. It is well known that the system $Ba(Pb,Bi)O_3$ becomes metallic only after a very large fraction of Bi is substituted by Pb. On the other hand, much less amount of K substitution for Ba induces the insulator to metal transition in $(Ba,K)BiO_3$. In this sense, the latter system may be much simpler than the former one, in which the local nature of the lattice distortion or the effect of the randomness seems to be very important.

References

1. L. F. Mattheiss, E. M. Georgy and D. W. Johnson Jr., Phys. Rev. B37, 3745 (1988).
2. R. J. Cava, B. Batlogg, J. J. Krajewski, R. C. Farrow, L. W. Rupp Jr., A. E. White, K. T. Short, W. F. Peck Jr. and T. Y. Kometani, Nature 332 814 (1988).
3. T. Kasuya and K. Takegahara, JJAP series 1 Superconducting Materials edited by S. Nakajima and H. Fukuyama p251.
4. D. E. Cox and A. W. Sleight, Solid State Commun. 19 969 (1976).
5. C. Chaillout, A. Santoro, J. P. Remeika, A. S. Cooper, G. P. Espinosa and M. Marezio, Solid State Commun. 65 1363 (1988).
6. S. Kondoh, M. Sera, K. Fukuda, Y. Ando and M. Sato, Solid State Commun. 67 879 (1988).
7. M. Sera, S. Kondoh and M. Sato, Solid State Commun. 68 647 (1988).
8. S. Kondoh, M. Sera, Y. Ando and M. Sato, Physica C 469 (1989).
9. P. M. Chaikin and J. F. Kwak, Rev. Sci. Inst. 46 218 (1975).
10. B. Batlogg, R. J. Cava, L. W. Rupp Jr., A. M. Mujsce, J. J. Krajewski, J. P. Remeika, W. F. Peck Jr., A. S. Cooper and G. P. Espinosa, Phys. Rev. Lett. 61 1670 (1988).
11. D. G. Hinks, D. R. Richards, B. Dabrowski, D. T. Marx and A. W. Mitchell, Nature 335 419 (1988).
12. L. F. Schneemeyer, J. K. Thomas, T. Siegrist, B. Batlogg, L. W. Rupp, R. L. Opila, R. J. Cava and D. W. Murphy, Nature 335 421 (1988).
13. G. Shirane, Y. Endoh, R. J. Birgeneau, M. A. Kastner, M. Oda and T. Murakami, Phys. Rev. Lett. 59 1613 (1987).
14. M. Sato, S. Shamoto, J. M. Tranquada, G. Shirane and B. Keimer, Phys. Rev. Lett. 61 1317 (1988).

15. Y. Ando, M. Sera, S. Yamagata, S. Kondoh, M. Onoda and M. Sato, Solid State Commun. 70 303 (1989).
16. Y. J. Uemura, B. J. Sternlieb, D. E. Cox, J. H. Brewer, R. Kadono, J. R. Kempton, R. F. Kiefl, S. R. Kreitzman, G. M. Luke, P. Mulhern, T. Riesman, D. L. Williams, W. J. Kossler, X. H. Yu, C. E. Stronach, M. A. Subramanian, J. Gopalakrishnan and A. W. Sleight, Nature 335 151 (1988).
17. M. Shirai, N. Suzuki and K. Mochizuki, preprint.

Flux Creep and the Crossover to Flux Flow in the Resistivity of High-T_c Superconductors

A.P. Malozemoff, T.K. Worthington, E. Zeldov, N.C. Yeh, M.W. McElfresh, and F. Holtzberg*

IBM Thomas J. Watson Research Center, Yorktown Hights, NY 10598, USA
*On leave from the Department of Electrical Engineering and Solid State Institute, Technion-Israel Institute of Technology, Haifa 32000, Israel

Abstract. A model for the resistive behavior of high temperature superconductors in a field is presented, including both flux flow and the possibility of long jumps or cascades of flux lines in flux creep. These mechanisms help to understand characteristic features of the resistivity, reported earlier in YBaCuO crystals and films, especially the large resistive prefactors in the activated regime. The importance of including the temperature dependence of the activation energy in the data analysis is emphasized. The crossover between flux creep and flux flow may explain the knee in the resistivity-vs.-temperature curves observed in much earlier work and particularly clearly in new work on YBaCuO crystals reported here.

1. Introduction

Because of the relatively low pinning barriers and high temperatures, flux creep and flow effects are unusually large in the magnetic and transport properties of the new high temperature superconductors.[1,2] We discuss here some recent developments in the interpretation of these effects, particularly the resistivity ρ versus temperature in a magnetic field.

Tinkham[3] first presented a detailed theory for the unusual broadening of the resistivity versus temperature in a magnetic field. He exploited the Ambegaokar and Halperin[4] (AH) model of a single overdamped Josephson junction. This provides perhaps the only analytical treatment of thermal excitation over a potential barrier which spans regimes of both low and high driving forces. Tinkham found reasonable initial comparison with resistance measurements in YBaCuO crystals, when combined with a simple scaling form[1] for the barrier energy with field H and reduced temperature $t = T/T_c$. Here we will use the scaling form

$$U_0 = A(1 - t^2)^{3/2}/H \ , \qquad (1)$$

(with $1 - t^2$ rather than $1 - t$ to permit extension to lower temperatures). In the limit of low resistivity and driving force, Tinkham's model predicts an activated form

$$\rho = \rho_n(U_0/kT) \exp(-U_0/kT) \ , \qquad (2)$$

where ρ_n is the normal-state resistivity. At temperatures comparable to U_0/k, the model predicts that ρ smoothly approaches ρ_n.

Editors: H. Fukuyama · S. Maekawa · A.P. Malozemoff

Nevertheless, subsequent more detailed experimental work by a number of authors[5-9] has shown that the behavior at low resistivity (below about 1 $\mu\Omega$cm for $H \parallel c$ in YBaCuO for example), although activated as Eq. 2 predicts, does not properly extrapolate to the data at higher resistivities according to Tinkham's prediction. We will show here that this can be interpreted as due to the very large prefactors ρ_0 determined from fits to $\rho = \rho_0 e^{-U/kT}$ in the low resistivity region. In fact, some data[5,6,10] (though not all), especially for $H \parallel c$, even show a knee which separates these two regimes, with a typical crossover value in the range of 1-10 $\mu\Omega$cm for YBaCuO crystals. A particularly clear example is shown in our new data in Fig. 1 on an YBaCuO crystal grown by a technique previously described[11] (further details on this experiment will be given in Section III). While others have ascribed this knee to sample inhomogeneity,[10] we find this knee most pronounced in the crystals with the sharpest zero field transitions. Since Tinkham's model gives a smooth evolution from the activated region right up to ρ_n, it cannot account for this anomaly or for the low and high resistivity regions simultaneously. The low resistivity activated region plausibly represents flux creep.[12-14] We propose here that the resistivity region above the knee but below ρ_n is a flux flow regime,[14-17] the physics of which is not contained in the Tinkham or AH model.

In Section 2, we begin by reintroducing an old idea[18,19] for the attempt frequency in flux creep, namely that it could be a relaxation frequency of the flux line or bundle oscillating in its pinning well. We show that this gives a formula closely analogous to Tinkham's except for an enhancement factor which comes from the ratio of the jump distance to the well dimension or from the size of a vortex cascade, and which makes possible a flux creep resistivity prefactor much larger than the normal-state

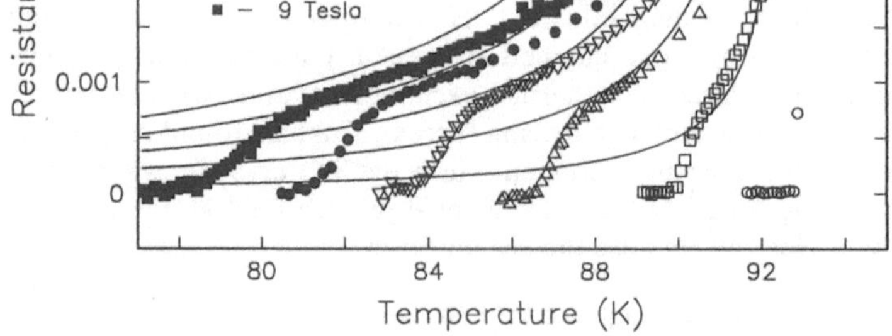

Fig. 1. Resistance versus temperature for an YBaCuO crystal at a set of fields applied parallel to the c-axis, determined at a current density of 5 A/cm^2. The data distinctly show a "knee" in the range 10 $\mu\Omega$cm. The line labeled R_n represents the extrapolated temperature-linear normal state resistivity. The superimposed curving lines represent the predictions of a simple Bardeen-Stephen flux-flow model (see text and Eq. 9).

resistivity. In Section 3 we analyze existing data as well as new data presented here, to determine and compare the activation energies and prefactors. The importance of analyzing the experimental data with a temperature-dependent activation energy is emphasized. In Section 4 we compare our intermediate resistivity data to the flux flow model and explore the crossover between the flux flow and flux creep.

While some puzzles in the earlier data can be clarified in this way, there is still much to understand, especially in the detailed current density dependence. Alternative pictures in terms of vortex glass freezing[20] and related scaling arguments, collective pinning[21–23], a pinning barrier distribution[24–25] or fluctuation effects[26] are also possible and should be examined. Therefore the ideas presented in this paper should not be viewed as a final explanation of the experimental phenomena but rather as an exploration of one particular model, namely the flux flow and flux creep model.

2. The flux creep resistivity in an overdamped oscillator model

First we consider the flux creep regime and introduce a hypothesis for the attempt frequency, which forms a basis for our subsequent (Section IV) comparison of the flux creep and flux flow regimes. The conventional formula[12–14] for flux creep resistivity ρ, representing thermally activated hopping of flux lines or bundles over a pinning barrier, can be written (in cgs units)

$$\rho \; = \; E/J \; = \; (2aNf_0B/Jc) \exp(-U_0/kT) \sinh(JBLV_c/2ckT) \; , \qquad (3)$$

where E is the electric field, J ($<<J_{c0}$) is the applied current density, a is the jump distance, N is the number of vortices in a cascade (see below), B is the magnetic induction or flux density, c is the speed of light, f_0 is the attempt frequency, U_0 is the pinning barrier height in the limit of zero current, L is the dimension of the pinning well defined by $U(x)=(-U_0/2)\cos(2\pi x/L)$, and V_c is the activation volume. In this sinusoidal model for the shape of the pinning well, the critical current J_{c0} in the absence of thermal activation is defined by the maximum force of the well, which leads to

$$J_{c0} \; = \; \pi U_0 c/BLV_c \; . \qquad (4)$$

It is important to note here that the net pinning barrier height U which is U_0 plus a driving force term due to the current, is in general a function of temperature and field, for example as in Eq. 1. U is also a non-linear function of J, as first suggested by Beasley et al.[13] and as found experimentally in the recent results of Zeldov et al.[8] on YBaCuO films. The form in Eq. 3 is only valid for J small compared to J_{c0}; this is why substitution of Eq. 4 into Eq. 3 gives a critical current density which appears to be too large by a factor of $\pi/2$ (for the sinusoidal pinning-well model).

Eq. 3 also introduces ad hoc the possibility that the jump or hopping length a could be much larger, than the characteristic well size L, causing an enhancement in the resistivity prefactor. This physical concept is not contained in the AH model and will be important in comparison to experiment. It implies collective pinning[21–23] or some other cooperative effect in the vortex motion. A further interesting point on Eq. 3 is that the resistivity does not depend on the number of vortices in a flux bundle. This is because, as long as the bundles are regarded as non-overlapping, the flux density B remains constant. However, if each single-vortex hop unleashes a cascade of N vortices, ρ is enhanced by a factor N.

Next we consider the attempt frequency f_0, for which many different models have been given in the literature. Here we assume that the motion of the flux line or bundle in its pinning potential well is overdamped,[18,19] in analogy to the AH model. In the Bardeen-Stephen approximation[15], the viscous drag force on a flux line per length can be written

$$F_d \equiv b\,dx/dt = (\Phi_0 B_{c2}/c^2\rho_n)dx/dt \ , \tag{5}$$

where b is the viscous damping coefficient, Φ_0 is the flux quantum, B_{c2} is the upper critical field, ρ_n is the normal state resistivity and x is the displacement of the flux line. Similarly the harmonic restoring force at low current densities can be written for the sinusoidal pinning well model as[18]

$$F_h = kx = (J_{c0}\Phi_0/c)(2\pi x/L) \ . \tag{6}$$

Thus the relaxation frequency is

$$f_0 = k/b = 2\pi c J_{c0}\rho_n/B_{c2}L \ . \tag{7}$$

We estimate the size of this frequency at 77 K to be about 2×10^{10} Hz if $J_{c0} = 10^5$ A/cm^2, $\rho_n = 50$ $\mu\Omega$cm, $B_{c2} = 26$ T and $L = 70\times10^{-8}$ cm (twice the 77 K ab-plane coherence length[27] in a single-vortex-hopping picture).

The key step now is to take this relaxation frequency to be the effective attempt frequency f_0. Thus we find

$$\rho = \rho_0 \exp(-U_0/kT) \sinh(\pi U_0 J/2kTJ_{c0}) \ , \tag{8a}$$

$$\rho_0 = (\rho_n B/B_{c2})(4\pi Na/L)(J_{c0}/J) \equiv (2/\pi)\rho_f \eta J_{c0}/J \ . \tag{8b}$$

Here ρ_f is the simplest form of the Bardeen-Stephen flux flow resistivity[14,15]

$$\rho_f = \rho_n B/B_{c2} \ , \tag{9}$$

and η is an enhancement factor which measures the ratio of the hopping distance to the well dimension and the size of the vortex cascade.

$$\eta = 2\pi^2 Na/L \ . \tag{10}$$

The resistivity of Eq. 8 is an effective resistivity E/J, not a true ohmic resistivity, because of the evident nonlinear dependence on J. The factors of 2 and π in these equations reflect the specific sinusoidal well shape and are valid for small J (although the argument of the sinh in Eq. 8a can still in principle be greater than 1).

In the limit of sufficiently small J, the sinh in Eq. 8a can be linearized to give a true ohmic resistivity

$$\rho_l = \rho_{0l} \exp(-U_0/kT) \ , \tag{11a}$$

$$\rho_{0l} = \rho_f \eta U_0/kT \ . \tag{11b}$$

It is of interest to compare these expressions to those derived earlier in the literature. Fisher[19] also used the Bardeen-Stephen model to derive an ohmic resistivity comparable to that of Eq. 11, obtaining $\rho_l \simeq \rho_f \exp(-U_0/kT)$. In his treatment he ignored the jump enhancement effect which we parametrize by η, and it is not clear if he has simply neglected U_0/kT in his prefactor. Tinkham,[3] by contrast, finds $\rho_l = \rho_n(U_0/kT)e^{-U_0/kT}$, which includes U_0/kT in the prefactor but of course does not have the flux flow resistivity or the jump enhancement factors. A related derivation has also been given by Yeh.[23]

The important result of these simple expressions is that the resistivity prefactors in Eqs. 8 and 11 can be larger than either ρ_f or even ρ_n, if the enhancement factor η and U_0/kT are sufficiently large. Physically one can picture the large dissipation coming from a long jump as arising during the flux line equilibration in its new well while it waits for the next jump.

3. Detailed analysis of flux creep resistivity experiments

Because different authors have analyzed their resistivity data in different ways, we reanalyze these data here in a consistent way. We start with the low-resistivity data on YBaCuO films ($H \parallel c$) of Zeldov et al.[8]. They used a simplified flux creep equation of the form

$$\rho = \rho_0 \exp(-U/kT) \ , \tag{12}$$

with a temperature-, field- and current-density-dependent activation energy U of the form

$$U(T,H,J) = (A_J/\sqrt{2})(1-t^2)(1-t^4)^{1/2}/B^m \ . \tag{13}$$

They found the power m to be close to though slightly weaker than 1; so that Eq. 13 reduces near T_c to the form of Eq. 1, with a J-dependent coefficient A_J. An implicit assumption here is that the current driving-force term has the same temperature and field-dependence as the barrier energy U_0. This is not easily justified but significantly simplifies the initial analysis of the data.

Most important were the current density results of Zeldov et al. For example, between $J = 4.5\mathrm{x}10^4$ and $3\mathrm{x}10^5$ A/cm^2 at 20 kG, they found A_J between 38 and 20 eV-kG respectively. A simple linear extrapolation from this region, based on $U = U_0(1 - J/J_{c0})$, gives a zero-current barrier parameter $A_0 \simeq 42$ eV-kG. However Zeldov et al. found that at lower current density, the effective barrier dependence on J curved upwards, for example reaching $A \simeq 51$ eV-kG at the $1.5\mathrm{x}10^4$ A/cm^2 . They could fit this dependence to a logarithmic form $U \propto \ln(J_0/J)$, with J_0 a constant. Although this specific form does not emerge from our treatment of a sinusoidal barrier in Section II, even that simple model predicts an upward-curving nonlinearity in the current density dependence. The average resistivity prefactor extracted from Eq. 12 is of order 1000 $\mu\Omega$cm, well above the normal-state resistivity of about 65 $\mu\Omega$cm just above T_c.

In closely related work, Hettinger et al.[9] analyzed their low-resistivity data ($H \parallel c$) on an YBaCuO film, though with a slightly different form of the activation energy

$$U = (A_J/\sqrt{2})(1 - t^2)^2(1 - t)^{-1/2}/B \ . \tag{14}$$

Here the coefficient A_J, as in Eq. 13, has been written to give the consistent form of Eq. 1 for the barrier energy near T_c. They found a barrier approximately linearly dependent on current density, though their data covered a narrower range of J than in the work of Zeldov et al., and an upward curvature in U(J) cannot be excluded. Their linear extrapolation to zero current density gave $A_0 \simeq 29$ eVkG, close to the linear extrapolation of the Zeldov data from the same range of J. An estimate of their prefactor from Eq. 14 and their 12 T data at 0.1 $\mu\Omega$cm gives $\log \rho_0 = 4.5 \pm 1$ $\mu\Omega$cm, which is several orders of magnitude larger than the normal-state resistivity.

In another study, Sun et al.[7] analyzed resistivity data on YBaCuO films in a similar way but obtained noticeably smaller values of A and ρ_0. We attribute this discrepancy to the fact that their data did not probe the low resistivity range (below $\simeq 1 \mu\Omega$cm), and so we do not consider their results further here.

These thin film studies can be compared to Palstra et al.'s[6] and our YBaCuO crystal studies, which typically use much lower current densities. Starting with the work of Palstra et al.,[6] we note that their reported activation energies should not be directly compared to the numbers quoted above because they initially analyzed their data ignoring the temperature dependence of U, thus obtaining anomalously large values as will be shown below. Obviously troubling in their approach is that linear extrapolations on a plot of log ρ vs 1/T give prefactors ρ_o of 10^{50} $\mu\Omega$cm or greater (see Table I)!

Table I. Reanalysis of YBaCuO crystal resistivity data of Palstra et al.[6] Magnetic fields B and temperatures T are determined at $\rho = 0.1$ $\mu\Omega$cm for B $\parallel$ c with $T_c = 88$ K and n=1.5. U_{eff} and $\rho_{0,eff}$ are the effective activation energy and prefactor determined by Palstra et al. ignoring the temperature dependence of U, $1 + (2nt^2/1 - t^2)$ is the enhancement factor for U, U and ρ_0 are the corrected values, and A_J is the current-density-dependent coefficient in Eq. 1.

B(T)	T(K)	$U_{EFF}(K)$	$\log \rho_{0,EFF}$	$1 + \frac{2nt^2}{1-t^2}$	U(K)	$\log \rho_0$	$A_J(eVkG)$
1	84.5	46000	235	39	1150	5.0	51
2	83	28500	150	25	1150	5.0	54
5	77.5	17500	98	11.5	1550	7.7	63
7.5	74	13000	74	8	1600	8.3	62
10	70.5	10500	64	6.5	1650	9.3	66
12	68	8000	50	5.5	1450	8.3	59

While an analysis ignoring the temperature dependence of U might be appropriate for low temperature data such as that in their earlier work[5] on BiSrCaCuO, we argue here that the temperature dependence near T_c makes a substantial difference. To see this, we differentiate the logarithm of the resistivity (Eq. 12) with respect to 1/T, using the pinning barrier temperature dependence $U \propto (1 - t^2)^n$, to obtain an effective barrier energy

$$U_{eff} \equiv -\frac{d \ln \rho}{d(1/T)} = -\frac{d \ln \rho_0}{d(1/T)} + U_0(1 + \frac{2nt^2}{1 - t^2}) . \qquad (15)$$

Even when ρ_0 is temperature independent so that the first term can be ignored, the $2nt^2/(1 - t^2)$ term can give a large enhancement, making U_{eff} much larger than U. Some measure of the size of this effect is given in Table I, where the $1 + 2nt^2/(1 - t^2)$ factors for the Palstra et al. data are seen to range from 5 to 39, for the 12 to 0.5 T data respectively! In this reanalysis we have used n=3/2, as in Eq. 1, which describes the temperature shifts with field at constant resistivity reasonably well. Table I shows that the corrected activation energies U are essentially independent of field, in contrast to U_{eff}. This simply means that the temperatures are tracking with field according to the law $(1 - t^2)^{3/2} \propto B$, so that U in Eq. 1 remains constant. The coefficient A in Eq. 1 comes out on average to be 59 eV-kG and the prefactors $10^{7\pm1}$ $\mu\Omega$cm. Although there is some apparent field dependence in the prefactor, these variations are within the errors of the analysis, considering that we have ignored the temperature dependence of ρ_0 in Eq. 15.

We also report here new results on an YBaCuO crystal, 1 mm square and 0.02 mm thick. Gold wires were thermal-compression-bonded to small (0.05mm) gold pads which were evaporated through a mask at the corners of the crystal face. The contact resistance was ~10-20 Ω. The R vs T plot of Fig. 1 was measured at a current density of ~5 A/cm². The zero-field transition is centered at 92.8 K with a width of less than 0.3 K. The IV curves above the knee in Fig. 1 are linear while nonlinearity sets in at some point below the knee, even at the lowest current densities, in contrast to the claim of Palstra et al.[6] We note that the demonstration of linearity presented in their paper is limited to a relatively high resistivity range, where we also observe linearity. Further details on our IV curves will be presented elsewhere. Figure 1 also shows an extrapolation of the high-temperature normal-state resistivity, which is linear in temperature above about 120 K. Table II gives our activation energies and prefactors determined in the same way as we used above for the Palstra et al. data. The values are remarkably similar in the two studies, and the average values are summarized in Table III.

Finally, these results can be compared to earlier measurements of the frequency-dependent ac susceptibility.[2] Since the resistivity in the flux creep equation is proportional to the effective frequency of flux motion, Eq. 3 can be rewritten as

$$f = f_0 \exp(-U_0/kT) \sinh(\pi U_0 J/2kTJ_{c0}) . \qquad (16a)$$

Thus, for a given frequency in the ac susceptibility experiment, the loss peak or inductive transition will occur at the fields and temperatures specified implicitly in this equation. This field-temperature contour defines the "irreversibility line" and corresponds closely to the field-temperature contours defined by a given resistivity.

Table II. Analysis of YBaCuO crystal resistivity data (this work). Magnetic fields B and temperatures T are determined at $\rho = 0.1\ \mu\Omega$cm for B $\|$ c and J=5 A/cm^2, with T_c=92.8 K and n=1.5. Other entries as in Table I.

B(T)	T(K)	$U_{EFF}(K)$	$1 + \frac{2nt^2}{1-t^2}$	U(K)	log ρ_0	$A_J(eVkG)$
1	89.5	40000	41	1000	3.8	46
2	87.4	33000	24	1350	5.8	61
3	85.6	28000	18	1550	6.8	70
5	82.3	20000	12	1650	7.8	73
7	79.4	13000	9	1450	7.0	63
9	76.4	10000	7.5	1350	6.8	58

Table III. Activation energies and prefactors from resistive or (where noted) ac susceptibility data, determined as discussed in the text. The activation energy U is parametrized as $A_J(1-t)^{3/2}/B$ near T_c . Error bars indicate the range of values obtained for different applied magnetic fields.

A_J (eV-kG)	log ρ_0 ($\mu\Omega$ cm)	J-range (10^4 A/cm^2)	Comment	Authors	Reference
42±7	3±1	4.5-30	nonlinear	Zeldov et al.	8
29±2	4.5±1	3-24	nonlinear	Hettinger et al.	9
59±8	7.5±2	$\sim 10^{-4}$	linear ?	Palstra et al.	6
62±10	7±1	$\sim 10^{-4}$	nonlinear	this work	–
29±4	–	–	ac - χ	Malozemoff et al.	2

Linearizing the sinh (assuming a low current density limit), one can write

$$U_0 = kT \ln(f_{eff}/f) , \qquad (16b)$$

where we absorb the U_0J/kTJ_{c0}-factor in the logarithm into an effective attempt frequency f_{eff}. Differentiating with respect to the measurement frequency f, and assuming $U_0 \propto (1-t^2)^n$, one derives near T_c

$$U_0 = kT_c^2(1 - t^2)/2n(dT/d\ln f) \ . \tag{17}$$

Thus a measurement of the logarithmic derivative of the ac-loss-peak temperature directly gives U_0. Carrying out this procedure for the YBaCuO crystal data of Ref. 2 with H$\parallel$c and n=1.35, we find $A \simeq 29 \pm 3$ eV-kG, a value within a factor of two of the resistive determinations. A more detailed comparison of both resistive and ac-susceptibility data on the same crystal is in progress.

4. Crossover to flux flow and discussion

The results of all these studies are summarized in Table III, although not all the numbers are strictly comparable, even after the reanalysis, because of the different current densities. Nevertheless the activation energies described by the coefficient A_J defined in Eq. 1, are within a factor of two of each other. Such variations are certainly plausible considering possibly different pinning in different samples.

The resistivity prefactors are all orders of magnitude larger than the normal-state resistivity. We estimate the prefactor in the linear regime (for the data of Palstra et al.) from Eq. 11b. Since $B_{c2} \simeq b(T - T_c)$ depends linearly on temperature according to recent data of Welp et al.,[27] Eq. 11b can be written near T_c as

$$\rho_0/\rho_n = 2\sqrt{2}\,\eta(1 - t)^{1/2}A_J/kTT_cb \ . \tag{18}$$

With $b \simeq 20$ kG/K, $(1 - t)^{1/2} \simeq 0.4$, $A_J \simeq 60$ eV-kG and $T_c \simeq 90$ K, we find $\rho_0/\rho_n \simeq 5\eta$. Thus a resistivity enhancement ρ_0/ρ_n of 10^5 requires an η of over 10^4. According to Eq. 10, this requires a cascade of about 1000 vortices or a hopping distance of order 10 microns (1000 times the minimum well size $2\xi \simeq 75$Å at this temperature). Similar estimates can be made for the nonlinear prefactor of Eq. 8b. Obviously the large cascades or hopping distances imply a strongly co-operative behavior of the flux-line motions.

While qualitative, these notions of large cascades or hopping distances offer a possible explanation of the large flux-creep-resistivity prefactors. The large prefactors imply in turn that, extrapolated to T_c, the resistivity would become much larger than the normal state resistivity ρ_n. In fact, even before ρ_n is reached, the resistivity should plausibly be limited by damping of the vortex motion during its hop over the pinning barrier. This implies that the flux creep resistivity cannot become larger than the flux flow resistivity.

Thus the crossover between flux creep and flux flow is determined by the condition where the flux flow (Eq. 9) and flux creep (Eq. 8a) resistivities are approximately equal:

$$\rho_f = (2\rho_f\eta J_{c0}/\pi J)\exp(-U_0/kT)\sinh(\pi U_0 J/2kTJ_{c0}) \ . \tag{19}$$

In the regime where $\pi U_0 J/2kTJ_{c0} >> 1$, one can rewrite Eq. 19 in the form

$$J = (2J_{c0}/\pi)[1 - (kT/U_0)\ln(\eta J_{c0}/\pi J)] \ . \tag{20}$$

This gives the current density beyond which flux flow dominates. This can be compared to the flux creep formula for the critical current density

$$J_c = (2J_{c0}/\pi)[1 - (kT/U_0)\ln(\eta J_{c0}\rho_f/\pi J\rho_c)] , \qquad (21)$$

where ρ_c is the critical resistivity used as the experimental criterion for the critical current. As long as this criterion is well below ρ_f, the critical current density is always lower than the value for crossover to flux flow.

For data in the linear current density regime, the condition for the crossover becomes simply

$$\eta(U_0/kT)\exp(-U_0/kT) = 1 , \qquad (22)$$

defining a line in the H-T plane, which, if U_0 follows Eq. 1, will show the usual $(1-t)^{3/2} \propto B$ -relationship. But since B_{c2} is linear[27] in (1-t) near T_c, the resistivity at the crossover, from Eq. 9, goes as $B^{1/3}$.

Some indication for a crossover to the flux flow regime at higher resistivities comes from the YBaCuO crystal data in Fig. 1, which shows a pronounced knee. We compare the data above the knee to the simple Bardeen-Stephen form of Eq. 9, using the linearly extrapolated high-temperature resistivity and the linear temperature-dependence for B_{c2} recently determined by Welp et al..[27] (we use 2.5 T/K rather than their 1.9 T/K to optimize the fit, recognizing some corresponding uncertainty in the formula for the Bardeen-Stephen prefactor). The predicted resistivity is shown by the solid lines in Fig. 1. The weak dependence on B is in qualitative agreement with the $B^{1/3}$ -dependence predicted above.

Although the fit is obviously not perfect, it is remarkable how closely the flux-flow prediction approaches the data at intermediate resistivities and how the data mimics the predicted upward curvature. The deviations both at lower and higher resistivities can be plausibly understood. At lower resistivities, the crossover to flux creep is approached, and one would expect a smeared out downturn with the degree of smearing determined by the distribution of pinning barriers. Thus one can understand qualitatively why many data sets might not have observed the knee which is so evident in Fig. 1, if one assumes that our crystal has a more homogeneous distribution of defects. On the other hand, as the resistivity approaches the normal-state resistivity at higher temperatures, one expects deviations from the simple Bardeen-Stephen form because of fluctuation effects and because the isolated vortex core approximation used in the model breaks down.

An interesting further indication for the crossover from flux creep to flux flow in the regime of about 10 $\mu\Omega$cm comes from the photoresponse studies of Zeldov et al.[8] They found an enhanced resistivity due to light, but the nonbolometric contribution came only at resistivities below this value. They interpreted the enhancement as coming from photoinduced flux creep. Conversely the absence of the effect at higher resistivities is consistent with flux flow, where the light is no longer helpful in overcoming a pinning barrier.

In summary, the introduction of a flux flow regime and an overdamped attempt frequency permit a natural extension of Tinkham's original model for the resistive behavior of the high temperature superconductors in a field. This is all, of course, in the context of a flux creep model. Rather large ratios of hopping distance to well size, or else large vortex cascades, are required to explain the large observed prefactors. Further work is also required to test the model with a more detailed study of the current-voltage characteristics and to resolve puzzling discrepancies of the low and high temperature or low and high current density activation energies.[8] These difficulties may require consideration of alternative models, involving either distributions of barriers,[24,25] collective flux pinning[21–23], fluctuation effects[26] or especially the new co-operative concepts like vortex glass freezing.[20]

The authors appreciate key conversations with J. R. Clem, M. P. A. Fisher, B. Batlogg, T. Palstra and N. Amer.

References

1. Y. Yeshurun and A. P. Malozemoff, Phys. Rev. Lett. **60**, 2202 (1988); Y. Yeshurun, A. P. Malozemoff, F. Holtzberg and T. Dinger, Phys. Rev. B **38**, 11828 (1988).
2. A. P. Malozemoff, T. K. Worthington, Y. Yeshurun, F. Holtzberg and P. H. Kes, Phys. Rev. B **38**, 7203 (1988); T. K. Worthington, Y. Yeshurun, A. P. Malozemoff, R. Yandrofski, F. Holtzberg and T. Dinger, J. de Physique, Colloque C8, Suppl. 12, 2093 (1989).
3. M. Tinkham, Phys. Rev. Lett. **61**, 1658 (1988).
4. V. Ambegaokar and B. I. Halperin, Phys. Rev. Lett. **22**, 1364 (1969).
5. T. T. M. Palstra, B. Batlogg, L. F. Schneemeyer and J. V. Waszczak, Phys. Rev. Lett. **61**, 1662 (1988).
6. T. T. M. Palstra, B. Batlogg, R. B. van Dover, L. F. Schneemeyer and J. V. Waszczak, Appl. Phys. Lett. **54**, 763 (1989).
7. J. Z. Sun, K. Char, M. R. Hahn, T. H. Geballe and A. Kapitulnik, Appl. Phys. Lett. **54**, 663 (1989).
8. E. Zeldov, N. M. Amer, G. Koren and A. Gupta, Phys. Rev. B **39**, 9712 (1989); E. Zeldov, N. M. Amer, G. Koren, A. Gupta, R. J. Gambino and M. W. McElfresh, Phys. Rev. Lett., June 26, 1989.
9. J. D. Hettinger, A. G. Swanson, W. J. Skocpol, J. S. Brooks, J. M. Graybeal, P. M. Mankiewich, R. E. Howard, B. L. Straughn and E. G. Burkhardt, to be published.
10. M. Hikita and M. Suzuki, Phys. Rev. B **39**, 4756 (1989).
11. D. L. Kaiser, F. Holtzberg, M. F. Chisholm and T. K. Worthington, Journal of Crystal Growth **85**, 593 (1987).
12. P. W. Anderson and Y. B. Kim, Rev. Mod. Phys. **36**, 39 (1964).
13. M. R. Beasley, R. Labusch and W. W. Webb, Phys. Rev. **181** , 682 (1969).
14. M. Tinkham, **Introduction to Superconductivity** (McGraw-Hill Inc., New York 1975).
15. J. Bardeen and M. J. Stephen, Phys. Rev. **140** A1197 (1965).
16. R. B. van Dover, L. F. Schneemeyer, E. M. Gyorgy and J. V. Waszczak, Phys. Rev. B **39**, 4800 (1989).

17. R. J. Soulen and S. A. Wolf, Bull. Am. Phys. Soc. **34**, 846 (1989).

18. J. I. Gittleman and B. Rosenblum, Phys. Rev. Lett. **16**, 734 (1966).

19. D. S. Fisher, Phys. Rev. B **22**, 1190 (1980).

20. M. P. A. Fisher, Phys. Rev. Lett. **62** 1415 (1989); R. H. Koch, V. Foglietti, W. J. Gallagher, G. Koren, A. Gupta and M. P. A. Fisher, to be published.

21. A. I. Larkin and Yu. N. Ovchinnikov, J. Low Temp Phys. **34**, 409 (1979).

22. P. H. Kes, J. Aarts, J. van den Berg, C. J. van der Beek and J. A. Mydosh, Superconductor Science and Technology **1**, 242 (1989).

23. N. C. Yeh, Phys. Rev. B, to be published.

24. A. P. Malozemoff, T. K. Worthington, R. M. Yandrofski and Y. Yeshurun, International Journal of Modern Physics B **1**, 1293 (1988).

25. C. W. Hagen and R. Griessen, in **Studies of High Temperature Superconductors**, ed. A. V. Narlikar (Nova Science Publishers, New York 1988), to be published.

26. K. Kitazawa, S. Kambe and M. Naito, in **Strong Correlation and Superconductivity**, ed. H. Fukuyama, S. Maekawa and A. P. Malozemoff, Springer Ser. Solid State Sci. (Springer Verlag, Berlin, Heidelberg, 1989), to be published.

27. U. Welp, W. K. Kwok, G. Crabtree, K. G. Vandervoort and J. Z. Liu, Phys. Rev. Lett. **62**, 1908 (1989).

Giant Fluctuation Model and the Field-Induced Broadening of the Superconducting Transition in Oxides

K. Kitazawa[1], *S. Kambe*[1], *and M. Naito*[2]

[1]Department of Industrial Chemistry, Faculty of Engineering, University of Tokyo, 7-3-1 Hongo, Bunkyo-ku, Tokyo 113, Japan

[2]Basic Research Laboratories, NTT, 3-9-1 Midori-cho, Musashino-shi, Tokyo 180, Japan

The concept of the giant fluctuation is proposed with a preliminary model which assumes the field-induced broadening of superconducting transition to be intrinsic to the highly anisotropic superconductor and to be resulting from the state of dynamical mixture of superconductive and non-superconductive microscopical regions of which size is limited by the magnetic field and the coherence length. Recent experimental results are referred to examine the likeliness of this model and to delineate the phenomenon. It is pointed out from the experimental results that essentially the entire transition region is governed by a single mechanism and that its broadening under the magnetic field follows a scaling law $\Delta T^{3/2}/H$. A prediction is made that the remarkable broadening observed for H//c-axis configuration is solely due to the extremely short coherence length along c-axis of the cuprate superconductors and that it cannot be altered by the introduction of the pinning centers.

Introduction

It has been known that the temperature range of the resistive transition is broadened remarkably under a strong magnetic field especially when it is applied perpendicular to the CuO_2 basal plane[1]. There have been proposed various mechanisms in order to explain the broadening phenomenon including the superconducting glassy state model[2] and the giant flux creep model[3]. we have demonstrated that the extent of the broadening is essentially independent of the relative orientation between the field and the current and hence that the Lorentz force cannot be involved in the major mechanism[4], employing a novel measurement configuration of resistivity on the single crystal $(La,Sr)_2CuO_4$ which has the long c-axis dimension grown by the TSFZ method[5]. Based on this fact, we have argued that the broadening cannot be explained by the mechanisms mentioned above but it is rather due to the manifestation of the "giant fluctuation" which may be equivalent to assume the significant widening of the critical region of the superconducting phase transition[4,6]. Iye et al. has performed a precise measurements to show that the broadening is independent of the orientation of the magnetic field when the current is passed in the basal plane in a thin film of $Bi_2Sr_2CaCu_2O_y$[7].

In the present paper, we argue further that the whole transition region where the finite resistivity is observed is governed by a single mechanism and that it is the "giant fluctuation".

Non-involvement of the Lorentz force in the broadening mechanism

Figs. 1 to 4 show the resistivity-T-H relationship for the four orientations[4]. Comparisons of the Figs. of the sets 1:2 vs. 3:4 show that the field-induced broadening of the resistive transition is much more remarkable for H//c-axis than for H//ab-plane. While it is understood by comparing the two sets, 1, 4 and 2, 3 that the broadening effect is essentially unchanged for H//I and H⊥I. Therefore, we can conclude that the effect is independent of the relative orientation of H vs. I but is solely determined by the relative orientation of H vs. c-axis. Consequently, the involvement of the Lorentz force, which is maximized for H⊥I and minimized for H//I, is excluded from the possible motive force to raise the finite resistivity in the transition region. This is the reason why the fluctuation mechanism is more favored to the flux creep or the glassy state mechanism.

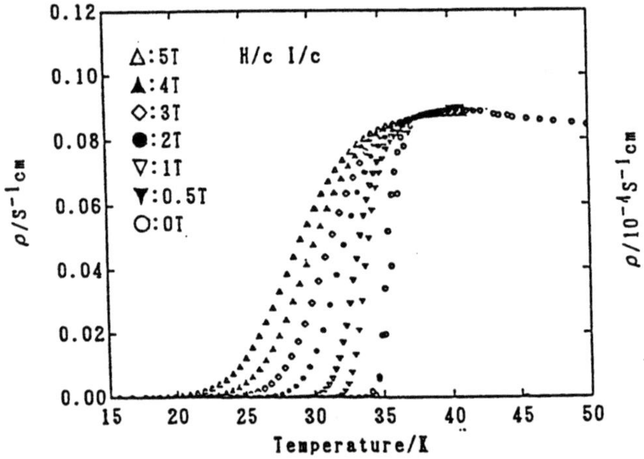

Fig. 1. Resistivity vs. temperature curves under various magnetic fields in the measurement configuration. H// I// c-axis for $(La_{0.93}Sr_{0.07})_2CuO_4$ single crystal.

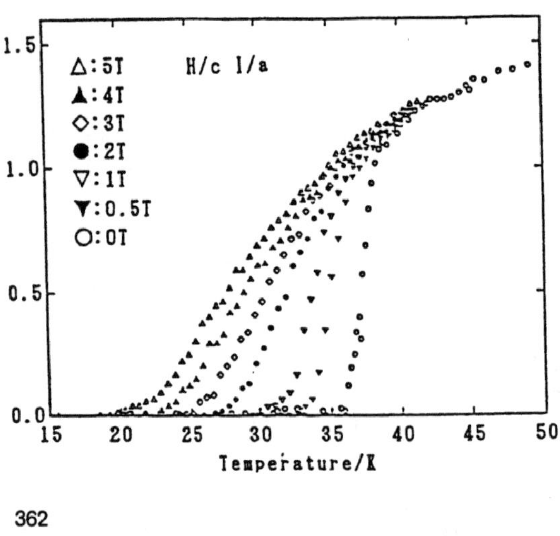

Fig. 2. Resistivity vs. temperature curves in configuration H//c-axis and I//a-axis for LSCO.

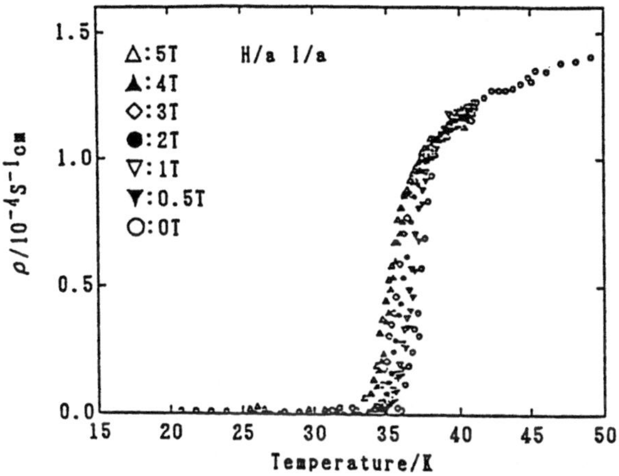

Fig. 3. Resistivity vs. temperature curves in configuration H/a-axis and I/a-axis for LSCO.

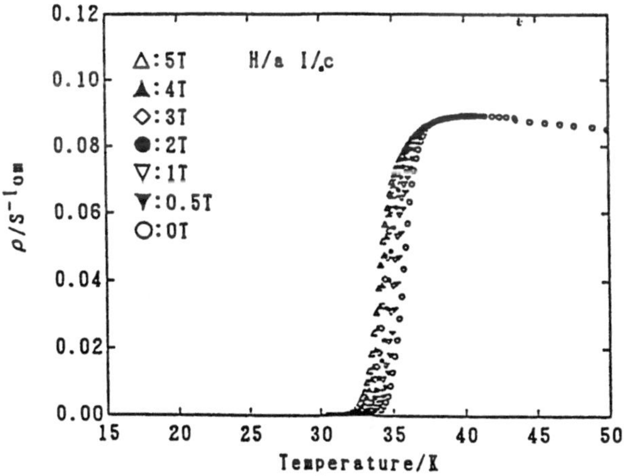

Fig. 4. Resistivity vs. temperature curves in configuration H/a-axis and I/c-axis for LSCO.

Besides, the transition range can be as wide as 20K in LSCO under the field of 5T. And the whole range seems to be governed by the single mechanism as will be described below. Hence this fluctuation phenomenon should better be termed as "giant fluctuation" in order to discriminate it from the classical gaussian fluctuation. As seen in Figs. 1 to 4, the resistivity-T curves show S-shaped structures rather than sharpened transition as the resistivity approaches toward zero. The latter behavior is expected from the conventional fluctuation theories which assume the occurrence of the clear phase transition. The S-shaped structure suggests that the clear phase transition in which the coherence length is assumed to develop infinitely large at the transition, may not even be present.

Non-GL scaling law in the transition region

As illustrated in Figs. 5 to 7, it is shown that $\Delta T(\rho,H)=T(\rho,0)-T(\rho,H)$ can be scaled as $\Delta T^{3/2}/H$ at any resistivity values in the transition region[8], which is quite distinctive from the linear scaling law expected from the mean field GL theory. This scaling law seems to hold in the whole transition range. This scaling law has been pointed out by Tinkham in analyzing Iye et al.'s ρ-T-H results on the $Ba_2YCu_3O_y$ single crystal[9], although he has attributed this scaling law to the flux creep mechanism. Ikeda et al. has shown that the dependence on H

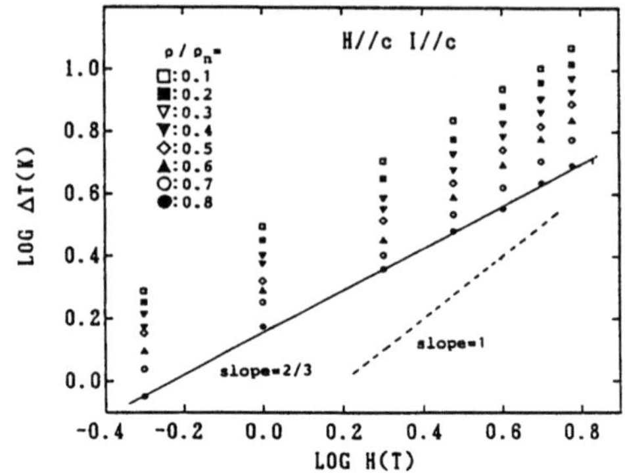

Fig. 5. Relationship between the field-induced broadening $\Delta T(\rho,H)$ and the field H at various fixed resistivity values in configuration H//I//c-axis.

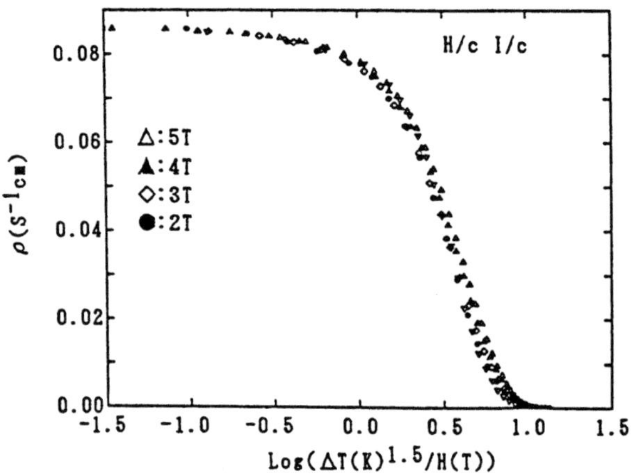

Fig. 6. Resistivity vs. Log($\Delta T^{1.5}/H$) curves under various magnetic fields in configuration H//I//c-axis for LSCO.

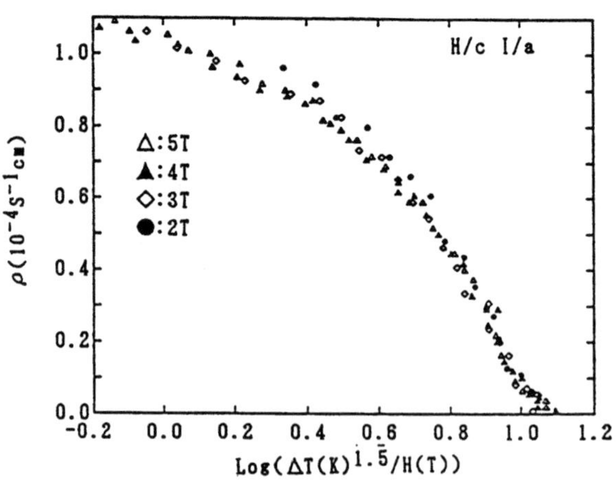

Fig. 7. Resistivity vs. Log($\Delta T^{1.5}/H$) curves under various magnetic fields in configuration H//c-axis and I c-axis for LSCO.

and T of any quantity resulting from fluctuation should be through the scaling variable $\Delta T^{3/2}/H$[10]. We presume that this field-induced scaling law is the one observed in such a wide range of the transition region as seen in Figs. 5 to 7, because the foot of the S-shaped ρ-T curves is not subjected to the Lorentz force-induced flux motion.

However, fluctuation theories have not been fully developed so that they can quantitatively analyze the results of such a wide and S-shaped transition. Therefore, an attempt is made to semiquantitatively explain the ρ-T-H behavior as the manifestation of the giant fluctuation based on the concept of the formation of point defects in a crystal[6].

Giant fluctuation model

We simply view the critical region as the microscopical and dynamical mixture of the superconducting(SC) domains and the non-superconducting(NSC) domains. We only treat the case H//c-axis which is the configuration for the most significant broadening. We assume the superconductivity is two-dimensional for simplicity. Then the minimum domain size may be taken as $\xi_c(0)$ along c-axis and as the ground-state Landau diameter $r_H=(\Phi_0/2\pi H)^{1/2}$ in the basal plane. We also assume that the smallest possible size of domains contributes dominantly to the free energy of the mixture system because the smallest difference is expected for the "defect formation energy", i.e. the difference between the SC and NSC energy values per domain and because the greatest stabilization is expected for the entropy term due to the dynamical mixing of the largest numbers of SC and NSC domains. Although the classical GL theory should not hold in the critical region, we tentatively adopt the condensation energy as a function of temperature as approximated by GL, i.e. $\varepsilon_c=[H_c(T)^2/8\pi][1-H/H_{c2}(T)]$ where the thermodynamical critical field $H_c(T)=H_c(0)[1-T/T_c]$ and $H_{c2}(T)=(\Phi_0/2\pi\xi_{ab}(0)^2)(1-T/T_c)$.

Then the defect formation energy per smallest coherence volume may be expressed as $\Delta H_d=\pi\varepsilon_c r_H^2\xi_c(0)$. Taking the volume fraction of the NSC region as 1-f, the enthalpy change per unit volume in the formation of the dynamically mixed NSC region is $\Delta H_d(1-f)(\pi r_H^2\xi_c(0))^{-1}$. On the other hand, the entropy of mixing is expressed as $\Delta S_m=k(\pi r_H^2\xi_c(0))^{-1}(f\ln f+(1-f)\ln(1-f))$. Minimizing the free energy against f, we obtain the thermodynamical equilibrium value of the NSC volume fraction, $1-f=(1+\exp(\Delta H_d/kT))^{-1}$. Fig. 8 shows the 1-f vs. T curves under various magnetic fields when the following values are adopted: $H_c(0)=1.5T$, $\xi_c(0)=8A$, $\xi_{ab}(0)=36A$. In the figure, it is noticed that each of the curves show anomalous shoulder at the GL phase transition point $T_c(H)$. This is caused because of the employment of the GL approximation. As a matter of fact, in the present giant fluctuation model, there should not be any clear

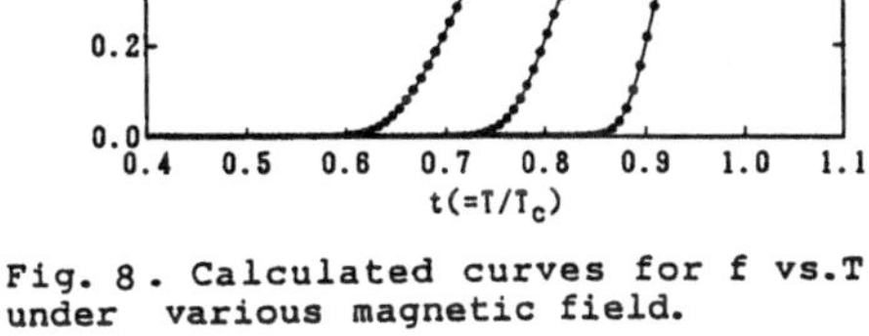

Fig. 8. Calculated curves for f vs.T under various magnetic field.

phase transition. Therefore, the anomalous small structure in the curves should not be taken up seriously.

The significance of the results, however, is the great similarity of the curves in Fig. 8 as the ones in Figs. 1 and 2, which tempts one to assume that the resistivity may be roughly proportional to the volume fraction of the NSC regions in the giant fluctuation range, i.e. roughly the range over which the scaling law $\Delta T^{3/2}/H$ is observed as mentioned above. Although we still cannot explain clearly why the resistivity could be proportional to 1-f, we presume that this should be related with the relative time constant of the dynamical fluctuation vs. electrical conduction in the mixture.

Significance of the giant fluctuation

The above proposed picture is still a preliminary one, but since it is based just on the thermodynamical consideration, we believe that it should be describing the essential phenomenon of the field-induced broadening. As a matter of fact, because of the extremely small SC domain size, one should directly expect the stabilization of the free energy as the result of manifestation of the dynamically mixing state of the SC and NSC domains as far as the condensation energy per domain is comparable with the thermal energy kT. Therefore, the reason to have the giant fluctuation in the cuprate superconductors is because of the nearly two-dimensional nature as well as of the high T^c which leads to the high thermal energy in the region of interest.

If the above picture is correct, the width of the broadened transition region under magnetic field cannot be lessened even by introducing some effective pinning centers in the material. But rather it is suggested that the coherence length along the c-axis should be the most crucial material parameter that determines the extent of the broadening. In this regard, it is noteworthy that the $\xi_c(0)$ seems to be longer for BYCO than for LSCO and shorter for BSCCO[11]. As seen in Figs. 1 and 2 the broadening is as wide as 20K under 5T in the present LSCO, while it is about 10K in BYCO in spite of the higher T_c of the latter. The broadening on the other hand, is much more serious in BSCCO, more than 50K as is expected in this context.

However, when the fluctuation becomes small as T and H are reduced, then the situation continuously shifts into the flux creep regime. Then it will be the regime where the flux pinning becomes important. In this sense, BYCO seems to be nearly on the border line for the practical usage of it under high magnetic field at liquid nitrogen temperature without being disabled by the giant fluctuation.

Conclusion

The above discussions leads to the conclusion that the field-induced broadening is the manifestation of the giant fluctuation which is a thermodynamical phenomenon, and hence cannot be suppressed either by introducing more effective pinning centers or by making the specimens in a more homogeneous state. Therefore, new compositions or new materials with longer coherence lengths along c-axis must be searched in order to further improve the superconducting performance at 77K of the cuprate superconductors under magnetic fields.

References

[1] Y. Iye, T. Tamegai, T. Sakakibara, T. Goto, N. Miura, H. Takeya and H. Takei, Physica C153-155, 26(1988).
[2] K. A. Muller, M. Takashige and J. G. Bednorz, Phys. Rev. Lett. 58, 1143(1987).
[3] Y. Yeshurun and A. P. Malozemoff, Phys. Rev. Lett. 60, 2202(1988).
[4] K. Kitazawa, S. Kambe, M. Naito, I. Tanaka and H. Kojima, Jpn. J. Appl. Phys. 28, 555(1989).
[5] I. Tanaka and H. Kojima, Nature 337, 21(1989).
[6] K. Kitazawa, S. Kambe and N. Okazaki, Proc. Taiwan Int. Symp. on Superconductivity, Hsinchu, Taiwan, Apr. 17-19, 1989 to be published by World Scientific.
[7] Y. Iye, S. Nakamura and T. Tamegai, submitted to Physica C.
[8] S.Kambe, M. Naito, K. Kitazawa, I. Tanaka and H. Kojima, submitted to Phys. Rev. B.
[9] M. Tinkham, Phys. Rev. Lett. 61, 1658(1988).
[10] I. Ikeda, T. Ohmi, T. Tsuneto, J. Phys. Soc. Jpn. 58, 1377(1988).
[11] Y. Iye, "Studies of High Temperature Superconductors" Ed. by A. V. Narlikar (NOVA Science Publishers inc.)to be published.

Flux Dynamics and Electronic Anisotropy in High-T_c Superconductors

B. Batlogg, T.T.M. Palstra, L.F. Schneemeyer, and J.V. Waszczak

AT&T Bell Laboratories, Murray Hill, NJ 07974, USA

The dissipation below T_c in external fields up to 12T has been studied in five groups of high temperature cuprates. The dissipation is due to Lorentz-force driven flux motion and is least pronounced in the compound with the smallest electronic anisotropy ($Ba_2YCu_3O_7$). A crossover temperature T_x is defined at which the effective activation energy for flux motion is ~kT. This T_x is material specific. An H-T phase diagram is presented for $Ba_2YCu_3O_7$ and $Bi_2Sr_2CaCu_2O_8$ demonstrating the qualitative similarities and quantitative differences in the dissipation behavior in these two compounds.

A wealth of experimental studies focused on the transport and magnetic properties of cuprate superconductors below T_c, both to elucidate the state in the presence of a magnetic field and to estimate the performance of these promising materials in practical applications. Experimental techniques include: time dependent magnetization [1-3], dc resistivity [4-9], ac susceptibility [10,11], mechanical oscillator [12,13] and decoration experiments [14,15]. Indeed, accumulating data allow us to have a clearer understanding of apparently intrinsic limitations and to point to the unresolved problems.

This is a short discussion of selected points brought up by resistance measurements in high magnetic fields. We first demonstrate that dissipative flux motion is driven by Lorentz-force. Then we propose the existence of a "crossover" line which separates distinctly different areas in the H-T plane. Below the crossover temperature T_x, the dissipation grows with T in a thermally activated fashion, until the characteristic energies become of the order k_BT at the crossover temperature T_x. This crossover temperature is characteristic for each compound and we suggest it to be related to the electronic anisotropy. The more anisotropic the superconductor, the lower T_x. Among the five major classes of cuprates studied, $Ba_2YCu_3O_7$ has the lowest dissipation levels.

1. Lorentz-Force Driven Dissipation

A basic underlying assumption in the discussion of flux motion is that flux lines (or bundles of lines) can move under the influence of the Lorentz-force. This force is largest when the applied field H and the current I are perpendicular to each other. [Not so widely appreciated is the fact that a (much smaller) Lorentz-force exists even when current and field are parallel, because the flux line ceases to be straight.] We have checked the Lorentz-force assumption in a single crystal of $Ba_2YCu_3O_7$. The current flows in the a,b plane and the external field is also applied in the a,b plane, either parallel or perpendicular to the current. The resistive transition $\rho(T)$ is much

Springer Series in Solid-State Sciences, Vol. 89 Editors: H. Fukuyama · S. Maekawa · A.P. Malozemoff **Strong Correlation and Superconductivity**

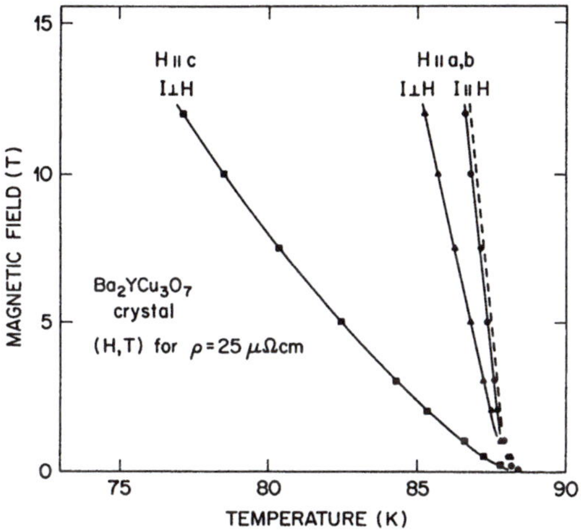

Fig. 1 Evidence for Lorentz-force driven dissipation below T_c in single crystal $Ba_2YCu_3O_7$. The symbols give (H,T) for a resistivity of 25μΩcm, corresponding to ~1/2 of the normal state value ($T_{1/2}$). When the current flows parallel to the magnetic field (round symbols) the broadening is the smallest and $T_{1/2}$ is suppressed at the same rate as the magnetically measured T_c (Ref 22).

more narrow in the H∥I geometry than for H⊥I. This is quantitatively shown in Fig. 1 where the ρ(T) curves are simply parametrized by the $\rho_n/2$ temperature (i.e. $\rho = 25\mu\Omega$cm). The I∥H points follow closely the dashed line which is the slope of the upper critical field H_{c2} determined by magnetic measurements in Ref.22. The slopes are the same within experimental uncertainties: 10±1 T/K. Thus, we have provided clear evidence for Lorentz-force induced dissipation in $Ba_2YCu_3O_7$ single crystals.

Recent studies have raised some doubt whether the Lorentz-force is involved in the broadening of the transitions, [9,17,18] but we suspect that the microscopic current path in those samples was not identical to the macroscopic straight-line geometry assumed in the interpretation. This could be caused in the thin film samples by slight misalignment of the grains and in the single crystals by stacking-fault type imperfections, which would allow the current to flow across the Cu-O layers.

2. Crossover Temperature T_x

When the resistive transition in a field is plotted on a linear scale, it appears broadened with a small "foot" at lowest resistivities. On an Arrhenius plot (log ρ vs 1/T), however, the thermally activated nature of the dissipation at lowest levels becomes apparent. We have reported details for $Ba_2YCu_3O_7$ and $Bi_2Sr_2CaCu_2O_8$ previously. (Several more recent studies are in essential agreement.) As the temperature is raised, the resistivity increases rapidly and would exceed the normal state value well below T_c if there were not a crossover regime. In the temperature interval between T_x and T_c, ρ approaches ρ_n gradually. While it is obvious in the Bi compound (see Fig. 1, Ref.4), it is equally pronounced in $Ba_2YCu_3O_7$ (top part of Fig. 2). In the following, we show how the "crossover"

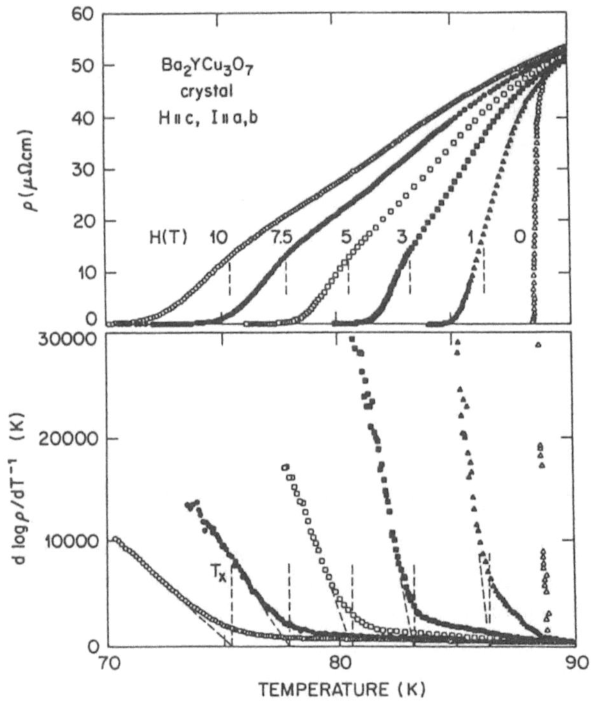

Fig. 2 Experimental definition of the crossover temperature T_x. Upper panel: ρ(T) on a linear scale. Lower panel: slope of the Arrhenius plot d(log ρ)/d(1/T). The slopes decrease rapidly and extrapolate to 0 at T_x, marking the vanishing of the effective activation energy which dominates the flux dynamics at lower temperatures. The "knees" in ρ(T) on the upper panel coincide with T_x.

temperature T_x can be defined unambiguously from experimental data in a way that reflects the underlying physical mechanism.

The activation energy in a thermally activated process is usually extracted from the slope of an Arrhenius plot d(log ρ)/d(1/T). If the activation energy U itself is temperature-dependent the value of the slope is renormalized and is not identical with U. Numerical examples for different functional dependences of U(T) can be found, e.g. in Refs.21 and 23. Thus, instead of representing the data on an Arrhenius plot, we show d(log ρ)/d(1/T) for $Ba_2YCu_3O_7$ in the lower part of Fig. 2. This local slope or apparent activation energy rapidly drops and extrapolates to zero as indicated by the dashed lines, defining T_x. This T_x marks the crossover from thermally activated dissipation and occurs when the activation energy vanishes on a scale of kT. Measurements of the kinetic inductance of $Ba_2YCu_3O_7$ thin films have been discussed in a similar spirit and a detailed comparison is given in Ref. 19,20. We also refer to a discussion on creep-flow crossover in these proceedings [21].

It is interesting to note that T_x is the temperature of the "knee" in the ρ(T) curves on a linear scale (upper part Fig. 2), a feature which is clearly observed in good quality materials. The resistivity at the crossover temperature is a sizeable fraction of the normal state resistivity.

3. Comparison Between $Bi_2Sr_2CaCu_2O_8$ and $Ba_2YCu_3O_7$

The similarities and differences between the two compounds ("Bi 2212" and "123") are best illustrated in an H-T plot (Fig. 3.) The temperature ranges over which thermally activated dissipation was measured are indicated by the hatched areas. The low temperature border is given by experimental limitations to measure very low resistivities, and might be extended by using SQUID-based voltmeters. The crossover temperature is limiting the thermally activated range at the high temperature and in $Ba_2YCu_3O_7$ T_x is operationally defined as described in the previous section. The same procedure cannot be applied to "Bi 2212" since the activation energies are not much larger than kT and the d log ρ/d(1/T) does not lead to the pronounced drop as in $Ba_2YCu_3O_7$. (See Ref.23). Therefore, we have not drawn the crossover line T_x(H) and we have also chosen a very restrictive criterion of 1% of ρ_n to delineate the thermally activated regime. In analogy to "123" where T_x coincides with the "knee" in the ρ(T) plots, we could rely on the same feature in "Bi 2212" to estimate T_x. Such a procedure then places the crossover line at significantly higher temperatures than the border of hatched area (by ~10-15K).

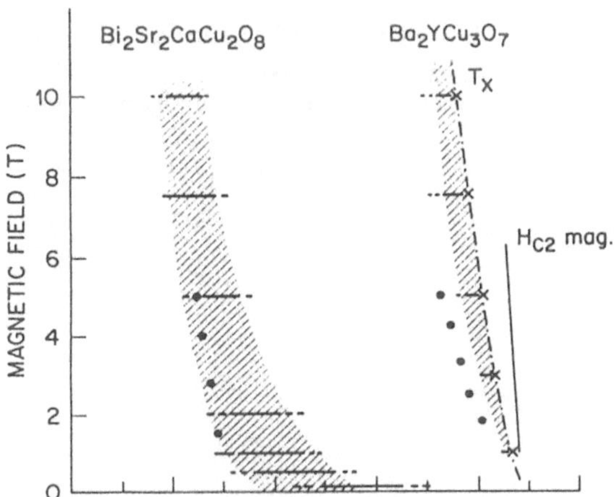

Fig. 3 Comparison of flux motion induced dissipation in $Ba_2YCu_3O_7$ and $Bi_2Sr_2CaCu_2O_8$. The crosses indicate the crossover temperature T_x(H), as defined in Fig. 2, and represent the high-temperature limit of thermally activated dissipation (hatched area). The main observations are: (1) there exists a region between T_x and the magnetically measured H_{c2} where the resistivity is close to the normal state value. (2) The crossover temperature is material specific. Round symbols: dissipation maxima in mechanical oscillator experiments [12].

Irrespective of how T_x is extracted from the "Bi 2212" data, the two major points of Fig. 3 are evident. First, there exists a temperature interval between the crossover temperature $T_x(H)$ and the magnetically measured $H_{c2}(T)$.[23] The resistivity in this range approaches the normal state value within an order of magnitude and renders practical applications unfeasible in this temperature and field range. Second, the crossover temperature is material specific and is lower in $Bi_2Sr_2CaCu_2O_8$ than in $Ba_2YCu_3O_7$. While variations in the preparation process or subsequent treatment might modify $\rho(H,T)$ in a given compound to some degree, we suggest that the crossover temperature cannot be changed significantly because it is associated with intrinsic microscopic properties, such as electronic anisotropies.

Dissipation maxima in mechanical oscillator experiments [12] (round symbols in Fig. 3) have been interpreted as due to flux lattice melting. Those temperatures are lower than T_x and in a range where the resistivity is by many orders of magnitude smaller than ρ_n. The microscopic relationship between the "melting" temperature and T_x remains to be elucidated, keeping in mind the different dissipation levels employed in the two experiments.

4. Comparison of Various Compound Families

The discussion in the previous part focussed on the two most intensively studied compounds. A comparison on a more qualitative level also can be made for other families of high T_c cuprates. As an example, we show log ρ vs T in a field of 5T (H||C) for five compounds. Starting with $Ba_2YCu_3O_7$ (T_c = 88K), the dissipation increases on going to $Ba_2YCu_3O_{6.7}$ (T_c = 60K) $Pb_2Sr_2RCu_3O_8$ ($T_c \approx 50K$), $Bi_2Sr_2CaCu_2O_8$ (T_c=84K) and $Tl_2Ba_2CaCu_2O_8$ (T_c = 90K). In analogy to the previous discussion, we would expect the crossover temperature to decrease (on a

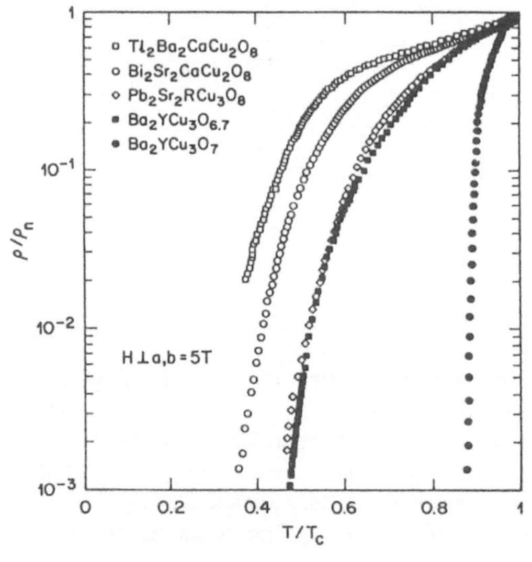

Fig. 4 Normalized resistance vs. T/T_c in a field of 5T for five high T_c cuprates. It is suggested that flux motion induced dissipation is more pronounced in the compounds with larger electronic anisotropy.

reduced temperature scale T/T_c) as the dissipation increases. This tendency is apparently correlated with the electronic anisotropy and suggests that detailed microscopic discussion of flux dynamics will have to address the possibility of "broken" flux lines due to very weak coupling between the Cu-O layers.

Acknowledgement

We gratefully acknowledge stimulating discussions with D. J. Bishop, S. N. Coppersmith, R. C. Dynes, A. T. Fiory, D. S. Fisher, P. L. Gammel, E. M. Gyorgy, A. F. Hebard, D. A. Huse, M. Inui, P. B. Littlewood, A.P. Malozemoff, A. P. Ramirez, J. M. Valles, C. M. Varma and T. K. Worthington.

[1] K. A. Müller, M. Takashige and J. G. Bednorz, *Phys. Rev. Lett.* **58,** 1143 (1987).

[2] Y. Yeshurun and A. P. Malozemoff, *Phys. Rev. Lett.* **60,** 2202 (1988).

[3] A. P. Malozemoff, L. Krusin-Elbaum, D. C. Cronemyer, Y. Yeshurun and F. Holtzberg, *Phys. Rev.* **B38,** 6940 (1988).

[4] T. T. M. Palstra, B. Batlogg, L. F. Schneemeyer and J. V. Waszczak, *Phys. Rev. Lett.* **60,** 1662 (1988).

[5] M. Tinkham, *Phys. Rev. Lett.* **60,** 1658 (1988).

[6] T. T. M. Palstra, B. Batlogg, R. B. van Dover, L. F. Schneemeyer and J. V. Wasczak, *Appl. Phys. Lett.* **54,** 763 (1989).

[7] J. D. Hettinger, A. G. Swanson, W. J. Skocpol, J. S. Brooks, J. M. Graybeal, P. M. Mankiewich, R. E. Howard, B. L. Straughn and E. G. Burkhardt, *Phys. Rev. Lett.* **62,** 2044 (1989).

[8] J. Z. Sun, K. Char, M. R. Hahn, T. H. Geballe and A. Kapitulnik, *Appl. Phys. Lett.* **54,** 663 (1989).

[9] T. K. Worthington, Y. Yeshurun, A. P. Malozemoff, R. M. Yandrowski, F. H. Holtzberg and T. R. Dinger, *J. de Physique* (to be published).

[10] P. H. Kes, J. Aarts, J. van den Berg, C. J. van der Beck, J. A. Mydosh, *Superconductor Science and Technology 1 242 (1989).*

[11] P. L. Gammel, L. F. Schneemeyer, J. V. Waszczak and D. J. Bishop, *Phys. Rev. Let.* **61,** 1666 (1988).

[12] S. Gregory, C. T. Rogers, T. Venkatesan, X. D. Wu, A. Inam and B. Dutta, *Phys. Rev. Lett.* **62,** 1548 (1989).

[13] P. L. Gammel, D. J. Bishop, G. J. Dolan, J. R. Kwo, C. A. Murray, L. F. Schneemeyer and J. V. Waszczak, *Phys. Rev. Lett.* **59,** 2592 (1987).

[14] G. J. Dolan, F. Holtzberg, C. Feild and T. R. Dinger, *Phys. Rev. Lett.* **62,** 2184 (1989).

[15] T. Matsushita, *J. Phys. Soc. Japn.* **57**, 3941 (1988).

[16] K. Kitazawa, S. Kambe, A. Fukuoka, I. Tanaka and H. Kojima (preprint, *Japn. J. Appl. Phys.)*

[17] N. Kobayashi, H. Iwasaki, H. Kawabe, K. Wakanabe, H. Yamane, H. Kurosawa, H. Masumoto, T. Hirai and Y. Muto (preprint).

[18] A. F. Hebard, P. L. Gammel, C. E. Rice and A. F. J. Levi (to be published).

[19] A. F. Hebard and T. T. M. Palstra (Proc. Stanford Conf. M^2S–HTSC).

[20] Y. Iye, S. Nakamura and T. Tamegai (to be published).

[21] A. P. Malozemoff, T. K. Worthington, E. Zeldov, N. C. Yeh, M. W. McElfresh and F. Hotzberg (this volume).

[22] U. Welp, W. K. Kwook, G. W. Crabtree, K. G. Vandervoort and J. Z. Liu, *Phys. Rev. Lett.* **62,** 1908 (1989).

Part IV

Summary

Summary (Experimental)

H.R. Ott

Laboratorium für Festkörperphysik, ETH-Hönggerberg,
CH-8093 Zürich, Switzerland

What I shall try to do in this very brief summary of what we have heard about experimental aspects in this very lively symposium is to mention striking or characteristic features and to formulate questions that emerge when contemplating over the vast amount of experimental facts that is now available. Many of these facts have been presented here by experts in various different experimental techniques and we may assume that we, that were lucky enough to be able to participate in this workshop, are as well informed about high-T_c superconductors as one can possibly be.

Materials

It seems that still a lot can be learned from just studying the systematic variation of physical properties of different materials. The most outstanding class of materials in this business are still the various copper oxides. For most of them it seems to be true that doping a parent compound which has an insulating and antiferromagnetically-ordered ground state [La_2CuO_4, Ln_2CuO_4 (Ln = Lanthanides), $Ba_2YCu_3O_6$] with suitable dopants leads to a suppression of magnetic order giving way to a conducting normal state and superconductivity at elevated temperatures. In some cases [$La_{2-x}M_xCuO_4$, $Ln_{2-x}M_xCuO_4$] further doping increases the conductivity in the normal state but, alas, kills the wonderful feature of high-T_c superconductivity. The competition between antiferromagnetism and superconductivity is particularly intriguing and very well documented for the series $La_{2-x}M_xCuO_4$, $Ln_{2-x}M_xCuO_4$ and $Ba_2YCu_3O_{6+x}$. Of course, this feature is an important basis for all theoretical models trying to describe high-T_c superconductivity. Also for the Bi- and Tl-based copper oxides, magnetically-ordering phases have been found in materials where Ca has been replaced by Y. Detailed work indicates that for each doping concentration in $La_{2-x}M_xCuO_4$ and $Ba_2YCu_3O_{6+x}$, homogeneous material may be prepared. I am still not quite sure that this is so in each case and this problem deserves further careful investigations.

Springer Series in Solid-State Sciences, Vol. 89 **Strong Correlation and Superconductivity**
Editors: H. Fukuyama · S. Maekawa · A.P. Malozemoff

Doping often also triggers or suppresses structural instabilities that lead to significant changes in the charge distribution within these substances. Unfortunately there seems to be no one-to-one correspondence between symmetries of crystal structures and the occurrence of magnetic order or superconductivity. Nevertheless, materials adopting the tetragonal T, T' or T* structures seem to be a new and interesting playground in our game.

While the first-generation high-T_c superconductors were all claimed to be p-type conductors, more recent work has claimed superconductivity between 20 and 30 K in n-type conductors [$Ln_{2-x}(Ce$ or $Th)_xCuO_4$ or $Ln_2CuO_{4-y}F_y$]. At this conference it has been reported that also in compounds of the type $(Ln_{1-x}Ce_x)_2$ $(Ba_{1-y}Ln_y)_2$ $Cu_3O_{10-\delta}$, where Ln = Nd, Sm or Eu, superconductivity may set in at temperatures exceeding 40 K. In all these Ln compounds, which also have magnetically-ordering parents, the preparation of materials seems to be delicate and it appears that the homogeneity problems are not quite under control yet.

Not quite so spectacular with respect to T_c but nevertheless interesting are the substances of the type $Ba(Pb,Bi)O_3$, $Ba(Pb,Sb)O_3$ and $(Ba,K)BiO_3$. A significant difference to the copper oxides is the missing competition of superconductivity with magnetic order, but structural instabilities and metal-insulator transitions are obviously also present. Again, preparation problems are formidable and it appears that the occurrence of superconductivity is limited to a rather narrow range of chemical composition. As we shall see below, the Ba-oxide compounds are most likely not in the same class of materials as the Cu oxides, although critical temperatures of almost 30 K have been reported.

Normal-State Properties

It is not surprising that at least 50% of all discussions at this symposium dealt with normal-state properties because the conducting compounds can safely be characterized as anomalous metals. An outstanding feature is certainly the linear temperature dependence of the electrical resistivity that is now observed in all good-quality samples of Cu-oxides if measured parallel to the Cu-O planes. As a side remark I should like to point out that $\rho(T)$ linear in T has also been found in heavy-electron materials at low temperatures just above the range where the T^2 Fermi-liquid behaviour is observed. With the exception of a report on $Ln_{2-x}Ce_xCuO_4$ samples, no tendency to temperature independence of ρ is observed above T_c and in this latter case, it remains to be seen whether improvement of sample quality may get rid of this effect. In other Cu-O compounds where still an overall "semiconducting" behaviour is found in the normal state, the sample quality has probably not yet reached optimal conditions. Distinctly different with respect to the electrical resistivity are the above mentioned Ba-oxides where $\rho(T)$ is also large, at most temperature independent and usually found to be increasing with decreasing temperature.

Not so unambiguous are the claims for the resistivity behaviour of the Cu-oxides for currents perpendicular to the planes. Here the conductivities are far beyond the Mott limit and still, in the best characterized samples, the resistivity decreases with decreasing temperature below room temperature. At lower temperatures, trends to saturation or even upturns of ρ with decreasing temperature are observed. Whether contributions varying proportional to T^{-1} can be definitely identified in general remains to be seen in future experiments.

Where reliable Hall-effect data from single crystals are available, considerable temperature dependences of R_H below room temperature question the simple free-electron one-band intepretation of this quantity as being inversely proportional to the carrier concentration. This kind of analysis, however, seems to indicate correctly the trend from less to more metallic behaviour consistent with the conductivity and magnetic-susceptibility measurements. Very exotic are also, where known, the temperature dependences and the anisotropies of other transport properties (thermal conductivity, thermopower) where simple interpretations valid for common metals again seem to fail.

Most animated and often heated discussions were held about the electronic structure of some materials, mostly based on results from spectroscopic experiments on different copper oxides. Optical data on oriented films of parent-compound insulators reveal the expected gap in the electronic energy excitation spectrum. Spectra of doped material were claimed to be consistent with Drude behaviour while for far-infrared reflectivity results, this interpretation does not seem to be so successful. Quite intriguing is the temperature dependence of the spin-lattice relaxation times that were reported from NMR experiments, where different lattice sites and hence internal variations of interactions may be probed. Non-superconducting but metallic Cu oxides show classical Korringa (T_1T = const.) behaviour at low temperatures and fairly low relaxation rates. Superconducting materials reveal in general fairly high relaxation rates and no classical Korringa behaviour in the normal state. Such dramatic enhancements of T_1^{-1} and distinct deviations from Korringa behaviour were previously also observed in heavy-electron compounds. In doped La_2CuO_4-based material, an enhanced suppression of the relaxation rates is observed far above T_c and may well be correlated with the decrease of magnetic spin-fluctuation intensity that is noted in respective inelastic neutron-scattering experiments in the same temperature range. T_1^{-1} may indeed be correlated with exchange frequencies implied from a local-moment fluctuation scheme and most surprisingly the frequencies deduced in this manner scale with the observed critical temperatures for superconductivity of the respective substances. Of course, NMR data also give information on internal charge redistributions with varying doping, an important ingredient for model calculations.

In connection with the electronics of these materials we also learned how photoemission spectroscopy can be used to determine model parameters like energy-band widths, correlation energies and exchange parameters. This, of course, is invaluable for deciding which approximations in the model calculations are meaningful.

Fermi-liquid or not was clearly a major issue as far as descriptions of the electronic behaviour of superconducting material are concerned. New results of angle-resolved photoemission measurements were valued by some as unambiguous proof that suitable band theory is sufficient to describe these materials. Others seriously question this intepretation and point out that there is so far no experimental proof for the existence of a well defined Fermi surface in momentum space. It is true that one of the possibilities to settle this question, positron-annihilation experiments, have so far not been successful, mainly because of inadequate positron beam intensities. Meanwhile, the energy dependence of the conductivity at low energies as obtained from tunneling data and a strange source of scattering by an unidentified continuum of states revealed by Raman scattering experiments are put forward as indications that Fermi-liquid-type descriptions of the normal-state behaviour of these materials are inadequate. Photoemission data with high resolution show the energy cut-off at E_F and a linear rise of intensity with increasing binding energy. This intensity cannot, however, be directly compared with the energy dependence of the conductivity and it is also clear that, in spite of tremendous advances, the sensitivity of the emission technique cannot yet match the one of tunnel spectroscopy.

Magnetic interactions play a dominant role in these substances and hence their manifestations were also thoroughly discussed. The magnetically-ordering parent compounds have been studied in some detail and for some of them, distinct 2D-behaviour has been claimed. For $La_{2-x}M_xCuO_4$-type compounds, inelastic neutron-scattering reveals magnetic excitations that persist in superconducting material but whose q-vector becomes incommensurate or modulated in superconducting samples. Again in superconducting material the intensity of this signal weakens considerably with decreasing temperature far above T_c, approximately at the same temperatures where the spin-lattice relaxation rate mentioned above decreases rapidly.

In spite of distinct differences in crystal structure between La_2CuO_4 and Ln_2CuO_4, the characteristics of the magnetic order among Cu moments are very similar, at least for La_2CuO_4 and Pr_2CuO_4. More complicated is the temperature dependence of magnetic order in Nd_2CuO_4, indicating that the role of the rare-earth ions with respect to interactions is not completely negligible. It is clear that magnetic interactions are altered in superconducting material and it remains to be seen, how this interplay may give hints with respect to possible mechanisms for superconductivity. It would certainly be helpful if studies on magnetic excitations in $Ba_2YCu_3O_{6+x}$ turned out to be successful.

Superconductivity

An important question is still the existence, magnitude and anisotropy of the energy gap that develops with the onset of the superconducting state. Various contributions here discussed this problem. Tunneling seems to be possible with the current both parallel and perpendicular to the Cu-O planes. Anisotropies of the order of 1.5 to 2 are deduced. The evaluation of the important ratio $\Delta(0)/k_BT_c$ is somewhat hampered by the uncertainties of T_c's as they follow from the temperature dependence of the experimental gap structure. In some measurements it appears that also defining the gap from the experimental I/V-curves is not trivial. Other data that are consistent with a gap formation are obtained from far-infrared studies. Here the main obstacle is the lack of comparison between spectra of the normal- and the superconducting state at the same temperature because of the very high critical fields of the copper oxides. The situation is much better for Ba oxides, where the superconducting state can be suppressed with accessible magnetic fields and for the ratio $2\Delta(0)/k_BT_c$ a rather normal value between 3.5 and 4 is found.

Clear manifestations of an energy gap are also observable in measurements of the spin-lattice relaxation time and the Knight shift with NMR. Here one has a possibility to study the internal anisotropy of the gap by studying these effects on nuclei at different lattice sites. It appears that the temperature dependence of the Knight shifts of nuclei on the Cu(2) and Cu(1) sites in $Ba_2YCu_3O_7$ can be described on the basis of classical Yoshida functions with different values of the ratio $2\Delta(0)/k_BT_c$. Somewhat less straightforward seems the interpretation of $T_1(T)$ below T_c.

Another quantity that is intimately related with the superconducting state per se, is the London penetration depth λ_L for external magnetic fields. Both the zero-temperature value $\lambda_L(0)$ and the temperature dependence $\lambda_L(T)$ are of interest and various different attempts to measure these values have been made. At present all data seem to indicate that a two-fluid model can describe the temperature dependence of λ_L and the absolute values for $\lambda_L(0)$ are in the range of a few thousand Å. So far there is no overwhelming experimental proof for anomalous temperature dependences at low T. The anisotropies for fields either parallel or perpendicular to the Cu-O planes are of the order of 5 in these more or less direct measurements but it appears that calculated anisotropies using classical correlations between λ_L and the conductivity are found to be distinctly larger. Time dependent magnetization measurements in low fields were used to determine $H_{c1}(T)$ parallel and perpendicular to the Cu-O planes of $Ba_2YCu_3O_7$. It was shown here that calculations of λ_L from these data give very good agreement with results that were obtained with other experimental techniques. Another characteristic length, the correlation length ξ has been confirmed to be short (in the range of a few Å) and torque measurements indicate an anisotropy here of the order of 6 for $Ba_2YCu_3O_7$. Amazing are still the extreme anisotropies that are found for Bi-copper oxides.

A considerable influence in T_c by different isotopes of oxygen in $(Ba,K)BiO_3$ has been reported here. Even when considering the experimental difficulties and uncertainties, this is probably not surprising and supports, in some way, the impression that this kind of oxides is different from the copper-oxide variety.

Quite important, also in view of technical applications, is the response of these high-T_c superconductors to external magnetic fields H. The most intriguing manifestations of this response are the logarithmic time dependence of induced magnetization, the frequency dependence of anomalies in χ_{ac}, the temperature dependence of the resistivity in external magnetic fields, and mechanically probed changes of the flux-lattice behaviour by varying T and H. Here we were informed in detail on features of the resistivity in low fields where, with improved quality of the investigated samples, the temperature dependence of ρ in a fixed field is interpreted to indicate a transition from the flux-creep to the flux-flow regime by a distinct break in $\rho(T)$. Most available data are consistent with at least three different regimes in the [H,T]-phase diagram, distinguishing between a defined rigid flux lattice at the lowest temperatures and some thermally-activated motion of the flux lattice above a certain temperature which then transforms into an unidentified cross-over behaviour before $H_{c2}(T)$ is reached. The temperature range for activated flux motion is rather narrow for $Ba_2YCu_3O_7$ and seems to be largest for Tl-based copper oxides. These interpretations rely on the assumption that the Lorentz force exerted on individual flux lines is essential. Indeed, it has been found that the relevant effects on the resistivity in the Cu-O planes are most pronounced when the current direction is perpendicular to the orientation of the external field. This observation is, however, not corroborated by others in experiments probing the resistivity perpendicular to the planes in samples with large dimensions in the c-direction. From these latter experiments it was concluded that flux-creep and flux-flow are not responsible for the apparent broadening of the resistive transitions in external fields. Model calculations of ρ in the transition region invoking fluctuations and dynamical mixtures of superconducting and normal regions within the samples give a remarkable coincidence with observations.

For applications, the principal remaining concern about these effects will be the non-zero resistance of these materials in high external magnetic fields at liquid-nitrogen temperature. A more positive message with respect to applications was the report that proximity-induced superconductivity might persist in noble metals like Au in closed contact with copper oxides at elevated temperatures. This possibility clearly deserves future research efforts.

In conclusion we still have to admit that we have to take the superconductivity in these copper oxides at high temperatures as a godsent present. In spite of remarkable progress we cannot claim to know what the pairing mechanism is, whether weak or strong-coupling schemes apply, nor are we certain as to what kind of superconducting state is adopted. To find an answer to all these questions, on the experimental side, much will depend on how well we can characterize our substances and samples and whether we shall be able to recognize the essential parameters and their influence on the superconducting properties. As usual in our business, all the fun is getting there.

Acknowledgement

Particularly on this occasion I should like to thank the organizers of this seminar for hosting such a fruitful, pleasant and memorable meeting.

Summary Talk: Theory

P.A. Lee

Department of Physics, Massachusetts Institute of Technology,
Cambridge, MA 02139, USA

In the past four days we have heard many excellent theory talks. Here I would take a rather broad view and attempt to place some of these talks in the context of where we stand in our common goal of achieving an understanding of the oxide superconductors.

The first task of the theorist is to establish a model which he believes contains the essential physics of the problem. Since the initial suggestion of Anderson [1], there is now a growing concensus that the Hubbard model with sufficient correlation to open up a Mott-Hubbard gap at half filling is a good starting point. It is generally believed that with hole doping, the holes reside on the oxygen p σ- orbital. Zhang and Rice [2] have argued that it is energetically favorable to first form a symmetrized Wannier orbital about the copper site and then form a spin singlet between the oxygen hole and the copper local moment. The resulting singlet complex behaves as a hole in a one band model. At this meeting Sawatzky and Kuramoto have presented detailed cluster calculations which support this point of view in that a singlet state is found to split off cleanly as a low lying state. There is also general agreement on the parameters in a Hubbard type model with $t_{pd} \approx 1.5$ eV, $\varepsilon_p - \varepsilon_d \approx 2.5$ to 3.5 eV, leading to a copper-copper exchange J in rough agreement with the experiment value of 0.12 eV and an effective hopping t in a one band t - J model of roughly 0.4 eV. Fukuyama has carried out this point of view further by extending the model to include various higher order terms, and he found that the further neighbor hopping terms may be substantial. I have emphasized in my talk that second and third neighbor hops which connect the same sublattice may be beneficial to superconductivity.

The cluster calculation still does not address the lingering doubt concerning the Zhang-Rice mapping in that while a singlet-triplet splitting exists in the cluster, the admixture between the singlet once it is allowed to hop is expected to lead to a band width of the same order of magnitude as the gap. The discovery of the electron doped copper oxide increases our confidence that the one band t-J type model should contain sufficient physics to describe the high T_c superconductivity, in that the electron presumably resides on the copper. Maekawa spoke on the electronic model for the electron doped materials, and Kondo and Torrence addressed the question of stability of these compounds based on Madelung energy considerations. Experimentally Tokura has concluded from analyzing the spectral weight of his optical data that the effective mass for the carriers in the electron doped material is very small (of the order of 1/4 electron mass) so that the effective t is about four times larger than that for the hole doped materials. I find it remarkable that despite such a large change in the microscopic parameter, the transition temperature remains pretty much the same. This perhaps is a consequence of the remormalization of the coherent hopping matrix element downwards from t to J but it will be very interesting to check this experimentally by measuring the spectral weight of the Drude peak in the electron-doped materials.

During this meeting we had much heated debate on whether the normal state properties can be described by a Ferni liquid theory. I think everyone would agree that with sufficiently high dopping, a Fermi liquid state would form pretty much as in the heavy Fermion electron materials such as $CeAl_3$. In our case, the local moment on the copper plays the role analogous to the Ce moment and is "liberated" to become part of the Fermi surface by strong coupling with the conduction electrons introduced by doping on the oxygen orbital. The important point of this state is not just that a sharp Fermi surface exists, but that Luttinger Theorem is obeyed and the Fermi surface has a volume given by $1 + \delta$ holes where δ is the number of doped hole per copper. At the same time it seems intuitively clear that this state cannot survive in the limit of very small δ. This is an extreme case of the question raised by Nozieres [3] in the context of the heavy Fermion problem: how can there be sufficient conduction electrons around to compensate the local moments when $\delta \ll 1$. Thus a transition between Fermi liquid and non-Fermi liquid behavior must occur as a function of δ. A reasonable guess may be that the criterion is $t\,\delta \approx J$. The debate we had is whether the superconducting state with $\delta \approx 0.15$ is on the Fermi liquid side. The Fermi liquid theory was presented by Newns and Kuroda at this meeting and Newns argued strongly, based on the photo-emission data of Takahashi and others, for the Fermi liquid state.

At this meeting a number of experiments were presented which suggest that there is a qualitative change in the physical properties of $La_{2-x}Sr_xCuO_4$ when x changes from 0.15 to 0.30, when the material is no longer superconducting. The Hall number for $\delta \lesssim 0.1$ corresponds to a hole concentration of δ while the Hall number becomes large and negative for $\delta \approx 0.3$. Uchida showed that very similar behavior occurs with electron doping. Yashuoka showed that $1/T_1$ for $\delta \approx 0.3$ obeys the Korringa law expected of a Fermi liquid, in contrast to the unusual large relaxation rate in the normal state of the superconducting samples. These experiments suggest that a transition to the Fermi liquid state occurs around $\delta \approx 0.3$ concommitant with the disappearance of superconductivity. On the other hand, the photo-emission data are taken to indicate a band moving through the Fermi surface at a point expected by Luttinger Theorem and in agreement with the calculation of Newns and Pattnaik. The debate then centered on whether the data really showed a quasi-particle peak or simply an edge at the Fermi energy. It is clear that more and better data are needed. At the same time, it will be helpful if proponents of a non-Fermi liquid description can come up with a description of what the pho-emission should look like. One possibility is that there exist hole pockets which contain δ holes. Exact diagonalization of small clusters in fact suggest that these pockets are located at the $(\pi/2,\pi/2)$ point in the Brilloin zone, which incidentally, lies on the Fermi surface of the Luttinger Fermi liquid in the half filled limit. However, even these hole pockets may not behave as well defined quasi-particle, in that the Landau criterion that $\tau^{-1} \ll \varepsilon$, where τ^{-1} is the inelastic scattering rate of a quasi-particle of energy ε, may not be obeyed. If $\tau^{-1} \gg \varepsilon$, a quasi particle rapidly decays into multi-particle-hole pairs and loses its identity, and one does not expect well defined peak structures in photo-emission. There is in fact evidence for an inelastic rate τ^{-1} or order max (kT, ω) from DC and optical conductivity measurements. Additional evidence for this is provided by Klein at this conference, who showed data for the electronic background continuum in Raman scattering which is very different from Fermi liquid behavior, but can be modelled by an inverse lifetime which is of order max (kT,ω).

I mentioned at the beginning that theorists at this conference have converged on the one band t-J model or its extension as a starting model. The next job is to solve this model, which is recognized as a very difficult task. Much work has gone into the simpler limit of the half filled case where the problem reduces to the S = 1/2 Heisenberg model. It is generally believed that the model has a Neel ordered ground state. You may wonder why so much effort has been spent on discussing various forms of disordered ground state. The philosophy here is one borrowed from the successful experience in solving the critical phenomenon problem: when faced with a difficult problem, one first extend the problem into various unphysical space where the problem can be solved. By classifying different possibilities, one hopes that one or more of the possibilities are connected to the physical problem at hand. At this meeting Read described an extension of the S = 1/2 SU(2) Heisenberg model to $2S = N_c$ and SU(N). The large N limit can be solved by boson or Fermion representation. These correspond to different large N theories so that the boson representation corresponds to N_c/N constant, $N \rightarrow \infty$ while the Fermion theory correspnds to $N_c = 1$, $N \rightarrow \infty$. Read showed that the boson large N limit corresponds to a dimerized ground state except for N_c even, resulting in a short range ordered state with a gap in the spin excitation. In contrast, earlier work by Affleck and Marston [4] showed that the Fermion large N limit corresponds to the so called flux phase with gapless excitations. This is the best mean field description of the RVB picture pioneered by Anderson. Yet another possible state was introduced by Laughlin, who presented very physical arguments in analogy with the fractional quantum Hall problem for a ground state with a gap in the spin spectrum and neutral spin 1/2 excitation which obey fractional (1/2) statistics. Laughlin's picture is particularly appealing in that he has shown that the fractional statistics particles (anions) automatically form a superconducting ground state. Laughlin reviewed the recent work of Wen, Wilczek and Zee who were able to make connection betweeen the Laughlin state with the RVB development, in that they showed that a flux phase with an additional chiral symmetry breaking $< \vec{s_1} \cdot \vec{s_2} \times \vec{s_3} > \neq 0$ produces a gap in the spin spectrum and fractional statistics exactly as in the Laughlin state. Just as the simple dimer solid is a strong coupling paradigm for the boson large N limit of Read, so is the Laughlin state a strong coupling paradigm for the fermion large N limit with chiral symmetry breaking. Note that the chiral state breaks time reversal symmetry and parity separately and a number of unusual properties of this state have been predicted.

Thus out of the half filled generalized Heisenberg spin model grows a number of possibilities as finite dopping δ is introduced. The original RVB suggestion essentially grows out of the flux phase corresponding to the fermion large N limit. In my talks I described a t' - J model where a superconducting state emerges upon doping the boson large N limit. The Laughlin,Wen, Wilczek, Zee state grows out of the flux phase with additional chiral symmetry breaking. Recently, Anderson [5] and Wiegmann [6] have proposed a generalized flux phase and Rice has described a renormalized mean field theory for this state which he argued to be superconducting. Keeping in mind that the physical problem is the t-J model with finite δ, it is unlikely that theory can tell us which of these possible states, if any, are connected with the physical problem. Utlimately the question must be settled by experiment. We note that all these possible states (with the exception of the original RVB state) predict a spin gap, in agreement with the neutron scattering data described by Shirane.

Another possible way of distinguishg these different possibilities is by Monte Carlo calculations. In the past year, there has been significant advances in the algorithm for Monte Carlo simulation of quantum systems and at this meeting, Sorella and Imada presented impressive results which encourages me to believe that Monte Carlo calculations are reaching sufficient large size and low temperature that they will soon have a real impact on the problem. In particular, Imada's results showed that the one band Hubbard model produces a spiral state, in agreement with the neutron scattering data described by Shirane and in qualitative agreement with the mean field solution described by Yoshioka. At this point, there is still no indication of superconducting instability in the Monte Carlo simulation of the one band Hubbard model.

In conclusion, I believe that many theoretical issues are being clarified. At the same time, the experimental data, particularly microscopic probes such as neutron scattering, NMR relaxation, Raman scattering and optical measurements, are defining the problem sufficiently so that I am hopeful for the next stage when the various theoretical possibilities must be confronted with the data.

REFERENCES

1. P. W. Anderson, Science 235, 1196 (1987).
2. F.C. Zhang and T.M. rice, Phys. Rev. B 37, 3759 (1988).
3. P. Nozieres, Ann. de Phys. 10, 19 (1985).
4. U. Affleck and J.B. Marston, Phys. Rev. B 37, 3774 (1988).
5. P.W. Anderson, Nobel Symposium (1988), to be published.
6. P. Wiegmann, Nobel Symposium (1988), to be published.

Part V

Abstracts

Recent Numerical Results on the Two-Dimensional Hubbard Model

S. Sorella, S. Baroni, R. Car, M. Parrinello, A. Parola, and E. Tosatti

SISSA, Strada Costiera, 11, I-34014 Trieste, Italy

A new method for simulating strongly correlated fermionic systems has been applied to the study of the ground state properties of the 2D Hubbard model at various fillings. Comparison has been made with exact diagonalization in the 2x2, 3x3 and 4x4 lattices where very good agreement has been verified. Numerical results concerning a)magnetization, b)momentum distribution, c)charge, spin and pairing correlations have been obtained with this method for sizes up to 16x16 sites and several fillings. The 2D antiferromangetic order is destroyed by the presence of few holes and is very sensitive to the boundary conditions. The momentum distribution of the doped system does not have a clear Fermi surface. This is also supported by the exponential decay of the density matrix at large enough distances. Particular attention will be paid to the single hole properties and hole-hole correlations.

Springer Series in Solid-State Sciences, Vol. 89 **Strong Correlation and Superconductivity**
Editors: H. Fukuyama · S. Maekawa · A.P. Malozemoff

Valence Bond and Spin-Peierls Ground States of Low-Dimensional Quantum Antiferromagnets

*N. Read**

Section of Applied Physics, Becton Center, Yale University,
P.O. Box 2157 Yale Station, New Haven, CT06520, USA

The nature of the ground state in the non-Neel phase of Heisenberg Hamiltonians on bipartite lattices will be discussed using a large -N method. Topological effects produce different spin-Peierls or valence bond solid states depending periodically on the magnitude of the spin, with periodicity given by the co-ordination number of the lattice. If time allows, topological effects and statistics of excitations for other Heisenberg and Hubbard models will also be discussed.

* In collaboration with S. Sachdev; Phys. Rev. Letter 62, 1694 (1989).

Holes in Copper Oxides: Where Are They? How Do They Conduct? How Do They Affect Superconductivity?

J.B. Torrance

IBM Research Division, Almaden Research Center, San Jose, CA 95120, USA

Samples of $La_{2-x}Sr_xCuO_4$ and $Gd_{2-x}Sr_xCuO_4$ have been prepared with hole concentrations as high as p = 0.4 per copper by annealing them under high oxygen pressure (100 bar) in order to eliminate the oxygen vacancies which would otherwise compensate these holes. As x and the hole concentration are increased beyond the usual $p \sim x \sim 0.15$ region of the La compounds, the conductivity remains metallic and continues to increase in magnitude. The superconducting transition temperature, which had increased up to 36 K from lower concentrations, levels off at this value between p = 0.13 and p ~ 0.24, beyond which T_c falls and superconductivity surprisingly disappears for $p > 0.32$. The metallic conductivity in $La_{2-x}Sr_xCuO_4$ is contrasted with the semiconducting behavior observed in $Gd_{2-x}Sr_xCuO_4$. Both systems have sheets of CuO_2 and comparable hole concentrations. A possible explanation for their different behavior is presented as part of a broad overview of the relative energies of holes on copper and oxygen sites in a large variety of copper oxides. It is proposed that the holes in $Gd_{2-x}Sr_xCuO_4$ are self-trapped on oxygen sites.

* Work carried out in close collaboration with Y. Tokura, A.I. Nazzal, A. Bezinge, R.M. Metzger (Alabama), T.C. Huang, and S.S.P. Parkin.

Springer Series in Solid-State Sciences, Vol. 89 **Strong Correlation and Superconductivity**
Editors: H. Fukuyama · S. Maekawa · A.P. Malozemoff

Optical Excitations in CuO_2 Sheets With and Without Electron Doping: Relevance to Superconductivity Produced by Electron Doping

Y. Tokura[1], *H. Takagi*[2], *S. Kosihara*[1], *and S. Uchida*[2]

[1]Department of Physics, University of Tokyo, Tokyo 113, Japan
[2]Engineering Research Institute, University of Tokyo, Tokyo 113, Japan

Optical excitations, in particular charge-transfer(CT) excitons, plasmons and two-magnon excitations, have been investigated spectroscopically for single crystals of prototypical cuprate compounds with single-layer Cu-O sheets and chains. In all crystals investigated (for example Nd_2CuO_4, $CaCuO_2$, Ca_2CuO_3, $Sr_2CuO_2Cl_2$ and etc.), the strong CT transtion from O-2p to Cu-3d is observed at 1.5-1.9eV and its anisotropy depending on the Cu-O network is quite evident. Together with results of two-magnon Raman scattering and magntetic susceptibility, analysis of these optical data enables us to evaluate the values of fundamental physical parameters for Cu-O sheets(e.g. transfer energy, CT gap and exchange energy) as well as their depedence on the Cu-O network and dimension. To investigate change of electronic states upon electron-doping in CuO_2 and its relevance to electron-superconcductivity, reflectance spectra of single crystals of Ce-doped Nd_2CuO_4 have been measured in the wide electron-concentration range where the compound undergoes the phase change from the CT insulator to nonsuperconducting metal through the high-T_c superconductor.

Experimental Observation on Superconducting Single Crystals

Z.X. Zhao

Institute of Physics, Chinese Academy of Sciences,
P.O. Box 603, Beijing 100080, People's Rep. of China

Single Crystals of Bi-system have been grown by self-flux method. Incommensurate structure in 2212, 2223 structure and anti-phase boundary have been observed. The optical, magnetic, thermal and transport properties have been measured. For example, the resistivity along c axis shows semiconducting behavior just above T_c with a minimum between 140 - 180 K and in the a-b plane is metallic. At 250 K, $\rho_c/\rho_{ab} \gtrsim 1000$ indicates the even stronger anisotropy in $Bi_2Sr_2CaCu_2O_y$ than in YBCO. The thermopower measurement shows in a-b plane the sign of carriers is hole-like, but cross at 250 K. The spin fluctuation has been observed in Bi-Sr-Ca-Cu-O system. The center frequencies of the excitations are about 1900 - 2300 cm^{-1}, which correspond to the value of the exchange constant $J \approx 690 - 840\ cm^{-1}$. The measurements of superconducting transition under high pressure show that the transition width and the T_c change. The T_c maximum is related to the minimum of resistivity under pressure. Some new results in details will be given in the talk.

Springer Series in Solid-State Sciences, Vol. 89 **Strong Correlation and Superconductivity**
Editors: H. Fukuyama · S. Maekawa · A.P. Malozemoff

Index of Contributors

K. Kamimura, University of Tokyo; **A. Oshiyama**, NEC Corporation, Kawasaki, Japan (Eds.)

Mechanisms of High Temperature Superconductivity

Proceedings of the 2nd NEC Symposium, Hakone, Japan, October 24–27, 1988

1989. X, 343 pp. 203 figs. (Springer Series in Materials Science, Volume 11)
ISBN 3-540-50726-4

Since the discovery by Bednorz and Müller of Cu-O alloys displaying high temperature superconductivity, great energy has been put into research in this field. One of the most important and interesting issues, and the subject of this volume, is the clarification of the microscopic origin and mechanism of high temperature superconductivity. This book discusses the latest experimental results on magnetic, optical, electrical, thermal and mechanical properties of the Cu-O and Bi-O superconductors, as well as proposed theoretical models of the mechanisms. The participants in the symposium agreed that for the high T_c Cu-O superconductors electron correlation effects are of central importance. For the Bi-O superconductors the main topic was whether the mechanism of superconductivity is the same as that of high T_c Cu-O superconductors. What was and what was not resolved at the symposium is summarized at the end of the volume.

K. Kitazawa, University of Tokyo; **T. Ishiguro**, Kyoto University, Kyoto, Japan (Eds.)

Advances in Superconductivity

Proceedings of the 1st International Symposium on Superconductivity (ISS '88), August 28–31, 1988, Nagoya

1989. XXIII, 920 pp. 938 figs. ISBN 3-540-70039-0

Intense recent activity in the field of high-temperature superconductivity both in Japan and in the rest of the world was discussed at the First International Symposium on Superconductivity held in Nagoya in August 1988.
Current research and development efforts by major Japanese companies in the field of high-temperature superconductivity are reported by leading company scientists, to give an overview of the high level of activity in the area. Progress in the development of new materials and recent theoretical work is reported both from Japanese and international researchers. Contributions are organized by topic, with such topics as crystal chemistry and electronic structure, processing and microstructure, tapes and thick films, wires and coils, and thin film processing and properties. Future applications of superconductivity including magnetic levitation vehicles, electronics based on Josephson junctions, power delivery, energy storage, ship propulsion and magnetic resonance imaging are particularly stressed.

Springer-Verlag
Berlin Heidelberg New York London Paris Tokyo Hong Kong

Springer

MIX
Papier aus verantwortungsvollen Quellen
Paper from responsible sources
FSC® C105338

If you have any concerns about our products,
you can contact us on
ProductSafety@springernature.com

In case Publisher is established outside the EU,
the EU authorized representative is:
Springer Nature Customer Service Center GmbH
Europaplatz 3, 69115 Heidelberg, Germany

Printed by Libri Plureos GmbH
in Hamburg, Germany